VOLUME ONE HUNDRED AND SEVENTY TWO

PROGRESS IN MOLECULAR BIOLOGY AND TRANSLATIONAL SCIENCE

Autophagy in Health and Disease

VOLUME ONE HUNDRED AND SEVENTY TWO

Progress in Molecular Biology and Translational Science

Autophagy in Health and Disease

Edited by

AITZIBER BUQUÉ MARTINEZ
Department of Radiation Oncology, Weill Cornell Medical College, New York, NY, United States

LORENZO GALLUZZI
Department of Radiation Oncology, Weill Cornell Medical College; Sandra and Edward Meyer Cancer Center; Caryl and Israel Englander Institute for Precision Medicine, New York, NY; Department of Dermatology, Yale University School of Medicine, New Haven, CT, United States; Université de Paris, Paris, France

ELSEVIER

ACADEMIC PRESS
An imprint of Elsevier

Academic Press is an imprint of Elsevier
50 Hampshire Street, 5th Floor, Cambridge, MA 02139, United States
525 B Street, Suite 1650, San Diego, CA 92101, United States
The Boulevard, Langford Lane, Kidlington, Oxford OX5 1GB, United Kingdom
125 London Wall, London EC2Y 5AS, United Kingdom

First edition 2020

ISBN: 978-0-12-822021-4
ISSN: 1877-1173

For information on all Academic Press publications
visit our website at https://www.elsevier.com/books-and-journals

Publisher: Zoe Kruze
Acquisitions Editor: Ashlie M. Jackman
Editorial Project Manager: Leticia M. Lima
Production Project Manager: James Selvam
Cover Designer: Matthew Limbert

Typeset by SPi Global, India

Contents

Contributors

Mahmoud Abdellatif
Department of Cardiology, Medical University of Graz, Graz, Austria

Divya Bhatia
Division of Nephrology and Hypertension, Joan and Sanford I. Weill Department of Medicine, NewYork-Presbyterian Hospital, Weill Cornell Medicine, New York, NY, United States

Augustine M.K. Choi
Division of Pulmonary and Critical Care Medicine, Joan and Sanford I. Weill Department of Medicine; NewYork-Presbyterian Hospital, Weill Cornell Medicine, New York, NY, United States

Mary E. Choi
NewYork-Presbyterian Hospital; Division of Nephrology and Hypertension, Joan and Sanford I. Weill Department of Medicine, Weill Cornell Medicine, New York, NY, United States

Tamás Csizmadia
Department of Anatomy, Cell and Developmental Biology, Eötvös Loránd University, Budapest, Hungary

Donna Denton
Centre for Cancer Biology, University of South Australia and SA Pathology, Adelaide, SA, Australia

Nilima Dinesh Kumar
Department of Biomedical Sciences of Cells & Systems; Department of Medical Microbiology and Infection Prevention, University of Groningen, University Medical Center Groningen, Groningen, The Netherlands

Christopher Eickhorst
Institute of Biochemistry and Molecular Biology, ZBMZ, Faculty of Medicine; Faculty of Biology; Spemann Graduate School of Biology and Medicine (SGBM), University of Freiburg, Freiburg, Germany

Zvulun Elazar
Department of Biomolecular Sciences, The Weizmann Institute of Science, Rehovot, Israel

Milana Fraiberg
Department of Biomolecular Sciences, The Weizmann Institute of Science, Rehovot, Israel

Gábor Juhász
Department of Anatomy, Cell and Developmental Biology, Eötvös Loránd University, Budapest; Institute of Genetics, Biological Research Centre, Szeged, Hungary

Claudine Kraft
Institute of Biochemistry and Molecular Biology, ZBMZ, Faculty of Medicine; CIBSS—Centre for Integrative Biological Signalling Studies, University of Freiburg, Freiburg, Germany

Anita V. Kumar
Department of Molecular Biology, Cell Biology, and Biochemistry, Brown University, Providence, RI, United States

Sharad Kumar
Centre for Cancer Biology, University of South Australia and SA Pathology, Adelaide, SA, Australia

Louis R. Lapierre
Department of Molecular Biology, Cell Biology, and Biochemistry, Brown University, Providence, RI, United States

Mariya Licheva
Institute of Biochemistry and Molecular Biology, ZBMZ, Faculty of Medicine; Faculty of Biology, University of Freiburg, Freiburg, Germany

Senka Ljubojevic-Holzer
Department of Cardiology, Medical University of Graz; BioTechMed Graz, Graz, Austria

Frank Madeo
BioTechMed Graz; Institute of Molecular Biosciences, NAWI Graz, University of Graz, Graz, Austria

Guillermo Mariño
Laboratorio "Autofagia y Metabolismo", Instituto de Investigación Sanitaria del Principado de Asturias; Departamento de Biología Funcional, Universidad de Oviedo, Oviedo, Spain

Gemma G. Martínez-García
Laboratorio "Autofagia y Metabolismo", Instituto de Investigación Sanitaria del Principado de Asturias; Departamento de Biología Funcional, Universidad de Oviedo, Oviedo, Spain

Joslyn Mills
Department of Molecular Biology, Cell Biology, and Biochemistry, Brown University, Providence, RI, United States

Christian Münz
Viral Immunobiology, Institute of Experimental Immunology, University of Zürich, Zürich, Switzerland

Louise O'Keefe
Department of Molecular and Biomedical Science, School of Biological Sciences, The University of Adelaide; Hopwood Centre for Neurobiology, South Australian Health and Medical Research Institute, Adelaide, SA, Australia

Margarita-Elena Papandreou
Institute of Molecular Biology and Biotechnology, Foundation for Research and Technology—Hellas; Department of Basic Sciences, Faculty of Medicine, University of Crete, Heraklion, Greece

Alexandra C. Racanelli
Division of Pulmonary and Critical Care Medicine, Joan and Sanford I. Weill Department of Medicine; NewYork-Presbyterian Hospital, Weill Cornell Medicine, New York, NY, United States

Fulvio Reggiori
Department of Biomedical Sciences of Cells & Systems, University of Groningen, University Medical Center Groningen, Groningen, The Netherlands

Natalia Rodríguez-Muela
Deutsche Zentrum für Neurodegenerative Erkrankungen e.V, DZNE/German Center for Neurodegenerative Diseases; Technische Universität Dresden, CRTD/Center for Regenerative Therapies Dresden; Max Planck Institute for Molecular Cell Biology and Genetics, Dresden, Germany

Simon Sedej
Department of Cardiology, Medical University of Graz; BioTechMed Graz, Graz, Austria; Institute of Physiology, Faculty of Medicine, University of Maribor, Maribor, Slovenia

Jolanda M. Smit
Department of Medical Microbiology and Infection Prevention, University of Groningen, University Medical Center Groningen, Groningen, The Netherlands

Flavie Strappazzon
IRCCS Fondazione Santa Lucia, Rome, Italy

Nektarios Tavernarakis
Institute of Molecular Biology and Biotechnology, Foundation for Research and Technology—Hellas; Department of Basic Sciences, Faculty of Medicine, University of Crete, Heraklion, Greece

Andrew Thorburn
Department of Pharmacology, University of Colorado School of Medicine, Aurora, CO, United States

Shi Quan Wong
Department of Molecular Biology, Cell Biology, and Biochemistry, Brown University, Providence, RI, United States

Preface—Autophagy: An old mechanism with new challenges ahead

Aitziber Buqué Martinez[a,*], Lorenzo Galluzzi[a,b,c,d,e]
[a]Department of Radiation Oncology, Weill Cornell Medical College, New York, NY, United States
[b]Sandra and Edward Meyer Cancer Center, New York, NY, United States
[c]Caryl and Israel Englander Institute for Precision Medicine, New York, NY, United States
[d]Department of Dermatology, Yale University School of Medicine, New Haven, CT, United States
[e]Université de Paris, Paris, France
[*]Corresponding author: e-mail addresses: abm2013@med.cornell.edu; deadoc80@gmail.com

Autophagy is a cellular mechanism common to all existing eukaryotes that ensures the lysosomal degradation of potentially cytotoxic or disposable cytosolic material in support of physiological homeostasis or adaptation to stress.[1,2] The modalities through which autophagy substrates are delivered to lysosomes for degradation identify three major variants of the process: (1) microautophagy, which involves the direct invagination of late endosomal or lysosomal membranes[3]; (2) chaperone-mediated autophagy (CMA), which involves the translocation of cytosolic proteins chaperoned by heat shock protein family A (Hsp70) member 8 (HSPA8) across lysosomal membranes via oligomers of a specific splicing isoform of lysosomal associated membrane protein 2 (LAMP2), namely LAMP2A[4]; and (3) macroautophagy, which is characterized by the formation of a novel double-membraned organelle (the autophagosome) that delivers autophagy substrates to lysosomes upon fusion.[5] Macroautophagy (from here onward referred to as autophagy) operates at baseline levels in all eukaryotic cells as a housekeeping mechanism.[6–9] Moreover, the autophagic flux (i.e., the rate of lysosomal degradation of autophagy substrates) is responsive to a variety of stress conditions, including nutritional cues,[10,11] viral infection,[12,13] as well as exposure to toxic xenobiotics[14] or radiation.[15] In line with a key role for autophagy in organismal development, mice lacking key components of the molecular machinery for autophagy die in utero or shortly after birth.[5] Similarly, the homozygous deletion of essential autophagy genes from the adult mice is often associated with fatal, rapid-onset neurodegeneration,[16,17] suggesting that the central nervous system is especially reliant on autophagy for physiological activity in the adult.

Substrate selection can be rather nonspecific, especially when autophagy is driven by an increased demand for its products or functions, as well as relatively targeted, especially when autophagy is initiated by an excess of its substrates.[18] Thus, permeabilized mitochondria are rapidly taken up by autophagosomes and delivered to lysosomes for degradation, a process that is commonly referred to as mitophagy.[19] A growing list of neologisms has been introduced to indicate variants of autophagy specific for nuclear fragments (nucleophagy),[20] portions of the endoplasmic reticulum (reticulophagy),[21] dysfunctional lysosomes (lysophagy),[22] invading bacteria (xenophagy),[23] secretory granules (crinophagy),[24] etc. Importantly, some components of the molecular machinery for autophagy can also mediate cytotoxic effects,[25,26] at least in selected circumstances (including numerous developmental scenarios in lower eukaryotes).[27,28] Thus, the molecular apparatus for autophagy can mediate a specific form of regulated cell death that has been dubbed "autophagy-dependent cell death" (ACDC).[29,30]

Reflecting the key functions that autophagy mediate in the preservation of cellular and organismal homeostasis,[31] autophagy defects have been implicated in the etiology of numerous human pathologies, encompassing (but not limited to): neurodegenerative conditions,[16,17] cardiovascular disorders,[19,32–35] cancer,[36–39] liver disease,[40,41] obesity,[42] kidney conditions,[43] and viral infection.[12] However, targeting autophagy with pharmacological modulators remains challenging,[14,42,44–49] largely reflecting: (1) the reduced specificity of currently available molecules (e.g., the lysosomal inhibitors chloroquine and hydroxychloroquine),[50,51] (2) the elevated redundancy that characterizes the molecular machinery for autophagy,[5] and (3) the key role of autophagy in nondiseased tissues.[52,53] Nonetheless, we surmise that the development of novel, highly specific pharmacological agents may unlock the hitherto unrealized potential of autophagy modulation for the treatment of numerous pathologies.

In *Autophagy in Health and Disease* (a special issue of the *Progress in Molecular Biology and Translational Science* series), a large group of experts in the field discusses autophagy from both mechanistic and pathophysiological perspectives. Altogether, the 15 chapters of *Autophagy in Health and Disease* present specific features of autophagy as an increasingly characterized molecular pathway, dissect the links between autophagy deregulation and human disorders, and introduce model organisms that are suitable for autophagy research. The book is equally addressed to established researchers with expertise in autophagy and to students or other neophytes in this rapidly evolving field of investigation.

Acknowledgments

The LG lab is supported by a Breakthrough Level 2 grant from the US Department of Defense (DoD), Breast Cancer Research Program (BRCP) [#BC180476P1], by a pre-SPORE grant from the Meyer Cancer Center at Weill Cornell Medicine (New York, NY, USA), by a startup grant from the Department of Radiation Oncology at Weill Cornell Medicine (New York, NY, USA), by industrial collaborations with Lytix (Oslo, Norway) and Phosplatin (New York, NY, USA), as well as by donations from Phosplatin (New York, NY, USA), the Luke Heller TECPR2 Foundation (Boston, MA, USA) and Sotio a.s. (Prague, Czech Republic).

Disclosures

L.G. provides remunerated consulting to OmniSEQ (Buffalo, NY, USA), Astra Zeneca (Gaithersburg, MD, USA), Inzen (New York, NY, USA) and the Luke Heller TECPR2 Foundation (Boston, MA, USA), and he is member of the Scientific Advisory Committee of OmniSEQ (Buffalo, NY, USA).

References

1. Ballabio A, Bonifacino JS. Lysosomes as dynamic regulators of cell and organismal homeostasis. *Nat Rev Mol Cell Biol.* 2020;21:101–118.
2. Galluzzi L, et al. Molecular definitions of autophagy and related processes. *EMBO J.* 2017;36(13):1811–1836.
3. Li WW, et al. Microautophagy: lesser-known self-eating. *Cell Mol Life Sci.* 2012;69(7):1125–1136.
4. Kaushik S, Cuervo AM. The coming of age of chaperone-mediated autophagy. *Nat Rev Mol Cell Biol.* 2018;19(6):365–381.
5. Dikic I, Elazar Z. Mechanism and medical implications of mammalian autophagy. *Nat Rev Mol Cell Biol.* 2018;19(6):349–364.
6. Pohl C, Dikic I. Cellular quality control by the ubiquitin-proteasome system and autophagy. *Science.* 2019;366(6467):818–822.
7. Lopez-Otin C, Kroemer G. Decelerating ageing and biological clocks by autophagy. *Nat Rev Mol Cell Biol.* 2019;20(7):385–386.
8. Hansen M, et al. Autophagy as a promoter of longevity: insights from model organisms. *Nat Rev Mol Cell Biol.* 2018;19(9):579–593.
9. Clarke AJ, Simon AK. Autophagy in the renewal, differentiation and homeostasis of immune cells. *Nat Rev Immunol.* 2019;19(3):170–183.
10. Anderson CM, Macleod KF. Autophagy and cancer cell metabolism. *Int Rev Cell Mol Biol.* 2019;347:145–190.
11. Galluzzi L, et al. Metabolic control of autophagy. *Cell.* 2014;159(6):1263–1276.
12. Choi Y, et al. Autophagy during viral infection—a double-edged sword. *Nat Rev Microbiol.* 2018;16(6):341–354.
13. Zhang L. Autophagy in hepatitis B or C virus infection: an incubator and a potential therapeutic target. *Life Sci.* 2019;242:117206.
14. Levy JMM, et al. Targeting autophagy in cancer. *Nat Rev Cancer.* 2017;17(9):528–542.
15. Galluzzi L, et al. Activating autophagy to potentiate immunogenic chemotherapy and radiation therapy. *Nat Rev Clin Oncol.* 2017;14(4):247–258.
16. Boland B, et al. Promoting the clearance of neurotoxic proteins in neurodegenerative disorders of ageing. *Nat Rev Drug Discov.* 2018;17(9):660–688.
17. Menzies FM, et al. Compromised autophagy and neurodegenerative diseases. *Nat Rev Neurosci.* 2015;16(6):345–357.

18. Sica V, et al. Organelle-specific initiation of autophagy. *Mol Cell.* 2015;59(4):522–539.
19. Picca A, et al. Mitochondrial quality control mechanisms as molecular targets in cardiac ageing. *Nat Rev Cardiol.* 2018;15(9):543–554.
20. Papandreou ME, Tavernarakis N. Nucleophagy: from homeostasis to disease. *Cell Death Differ.* 2019;26(4):630–639.
21. Nakatogawa H, Mochida K. Reticulophagy and nucleophagy: new findings and unsolved issues. *Autophagy.* 2015;11(12):2377–2378.
22. Papadopoulos C, et al. Repair or lysophagy: dealing with damaged lysosomes. *J Mol Biol.* 2020;432(1):231–239.
23. Sharma V, et al. Selective autophagy and xenophagy in infection and disease. *Front Cell Dev Biol.* 2018;6:147.
24. Csizmadia T, et al. Molecular mechanisms of developmentally programmed crinophagy in drosophila. *J Cell Biol.* 2018;217(1):361–374.
25. Tompkins KD, Thorburn A. Regulation of apoptosis by autophagy to enhance cancer therapy. *Yale J Biol Med.* 2019;92(4):707–718.
26. Galluzzi L, Green DR. Autophagy-independent functions of the autophagy machinery. *Cell.* 2019;177(7):1682–1699.
27. Berry DL, Baehrecke EH. Growth arrest and autophagy are required for salivary gland cell degradation in Drosophila. *Cell.* 2007;131(6):1137–1148.
28. Denton D, et al. Autophagy, not apoptosis, is essential for midgut cell death in drosophila. *Curr Biol.* 2009;19(20):1741–1746.
29. Galluzzi L, et al. Molecular mechanisms of cell death: recommendations of the nomenclature committee on cell death 2018. *Cell Death Differ.* 2018;25(3):486–541.
30. Spetz J, et al. T cells and regulated cell death: kill or be killed. *Int Rev Cell Mol Biol.* 2019;342:27–71.
31. Galluzzi L, et al. Linking cellular stress responses to systemic homeostasis. *Nat Rev Mol Cell Biol.* 2018;19(11):731–745.
32. Bravo-San Pedro JM, et al. Autophagy and mitophagy in cardiovascular disease. *Circ Res.* 2017;120(11):1812–1824.
33. Delbridge LMD, et al. Myocardial stress and autophagy: mechanisms and potential therapies. *Nat Rev Cardiol.* 2017;14(7):412–425.
34. Henning RH, Brundel B. Proteostasis in cardiac health and disease. *Nat Rev Cardiol.* 2017;14(11):637–653.
35. Bonora M, et al. Targeting mitochondria for cardiovascular disorders: therapeutic potential and obstacles. *Nat Rev Cardiol.* 2019;16(1):33–55.
36. Rao S, et al. Cancer immunosurveillance by T cells. *Int Rev Cell Mol Biol.* 2019;342:149–173.
37. Rybstein MD, et al. The autophagic network and cancer. *Nat Cell Biol.* 2018;20(3):243–251.
38. Martinez-Outschoorn UE, et al. Cancer metabolism: a therapeutic perspective. *Nat Rev Clin Oncol.* 2017;14(1):11–31.
39. Pietrocola F, Kroemer G. Caloric restriction promotes the stemness and antitumor activity of T lymphocytes. *Onco Targets Ther.* 2019;8(10):e1616153.
40. Ueno T, Komatsu M. Autophagy in the liver: functions in health and disease. *Nat Rev Gastroenterol Hepatol.* 2017;14(3):170–184.
41. Hazari Y, et al. Autophagy in hepatic adaptation to stress. *J Hepatol.* 2020;72(1):183–196.
42. Zhang Y, et al. Targeting autophagy in obesity: from pathophysiology to management. *Nat Rev Endocrinol.* 2018;14(6):356–376.
43. Ye X, et al. Autophagy in immune-related renal disease. *J Immunol Res.* 2019;2019:5071687.
44. Aghaei M, et al. Targeting autophagy in cardiac ischemia/reperfusion injury: a novel therapeutic strategy. *J Cell Physiol.* 2019;234(10):16768–16778.

45. Ren J, Zhang Y. Targeting autophagy in aging and aging-related cardiovascular diseases. *Trends Pharmacol Sci.* 2018;39(12):1064–1076.
46. Soria LR, Brunetti-Pierri N. Targeting autophagy for therapy of hyperammonemia. *Autophagy.* 2018;14(7):1273–1275.
47. Chude CI, Amaravadi RK. Targeting autophagy in cancer: update on clinical trials and novel inhibitors. *Int J Mol Sci.* 2017;18(6).
48. Galluzzi L, et al. Pharmacological modulation of autophagy: therapeutic potential and persisting obstacles. *Nat Rev Drug Discov.* 2017;16(7):487–511.
49. Vanpouille-Box C, et al. Pharmacological modulation of nucleic acid sensors—therapeutic potential and persisting obstacles. *Nat Rev Drug Discov.* 2019;18(11): 845–867.
50. Zhang L, Liu B. Targeting autophagy with small-molecule modulators in immune-related diseases. *Adv Exp Med Biol.* 2019;1209:181–203.
51. Montrose DC, Galluzzi L. Drugging cancer metabolism: expectations vs. reality. *Int Rev Cell Mol Biol.* 2019;347:1–26.
52. Liu K, et al. Autophagy in normal stem cells and specialized cells. *Adv Exp Med Biol.* 2019;1206:489–508.
53. Levine B, Kroemer G. Biological functions of autophagy genes: a disease perspective. *Cell.* 2019;176(1–2):11–42.

CHAPTER ONE

Nucleophagy mediators and mechanisms

Margarita-Elena Papandreou[a,b], **Nektarios Tavernarakis**[a,b,*]

[a]Institute of Molecular Biology and Biotechnology, Foundation for Research and Technology—Hellas, Heraklion, Greece
[b]Department of Basic Sciences, Faculty of Medicine, University of Crete, Heraklion, Greece
*Corresponding author: e-mail address: tavernarakis@imbb.forth.gr

Contents

Abstract

Nuclear recycling is essential for cell and organismal homeostasis. Nuclear architecture perturbations, such as nuclear loss or nuclear enlargement, have been observed in several pathological conditions. Apart from proteasomal components which reside in the nucleus, specific autophagic proteins also shuttle between the nucleus and the cytoplasm. Until recently, only the microautophagic degradation of nuclear components had been described. Recent studies, dissecting nuclear material recycling in organisms ranging from yeast to mammals, provide insight relevant to other forms of nucleophagy and the mediators involved. Nucleophagy has also been implicated in pathology. Lamins are degraded in cancer through direct interaction with LC3 in the nucleus. Similarly, in neurodegeneration, Golgi-associated nucleophagy is exacerbated. The physiological role of nucleophagy and its contribution to other pathologies remain to be elucidated. Here we discus recent findings that shed light into the molecular mechanisms and pathways that mediate the autophagic recycling of nuclear material.

Progress in Molecular Biology and Translational Science, Volume 172
ISSN 1877-1173
https://doi.org/10.1016/bs.pmbts.2020.01.003

Abbreviations

ALS amyotrophic lateral sclerosis
AMPK AMP-activated protein kinase
BNIP3 BCL2 and adenovirus E1B 19-kDa interacting protein 3
ER endoplasmic reticulum
FUNDC1 FUN-14 domain containing protein 1
GAPDH glyceraldehyde 3-phosphate dehydrogenase
HDAC histone deacetylase
Hmo1 high-mobility group protein 1
LN late nucleophagy mTOR mammalian target of rapamycin
NPC nuclear pore complexes
Nvj1 nucleus-vacuole junction
PE phospahtidylethanolamine
PEX5 pexophagy 5
PI3K phosphatidylinositol-3-kinase
PMN macronucleophagy
PNS peripheral nerve syndrome
RNP ribonucleoprotein complex
SAR selective autophagy receptor
SCA spinocerebellar ataxia
Sirt1 sirtuin 1
TDP-43 TAR DNA-binding protein 43
TORC1 target of rapamycin complex 1
UBC ubiquitin-C
UBLs ubiquitin like proteins
ULK1 Unc-51 like autophagy activating kinase
Vacp8 vacuolar protein 8

1. Introduction: General and selective autophagy

Autophagy from the Greek words "auto," self, and "phagy," eating, is a physiological homeostatic mechanism in all eukaryotes. It is the main catabolic pathway of the cell and breaks down macromolecules such as lipids, proteins and organelles, via the main degradative compartment, the lysosome. Although it was initially considered a bulk degradation pathway, recent evidence highlights its highly selective nature. It can be subdivided into three mechanisms: macroautophagy, thereby referred to as autophagy, microautophagy and chaperone-mediated autophagy.[1]

Autophagy entails the formation of a double membranous vesicle, the autophagosome, which encircles the cargo to be degraded. Specificity is accomplished through recruitment of selective autophagy receptors (SARs).

Yeast Atg19 and mammalian p62/sequesterome are examples of SARs which mediate the interaction between core autophagic components and substrates.[2] Briefly, initiation of autophagy involves yeast Atg1/Unc-51-like, autophagy activating kinase (ULK1), ATG13 and FIP200 which trigger phagophore formation followed by vesicle nucleation through class-III phosphatidylinositol-3-kinase (PI3K) complex which comprises VPS34, VPS15, Beclin-1.[3] SARs interact with ubiquitin like proteins (UBLs), small globular proteins, such as the yeast Atg8/mammalian LC3, required for autophagosomal biogenesis and recruited to the autophagosomal membranes through phosphatidylethanolamine (PE) conjugation, which causes the conversion of LC3-I to LC3-II. The Atg5/Atg12 complex also contributes to the maturation of the autophagosome. Autophagy of specific organelles, lipids or proteins is coordinated by multiple signaling pathways under various environmental stresses. An Atg8-interacting motif (AIM)/LC3-interacting region (LIR) is found in SARs and targeted by Atg8/LC3. The general core motif, which includes W/F/Y-X-X-L/I/V, with negatively charged residues upstream of the motif triggers a higher affinity interaction as well as post-translational modifications, such as phosphorylation.[4] In addition, these adaptor proteins contain domains such as the UBA motif, which recognizes ubiquitinated proteins for degradation. K27-linked mono-ubiquitination and K63 poly-ubiquitination are characteristic examples of tagging for autophagosomal recycling.[5] This allows tight regulation of transport of autophagic substrates for lysosomal degradation.

Different selective autophagy receptors exist depending on the substrate to be degraded. Protein aggregate degradation, aggrephagy is mediated by SARs such as p62, optineurin, NBR1 and NDP52.[6–8] Organellophagy, the autophagic recycling of organelles can occur with the concerted action of the SARs referred to above, and more specific ones such as the BCL2 and adenovirus E1B 19-kDa interacting protein 3 (BNIP3), and FUN-14 domain containing protein 1 (FUNDC1) for mitophagy.[9,10] Mutations in SARs can lead to severe phenotypes such as metabolic disorders, neurodegenerative and age-related disease.[11] In mitophagy, there are several receptors modulating mitochondrial degradation. Mutations in the PINK1 and Parkin proteins involved in the ubiquitination of outer mitochondrial membrane proteins that are then recognized by SARs such as p62, OPTN and NDP52, cause Parkinson's disease.[12] Optineurin mutations have been associated with amyotrophic lateral sclerosis (ALS).[13] In addition, mutations in the PEX5 gene cause the fatal cerebro-hepato-renal Zellweger syndrome. PEX5 encodes a pexophagy-specific receptor, which when ubiquitinated interacts with p62, and is transported to the autophagosome.[14,15] Therefore, different protein receptors/adaptors found either on the organelle, or recruited under stress, are required to efficiently regulate autophagic degradation.

1.1 The nucleus

Nucleophagy, the recycling of the nucleus by autophagy, has not been investigated in depth although the nucleus is the most prominent organelle of the cell. To discern potential receptors and pathways, which control degradation of specific nuclear components and compartments, a meticulous study of nuclear structure is required. The phospholipid bilayer of the nuclear membrane is a barrier tightly controlling communication between the cytoplasm and the nucleoplasm. The nuclear membranes are continuous with those of the endoplasmic reticulum (ER) but significantly differ in protein composition. Nucleoporins form large nuclear pore complexes (NPCs) for bidirectional movement of ribonucleoprotein complexes (RNPs), RNA and proteins between the nucleus and the cytoplasm.[16] The inner nuclear membrane of the nuclear envelope consists of integral proteins that interact with the nuclear lamina (nucleoskeleton) and chromatin, thus, influencing chromatin localization and gene expression. Examples include emerin, SUNs, lamin B receptors (LBRs) that interact with B-type lamins and heterochromatin protein 1 (HP1), which, in turn, interacts with chromatin. SUN proteins interact with nesprins, the outer nuclear membrane proteins (ONMs) that connect the nucleus with the cytoskeleton, and together form the linker of the nucleoskeleton-cytoskeleton complex (LINC). ONMs share a small Klarsicht, ANC-1, Syne Homology (KASH) domain. They regulate nuclear migration, cell polarization and maintain nuclear shape. However, many of the nuclear envelope proteins remain largely uncharacterized. Inner nuclear components include the nucleolus, nuclear bodies, such as the promyelocytic leukemia PML bodies, DNA, RNA, and proteins that shuttle nucleocytoplasmically.

1.2 Nucleophagy in yeast

Yeast cells perform micronucleophagy as well as macronucleophagy. Nucleophagy was first described in *Saccharomyces cerevisiae*, in two forms, piecemeal macronucleophagy (PMN) and late nucleophagy (LN).[17,18] Apart from inducing general autophagy, TORC1 (target of rapamycin 1) inactivation triggers nucleophagy through the Nem1/Spo7-Pah1 axis, which properly localizes micronucleophagy component Nvj1 and nucleophagy receptor Atg39.[19]

PMN occurs under nutrient-rich conditions, after rapamycin treatment, or short nutrient and nitrogen deprivation. Mechanistically, the lytic vacuole and the nucleus initiate direct interaction with the formation of tight nuclear-vacuole junctions (NVJ). Subsequently, ONM and nuclear ER protrusions develop, which then become ER-derived vesicles encircled by the lytic vacuole.[20] Then, vesicles are pinched off the nucleus; nuclear-derived membranes are fused at multiple points with the vacuole membrane for the release of substrates to be degraded by lytic enzymes. The outer nuclear membrane nucleus to vacuole protein 1 (Nvj1p) and vacuolar protein 8 (Vacp8) mediate micronucleophagy, emphasizing the importance of autophagy selectivity, even in lower eukaryotes. Vacp8 together with Apg13p mediate cytoplasm-to-vacuole targeting pathway.[21] Although it is suggested that core autophagic genes are implicated in this process., recently a new study reveals that microautophagy does not require macroautophagic proteins, but endosomal sorting complex proteins instead such as Vph1, Pho8 and Vps27.[22] Of note, RNA non-selective degradation has been reported after nitrogen starvation.[23] However, it is not known whether nuclear RNA is directly degraded or whether cytoplasmic RNA is recycled by general autophagy after transported to the cytoplasm. Autophagic cargo of PMN includes nuclear envelope components, the granular nucleolus, containing pre-ribosomes, without nuclear pore complexes and spindle pole bodies, and RNA.[24] CLIP and cohibin are mediators of nucleolar protein degradation by spatially separating them from ribosomal DNA that is not degraded, tethering them on the ONM, and directing them proximal to NVJs.[25] The complex of condensin, Rpd3–Sin3 histone deacetylase (HDAC), and high-mobility group protein 1 (Hmo1) are essential for ribosomal DNA condensation.

PMN and LN are almost always mutually exclusive, both temporally and spatially. LN which was later identified as macronucleophagy, is only induced after prolonged (18–24 h) nitrogen starvation, and, in contrast to PMN, can trigger nuclear shape alterations, indicating accumulation and/or aggregation of nuclear cargo that is not degraded.[17] This can be distinguished by two differential reporters Nvj1p-EYFP, for the former and n-Rosella for the latter.[26] LN requires Nvj1p and Vac8p and other autophagy-related proteins, such as Vps34p, Vps15p, Atg6p, and Atg14p or Atg11p.

Recently, another mode of ERphagy and nucleophagy has been described in yeast.[27] Specifically, the Atg39 receptor, found in perinuclear ER is required for macronucleophagy, while Atg40 mediates ERphagy while contributing to macronucleophagy. Moreover, it was shown that both

Table 1 Nucleophagy factors and pathways.

Factors	Piecemeal nucleophagy	Macronucleophagy	Mammalian nucleophagy
Macroautophagy genes	–	*atg1, atg2, atg3, atg4, atg5, atg7, atg8, atg9, atg10, atg12, atg13, atg16, atg18, atg23, atg29, atg31*	*Atg7, Lc3*
Receptor	Nvj1p, Vac8p	Vac8p, Atg39, Atg40	–
Substrates	Nuclear envelope components, granular nucleolus, pre-ribosomes, spindle pole bodies, RNA	Nuclear envelope components, granular nucleolus, pre-ribosomes, spindle pole bodies, RNA, Hmg1, Src1, Nop1	Lamin A, lamin B2
Pathway/Inducer	Short nitrogen starvation, TORC1 inactivation, Nem1/Spo7-Pah1 axis	Prolonged nitrogen starvation, TORC1 inactivation, Nem1/Spo7-Pah1 axis	DNA damage, oncogenesis
Phenotype	–	Nuclear shape alterations	Cell death, neurodegeneration, senescence

Atg39 and Atg40 are substrates of macroautophagy as genetic inhibition of macroautophagy causes an increase of their levels. Substrate degradation by this key regulator includes ONM and INM proteins, Hmg1, Src1, and nucleolar protein Nop1. However, no exact homolog has been found in mammals although functionally similar SARs have been identified.[18] A comparison between PMN and LN is shown in Table 1.

1.3 Interplay between autophagy and the nucleus in mammals

The crosstalk between the autophagic machinery and nuclear components in mammals has recently drawn much attention. The main autophagosomal membrane protein LC3 is stored in the nucleus under basal conditions in an acetylated form.[28] Deacetylation occurs selectively under starvation at residues K49 and K51 by sirtuin 1 (Sirt1). LC3 is then transported to the cytoplasm by DOR, where it can be conjugated in preautophagosomal membranes through conversion to LC3-II. It is still unclear whether LC3 localization in the nucleus has other functional implications, apart from storage purposes.

Selective autophagy receptors and adaptors are also localized in the nucleus, the function of which remains largely unexplored (Fig. 1). p62, which continuously shuttles between the nucleus and the cytoplasm, together with large protein ALFY, synergize to target ubiquitinated proteins to PML bodies.[29] Moreover, upon stress, ALFY becomes localized with p62 to ubiquitin-positive bodies in the cytoplasm. In addition, nuclear ATG5 and ATG7 can regulate p53 activation and cell cycle progression as well as autophagy.[30,31]

Filamentous fungi contain multinucleate cells called hyphae that proliferate from the hyphal tip. Basal cells undergo nucleophagy of whole nuclei as well as pexophagy and mitophagy.[32] It has been proposed that autophagosomes for nuclei are specific as they are much larger. Colony growth is inhibited upon perturbation of autophagy, indicating that autophagy and potentially nucleophagy-derived catabolic products are

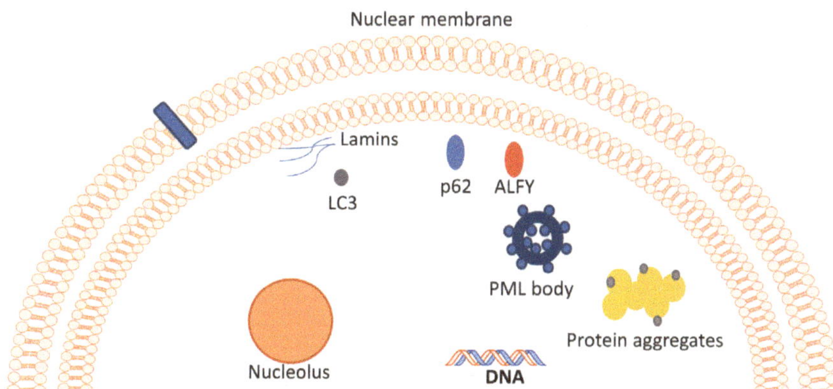

Fig. 1 Autophagy and the nucleus. LC3 (gray), when activated by deacetylase Sirt1, is transported to the cytoplasm. LC3 can transport cargo, such as lamin B2 (blue lines), to autophagosomes during oncogenesis. Similarly, Golgi-mediated lamin A degradation occurs during neurodegeneration. P62 (blue) and ALFY (red) synergize to transport ubiquitinated proteins (yellow) to PML bodies (blue circle). DNA is degraded in the lysosome by Dnase2a, but the path to degradation is still poorly understood. Outer nuclear components (blue rectangle) can also be degraded by interacting with core autophagic components localized in the cytoplasm. Parts of the nucleolus (orange) are degraded by microautophagy, in yeast. Whether this also occurs in higher organisms is unknown.

required for survival and proliferation under starvation, given that nuclei contain large amounts of phosphorus and nitrogen. Thus, nucleophagy is likely an energy source for survival, in this case. Nucleophagy has also been observed during differentiation of skin epidermal keratinocytes. As these cells migrate from the epidermis to the granular layer, they convert into corneocytes which are anucleated cells.[33] During this process, perinuclear LC3 positive vesicles containing p62 and HP1α are localized close to lamins.

Nucleophagy has been shown to play an important role in an oncogenic setting. Electron microscopy has revealed specific nuclear alterations in shape, size and patterns.[34] Chromosomes, protein, and PML bodies are relocalized, while the largest structure of the nucleus, the nucleolus, can become enlarged to meet the excessive protein synthesis needs of transformed cells.[35] Autophagy induces cellular senescence after administration of DNA damaging agents, such as etoposide. The first unequivocal indication of nucleophagy in mammals was obtained during oncogenesis by KRAS, in primary human cells.[36] In this setting, nuclear LC3 directly interacts with lamin B1 through a LIR motif. They are then transported with heterochromatic regions to the cytoplasm for lysosomal degradation (Fig. 2).

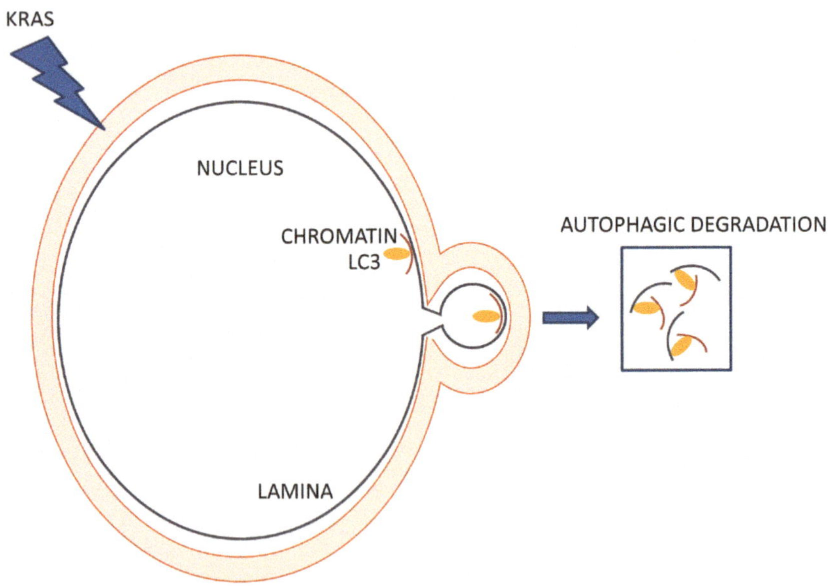

Fig. 2 Mammalian nucleophagy in oncogenesis. LC3 (yellow) interacts with the nuclear lamina (gray), specifically lamin B2 which contains a LIR motif, and transports it together with chromatin fragments (red). A nuclear bulge is formed, which is pinched off and transported to the cytoplasm for autophagic degradation.

Interestingly enough, starvation or mTOR inhibition does not elicit this response, underlining the specificity of this mechanism. Genetic inhibition of ATG7 or mutations in the LIR motif lead to failure of lamin B degradation and subsequent premature senescence, although the exact mechanism has not yet been elucidated. In photoaging, lamin B has been shown to increase and thus act as a biomarker after UV irradiation.[37] Even more recently, SUMOylation of laminA/C by nuclear ubiquitin-C 9 (UBC9) has been shown to be critical for DNA leakage and lamin degradation.[38] Thus, nucleophagy could serve a protective function to restrict DNA replication and proliferation of cancerous cells. Self DNA has been shown to cause autoimmune disorders when not cleared out by transport to the cytoplasm and degradation in the lysosome. Dnase2a or autophagy deficiency causes extra nuclear DNA accumulation and induces inflammation via the STING DNA sensing pathway.[39]

Defects in general and selective autophagy have been implicated in neurodegenerative disease, whereby protein and organelle aggregation is a direct result of defective autophagy. Polyglutamine (PolyQ) diseases are triggered by expansion of CAG repeats in the genes encoding PolyQ proteins. Dentatorubral-pallidoluysian atrophy which is characterized by ataxia, dementia and epilepsy, is caused by atrophin mutations, and manifests a pathological form of nucleophagy.[40,41] At the cellular level, canonical autophagy is stalled, while Golgi membrane degradation and nuclear breakdown occurs via nucleophagy based lamin B degradation and excretion. Concomitantly, there is reduction of LC3-I/II conversion, increase of p62, DNA damage, and senescence. Blockage of autophagy with simultaneous induction of alternative degradative pathways causes nuclear architecture disruption, nuclear collapse and cellular atrophy. Apart from specialized cases in fungi, nucleophagy has been shown to occur in pathological, rather than physiological conditions. Both in a cancerous and neurodegeneration setting, nucleophagy is recruited to clear out nuclear aggregates, damaged DNA, or protein aggregates. Whether, basal nucleophagy occurs, albeit at low, non-pathological levels, remains to be elucidated.

1.4 Nuclear morphology alterations and disease

Nuclear abnormalities and nuclear-derived pathology have been observed in disease. However, in many cases, the mechanisms underlying such phenotypes have not been delineated. Complete nuclear loss or reduction in DNA copy number in the intestine and other cells has been observed during aging, in the nematode *Caenorhabditis elegans*.[42] Long-lived *daf-2* mutant animals display delayed manifestation of such phenotypes, indicating at least partial regulation of the underlying processes by the insulin/IGFI signaling pathway.[43] It has been suggested that autophagy degrades nuclei during aging, possibly to supply nutrients. Nuclear size alterations, or nuclear malformations, which could be a direct or indirect consequence of nucleophagy, occur in pathological conditions such as malignancy which could be the result of DNA, nuclear protein accumulation or nuclear envelope recycling dysregulation.[44]

A multitude of nuclear envelopathies have been identified. Mutations in *LMNA* can affect different tissues causing a variety of pathologies. Examples include the Emery-Dreifuss muscular dystrophy, and dilated cardiomyopathy, which can also be caused by mutations in emerin and Nesprins.[45–47] Apart from striated muscle pathology, lipodystrophy manifests in patients with *LMNA* and *LMNB2* defects. Other diseases include peripheral nerve syndrome (PNS) disorders, bone abnormalities and progeria syndromes. Nesprin 1/2 mutations have also been linked to cardiomyopathies. Indeed, myocytes from such patients have enlarged nuclei. Senescence can also arise upon perturbation of Nesprin function, which could be attributed to the compartmentalization of the extracellular signal-regulated kinase 1/2 (ERK1/2) to PML bodies.[48]

Defects in nucleolar shape and size are prominent during aging and have been proposed as biomarkers of premature aging. Reproductively aged mice exhibit larger nucleoli, thus, lower oocyte quality, which leads to decreased fertility.[49] This extends to different tissues, both in cell cultures and in *C. elegans*. Nucleolar size is predictive of cell/organismal lifespan.[50,51] *daf-2*, *eat-2*, *ife-2* and *glp-1*, long-lived mutant strains display smaller nuclei compared to wild-type worms. Thus, these signaling pathways extend lifespan partially through restricting nucleolar expansion and ribosomal biogenesis. Complementarily, in Hutchison-Gilford progeria syndrome, cells contain larger nucleoli, indicating increased ribosome biogenesis and protein translation.[50] Nucleolar size is not merely a biomarker of aging. Reducing fibrillarin protein production, a key nucleolar protein, extends lifespan. In addition, mutant prelamin A, progerin, which causes progeria syndrome increases nucleolar size. Rapamycin, which induces macroautophagy, has been shown to ameliorate cellular phenotypes in Hutchison-Gilford progeria syndrome by degrading the mutant protein.[52]

Perturbed nucleophagy could be the underlying cause of a variety of nuclear inclusion body diseases. Accumulation of mutant huntingtin, ataxin and androgen receptor cause Huntington's disease, spinocerebellar ataxias

(SCAs), and spinobulbar muscular dystrophy, respectively. Multiple muscle atrophy is characterized by filamentous nuclear aggregates associated with the nucleolus or the inner nuclear membrane.[53] Inefficient clearance of these proteins causes neurological defects and neuronal cell death. Nuclear GAPDH, when associated with DNA, upon oxidative stress, blocks DNA damage repair and triggers cells death.[54] Importantly, the AMP-activated protein kinase (AMPK) activates GAPDH upon glucose starvation, which in turn activates Sirt1. Sirt1 has been shown to deacetylate nuclear LC3 triggering its cytoplasmic shuttling. This molecular pathway could be pharmacologically exploited to trigger nuclear autophagy. Identifying inducers of nucleophagy could potentially allow clearance of excess pathological aggregates, particularly because nuclear pores further restrict large protein transport, when protein aggregates are formed that can themselves perturb nuclear pore function.[55]

2. Conclusion and future perspectives

Selective autophagy of nuclear components occurs in organisms ranging from simple single cell eukaryotes to humans. Substrates can be RNA, diffuse proteins or components of the nuclear envelope, and the nucleolus. In yeast, different types of micronucleophagy exist under both basal and nutrient deprivation conditions. In mammals, although there are indications of nucleophagic events under physiological conditions, nucleophagy has mainly been associated with pathology. For example, during oncogenesis and neurodegeneration, nucleophagy targeting lamins is prominent, resulting in cellular senescence and neuronal death, respectively. Identification of potential instigators of macronucleophagy in physiology would spur new research directions in the field. Apart from neurodegeneration caused by excess nucleophagy, an additional open question relates to the consequences of nucleophagy deficiency. Molecular pathways selectively inducing this autophagic process could be therapeutic targets for a multitude of diseases linked to the formation of nuclear aggregates. Hence, given that the nucleus safeguards the genetic material in addition to controlling its expression via mRNA transcription and translation, it is important to better understand the contribution of nucleophagy in maintaining nuclear homeostasis, both in the context of physiology and pathology. Nucleophagy may well be a protective mechanism against age-related disease due to nuclear dysfunction and a potential therapeutic target.

Acknowledgments

Work in the authors' laboratory is funded by grants from the European Research Council (ERC), the European Commission Framework Programmes, and the Greek Ministry of Education. Margarita-Elena Papandreou was supported by the European Commission Marie Curie Initial Training Network (ITN) CODAGE.

Conflict of interest

The authors declare no financial or commercial conflict of interest.

References

1. Mizushima N. Autophagy: process and function. *Genes Dev*. 2007;21(22):2861–2873.
2. Kirkin V, Rogov VV. A diversity of selective autophagy receptors determines the specificity of the autophagy pathway. *Mol Cell*. 2019;76(2):268–285.
3. Kraft C, Martens S. Mechanisms and regulation of autophagosome formation. *Curr Opin Cell Biol*. 2012;24(4):496–501.
4. Birgisdottir AB, Lamark T, Johansen T. The LIR motif—crucial for selective autophagy. *J Cell Sci*. 2013;126(pt 15):3237–3247.
5. Shaid S, et al. Ubiquitination and selective autophagy. *Cell Death Differ*. 2013; 20(1):21–30.
6. Lamark T, et al. NBR1 and p62 as cargo receptors for selective autophagy of ubiquitinated targets. *Cell Cycle*. 2009;8(13):1986–1990.
7. Ryan TA, Tumbarello DA. Optineurin: a coordinator of membrane-associated cargo trafficking and autophagy. *Front Immunol*. 2018;9:1024.
8. von Muhlinen N, et al. NDP52, a novel autophagy receptor for ubiquitin-decorated cytosolic bacteria. *Autophagy*. 2010;6(2):288–289.
9. Zhang J, Ney PA. Role of BNIP3 and NIX in cell death, autophagy, and mitophagy. *Cell Death Differ*. 2009;16(7):939–946.
10. Chen M, et al. Mitophagy receptor FUNDC1 regulates mitochondrial dynamics and mitophagy. *Autophagy*. 2016;12(4):689–702.
11. Palikaras K, Lionaki E, Tavernarakis N. Coordination of mitophagy and mitochondrial biogenesis during ageing in *C. elegans*. *Nature*. 2015;521(7553):525–528.
12. Wang Y, Liu N, Lu B. Mechanisms and roles of mitophagy in neurodegenerative diseases. *CNS Neurosci Ther*. 2019;25(7):859–875.
13. Chu CT. Mechanisms of selective autophagy and mitophagy: implications for neurodegenerative diseases. *Neurobiol Dis*. 2019;122:23–34.
14. Eun SY, et al. PEX5 regulates autophagy via the mTORC1-TFEB axis during starvation. *Exp Mol Med*. 2018;50(4):4.
15. Jeong SJ, et al. p62/SQSTM1 and selective autophagy in cardiometabolic diseases. *Antioxid Redox Signal*. 2019;31(6):458–471.
16. Hetzer MW. The nuclear envelope. *Cold Spring Harb Perspect Biol*. 2010;2(3): a000539.
17. Mijaljica D, Prescott M, Devenish RJ. A late form of nucleophagy in Saccharomyces cerevisiae. *PLoS One*. 2012;7(6): e40013.
18. Mijaljica D, Devenish RJ. Nucleophagy at a glance. *J Cell Sci*. 2013;126(pt 19): 4325–4330.
19. Rahman MA, Mostofa MG, Ushimaru T. The Nem1/Spo7-Pah1/lipin axis is required for autophagy induction after TORC1 inactivation. *FEBS J*. 2018;285(10):1840–1860.
20. Krick R, et al. Piecemeal microautophagy of the nucleus requires the core macroautophagy genes. *Mol Biol Cell*. 2008;19(10):4492–4505.

21. Scott SV, et al. Apg13p and Vac8p are part of a complex of phosphoproteins that are required for cytoplasm to vacuole targeting. *J Biol Chem*. 2000;275(33):25840–25849.
22. Oku M, et al. Evidence for ESCRT- and clathrin-dependent microautophagy. *J Cell Biol*. 2017;216(10):3263–3274.
23. Huang H, et al. Bulk RNA degradation by nitrogen starvation-induced autophagy in yeast. *EMBO J*. 2015;34(2):154–168.
24. Millen JI, et al. Measuring piecemeal microautophagy of the nucleus in Saccharomyces cerevisiae. *Autophagy*. 2009;5(1):75–81.
25. Mostofa MG, et al. CLIP and cohibin separate rDNA from nucleolar proteins destined for degradation by nucleophagy. *J Cell Biol*. 2018;217(8):2675–2690.
26. Sargsyan A, et al. Rapid parallel measurements of macroautophagy and mitophagy in mammalian cells using a single fluorescent biosensor. *Sci Rep*. 2015;5:12397.
27. Mochida K, et al. Receptor-mediated selective autophagy degrades the endoplasmic reticulum and the nucleus. *Nature*. 2015;522(7556):359–362.
28. Huang R, et al. Deacetylation of nuclear LC3 drives autophagy initiation under starvation. *Mol Cell*. 2015;57(3):456–466.
29. Isakson P, Holland P, Simonsen A. The role of ALFY in selective autophagy. *Cell Death Differ*. 2013;20(1):12–20.
30. Simon HU, et al. ATG5 can regulate p53 expression and activation. *Cell Death Dis*. 2014;5: e1339.
31. Lee IH, et al. Atg7 modulates p53 activity to regulate cell cycle and survival during metabolic stress. *Science*. 2012;336(6078):225–228.
32. Shoji JY, et al. Macroautophagy-mediated degradation of whole nuclei in the filamentous fungus Aspergillus oryzae. *PLoS One*. 2010;5(12): e15650.
33. Akinduro O, et al. Constitutive autophagy and nucleophagy during epidermal differentiation. *J Invest Dermatol*. 2016;136(7):1460–1470.
34. Uhler C, Shivashankar GV. Nuclear mechanopathology and cancer diagnosis. *Trends Cancer*. 2018;4(4):320–331.
35. Galluzzi L, Bravo-San Pedro JM, Kroemer G. Autophagy mediates tumor suppression via cellular senescence. *Trends Cell Biol*. 2016;26(1):1–3.
36. Dou Z, et al. Mammalian autophagy degrades nuclear constituents in response to tumorigenic stress. *Autophagy*. 2016;12(8):1416–1417.
37. Wang AS, et al. Loss of lamin B1 is a biomarker to quantify cellular senescence in photoaged skin. *Sci Rep*. 2017;7(1):15678.
38. Li Y, et al. Nuclear accumulation of UBC9 contributes to SUMOylation of lamin A/C and nucleophagy in response to DNA damage. *J Exp Clin Cancer Res*. 2019;38(1):67.
39. Lan YY, et al. Dnase2a deficiency uncovers lysosomal clearance of damaged nuclear DNA via autophagy. *Cell Rep*. 2014;9(1):180–192.
40. Baron O, et al. Stall in canonical autophagy-lysosome pathways prompts nucleophagy-based nuclear breakdown in neurodegeneration. *Curr Biol*. 2017;27(23):3626–3642 e6.
41. Suzuki Y, Yazawa I. Pathological accumulation of atrophin-1 in dentatorubral-pallidoluysian atrophy. *Int J Clin Exp Pathol*. 2011;4(4):378–384.
42. Golden TR, et al. Dramatic age-related changes in nuclear and genome copy number in the nematode *Caenorhabditis elegans*. *Aging Cell*. 2007;6(2):179–188.
43. McGee MD, et al. Loss of intestinal nuclei and intestinal integrity in aging *C. elegans*. *Aging Cell*. 2011;10(4):699–710.
44. Jevtic P, Levy DL. Mechanisms of nuclear size regulation in model systems and cancer. *Adv Exp Med Biol*. 2014;773:537–569.
45. Bonne G, et al. Mutations in the gene encoding lamin A/C cause autosomal dominant Emery-Dreifuss muscular dystrophy. *Nat Genet*. 1999;21(3):285–288.
46. Schreiber KH, Kennedy BK. When lamins go bad: nuclear structure and disease. *Cell*. 2013;152(6):1365–1375.

47. Banerjee I, et al. Targeted ablation of nesprin 1 and nesprin 2 from murine myocardium results in cardiomyopathy, altered nuclear morphology and inhibition of the biomechanical gene response. *PLoS Genet.* 2014;10(2): e1004114.
48. Warren DT, et al. Nesprin-2-dependent ERK1/2 compartmentalisation regulates the DNA damage response in vascular smooth muscle cell ageing. *Cell Death Differ.* 2015;22(9):1540–1550.
49. Duncan FE, et al. Age-associated dysregulation of protein metabolism in the mammalian oocyte. *Aging Cell.* 2017;16(6):1381–1393.
50. Buchwalter A, Hetzer MW. Nucleolar expansion and elevated protein translation in premature aging. *Nat Commun.* 2017;8(1):328.
51. Tiku V, et al. Small nucleoli are a cellular hallmark of longevity. *Nat Commun.* 2017;8:16083.
52. Cao K, et al. Rapamycin reverses cellular phenotypes and enhances mutant protein clearance in Hutchinson-Gilford progeria syndrome cells. *Sci Transl Med.* 2011;3(89):89ra58.
53. Papp MI, Lantos PL. Accumulation of tubular structures in oligodendroglial and neuronal cells as the basic alteration in multiple system atrophy. *J Neurol Sci.* 1992;107(2):172–182.
54. Chang C, et al. AMPK-dependent phosphorylation of GAPDH triggers Sirt1 activation and is necessary for autophagy upon glucose starvation. *Mol Cell.* 2015;60(6):930–940.
55. Chou CC, et al. TDP-43 pathology disrupts nuclear pore complexes and nucleocytoplasmic transport in ALS/FTD. *Nat Neurosci.* 2018;21(2):228–239.

CHAPTER TWO

Scaffold proteins in bulk and selective autophagy

Christopher Eickhorst[a,b,c,†], Mariya Licheva[a,b,†], Claudine Kraft[a,d,*]
[a]Institute of Biochemistry and Molecular Biology, ZBMZ, Faculty of Medicine, University of Freiburg, Freiburg, Germany
[b]Faculty of Biology, University of Freiburg, Freiburg, Germany
[c]Spemann Graduate School of Biology and Medicine (SGBM), University of Freiburg, Freiburg, Germany
[d]CIBSS—Centre for Integrative Biological Signalling Studies, University of Freiburg, Freiburg, Germany
[*]Corresponding author: e-mail address: kraft@biochemie.uni-freiburg.de

Contents

Abstract

Autophagy is a crucial cellular degradation and recycling pathway. During autophagy double-membrane vesicles, called autophagosomes, encapsulate cellular components and deliver their cargo to the lytic compartment for degradation. Formation of autophagosomes is regulated by the Atg1 kinase complex in yeast and the homologous ULK1 kinase complex in mammals. While research on Atg1 and ULK1 has advanced our understanding of how these protein kinases function in autophagy, the other Atg1/ULK1 kinase complex members have received much less attention. Here, we focus on the functions of the Atg1 kinase complex members Atg11 and Atg17 as well as the ULK1 kinase complex member FIP200 in autophagy. These three proteins act as scaffolds in their respective complexes. Recent studies have made it evident that they have similar but also distinct functions. In this article, we review our current understanding of how these scaffold proteins function from autophagosome formation to fusion and also discuss their possible roles in diseases.

[†] Equal contribution, in alphabetical order.

Progress in Molecular Biology and Translational Science, Volume 172
ISSN 1877-1173
https://doi.org/10.1016/bs.pmbts.2020.01.009

1. Introduction

Autophagy and the ubiquitin-proteasome system are two major degradation pathways in eukaryotic cells. Highly conserved from budding yeast to mammals, autophagy represents an intracellular trafficking and recycling system, responsible for degradation of proteins and organelles as well as bulk cytoplasm turnover. Macroautophagy (hereafter referred to as autophagy) is a type of autophagy through which damaged organelles, aggregated or superfluous proteins and intracellular pathogens are taken up into a double-membrane vesicle, the autophagosome, and subsequently transported to the vacuole (in yeast) or lysosome (in mammals) for degradation. The resulting building blocks are released into the cytoplasm for reuse. Autophagy is a predominantly stress-induced process critical for intracellular homeostasis. In the last two decades a number of studies in higher eukaryotes demonstrated that defective autophagy is contributing to several pathologies of diseases such as cancer and neurodegeneration, exposing the necessity of fully functional autophagy in cellular homeostasis.[1,2]

Autophagy can be both a selective and non-selective process. Nonselective or bulk autophagy is primarily induced as a stress response to nutrient shortage and is characterized by the random engulfment of cytoplasmic material into an autophagosome. Selective autophagy involves the recognition of a specific cargo by dedicated receptor proteins. Once recognized by a receptor, the cargo is packed into an autophagosome and delivered to the lytic compartment for degradation.[3,4]

Genetic screens in yeast revealed a large group of autophagy-related genes (ATG), which are key to the underlying molecular mechanisms involved in autophagy.[5,6] To date more than 40 Atg proteins have been identified, carrying out a diverse range of functions during the different steps of autophagy. Many functional and structural homologs of these Atg proteins have also been found in higher eukaryotes.

In yeast the formation of the phagophore assembly site (PAS, also called preautophagosomal structure), which is in close vicinity to the vacuole and endoplasmic reticulum (ER), is deemed the first step of autophagy.[7,8] The PAS has been defined as the region where autophagy proteins assemble in a hierarchical manner to induce the formation of a double-membrane structure, the phagophore (also called isolation membrane). In mammals, the phagophore is believed to form throughout the cytoplasm at specific subdomains of the ER. During autophagy progression, both in yeast and in higher eukaryotes, the phagophore grows around the cargo destined for degradation and eventually closes to form an autophagosome. Once formed, autophagosomes mature and fuse with the lytic compartment, to expose the cargo to resident hydrolases for degradation.

The earliest factor recruited to the PAS is the Atg1 kinase complex in yeast and the homologous ULK1 kinase complex in mammals. Atg1 and ULK1 are serine-threonine protein kinases, whose kinase activity is essential for both selective and nonselective autophagy.[9] Atg1 assembles with trimeric Atg17-Atg31-Atg29 and Atg13 into a complex.[4,10,11] Also part of the complex is the scaffold protein Atg11.[12] Similar to other kinases, Atg1 multimerizes as part of its activation mechanism, allowing trans-autophosphorylation and subsequent kinase activation.[13–15] Kinase activity consequently leads to recruitment of additional Atg proteins and the maturation of the PAS into a site of phagophore formation. Similarly, the ULK1 kinase acts early in autophagosome formation in mammals. The ULK1 kinase complex consists of ULK1, FIP200, ATG13, and ATG101.[16] This complex is thought to be one of the earliest structures to assemble at the ER upon autophagy induction. At these sites omega shaped structures, so called omegasomes, form, from which the phagophore then emerges. Similar to Atg1, multimerization of ULK1 leads to activation of ULK1 via autophosphorylation, which in turn promotes recruitment of autophagy proteins. Concomitant with Atg1 and ULK1 complex activation, Atg9 vesicles are recruited to the site of autophagosome formation and aid in membrane nucleation. Also conserved among species is the subsequent recruitment of the VPS34 complex, the Atg2–18 complex, the Atg12-Atg5-Atg16 conjugation machinery and Atg8 proteins (for simplicity we use the yeast nomenclature for conserved proteins).[17,18] The VPS34 complex is important for the production of the phospholipid phosphatidylinositol 3-phosphate (PI3P), which is needed for the recruitment of the Atg12-Atg5-Atg16 ubiquitin like conjugation machinery as well as the Atg2-Atg18 complex. Atg12-Atg5-Atg16 catalyzes the conjugation of the ubiquitin-like Atg8 family proteins (Microtubule-associated protein 1 light chain 3 (LC3A, B and C) and GABA type A receptor associated proteins (GABARAP, GABARAPL1 and GABARAPL2) in mammals, Atg8 in yeast) to the lipid phosphatidylethanolamine. Lipidated Atg8 family proteins are an essential part and also a hallmark of autophagic membranes. Atg8s are considered highly relevant in cargo recognition, autophagosome maturation, closure and fusion with lysosomes.[19,20]

Selective and non-selective autophagy share many players and mechanistic steps. A striking difference between these pathways in yeast is the way the

PAS is initiated, which depends on the presence of two different scaffold proteins—Atg11 in selective and Atg17 in bulk autophagy. In higher eukaryotes FIP200 (FAK Family Kinase-interacting protein of 200 kDa) has been proposed to combine the functions of both these yeast scaffold proteins in selective and bulk autophagy.[16] In this review we summarize the current knowledge on Atg11, Atg17 and FIP200 in autophagy. We further compare and contrast the functional and structural characteristics of these scaffolds and their contribution during different stages of autophagy, as well as the role of FIP200 in disease.

2. Scaffold proteins during autophagy initiation

Autophagosome formation in yeast is initiated at the PAS. Depending on the type of autophagy that is executed, the PAS adopts different kinds of structures, influenced by the presence of the cargo and the respective scaffold proteins, Atg11 during selective and Atg17 during bulk autophagy (Fig. 1A).[21] In both cases the Atg1 kinase needs to be activated at the PAS. Clustering of Atg1 at the PAS induces its trans-autophosphorylation and subsequent kinase activation. This clustering is achieved by two different mechanisms in selective and bulk autophagy.

In selective autophagy, Atg11 acts as an adaptor protein to deliver the receptor-bound cargo to the vacuole. Atg11 has been suggested to form an elongated homodimer with a parallel coiled-coil architecture.[21] In a parallel but independent manner, Atg13 recruits Atg1 to the vacuole. The pathways converge on the vacuolar surface, where Atg1 binds the Cargo-Atg11 complex. The cargo complex provides a local clustering platform for Atg1, leading to trans-autophosphorylation and activation of the kinase. Then, further autophagy proteins are recruited, resulting in the initiation of autophagic membrane formation. This two-step process allows the cell to regulate Atg1 kinase activation in space and time, thereby preventing undesirable kinase activity and futile autophagosome formation (Fig. 2A).[13]

Different to selective autophagy, there is no cargo that could serve as a clustering platform in bulk autophagy, necessitating a different mechanism. Initiation of bulk autophagy is tightly connected to nutrient levels and the Target of rapamycin complex 1 (TORC1). Active TORC1 inhibits bulk autophagy by negatively regulating Atg1 kinase activity via phosphorylation of Atg13. TORC1 inactivation upon nutrient depletion leads to dephosphorylation of Atg13 and subsequent recruitment of Atg1 to the vacuole.[22–24] At the vacuole, dephosphorylated Atg13 binds to Atg17, thereby

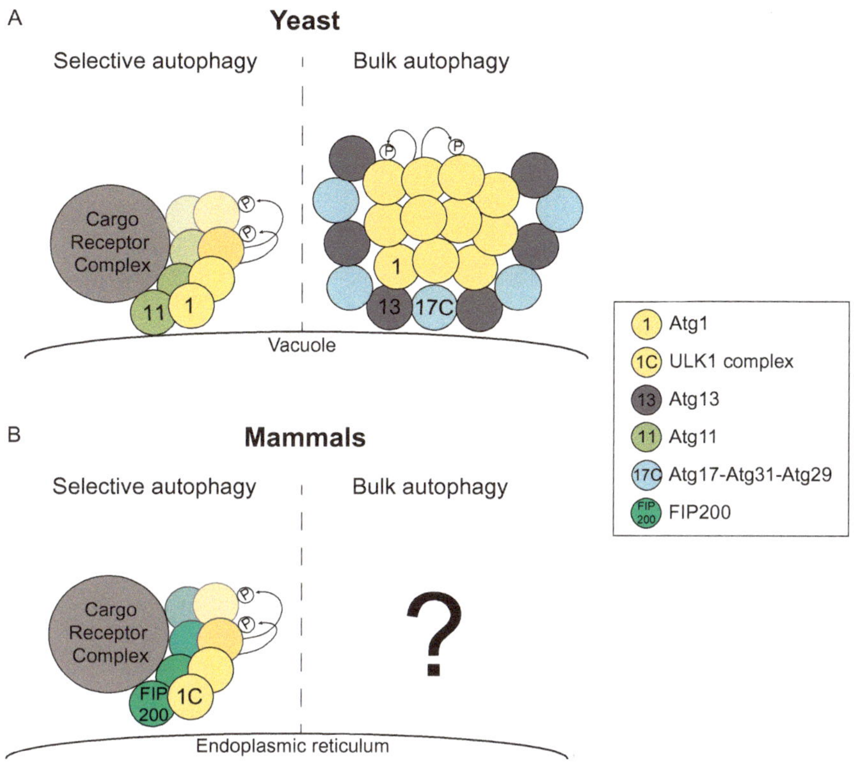

Fig. 1 Comparison of the site of autophagosome formation in different situations. The assembly of the autophagic machinery at a specific site to initiate autophagic membranes, is conserved from yeast to mammals. In all cases a prerequisite is the activation of the Atg1/ULK1 kinase at this site. The mechanistic steps between selective and non-selective autophagy, differ however. During selective autophagy in yeast, the cargo is bound to receptors, which in turn bind to the scaffold protein Atg11. Due to multiple binding, this results in clustering of Atg11 on the cargo. Atg11 anchors the cargo at the vacuole, where it serves as a platform for the assembly of further autophagy machinery proteins. Atg11 interacts with the Atg1 kinase complex, thereby facilitating the clustering of this complex and allowing its auto-phosphorylation, leading to kinase activation (A, left side). During bulk autophagy no cargo is available to act as a clustering platform. Here Atg17 fulfills the role of the scaffold protein. Atg17 forms a complex with Atg31 and Atg29 (17C), and binds Atg13 upon bulk autophagy initiation. Atg13 provides multiple intercomplex linkages between Atg17-Atg29-Atg31 complexes, leading to the supramolecular self-assembly of these proteins. Recent work suggests that the PAS is a liquid like condensate of Atg proteins rather than a rigid mesh.[24a] In both scenarios a concentration of Atg proteins at the PAS is achieved, leading to the local concentration of Atg1, which allows Atg1 kinase auto-activation (A, right side). In mammals, FIP200 acts as a scaffold protein and recruits the ULK1 kinase complex (1C) to a selective cargo. This may promote clustering and kinase activation, in a similar manner to Atg11 in yeast. How this structure forms and how the ULK1 kinase is activated without a specific cargo during bulk autophagy is unclear in mammals. Note that the ULK1 complex (1C) indicates the assembly of ULK1 with ATG13 and ATG101.

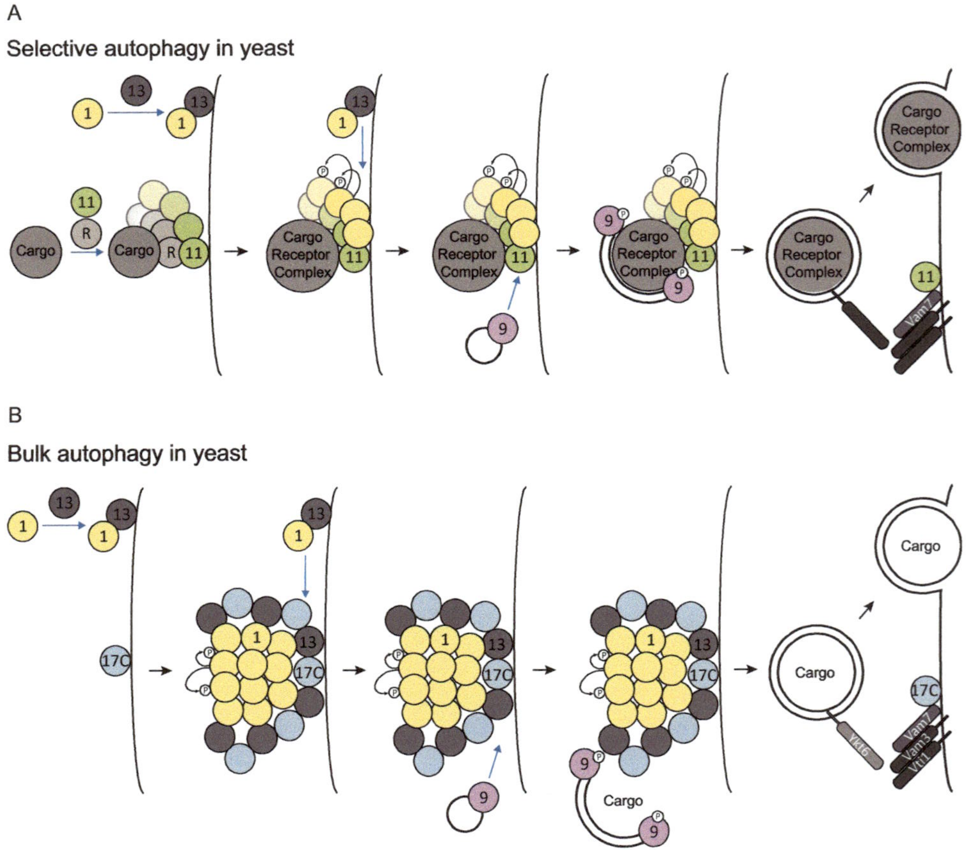

Fig. 2 See legend on next page.

linking multiple Atg17-Atg29-Atg31 complexes to one another. This has been proposed to result in the formation of a mesh like structure. Within this mesh like structure Atg1 molecules are brought into close proximity, triggering trans-autophosphorylation of Atg1 and thereby inducing kinase activation. The subsequent recruitment of further factors then result in the assembly of the PAS and autophagy initiation (Fig. 2B).[15] Recent work suggests that the PAS is a liquid like condensate of Atg proteins rather than a rigid mesh.[24a] In both scenarios a concentration of Atg proteins at the PAS is achieved, leading to Atg1 kinase molecules in close proximity, allowing their trans-activation.

It has been suggested that there are two different Atg1 kinase complexes formed depending on the selectivity of the autophagic process—one containing Atg17 (nonselective bulk autophagy), the other containing Atg11 (selective autophagy). However, a growing body of evidence is indicating that Atg11 and Atg17 act together to form a holo-complex with Atg31, Atg29, Atg1 and Atg13.[12] In the absence of other Atg proteins, the vacuolar localization of Atg17-Atg31-Atg29 is furthermore dependent on Atg11, thus demonstrating that Atg11 is necessary and sufficient to trigger PAS formation. It was also shown that phosphorylated Atg29 directly interacts with Atg11, indicating that the recruitment of the Atg1 kinase is achieved by an interplay between the Atg17-Atg31-Atg29 complex and Atg11.[12]

Within the Atg17-Atg29-Atg31 complex Atg17 is composed of a right-handed 3-helix bundle showing an antiparallel orientation with a dimerization domain in the C-terminus. Structural studies suggest two antiparallel Atg17 monomers to form an S-shaped homodimer. This antiparallel dimeric architecture of Atg17 is in contrast to the suggested parallel coiled-coil

Fig. 2 Selective and bulk autophagy in yeast and the involvement of scaffold proteins. (A) In yeast selective autophagy is initiated by the formation of cargo-receptor complexes, which are recruited to the vacuole by Atg11. In a parallel but independent manner, Atg13 recruits Atg1 to the vacuole. Atg1 can only interact with Atg11 at the vacuole, where it clusters on the cargo-receptor complex and subsequently activates via trans-autophosphorylation. Then Atg9 vesicles are recruited and Atg9 is phosphorylated by Atg1. This allows the recruitment of further autophagy proteins and outgrowth of the autophagic membrane around the cargo. Once the autophagosome is completed, it fuses with the vacuole. This requires SNARE proteins. Among these SNARE proteins is Vam7, which is recruited to the vacuole in an Atg17 and Atg11 dependent manner. (B) In yeast bulk autophagy is initiated by the formation of a supramolecular structure of Atg13 with the Atg17-Atg31-Atg29 complex (Atg17C). This results in clustering and autoactivation of Atg1. Subsequent steps, such as the recruitment and phosphorylation of Atg9, and the fusion of autophagosomes with the vacuole are believed to happen analogous to the selective pathway.

structure of Atg11. Dimerization of Atg17 is essential for PAS initiation and autophagy progression.[25] Although this model offers an elegant explanation for PAS formation, many molecular events surrounding this mesh assembly remain unknown. For example, it is unclear whether the Atg17-Atg31-Atg29 complex assembles at the vacuole prior to phagophore formation, or whether it forms in the cytoplasm and is recruited to the vacuole only after the complex has been assembled. If the latter is true, how the complex is transported to the vacuole and how the site of PAS formation is selected remain to be clarified.

FIP200, also known as RB1CC1, was first identified as an ULK1 complex interactor.[26] Of the ULK1 complex members only ATG13 directly interacts with FIP200. ATG13 also directly binds to ULK1 and ATG101 (Fig. 1B).[27–30] FIP200 has no clear sequence homologue in yeast, but contains an Atg11 homology region in its C-terminus. It is considered the mammalian functional counterpart to yeast Atg17 and Atg11.[31–33] Similar to yeast Atg11, FIP200 binds autophagy cargo receptors, e.g., NDP52, p62 and CCPG1, thus linking cargo to the ULK1 complex and acting as a scaffold protein. Binding of the ULK1 complex to a cargo is thought to activate ULK1 and facilitate recruitment of additional autophagy proteins (Fig. 1B).[34–37] The binding of FIP200 to these receptors is dependent on FIP200-interaction-regions on receptors. So far only the three autophagy receptors mentioned above have been identified to contain FIP200-interaction-regions and bind FIP200 directly. Whether other autophagy receptors also carry such regions and are capable of binding directly to FIP200 is unknown at this time.

The role of FIP200 in bulk autophagy is less studied. Inhibition of bulk autophagy initiation through phosphorylation of ULK1 and ATG13 by mammalian TORC1 (mTORC1) is however well established and demonstrates similarities in bulk autophagy regulation between mammals and yeast. Cells depleted of FIP200 fail to recruit ULK1 to the site of autophagosome formation or possibly fail in establishing such a site, resulting in impaired autophagy flux. This highlights the important role of FIP200 in autophagy. Several downstream autophagy proteins including ULK1, ATG16L1 and LC3 do not form puncta in FIP200 KO cells, supporting the notion of FIP200 as an essential scaffold in autophagy initiation.[38] If FIP200 multimerization is similar to the yeast scaffolds Atg11 or Atg17 is not known at this time.

A role of FIP200 in autophagy initiation upstream of mTORC1 regulation of the ULK1 complex has also been suggested. mTORC1 is negatively regulated by the Tuberous Sclerosis Complex 1 and 2 (TSC1/TSC2), which prevents mTORC1 association with the lysosome and thereby its activation.[39,40] FIP200 and WIPI3 were reported to interact with the middle region of TSC1 and colocalize with the TSC1/TSC2 complex at lysosomes.[41,42] This colocalization increased upon amino acid deprivation, indicating an involvement in amino acid dependent mTORC1 regulation.[42] From these findings one can speculate that FIP200 and WIPI3 aid in autophagy induction by promoting mTORC1 inhibition via their interaction with the TSC1/2 complex. As mTORC1 is crucial for bulk autophagy, understanding its regulation by FIP200 and WIPI3 is of considerable interest (Fig. 3).

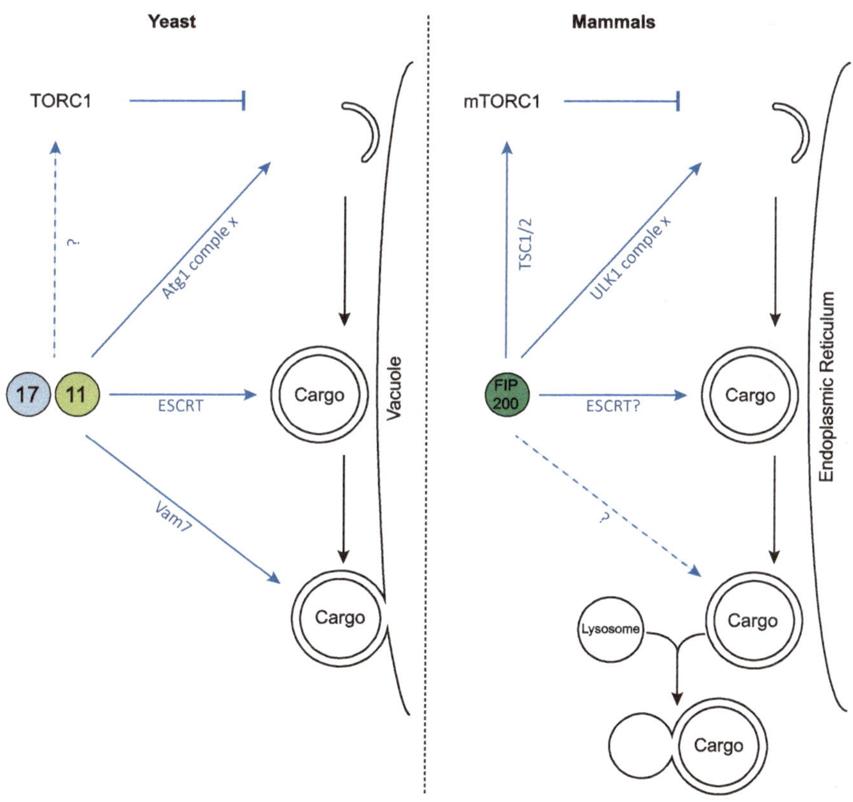

Fig. 3 Role of scaffold proteins throughout the autophagy pathway. Both in yeast and in mammals, the scaffold proteins Atg11, Atg17 and FIP200 have been describe to act throughout the entire autophagy process, possibly fulfilling similar functions across species.

Taken together, both in yeast and in mammals, the scaffold proteins Atg11, Atg17 and FIP200 elicit similar functions, by either connecting receptor bound or bulk cytoplasmic cargo to the autophagy machinery. These proteins act early in the process of autophagy initiation by providing a scaffold, thus aiding the assembly of the autophagic machinery at the PAS. Elucidating the functions of these scaffold proteins and understanding how FIP200 is able to combine the functions of Atg17 and Atg11 will need further analysis.

3. The role of scaffold proteins in phagophore expansion

Different levels of regulation are required to ensure the correct architecture and function of autophagosomes during the autophagic process. Although many steps within autophagosome formation, maturation and fusion are still insufficiently understood, it is becoming apparent that scaffold proteins play a critical role throughout the process of autophagy.

In yeast, activation of Atg1 drives autophagy progression and formation of the phagophore. Following Atg1 activation, Atg9 vesicles are recruited to the PAS in an Atg11 and Atg17 dependent manner. Atg1 directly phosphorylates Atg9 at the PAS, prompting the recruitment of further Atg proteins.[43] A small number of Atg9-vesicles, derived from the Golgi apparatus fuse, which nucleates the formation of the phagophore at the PAS (Fig. 2). These Atg9 vesicles become part of the autophagic membrane, however Atg9 eventually gets retrieved.[44] They are however not the sole membrane source.

The transition from nucleation to formation of a cup-shaped phagophore was also studied using a computational and biophysical simulation model.[45] This simulation revealed that membrane-associated Atg17 dimers play a critical role in the escape from a metastable tubular structure to phagophore formation by stabilizing the transition state from tube to disk shape on the one hand and further progression into a cup-shape structure on the other. Furthermore, Atg17 dimers lower the energy of these transition steps by interacting with the forming membrane.[45] However, an earlier study suggested that Atg17 does not directly bind membranes.[25] Thus, how Atg17 shapes phagophores and how Atg9 contributes, will need to be addressed in more detail.

Atg11 and Atg17 have been reported to interact with Atg9 in vivo and in vitro.[46–48] The role of Atg9 is proving more elusive in mammals and is not yet adequately explained. The ULK1 complex is known to first interact with phosphatidylinositol synthase-enriched ER subdomains, and then to translocate to Atg9a-positive autophagosome precursors in a PI3P-dependent manner.[19,34] However, no direct interaction between Atg9 and FIP200, or any other member of the ULK1 complex, has been shown to this date, despite Atg9 being a substrate of ULK1.[49] Also, Atg9 vesicles do not seem to become part of autophagic membranes in mammals, in contrast to their yeast counterparts, although conflicting results exist.[19,34,50]

The Atg8 subfamilies LC3 and GABARAP are proposed to be critical in autophagosome formation, recruitment of autophagy factors and selective sequestration of substrates via cargo receptors.[51,52] The GABARAP subfamily members GABARAP and GABARAPL2 were recently observed to efficiently mediate clustering of liposomes as well as mixing vesicular contents, suggesting a role in autophagosome maturation and fusion. In contrast, the LC3 subfamily did not further membrane fusion as efficiently, but rather promoted changes in vesicle shape to more elongated structures.[52–55] Recruitment of Atg8s to autophagosomal membranes requires several autophagy factors, including WIPI2 and the ATG12-ATG5-ATG16L1 ubiquitin like conjugation machinery. Several studies have reported a direct interaction of FIP200 with ATG16L1 which is downstream of WIPI2 binding.[56–59] In fact, WIPI2 and FIP200 bind to adjacent but distinct regions of ATG16L1. It was also shown that a FIP200 binding deficient ATG16L1 rescued autophagy in ATG16L1 −/− cells to similar levels as WT ATG16L1, while a WIPI2 binding deficient ATG16L1 did not.[59] In line with this finding another study reported that the FIP200-interaction site on Atg16L1 is not essential for autophagy, but is needed for accurate ATG16L1 targeting to autophagosomal membranes. It was also hypothesized that this interaction may facilitate the assembly of a large complex in the cytoplasm consisting of the ULK1 and ATG12-ATG5-ATG16L1 complexes.[57] These studies provide evidence that the FIP200-ATG16L1 interaction is nonessential for autophagy but plays a role in ATG16L1 membrane targeting. How FIP200 and the ULK1 complex execute a targeting function on the ATG12-ATG5-ATG16L1 complex remains unexplained.

Binding to proteins of the Atg8 family is facilitated by sequence motifs called LC3-interacting regions (LIRs, or also AIM for Atg8 interacting motif in yeast). Of the ULK1 complex members ULK1, ATG13 and FIP200 have such LIR motifs and are able to bind ATG8s.[60,61] The LIR motif containing ULK1 complex members all share a preference for binding to the GABARAP subfamily. FIP200 binds strongest to GABARAP and

GABARAPL1 and relatively weaker to GABARAPL2, LC3-C and LC3-A, and very weakly to LC3-B.[60,62,63] Due to the involvement of GABARAPs in autophagosome maturation and fusion, the preferential binding of ULK1 complex members to GABARAPs is thought to be an indication for a role of the ULK1 complex also in late autophagy steps.[53,60,61,64] The function of FIP200, ULK1 and ATG13 binding to ATG8s as well as their preference to the GABARAP subfamily is however still unclear. Further research is needed to fully characterize the role of ULK1 complex binding to ATG8s.

Both Atg1 and Atg13 have been found to bind Atg8 in a LIR-dependent manner in yeast, thereby associating with forming and/or mature autophagosomes.[61] Also plant ATG1 interacts with ATG8 in a LIR-dependent manner and thereby links the ATG1/13 complex to ATG8-decorated autophagic vesicles. Interestingly, plant ATG13 does not bind ATG8, but plant ATG11 contains a LIR motif for ATG8 interaction, thus providing a second contact site between the ATG1/13 complex and ATG8 membranes.[33]

Little is known about the involvement of scaffold proteins in phagophore expansion in yeast. While Atg1 is required for phagophore expansion by recruiting downstream factors such as Atg18, it is unclear if the scaffold proteins are directly involved.[43] Atg1 was found to not always assemble with its complex members. During phagophore expansion, Atg1 spreads along the expanding membrane, while Atg17 and Atg13 only localize to distinct spots at contact sites of the phagophore with the vacuole.[8] Therefore, a scenario in which scaffold proteins act independently and outside of the Atg1/ULK1 kinase complex during certain steps in autophagy seems plausible.

A prerequisite for the formation of autolysosomes in mammals or uptake of autophagosomes by the vacuole in yeast, is the closure of the phagophore. It was recently reported that CHMP2A, which is part of the endosomal sorting complexes required for transport (ESCRT)-III, is a critical regulator of phagophore closure in mammals (Fig. 3). Depletion of CHMP2A lead to an accumulation of unclosed phagophores in cells.[65] Similarly, depletion of the yeast ESCRT complex resulted in accumulation of unclosed autophagosomes. The interaction of Atg17 with the ESCRT protein Snf7 was found important for ESCRT recruitment to unclosed autophagosomes, where ESCRT catalyzes subsequent autophagosome closure (Fig. 3).[66] These findings raise the question whether FIP200 plays a comparable role to yeast Atg17 in recruiting ESCRT-III to autophagosomes in mammals. Studying closed vs unclosed phagophores could be a useful tool to quantitatively analyze autophagosome biogenesis and maturation in both yeast and mammals.

4. The role of scaffold proteins in the fusion of autophagosomes with the lytic compartment

The common concensus on autophagy scaffold proteins predominantly participating in early autophagy steps is being challenged by recent publications suggesting functions in later autophagy steps such as the fusion of autophagosomes with the lytic compartment. Both Atg17 and Atg11 have been reported to interact with the SNARE protein Vam7. A model has been proposed where Atg17 and Atg11 coordinate the recruitment of Vam7 to the PAS, and regulate timely autophagosome-vacuole fusion by blocking the premature interaction of Vam7 with other SNARE proteins.[67] Vam7 bundles with Vam3, Vti1 and Ykt6 during autophagosome-vacuole fusion.[68] If scaffold proteins also affect the recruitment and function of these SNARE proteins remains unknown (Fig. 2A and B).

Little is known about the role of FIP200 in autophagosome-lysosome fusion. FIP200 KO cell lines abolish the binding of ULK1 complex members to cargo receptors, thus impairing autophagy initiation. The function of FIP200 in fusion can therefore not be studied in such a setup. A modification of ULK1 important for autophagosome-lysosome fusion but irrelevant for autophagy initiation was recently discovered. ULK1 dephosphorylation at S423 allows recruitment of STX17.[69] STX17, a SNARE protein that is translocated to autophagosomes, forms a trimeric complex with SNAP29 and the endosomal/lysosomal SNARE VAMP8, and together mediate fusion of autophagosomes with lysosomes.[70,71] As ULK1 is in complex with ATG13, FIP200 and ATG101 at autophagy initiation, it is reasonable to speculate that this is also the case during fusion. Although the ULK1 S423D phosphomimicking mutant did not affect autophagy initiation, it is unknown whether dephosphorylation of endogenous ULK1 at S423 occurs before or after autophagy initiation and which phosphatase is responsible. FIP200 is an essential member of the ULK1 kinase complex and necessary for ULK1 activation. It is therefore possible that FIP200 plays a role in autophagosome-lysosome fusion in this setting (Fig. 3). Although STX17 function is mostly discussed in fusion events, a recent report describes STX17 and TBK1 playing a role in the formation of FIP200-ATG13 structures during autophagy initiation.[72]

5. Autophagy scaffold proteins in disease

The cytoprotective function of autophagy is imperative to cellular health, which is achieved and maintained by the removal of damaged, aggregated or undesirable cellular components. Dysfunctional autophagy is linked to several diseases. For this reason there has been increasing interest in autophagy as a potential target for medical treatments. Proteins involved in autophagy initiation are of particular interest, which has put a focus on the ULK1 complex.

Retinoblastoma 1 (RB1) was one of the first molecularly described tumor suppressors and is mutated in most retinoblastomas, osteosarcomas, small-cell lung cancers, and in a variety of other cancers at a lower frequency.[73] FIP200 was described as a regulator of RB1 in several tissues and cell lines.[74] A screen of primary breast cancers revealed that 20% of the screened cancers contained deletion mutations of the FIP200 gene.[75] In line with these findings, analyses of breast cancer cells suggest a tumor suppressor role of FIP200 through its interaction with, and consequent stabilization of p53, leading to an increase of p21 and decrease of cyclin D1.[76] Recent studies also describe a role of FIP200 in brain metastases of breast cancer patients and in naïve T cell survival in tumor immunity.[77,78]

Persistent high-risk human papillomavirus (hrHPV) infection has been closely linked to cervical cancer. Autophagy has been suggested to inhibit viral infections.[79–81] A miRNA microarray analysis on hrHPV infected cervical cancer tissue and cell lines revealed miR-224-3p upregulation. In addition, miR-224-3p negatively regulated autophagy via direct inhibition of FIP200 expression.[82] Taken together these studies clearly demonstrate a role of FIP200 in several types of cancer. Whether FIP200s involvement in these diseases is connected to autophagy or independent of its role in autophagy remains unknown. How FIP200 evokes anti-tumor functions and immunity, or influences tumor growth needs further analysis.

FIP200 plays a crucial role in the elimination of invading bacteria such as *S. typhimurium* by binding to the autophagy receptor NDP52, thereby recruiting the ULK1 complex and initiating engulfment of the bacterium.[83–85] It was recently demonstrated that xenophagy of cytosol invading *S. typhimurium* only progresses when NDP52 binds FIP200 and the TBK1 adaptor protein Sintbad/NAP1. Both FIP200 and Sintbad/NAP1 bind NDP52 on opposite faces in the SKICH domain of NDP52, forming a proposed trimeric complex. Binding of FIP200 to Sintbad and NAP1 was also shown, but was not required for efficient xenophagy of *S. typhimurium*.[85] The necessity of FIP200 to bind to the SKICH domain of NDP52 and recruiting the ULK1 complex was also demonstrated in a different study in mitophagy. Even though Sintbad/NAP1 was not analyzed in this study, TBK1 knockout revealed that it was crucial in fostering the association between NDP52 and the ULK1 complex.[36] Therefore recruitment of FIP200 and TBK1 to NDP52 happens in a similar manner in mitophagy and xenophagy. The SKICH domain of the autophagy receptor TAX1BP1 harbors a strong structural homology to the SKICH domain of NDP52. Indeed it was shown that TAX1BP1 can interact with NAP1, and is also predicted to interact with Sintbad.[83,85,86] As a result it may be hypothesized that like NDP52, TAX1BP1 binds FIP200 and Sintbad/NAP1 to form a trimeric complex. It should be noted that although it seems likely that the remaining ULK1 complex members and TBK1 are recruited to a FIP200-NDP52-Sintbad/NAP1 trimeric complex at the same time, this has not yet been shown. TBK1 has emerged as an important player in inflammation and autophagy over the past few years. Mutations in TBK1 or its pathways are linked to defective autophagy and neurodegenerative diseases such as amyotrophic lateral sclerosis and frontotemporal dementia.[87] The recently published data on simultaneous binding of FIP200 and Sintbad/NAP1 to NDP52, and the importance of TBK1 phosphorylation events in selective autophagy initiation, indicate towards a yet unexplained molecular relationship between TBK1 and the ULK1 complex.[36,85]

Autophagy is critical for neuronal homeostasis and impaired autophagy can lead to devastating neurodegenerative diseases such as Huntington's disease.[88] Huntington's disease is caused by mutations in the Huntingtin (htt) gene. Accumulation of mutant Htt is thought to impair autophagic flux. Cells afflicted with Huntington's disease form and successfully remove autophagosomes via lysosomes, but fail to efficiently recognize and engulf cytosolic cargo for degradation.[89] Wild-type Htt physically interacts with the autophagy receptor p62 and facilitates its association with LC3 and ubiquitin-modified substrates. Complete activation of ULK1 in selective autophagy is achieved through Htt interaction with ULK1, followed by its release from mTOR inhibition. In bulk autophagy Htt was shown to be dispensable.[90] Overall Htt shows unmistakable functional similarities to FIP200. Both bind p62-ubiquitin condensates and interact with ULK1 in selective autophagy.[18,29,30,37,91,92] Whether Htt interacts directly with ULK1 or through another protein, akin to FIP200 and ATG13, is unknown. It is also unclear why an additional protein capable of binding p62 and ULK1 is needed in selective autophagy.

Although interactions of Htt with autophagy proteins have been demonstrated, the exact role of Htt in autophagy remains uncertain. Intriguingly, both Htt and FIP200 C-termini contain weak similarity to the C-terminus of yeast Atg11, which is the region needed for autophagy receptor binding in all three proteins.[32,35,37,93,94] This further supports the notion that Htt and FIP200 play similar roles in selective autophagy. Ultimately more research is needed to unravel the similarities and differences between Atg11, FIP200 and Htt as well as their respective functions.

6. Concluding remarks

Over recent years research has advanced our understanding of how the Atg1 and ULK1 kinases function in autophagy. The other Atg1/ULK1 kinase complex members have received significantly less attention and are therefore also less understood. Recent publications show that scaffold proteins (in particular Atg11, Atg17 and FIP200) are at least important. It has become evident that these scaffold proteins carry out several functions throughout the autophagy pathway and that their function as scaffolds is crucial for Atg1/ULK1 activation and autophagy induction. The interplay between the scaffold proteins and other autophagy factors as well as the mechanisms involved in their functions are still largely unknown. Our current knowledge of the autophagy scaffold proteins in yeast and mammals suggest they function in similar ways, but also have exclusive functions, including functions outside the Atg1/ULK1 complex (Fig. 3). In general, many players in yeast and mammalian autophagy have been identified, now we need to work on understanding how, where and when they act in autophagy at a molecular level.

Both in yeast and mammals, the scaffold proteins act in autophagy initiation. They are part of the Atg1/ULK1 complex and participate in local clustering of the complex allowing its kinase activation. Recent reports suggest that the yeast scaffold proteins act during autophagosome completion, by interacting with the ESCRT machinery. Whether FIP200 acts in a similar manner in mammals is presently not known. Atg11 and Atg17 both coordinate the recruitment of the SNARE protein Vam7 to regulate autophagosome-vacuole fusion in yeast. Whether FIP200 also acts during autophagosome-lysosome fusion in mammals is unknown. FIP200 has also been reported to play a role in autophagy initiation upstream of mTORC1 signaling. If the yeast scaffolds also interfere with TORC1 signaling has not been addressed.

Acknowledgments

We thank Sharon Tooze for input on the manuscript. The Kraft laboratory has received funding from the Vienna Science and Technology Fund (WWTF, VRG10-001) from the FWF Austrian Science Fund (grant numbers P25522-B20, P28113-B28 and W1261), from the European Research Council (ERC) under the European Union's Horizon 2020 research and innovation programme under grant agreement No 769065, from the European Union's Horizon 2020 research and innovation programme under grant agreement No 765912, from the Deutsche Forschungsgemeinschaft (DFG, German Research Foundation)—Project-ID 403222702—SFB 1381, and Project ID 259130777—SFB 1177, and from the EMBO Young Investigator Program. We thank the SGBM graduate school for supporting our students. This work reflects only the authors' view and the European Union's Horizon 2020 research and innovation programme is not responsible for any use that may be made of the information it contains. The authors declare no competing financial interests.

References

1. Kraft C, Martens S. Mechanisms and regulation of autophagosome formation. *Curr Opin Cell Biol*. 2012;24:496–501.
2. Mizushima N, Levine B, Cuervo AM, Klionsky DJ. Autophagy fights disease through cellular self-digestion. *Nature*. 2008;451:1069–1075.
3. Nakatogawa H, Suzuki K, Kamada Y, Ohsumi Y. Dynamics and diversity in autophagy mechanisms: lessons from yeast. *Nat Rev Mol Cell Biol*. 2009;10:458–467.
4. Kraft C, Reggiori F, Peter M. Selective types of autophagy in yeast. *Biochim Biophys Acta*. 2009;1793:1404–1412.
5. Yang Z, Klionsky DJ. An overview of the molecular mechanism of autophagy. *Curr Top Microbiol Immunol*. 2009;335:1–32.
6. Klionsky DJ, et al. A unified nomenclature for yeast autophagy-related genes. *Dev Cell*. 2003;5:539–545.
7. Graef M, Friedman JR, Graham C, Babu M, Nunnari J. ER exit sites are physical and functional core autophagosome biogenesis components. *Mol Biol Cell*. 2013;24: 2918–2931.
8. Suzuki K, Akioka M, Kondo-Kakuta C, Yamamoto H, Ohsumi Y. Fine mapping of autophagy-related proteins during autophagosome formation in *Saccharomyces cerevisiae*. *J Cell Sci*. 2013;126:2534–2544.
9. Papinski D, Kraft C. Regulation of autophagy by signaling through the Atg1/ULK1 complex. *J Mol Biol*. 2016;428:1725–1741.
10. Kawamata T, Kamada Y, Kabeya Y, Sekito T, Ohsumi Y. Organization of the pre-autophagosomal structure responsible for autophagosome formation. *Mol Biol Cell*. 2008;19:2039–2050.
11. Kabeya Y, et al. Atg17 functions in cooperation with Atg1 and Atg13 in yeast autophagy. *Mol Biol Cell*. 2005;16:2544–2553.
12. Mao K, et al. Atg29 phosphorylation regulates coordination of the Atg17-Atg31-Atg29 complex with the Atg11 scaffold during autophagy initiation. *Proc Natl Acad Sci USA*. 2013;110:E2875–E2884.
13. Torggler R, et al. Two independent pathways within selective autophagy converge to activate Atg1 kinase at the vacuole. *Mol Cell*. 2016;64:221–235.
14. Kamber RA, Shoemaker CJ, Denic V. Receptor-bound targets of selective autophagy use a scaffold protein to activate the Atg1 kinase. *Mol Cell*. 2015;59:372–381.
15. Yamamoto H, et al. The intrinsically disordered protein Atg13 mediates supramolecular assembly of autophagy initiation complexes. *Dev Cell*. 2016;38:86–99.
16. Mizushima N. The role of the Atg1/ULK1 complex in autophagy regulation. *Curr Opin Cell Biol*. 2010;22:132–139.
17. Hurley JH, Young LN. Mechanisms of autophagy initiation. *Annu Rev Biochem*. 2017;86:225–244.

18. Mercer TJ, Gubas A, Tooze SA. A molecular perspective of mammalian autophagosome biogenesis. *J Biol Chem*. 2018;293:5386–5395.
19. Karanasios E, et al. Autophagy initiation by ULK complex assembly on ER tubulovesicular regions marked by ATG9 vesicles. *Nat Commun*. 2016;7:12420.
20. Zachari M, Ganley IG. The mammalian ULK1 complex and autophagy initiation. *Essays Biochem*. 2017;61:585–596.
21. Suzuki H, Noda NN. Biophysical characterization of Atg11, a scaffold protein essential for selective autophagy in yeast. *FEBS Open Bio*. 2018;8:110–116.
22. Funakoshi T, Matsuura A, Noda T, Ohsumi Y. Analyses of APG13 gene involved in autophagy in yeast, *Saccharomyces cerevisiae*. *Gene*. 1997;192:207–213.
23. Matsuura A, Tsukada M, Wada Y, Ohsumi Y. Apg1p, a novel protein kinase required for the autophagic process in *Saccharomyces cerevisiae*. *Gene*. 1997;192:245–250.
24. Kamada Y, et al. Tor-mediated induction of autophagy via an Apg1 protein kinase complex. *J Cell Biol*. 2000;150:1507–1513. (a) Fujioka Y, et al. Phase separation organizes the site of autophagosome formation. *Nature*. 2020;578:301–305.
25. Ragusa MJ, Stanley RE, Hurley JH. Architecture of the Atg17 complex as a scaffold for autophagosome biogenesis. *Cell*. 2012;151:1501–1512.
26. Hara T, et al. FIP200, a ULK-interacting protein, is required for autophagosome formation in mammalian cells. *J Cell Biol*. 2008;181:497–510.
27. Chen S, et al. Distinct roles of autophagy-dependent and -independent functions of FIP200 revealed by generation and analysis of a mutant knock-in mouse model. *Genes Dev*. 2016;30:856–869.
28. Qi S, Kim DJ, Stjepanovic G, Hurley JH. Structure of the human Atg13-Atg101 HORMA heterodimer: an interaction hub within the ULK1 complex. *Structure*. 2015;23:1848–1857.
29. Jung CH, et al. ULK-Atg13-FIP200 complexes mediate mTOR signaling to the autophagy machinery. *Mol Biol Cell*. 2009;20:1992–2003.
30. Hosokawa N, et al. Nutrient-dependent mTORC1 association with the ULK1-Atg13-FIP200 complex required for autophagy. *Mol Biol Cell*. 2009;20:1981–1991.
31. Hara T, Mizushima N. Role of ULK-FIP200 complex in mammalian autophagy: FIP200, a counterpart of yeast Atg17? *Autophagy*. 2009;5:85–87.
32. Steffan JS. Does huntingtin play a role in selective macroautophagy? *Cell Cycle*. 2010;9:3401–3413.
33. Li F, Chung T, Vierstra RD. AUTOPHAGY-RELATED11 plays a critical role in general autophagy- and senescence-induced mitophagy in Arabidopsis. *Plant Cell*. 2014;26:788–807.
34. Nishimura T, et al. Autophagosome formation is initiated at phosphatidylinositol synthase-enriched ER subdomains. *EMBO J*. 2017;36:1719–1735.
35. Smith MD, et al. CCPG1 is a non-canonical autophagy cargo receptor essential for ER-phagy and pancreatic ER proteostasis. *Dev Cell*. 2018;44:217–232.e11.
36. Vargas JNS, et al. Spatiotemporal control of ULK1 activation by NDP52 and TBK1 during selective autophagy. *Mol Cell*. 2019;74:347–362.e6.
37. Turco E, et al. FIP200 claw domain binding to p62 promotes autophagosome formation at ubiquitin condensates. *Mol Cell*. 2019;74:330–346.e11.
38. Itakura E, Mizushima N. Characterization of autophagosome formation site by a hierarchical analysis of mammalian Atg proteins. *Autophagy*. 2010;6:764–776.
39. Demetriades C, Doumpas N, Teleman AA. Regulation of TORC1 in response to amino acid starvation via lysosomal recruitment of TSC2. *Cell*. 2014;156:786–799.
40. Menon S, et al. Spatial control of the TSC complex integrates insulin and nutrient regulation of mTORC1 at the lysosome. *Cell*. 2014;156:771–785.
41. Gan B, Melkoumian ZK, Wu X, Guan K-L, Guan J-L. Identification of FIP200 interaction with the TSC1-TSC2 complex and its role in regulation of cell size control. *J Cell Biol*. 2005;170:379–389.

42. Bakula D, et al. WIPI3 and WIPI4 β-propellers are scaffolds for LKB1-AMPK-TSC signalling circuits in the control of autophagy. *Nat Commun.* 2017;8:15637.
43. Papinski D, et al. Early steps in autophagy depend on direct phosphorylation of Atg9 by the Atg1 kinase. *Mol Cell.* 2014;53:471–483.
44. Yamamoto H, et al. Atg9 vesicles are an important membrane source during early steps of autophagosome formation. *J Cell Biol.* 2012;198:219–233.
45. Bahrami AH, Lin MG, Ren X, Hurley JH, Hummer G. Scaffolding the cup-shaped double membrane in autophagy. *PLoS Comput Biol.* 2017;13:e1005817.
46. He C, Klionsky DJ. Atg9 trafficking in autophagy-related pathways. *Autophagy.* 2007;3:271–274.
47. Chang C-Y, Huang W-P. Atg19 mediates a dual interaction cargo sorting mechanism in selective autophagy. *Mol Biol Cell.* 2007;18:919–929.
48. Rao Y, Perna MG, Hofmann B, Beier V, Wollert T. The Atg1-kinase complex tethers Atg9-vesicles to initiate autophagy. *Nat Commun.* 2016;7:10338.
49. Zhou C, et al. Regulation of mATG9 trafficking by Src- and ULK1-mediated phosphorylation in basal and starvation-induced autophagy. *Cell Res.* 2017;27:184–201.
50. Orsi A, et al. Dynamic and transient interactions of Atg9 with autophagosomes, but not membrane integration, are required for autophagy. *Mol Biol Cell.* 2012;23:1860–1873.
51. Shpilka T, Weidberg H, Pietrokovski S, Elazar Z. Atg8: an autophagy-related ubiquitin-like protein family. *Genome Biol.* 2011;12:226.
52. Nguyen TN, et al. Atg8 family LC3/GABARAP proteins are crucial for autophagosome-lysosome fusion but not autophagosome formation during PINK1/Parkin mitophagy and starvation. *J Cell Biol.* 2016;215:857–874.
53. Weidberg H, et al. LC3 and GATE-16/GABARAP subfamilies are both essential yet act differently in autophagosome biogenesis. *EMBO J.* 2010;29:1792–1802.
54. Landajuela A, et al. Lipid geometry and bilayer curvature modulate LC3/GABARAP-mediated model autophagosomal elongation. *Biophys J.* 2016;110:411–422.
55. Abdollahzadeh I, Schwarten M, Gensch T, Willbold D, Weiergräber OH. The Atg8 family of proteins-modulating shape and functionality of autophagic membranes. *Front Genet.* 2017;8:109.
56. Gammoh N, Florey O, Overholtzer M, Jiang X. Interaction between FIP200 and ATG16L1 distinguishes ULK1 complex-dependent and -independent autophagy. *Nat Struct Mol Biol.* 2013;20:144–149.
57. Nishimura T, et al. FIP200 regulates targeting of Atg16L1 to the isolation membrane. *EMBO Rep.* 2013;14:284–291.
58. Fujita N, et al. Recruitment of the autophagic machinery to endosomes during infection is mediated by ubiquitin. *J Cell Biol.* 2013;203:115–128.
59. Dooley HC, et al. WIPI2 links LC3 conjugation with PI3P, autophagosome formation, and pathogen clearance by recruiting Atg12-5-16L1. *Mol Cell.* 2014;55:238–252.
60. Alemu EA, et al. ATG8 family proteins act as scaffolds for assembly of the ULK complex: sequence requirements for LC3-interacting region (LIR) motifs. *J Biol Chem.* 2012;287:39275–39290.
61. Kraft C, et al. Binding of the Atg1/ULK1 kinase to the ubiquitin-like protein Atg8 regulates autophagy. *EMBO J.* 2012;31:3691–3703.
62. Rogov VV, et al. Structural and functional analysis of the GABARAP interaction motif (GIM). *EMBO Rep.* 2017;18:1382–1396.
63. Wirth M, et al. Molecular determinants regulating selective binding of autophagy adapters and receptors to ATG8 proteins. *Nat Commun.* 2019;10:2055.
64. Joachim J, et al. Activation of ULK kinase and autophagy by GABARAP trafficking from the centrosome is regulated by WAC and GM130. *Mol Cell.* 2015;60:899–913.
65. Takahashi Y, et al. An autophagy assay reveals the ESCRT-III component CHMP2A as a regulator of phagophore closure. *Nat Commun.* 2018;9:2855.

66. Zhou F, et al. Rab5-dependent autophagosome closure by ESCRT. *J Cell Biol.* 2019;218:1908–1927.
67. Liu X, et al. The Atg17-Atg31-Atg29 complex coordinates with Atg11 to recruit the Vam7 SNARE and mediate autophagosome-vacuole fusion. *Curr Biol.* 2016;26:150–160.
68. Bas L, et al. Reconstitution reveals Ykt6 as the autophagosomal SNARE in autophagosome-vacuole fusion. *J Cell Biol.* 2018;217:3656–3669.
69. Wang C, et al. Phosphorylation of ULK1 affects autophagosome fusion and links chaperone-mediated autophagy to macroautophagy. *Nat Commun.* 2018;9:3492.
70. Itakura E, Kishi-Itakura C, Mizushima N. The hairpin-type tail-anchored SNARE syntaxin 17 targets to autophagosomes for fusion with endosomes/lysosomes. *Cell.* 2012;151:1256–1269.
71. Matsui T, et al. Autophagosomal YKT6 is required for fusion with lysosomes independently of syntaxin 17. *J Cell Biol.* 2018;217:2633–2645.
72. Kumar S, et al. Phosphorylation of syntaxin 17 by TBK1 controls autophagy initiation. *Dev Cell.* 2019;49:130–144.e6.
73. Dyson NJ. RB1: a prototype tumor suppressor and an enigma. *Genes Dev.* 2016;30:1492–1502.
74. Chano T, Ikegawa S, Kontani K, Okabe H, Baldini N, Saeki Y. Identification of RB1CC1, a novel human gene that can induce RB1 in various human cells. *Oncogene.* 2002;21:1295–1298.
75. Chano T, Kontani K, Teramoto K, Okabe H, Ikegawa S. Truncating mutations of RB1CC1 in human breast cancer. *Nat Genet.* 2002;31:285–288.
76. Melkoumian ZK, Peng X, Gan B, Wu X, Guan J-L. Mechanism of cell cycle regulation by FIP200 in human breast cancer cells. *Cancer Res.* 2005;65:6676–6684.
77. Hashemi-Sadraei N, et al. Expression of LC3B and FIP200/Atg17 in brain metastases of breast cancer. *J Neurooncol.* 2018;140:237–248.
78. Xia H, et al. Suppression of FIP200 and autophagy by tumor-derived lactate promotes naïve T cell apoptosis and affects tumor immunity. *Sci Immunol.* 2017;2.
79. Kudchodkar SB, Levine B. Viruses and autophagy. *Rev Med Virol.* 2009;19:359–378.
80. Tallóczy Z, et al. Regulation of starvation- and virus-induced autophagy by the eIF2alpha kinase signaling pathway. *Proc Natl Acad Sci USA.* 2002;99:190–195.
81. Walboomers JM, et al. Human papillomavirus is a necessary cause of invasive cervical cancer worldwide. *J Pathol.* 1999;189:12–19.
82. Fang W, et al. miR-224-3p inhibits autophagy in cervical cancer cells by targeting FIP200. *Sci Rep.* 2016;6:33229.
83. Thurston TLM, Ryzhakov G, Bloor S, von Muhlinen N, Randow F. The TBK1 adaptor and autophagy receptor NDP52 restricts the proliferation of ubiquitin-coated bacteria. *Nat Immunol.* 2009;10:1215–1221.
84. Kageyama S, et al. The LC3 recruitment mechanism is separate from Atg9L1-dependent membrane formation in the autophagic response against Salmonella. *Mol Biol Cell.* 2011;22:2290–2300.
85. Ravenhill BJ, et al. The cargo receptor NDP52 initiates selective autophagy by recruiting the ULK complex to cytosol-invading bacteria. *Mol Cell.* 2019;74:320–329.e6.
86. Fu T, et al. Mechanistic insights into the interactions of NAP1 with the SKICH domains of NDP52 and TAX1BP1. *Proc Natl Acad Sci USA.* 2018;115:E11651–E11660.
87. Ahmad L, Zhang S-Y, Casanova J-L, Sancho-Shimizu V. Human TBK1: a gatekeeper of neuroinflammation. *Trends Mol Med.* 2016;22:511–527.
88. Croce KR, Yamamoto A. A role for autophagy in Huntington's disease. *Neurobiol Dis.* 2019;122:16–22.
89. Martinez-Vicente M, et al. Cargo recognition failure is responsible for inefficient autophagy in Huntington's disease. *Nat Neurosci.* 2010;13:567–576.

90. Rui Y-N, et al. Huntingtin functions as a scaffold for selective macroautophagy. *Nat Cell Biol.* 2015;17:262–275.
91. Hosokawa N, et al. Atg101, a novel mammalian autophagy protein interacting with Atg13. *Autophagy.* 2009;5:973–979.
92. Ganley IG, et al. ULK1.ATG13.FIP200 complex mediates mTOR signaling and is essential for autophagy. *J Biol Chem.* 2009;284:12297–12305.
93. Yorimitsu T, Klionsky DJ. Atg11 links cargo to the vesicle-forming machinery in the cytoplasm to vacuole targeting pathway. *Mol Biol Cell.* 2005;16:1593–1605.
94. Pfaffenwimmer T, et al. Hrr25 kinase promotes selective autophagy by phosphorylating the cargo receptor Atg19. *EMBO Rep.* 2014;15:862–870.

CHAPTER THREE

A global view of the miRNA-mitophagy connexion

Flavie Strappazzon*
IRCCS Fondazione Santa Lucia, Rome, Italy
*Corresponding author: e-mail address: f.strappazzon@hsantalucia.it

Contents

Abstract

Mitochondria are highly dynamics organelles that provide the necessary energy for cellular functions. However, when they are dysfunctional, they can, by contrast, be very harmful for the cell. Mitophagy ensures their recycling and preserves cell performance. This mechanism is particularly important in neurons because they use a lot of energy. Failed mitophagy can thus affect the development of neurons and lead to brain problems. In this regard, a tight regulation of this process is needed. In recent years microRNAs, as regulators of several biological processes, have attracted attention in the field of mitophagy. In this review, we focused on the studies that highlight the miRNAs implicated in the regulation of mitophagic pathways. In particular, we described the first study carried out 7 years ago, in the context of mitophagy during erythroid differentiation. Next, we have cited all the other works to date on microRNAs and mitophagy regulation. Finally, we have underlined the importance of these discoveries in order to define new therapeutic approaches in the context of age-related diseases involving mitochondrial dysfunctions, such as cancers and neurodegenerative diseases.

Progress in Molecular Biology and Translational Science, Volume 172
ISSN 1877-1173
https://doi.org/10.1016/bs.pmbts.2020.03.006

Abbreviations

AMBRA1	autophagy and beclin 1 regulator 1
Atg8	protein linked to autophagy 8
BNIP3L	BCL2 interacting protein 3 like
CALCOCO2	calcium binding and coiled-coil domain 2
FUNDC1	FUN14 domain containing 1
HEK293	human embryonic kidney 293
HMBG1	*high mobility group box 1*
I/R	ischemia/reperfusion
KRAB-ZFPs	KRAB-containing zinc finger proteins
LC3	proteins associated with microtubules 1A/1B light chain 3B
LIR	LC3 interaction region
OMM	outer mitochondrial membrane (OMM)
OPTN	optineurin
PD	Parkinson's disease
PINK1	PTEN-induced kinase 1
ROS	reactive oxygen species
TRAF3	TNF receptor-associated factor 3
UBD	ubiquitin binding domain
UTR	untranslated region
VDAC1	voltage-dependent anion channel 1

1. Introduction

Autophagy is one of the processes of cellular degradation that has been preserved in all eukaryotic organisms. It involves molecular machinery and specific degradation compartments, which in mammals are represented by lysosomes and by vacuoles in plants and yeast.

Three common forms of autophagic pathways are defined: micro-autophagy, chaperone-mediated autophagy (CMA) and macroautophagy. Microautophagy is mediated by direct engulfment of the cytoplasmic material that needs to be degraded. Cytoplasmic material is trapped in the lysosome/vacuole by a random process of membrane invagination. This pathway is crucial for survival of cells under conditions of starvation, nitrogen deprivation, or after treatment with rapamycin. CMA contributes to cellular quality control through the removal of damaged or malfunctioning proteins through the lysosome and this is performed specifically through a recognition motif in the amino acid sequences of proteins. Macroautophagy is instead considered the main form of autophagy and is characterized by a discernible morphological feature. Indeed, it involves the formation of membranes, called phagophores, which close by sequestering cytoplasmic content and form vesicles, named autophagosomes. These autophagosomes will then merge with the lysosome, their content being degraded due to the different enzymes present in this compartment. Many diseases such as neurodegenerative diseases and cancers have been associated with defects in autophagy.[1,2]

Autophagy has been long considered a mechanism used only for the random degradation of cytosolic components (proteins, organelles, ribosomes, etc.). However, numerous studies have now reported that autophagy degradation can be selective for specific substrates, such as ribosomes, peroxisome, nucleus, endoplasmic reticulum or mitochondria.[3] Because maintaining damaged mitochondria is harmful for cells, eukaryotic species have developed a specific mitochondrial degradation process called mitophagy. It is a crucial process for the specific removal of damaged or excess mitochondria by micro or macroautophagy.[4] Microautophagic pathways have been observed in yeast cells following nitrogen deprivation. In this context, one mitophagic receptor has been described as playing a key role and is called Uth1/taking part of the SUN family protein UTH1).[5] Like for macroautophagy, the macromitophagy is widespread in a variety of model organisms such as yeast, nematodes, flies and mammals, making it the best path molecularly described to date. This process is important to remove superfluous mitochondria that have no apparent defects but also damaged mitochondria.[6] The mitophagy process was introduced in 2005 by John Lemasters.[7] He observed, in rat hepatocytes placed under deficiency conditions, that some of the mitochondria were sequestered in autophagosomes.[8] To be degraded by mitophagy, certain proteins of the outer mitochondria membrane (OMM) must be labeled by ubiquitin. Mitochondria are then recognized by proteins that specifically bind ubiquitin or by specific mitophagy receptors present in the OMM. The interaction of these effectors with LC3 (microtubule-associated protein 1A/1B-light chain 3), a protein localized in the membranes of phagophores would allow the recruitment and sequestration of mitochondria into autophagosomes (Fig. 1).

2. Mammalian key proteins in mitophagy

At the physiological level, the macroautophagic removal of mitochondria was revealed to be crucial in two contexts: (i) during maturation of reticulocytes into erythrocytes. Indeed, mitochondrial clearance is required during this differentiation process in order to deplete erythrocytes from mitochondria. In such conditions, the NIX protein, also known as

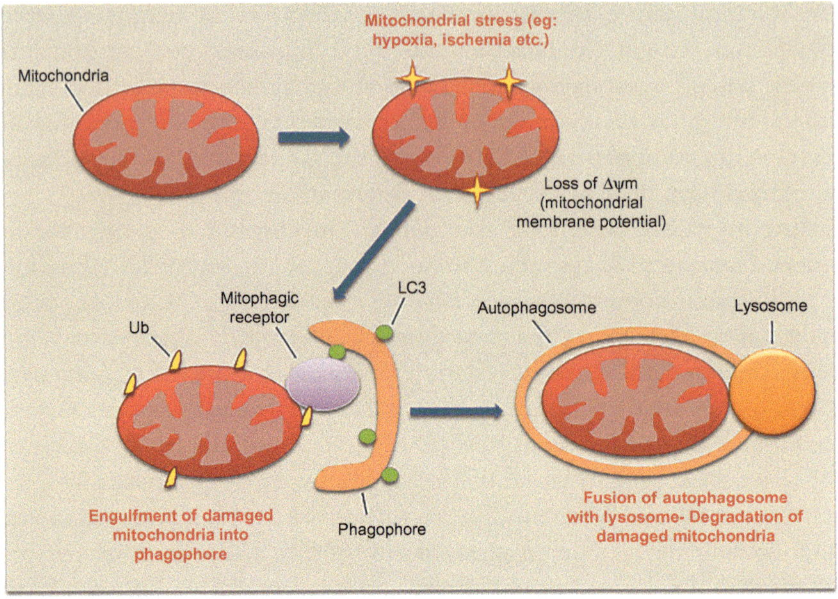

Fig. 1 The mitophagy mechanism in mammalian cells. The molecular pathway of mitophagy is illustrated. The principal major steps in the process are shown in red. Ub: Ubiquitin.

beclin-2 (BCL2)/adenovirus E1B 19 kDa protein interacting on proteins 3 (BNIP3L), increases and is responsible for the mitochondria elimination by directly binding LC3B thanks to its LC3 interaction region (LIR)[9]; (ii) during the first step of embryogenesis[10,11] when paternal mitochondria need to be removed. In this context, paternal mitochondrial which are fragmented and depolarised[10,12] are removed thanks to some molecular players such as endonuclease G in yeast,[13] prohibitin 2 (PHB2),[14] PTEN-induced putative kinase 1 (PINK1), and Parkin (PARK2) in mammals that contribute to tag mitochondria to be removed.

In mammals, mitophagy has been extensively studied since 2007 and, depending to the mitochondrial stress stimuli, various proteins that will be mentioned below, were found to regulate this process.

The main mitochondria degradation path described today occurs following depolarization of mitochondria. It is governed after the activation of the PINK1 kinase and the E3 Ubiquitin ligase, Parkin. After the loss of mitochondrial membrane potential, PINK1 accumulates in the OMM,[15] this in turn, allows PINK1-dependent recruitment of Parkin and its phosphorylation at Serine 65,[16] which promotes its dimerization and activation of its

E3 ubiquitin ligase activity. Activation of Parkin is followed by the ubiquitination of different OMM substrates. Afterwards, this phospho-ubiquitin signaling induces the recruitment of mitophagy cargo receptors such as SQSTM1 (also known as p62), Optineurin (OPTN) and CALCOCO2/Nuclear dot protein 52 (NDP52). These receptors are capable of binding ubiquitylated mitochondria through their UBD (Ubiquitin Binding domain) and thanks to a LIR (LC3-Interacting Region) motif, they bind Atg8 (protein linked to autophagy 8) /LC3 (proteins associated with microtubules 1A/1B light chain 3B) proteins on the membranes of phagophore in order to deliver mitochondria in autophagosomes.[17]

Other "alternative" mitophagy pathways, PINK1/Parkin-independent, have been described. For instance, in conditions of hypoxia, it has been shown that FUNDC1 (FUN14 domain containing 1), a protein of the OMM, activates mitophagy. In this context, FUNDC1 is mitophagic receptor that directly binds LC3 via its LIR domain.[18] Moreover NIX/Bcl-2-like 13 (BCL2L13), another mitophagic receptor of the OMM, is also known to favor mitophagy following hypoxia through a direct binding with LC3.[9] In addition, AMBRA1 is a cytosolic protein that can be located on damaged mitochondria after mitochondrial depolarization or ischemia, and that directly binds LC3 via its LIR motif, stimulating mitophagy.[19,20] This protein can be recruited to damaged mitochondria in contexts dependent[21] or independent of Parkin.[19] In this latest condition, AMBRA1 cooperates with the E3 Ubiquitin ligase HUWE1 in order to induce mitophagy.[19,22] Some members of the Fanconi Anemia (FA) protein family, Transglutaminase 2 (TG2), SMAD-specific E3 ubiquitin protein ligase 1 (SMURF1), peroxisomal biogenesis factor 3 (PEX3) and PEX13 also play a role in the mitophagy regulation. Their exact role, however, needs to be clarified.[23–26]

3. MicroRNAs

MicroRNAs (miRNAs or miRs), recently called "master regulators of gene expression,"[6] are a class of small non-coding RNAs (approximately 22 nucleotides long). They regulate the expression of genes via base complementarity between the seed region of the microRNA and the 3′-untranslated region (UTR) of the target mRNA.[27,28] miRNA binding induces degradation of the mRNA, translational repression, or both, depending on their level of complementarity.[29–31]

The post-transcriptional regulation of gene expression by microRNAs is huge since it could reach 60% of all genes encoding human proteins.[32] For this reason, miRNAs are involved in multiple cellular processes, such as proliferation, differentiation, apoptosis, carcinogenesis and energy metabolism.[33–39]

4. Mammalian mitophagy regulation by microRNAs

4.1 MicroRNAs and physiological mitophagy control

As said before, the mitophagy process is carried out by crucial proteins of the OMM or by proteins that translocate from the cytosol to mitochondria following mitochondrial stress. Hitherto it appears that mitophagy can be actively regulated by miRNAs in several contexts.

The first study which linked microRNAs to mitophagy was carried out in the context of erythroid differentiation, and therefore in a physiological mitophagic context. Indeed, in 2013, Barde et al. demonstrated that, during hematopoiesis, the mitophagy process is controlled through the induction of specific members of the large family of KRAB-containing zinc finger proteins (KRAB-ZFPs), which represses the expression of microRNAs targeting the transcripts of mitophagy effector genes.[40] Indeed, miR-351 and several other miRNAs, with predicted targets in the mitophagy pathway, are up-regulated in *kap1*-deleted murine erythroblasts. In human erythroblasts, the same authors demonstrated that KAP1 represses the NIX-targeting miR-125a-5p. These results underline a tight transcriptional regulation of mitophagy by microRNAs during hematopoiesis.

4.2 MicroRNAs that regulate mitophagy by directly targeting some mitophagic players

Here we cite the microRNAs that have been revealed as directly targeting some of the main players in mitophagy.

4.2.1 MicroRNAs and PINK1/Parkin-mediated mitophagy

Initially, miR-27a and miR-27b have been shown to regulate PINK1 expression at the translational level, which subsequently decreases the mitophagy efficiency. In particular, the authors of the study demonstrated that miR-27a,b expressions reduce ubiquitin phosphorylation, Parkin translocation, and LC3-II accumulation in damaged mitochondria.[41] The same year, miR-181a emerged as another inhibitor of mitophagy by targeting Parkin expression and consequently blocking the co-localization of mitochondria and autophagosomes. On the other hand, the authors demonstrated that the knockdown of miR-181 was sufficient to accelerate mitophagy in neuroblastoma cells.[42] More recently, miR-218 emerged as another Parkin targeting microRNA, able to delay mitochondrial clearance in HEK293 cells, by reducing Parkin expression to mitochondria, ubiquitinylation of OMM proteins and mitochondria co-localization with LC3-II.[43]

4.2.2 MicroRNAs and hypoxia-mediated mitophagy

In 2014, a study revealed a link between miR-137 and hypoxia-mediated mitophagy. miR-137 was found to be a novel hypoxia-responsive microRNA, which inhibits mitochondrial degradation by autophagy. The authors demonstrated that miR-137 directly downregulates the expression of the two main mitophagy receptors involved in hypoxia-induced mitophagy: FUNDC1 and NIX. They showed that the defect on mitophagy, due to the expression of miR-137, could be rescued by re-introduction of FUNDC1 and NIX expression constructs that lack the miR-137 pairing sites.[44]

4.2.3 MicroRNAs and starvation-mediated mitophagy

CREB1-driven expression of miR-320a was shown to promote mitophagy in HeLa cells following serum deprivation. In this context, miR-320a targets VDAC1 (voltage-dependent anion channel 1) expression and favors mitochondrial clearance.[45]

In summary, almost all mitophagy pathways that have been hitherto described can be regulated by microRNAs. This microRNAs/mitophagy interconnexion in mammalian cells is illustrated in Fig. 2.

4.3 MicroRNAs that modulate mitophagy in pathological condition

We report here some other microRNAs that can influence mitophagy without acting on a known "mitophagic target" or acting indirectly on mitophagy proteins.

4.3.1 MicroRNA and mitophagy mediated by injuries

Spinal cord injury. Liu and colleagues have postulated a protective role for mitophagy in spinal cord injury. In particular, they demonstrated, in rats with spinal cord ischemia-reperfusion (I/R), that the down-regulation of miR-124 (using antagomiR-124) protects against spinal cord injury, by

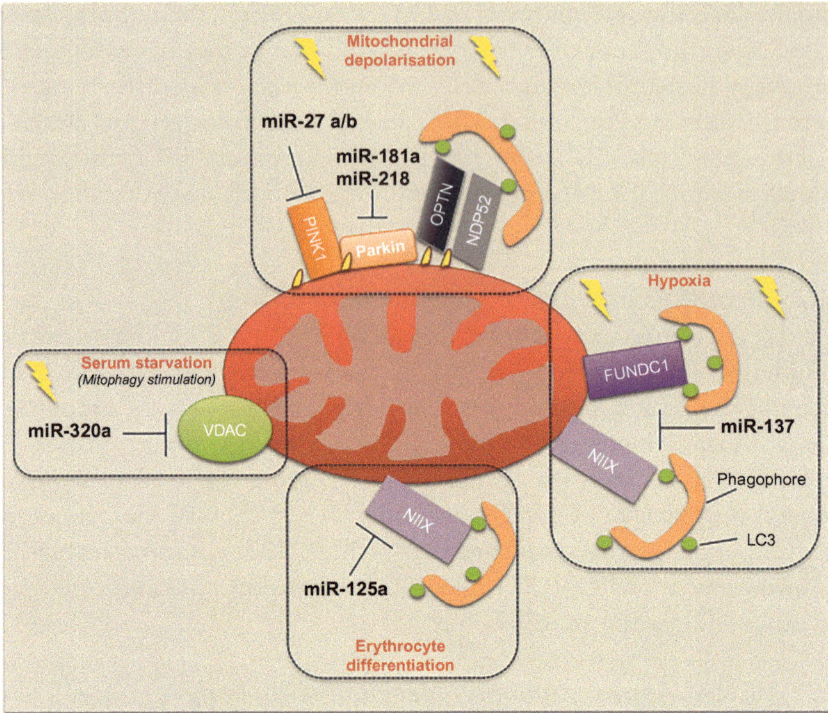

Fig. 2 Overview of the microRNAs controlling mammalian mitophagy pathways. Following mitochondrial depolarization, miR-27 a/b inhibit mitophagy through PINK1 down-regulation, whereas both miR-181a and miR-218 interfere with Parkin expression. During hypoxia, miR-137 inhibits both FUNDC1 and NIX expression, thus interfering with the main mitophagic receptor involved in hypoxia-mediated mitophagy. In the context of erythrocyte differentiation, miR-125 blocks NIX expression and thus mitophagy. Finally, following serum deprivation, miR-320 targets VDAC1 on the OMM, so favoring mitophagy. See text for more details.

increasing mitophagy.[46] Other studies are needed in order to understand the inhibitory mechanism of miR-124 on mitophagy since the authors did not analyze which target of miR-124 may be repressed following spinal cord injury.

Cardiac injury. Mitochondrial dysfunctions are critical in the cardiac system. Yang and coll. demonstrated that miR-410 inhibits mitophagy after cardiac I/R injury by modulating the heat shock protein B1 activity targeting HMGB1 (high-mobility group box 1 protein). This work opens thus new path for mitophagy modulation in heart injury related to I/R.[47]

Liver injury. 2,20,4,40-Tetrabromodiphenyl ether (BDE-47) induces mitochondrial dysfunction accumulation through a block of PINK1 mediated mitophagy in mouse liver and thus plays a role in liver injury. In this context, miR-34a-5p-mediates mitophagy impairment by targeting NAMPT expression reducing NAD production and favoring oxidation injury in mouse livers. miR-34a-5p-mediated mitophagy defect thus emerges as a potential target for the therapeutic intervention of BDE-47 toxicity.[48]

4.3.2 MicroRNA and mitophagy-related to inflammation

Adipose inflammation. In the context of obesity, miR-103 is a positive regulator of adipose inflammation by blocking mitophagy through TRAF3 (Tumor necrosis factor receptor-associated factor 3) down-regulation.[49] Further studies aiming to manipulate miR-103 and thus mitophagy could offer new opportunities for improving adipose inflammation in obesity.

Facial neuritis. Inflammatory response is involved in the development of facial neuritis. Liu and coll. explored the role of mitophagy in facial nerve damage induced by tumor necrosis factor α (TNFα). They indicated that TNFα treatment induces apoptosis in SH-SY5Y and suppressed mitophagy by reducing the expression of Bnip3. Functional studies have indicated that miR-145 inhibited Bnip3 transcription and expression, leading to mitophagy inhibition. In this context, miR-145/Bnip3/mitophagy axis may be considered as a potential target for the treatment of facial neuritis.[50]

Antiviral signaling pathways. During innate immune response, which plays a crucial role in the host defense against pathogens, miR-33/33* targets adenosine monophosphate activated protein kinase (AMPK), impeding the formation of active aggregates done by the mitochondrial adaptor mitochondrial antiviral-signaling protein (MAVS). This in turns inhibits mitophagy and generates an imbalance mitochondrial homeostasis, which is required for efficient MAVS activation. miR-33/33* thus emerges as a negative regulator of the RNA virus-triggered innate immune response.[51]

4.3.3 MicroRNA, mitophagy and cancers

Several connections between miRNAs, mitophagy and cancers have been discovered in the recent years, highlighting some potential new strategies in cancer prevention or treatments.

Mitochondrial dynamics is a highly deregulated process in cancer. Purhoit et al. have shown that miR-195 exerts pro-apoptotic effects in breast cancer cells by targeting mitofusin2. This in turn, favors mitochondria fission and cell death of breast cancer. Of note mitochondrial mass was decreased in this context but without increase of PINK1. Even if the precise mitophagy path induced in this system needs to be clarified, it appears that miR-195 expression could be used to fight breast cancer.[52]

Stem cell-based therapies are attractive for treating various human diseases, including cancer. Among the cell types that can be used for this end, mesenchymal stem cells (MSCs) are considered as promising source of stem cells. It has been shown that miR-155 disturbs mitophagy in MSCs.[53] In particular, miR-155 targets Bcl-2 associated athanogene 5 (BAG5) and this destabilizes the PINK1 kinase and consequently disrupted mitophagy. This study suggests a novel mechanism connecting mitochondrial dysfunction in stem cells through a miRNA-mediated mechanism and could offer new strategy for cancer therapies.

Also, because cancer stem cells are often resistant to anti-cancer treatment, modulation of their viability is becoming highly attractive. Zang et al. demonstrated that overexpression of miR-1 induces mitochondrial damage of cancer stem cells by targeting GPD2 (glycerol-3-phosphate dehydrogenase 2), MINOS1 (mitochondrial inner membrane organizing system 1) genes and interacting with LRPPRC (leucine-rich pentatrico peptide-repeat containing) protein. The role of mitophagy induction in this context is not well defined, but it suggests that miRNA-1-mediated regulation of mitochondria morphology may be a strategy to modulate cancer stem cells' viability.[54]

Another miRNA has been connected to mitophagy regulation and cancer, this in the context of lung tumors. In particular, in a mouse model of silicosis, it has been demonstrated that the expression of miR-1224-5p is increased both in lung tissues of silica-induced pulmonary fibrosis and fibroblasts exposed to TGF-1. Repression of miR-1224-5p expression attenuated silica-induced fibrotic progression in vivo and TGF-1-induced myofibroblast differentiation in vitro. miR-1224-5p facilitated silica-induced pulmonary fibrosis primarily by repressing one of target genes, Beclin1, thereby blocking Parkin translocation to mitochondria and inducing the accumulation of damaged mitochondria. These data indicate that miR-1224-5p exerts key functions in silica-induced pulmonary fibrosis and may represent a potential therapeutic target for lung tumors.[55]

Finally, because ROS production emerged as a strategy to kill cancer cells, Guo et al. proposed a novel strategy for cancer prevention or treatment through the use of miR-346.[56] Indeed, they found that following ER stress induced by thapsigargin treatment, the expression of miR-346 was increased. They next demonstrated that miR-346 reduced HeLa cell apoptosis following ER stress, this through a mechanism autophagy/mitophagy-dependent. In particular they found that miR-346 favors the expression of GSKβ, which disrupts Beclin1 (a well-known pro-autophagy protein) from its inhibitor Bcl-2, favoring like this the autophagy/mitophagy path and decreasing ROS production. The exact induction of autophagy and/or mitophagy in this context needs to be clarified but the reduction of Bcl-2 by miR-346 could be a therapeutic strategy for cancer prevention or treatment.

In summary, mitophagy processes induced by several pathological conditions such as spinal cord cardiac or liver injuries, inflammations or cancers are widely controlled by microRNAs. We illustrated these regulations in Figs. 3 and 4.

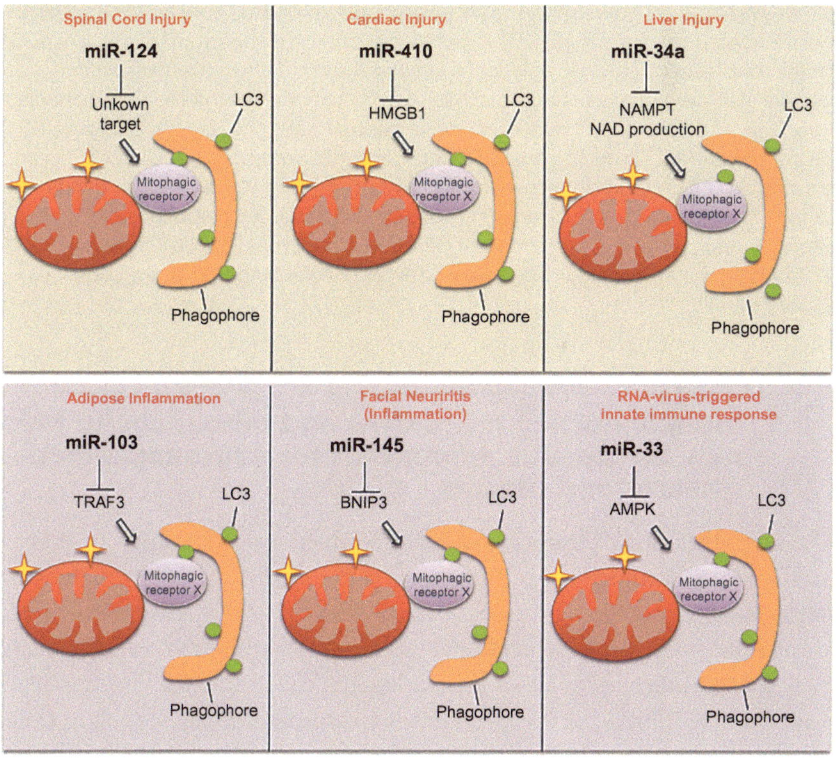

Fig. 3 Overview of microRNAs controlling mitophagy in pathological conditions. Following spinal cord, cardiac or liver injuries, protective mitophagy pathways are activated and can be repressed, respectively by miR-124, miR-410 and miR-34a. miR-410 targets HMBG1 which decreases its expression and inhibits mitophagy. In the context of obesity, miR-103 represses TRAF3, which in turn inhibits mitophagy and participates in adipose inflammation. Following BDE-47 treatment, miR-34a inhibits mitophagy and favors liver injury. See text for more details.

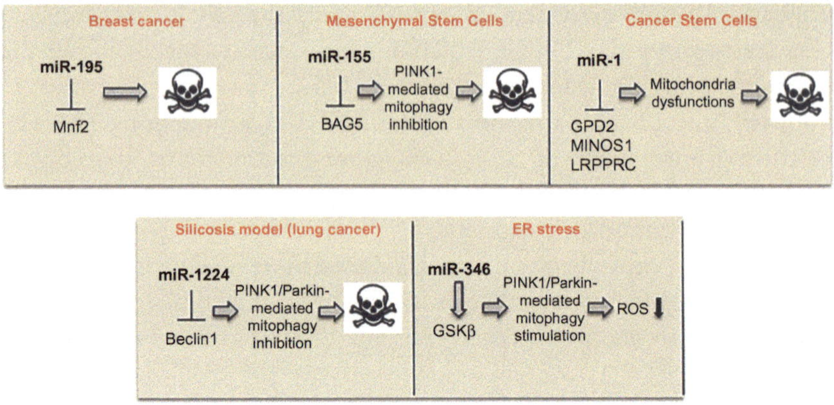

Fig. 4 Overview of microRNAs controlling mitochondria function/mitophagy in contexts related to cancers. Mir-195 expression induces cell death of breast cancer cells through mitofusin2 (Mfn2) down-regulation and thus mitochondrial dysfunction induction. miR-155 expression targets BAG5 which in turn inhibits PINK1-mediated mitophagy and favors cell death of mesenchymal stem cells. miR-1 targets GPD2, MINOS1 and LRPRRC, inducing mitochondrial dysfunctions and thus cell death of cancer stem cells. In the context of silicosis, miR-1224 targets Beclin1 and inhibits Parkin translocation to mitochondria, blocking mitophagy and favoring cell death. Following ER stress, miR-346 favors the expression of GSKB, which disrupts Beclin1 from its inhibitor Bcl2, favoring the autophagy/mitophagy path and decreasing ROS production. See text for more details.

5. Interconnecting microRNAs and mitophagy to favor new therapeutic approaches for neurodegenerative diseases and cancers

Mitochondrial clearance efficiency declines with age, leading to progressive accumulation of damaged and/or unwanted mitochondria, deterioration of cellular function, and cell death. For this reason, mitophagy modulation emerges as a potential therapeutic approach to tackle age-associated pathologies such as neurodegenerative diseases and cancers.[57] Because microRNAs are involved in several human diseases, such as cancer and disorders of the brain[58] their interconnection with the mitophagy process may offer novel attractive therapies.

5.1 Neurodegenerative diseases

PINK1 and Parkin are genetically associated with Parkinson's disease (PD) and play critical role in the mitophagy process. Endogenous miR-27a and- miR-27b regulate PINK1 expression in human dopaminergic-like M17 cells,[41] suggesting that miR27a/b may regulate PINK1 expression in dopaminergic neurons (main neurons involved in PD pathology). Endogenous miR-181 has been shown to be responsive to mitophagic agents in SH-SY5Y cells,[42] which are neuronal-like cells commonly used as an in vitro model of PD, suggesting a potential role of miR-181a in this pathology. In order to investigate potential contributions of these microRNAs in PD pathology, an accurate analysis of their specific brain expression should be investigated in the course of PD progression. Of note miR-181a, by targeting ATG5 mRNA and thus its protein level, has been recently described as an important inhibitor of macroautophagy induced by starvation and rapamycin in MCF-7, Huh-7 and K562 cells.[59] Moreover a key role of this microRNA has been demonstrated in mitochondrial homeostasis. Indeed, by favoring mitochondrial biogenesis and mitophagy, miR-181 inactivation delays neuronal death in two animal models of mitochondrial diseases (Micropthalmia with linear skin lesions and Leber's hereditary optic neuropathy).[60] This study thus highlights miR-181 as a therapeutic target for mitochondrial diseases.

Moreover, it is important to bear in mind that miR-218 is found enriched in motoneurons and mutant mice lacking miR-218 die neonatally and exhibit hallmarks of motoneuron diseases such as amyotrophic lateral sclerosis and spinal muscular atrophy.[61] In this regard, inhibition of mitophagy mediated by miR-218 may play an important role in controlling motoneurons homeostasis.

5.2 Cancers

Many studies have highlighted the role of autophagy in maintaining proper mitochondrial function. Concerning the role of mitophagy, some studies based on the analysis of the phenotypes in mice deleted for Parkin or PINK1 or for the BNIP3 and NIX mitophagy receptors have indicated a tumor suppressor function for mitophagy.[62,63] Preserving a correct mitophagy process is necessary for limiting reactive oxygen species (ROS) production, inflammation etc. By contrast, a block of mitophagy can lead to excessive ROS production and favor apoptosis. From this point of view, a general interest in mitophagy manipulation in order to improve cancer therapy is emerging. Indeed, taking advantage of the technological advances, chemotherapeutic agents can be administered with miRNAs to provide synergistic effects on patient. Interestingly, inhibition of autophagy/mitophagy enhanced drug sensitivity in several cases in which autophagosome formation or autophagosomes-lysosomes fusion events were inhibited.[64,65] The block of mitochondrial clearance by BNIP3L down-regulation enhanced doxorubicin sensitivity in colorectal cancer stem cells.[66] Even if macroautophagy inhibition has been already accepted as a therapeutic strategy in cancer, few tools that specifically block selective autophagy of mitochondria are available. Therefore, microRNAs that target mitophagy through PINK1, Parkin or mitophagic receptors down-regulation, are attractive. Interestingly, miR-218 was found significantly decreased in breast cancer tissues.[67] The authors demonstrated, in vitro, that miR-218 overexpression induces apoptosis. miR-218 mediated down-regulation of Parkin could be thus a potential mechanism that may contribute to apoptosis in breast cancer cells. Further studies are required to investigate this point. This microRNA is also strongly down-regulated in human glioma and it has been shown to work as a tumor suppressor in glioma cells.[68–70] Again, the role of mitophagy mediated by Parkin should be investigated in this context in order to understand whether mitophagy could regulate tumorigenesis in glioma cells. Interestingly, both miR-181a and Parkin siRNA sensitizes SH-SYSY cells to apoptosis in response to mitochondrial stresses.[42] miR-181a mediated down-regulation of Parkin, as well as other potential known targets (i.e., anti-apoptotic proteins) may thus contribute to apoptosis and underline miR-181a as a potential therapeutic strategy for neuroblastoma.

Finally microRNAs have emerged as key modulators of cellular senescence.[71,72] Since autophagy seems to be a causative agent for senescence, it would be interesting to evaluate the role of microRNA that control mitophagy in this context. Since autophagy induction is known to favor the immunogenic cell death (ICD) and thus to help in the success of chemotherapy,[73] it would also be interesting to test the use of miRNA's modulators of mitophagy in order to elucidate the role of mitophagy in ICD.

6. Conclusions

Since miRNAs/AntagoMirs deliverances can be easily performed through the use of nanoparticles, their application in cancer or neurodegenerative disorders has started to become highly attractive.[74] However, because miRNA-based therapeutic is quite recent, there are some challenges and limitations in their application. One major point is the fact that the modulation of miRNAs expression could have undesirable actions due to certain off-target effects (numerous mRNAs targets can be regulated by a single miRNA[75]). Further investigation in this regard is thus required in order to ensure a safe application of this kind of biotechnologies. Taken together, by linking microRNAs and mitophagy modulation, the studies cited in this review provide novel and promising therapeutic strategies for the treatment of mitochondrial-related diseases.

Acknowledgments

This work was supported by grants: ROCHE (Roche per la ricerca 2017) and 5XMILLE Italian Ministry of Health (2017) to Flavie Strappazzon. I thank M. Bennett for his proofreading.

References

1. Gallagher LE, Williamson LE, Chan EY. Advances in autophagy regulatory mechanisms. *Cells*. 2016;5:pii: E24.
2. Levine B, Packer M, Codogno P. Development of autophagy inducers in clinical medicine. *J Clin Invest*. 2015;125:14–24.
3. Stolz A, Andreas Ernst A, Dikic I. Cargo recognition and trafficking in selective autophagy. *Nat Cell Biol*. 2014;16:495–501.
4. Galluzzi L, Baehrecke EH, Ballabio A, et al. Molecular definitions of autophagy and related processes. *EMBO J*. 2017;13:1811–1836.
5. Kissova I, Salin B, Schaeffer J, Bhatia S, Manon S, Camougrand N. Selective and non-selective autophagic degradation of mitochondria in yeast. *Autophagy*. 2017;3:329–336.
6. Palikaras K, Tavernarakis N. Mitochondrial homeostasis: the interplay between mitophagy and mitochondrial biogenesis. *Exp Gerontol*. 2014;56:182–188.
7. Lemasters JJ. Selective mitochondrial autophagy, or mitophagy, as a targeted defense against oxidative stress, mitochondrial dysfunction, and aging. *Rejuvenation Res*. 2005;8:3–5.
8. Kim I, Rodriguez-Enriquez S, Lemasters JJ. Selective degradation of mitochondria by mitophagy. *Arch Biochem Biophys*. 2007;462:245–253.
9. Novak I, Kirkin V, McEwan DG, et al. Nix is a selective autophagy receptor for mitochondrial clearance. *EMBO Rep*. 2010;11:45–51.
10. Al Rawi S, Louvet-Vallee S, Djeddi A, et al. Postfertilization autophagy of sperm organelles prevents paternal mitochondrial DNA transmission. *Science*. 2010;334:1144–1147.
11. Sato M, Sato K. Degradation of paternal mitochondria by fertilization triggered autophagy in Caenorhabditis elegans embryos. *Science*. 2011;334:1141–1144.
12. Wang Y, Zhang Y, Chen L, et al. Kinetics and specificity of paternal mitochondrial elimination in Caenorhabditis elegans. *Nat Commun*. 2016;7:12569.
13. Zhou Q, Li H, Li H, et al. Mitochondrial endonuclease G mediates breakdown of paternal mitochondria upon fertilization. *Science*. 2016;353:394–399.
14. Wei Y, Chiang WC, Sumpter Jr R, Mishra P, Levine B. Prohibitin 2 is an inner-mitochondrial membrane mitophagy receptor. *Cell*. 2017;168:224–238.
15. Okatsu K, Oka T, Iguchi M, et al. PINK1 autophosphorylation upon membrane potential dissipation is essential for Parkin recruitment to damaged mitochondria. *Nat Commun*. 2012;3:1016.
16. Koyano F, Okatsu K, Kosako H, et al. Ubiquitin is phosphorylated by PINK1 to activate parkin. *Nature*. 2014;510(7503):162–166.

17. Lazarou M, Sliter DA, Kane LA, et al. The ubiquitin kinase PINK1 recruits autophagy receptors to induce mitophagy. *Nature*. 2015;524:309–314.
18. Chen M, Chen Z, Wang Y, et al. Mitophagy receptor FUNDC1 regulates mitochondrial dynamics and mitophagy. *Autophagy*. 2016;12:689–702.
19. Strappazzon F, Nazio F, Corrado M, et al. AMBRA1 is able to induce mitophagy via LC3 binding, regardless of PARKIN and p62/SQSTM1. *Cell Death Differ*. 2015;22:419–432.
20. Di Rita A, Peschiaroli A, D'Acunzo P, et al. HUWE1 E3 ligase promotes PINK1/PARKIN-independent mitophagy by regulating AMBRA1 activation via IKKalpha. *Nat Commun*. 2018;9:3755.
21. Van Humbeeck C, Cornelissen T, Hofkens H, et al. Parkin interacts with Ambra1 to induce mitophagy. *J Neurosci*. 2011;31:10249–10261.
22. Strappazzon F, Di Rita A, Peschiaroli A, et al. HUWE1 controls MCL1 stability to unleash AMBRA1-induced mitophagy. *Cell Death Differ*. 2019;21:1–14. https://doi.org/10.1038/s41418-019-0404-8.
23. Orvedahl A, MacPherson S, Sumpter Jr R, Talloczy Z, Zou Z, Levine B. Autophagy protects against Sindbis virus infection of the central nervous system. *Cell Host Microbe*. 2010;7:115–127.
24. Rossin F, D'Eletto M, Falasca L, et al. Transglutaminase 2 ablation leads to mitophagy impairment associated with a metabolic shift towards aerobic glycolysis. *Cell Death Differ*. 2015;22:408–418.
25. Lee MY, Sumpter Jr R, Zou Z, et al. Peroxisomal protein PEX13 functions in selective autophagy. *EMBO Rep*. 2017;18:48–60.
26. Sumpter Jr R, Sirasanagandla S, Fernandez AF, et al. Fanconi anemia proteins function in mitophagy and immunity. *Cell*. 2016;165:867–881.
27. Lee RC, Feinbaum RL, Ambros V. The C. elegans heterochronic gene lin-4 encodes small RNAs with antisense complementarity to lin-14. *Cell*. 1993;75:843–854.
28. Wightman B, Ha I, Ruvkun G. Posttranscriptional regulation of the heterochronic gene lin-14 by lin-4 mediates temporal pattern formation in C. elegans. *Cell*. 1993;75:855–862.
29. Behm-Ansmant I, Rehwinkel J, Doerks T, Stark A, Bork P, Izaurralde E. mRNA degradation by miRNAs and GW182 requires both CCR4:NOT deadenylase and DCP1:DCP2 decapping complexes. *Genes Dev*. 2006;20:1885–1898.
30. Behm-Ansmant I, Rehwinkel J, Izaurralde E. MicroRNAs silence gene expression by repressing protein expression and/or by promoting mRNA decay. *Cold Spring Harb Symp Quant Biol*. 2006;71:523–530.
31. Vella MC, Choi E-Y, Lin S-Y, Reinert K, Slack FJ. The C. elegans microRNA let-7 binds to imperfect let-7 complementary sites from the lin-41 3'UTR. *Genes Dev*. 2004;18:132–137.
32. Friedman RC, Farh KK-H, Burge CB, Bartel DP. Most mammalian mRNAs are conserved targets of microRNAs. *Genome Res*. 2009;19:92–105.
33. Roccaro AM, Sacco A, Thompson B, et al. MicroRNAs 15a and 16 regulate tumor proliferation in multiple myeloma. *Blood*. 2009;113:6669–6680.
34. Chen J-F, Tao Y, Li J, et al. microRNA-1 and microRNA-206 regulate skeletal muscle satellite cell proliferation and differentiation by repressing Pax7. *J Cell Biol*. 2010;190:867–879.
35. Yamakuchi M, Ferlito M, Lowenstein CJ. miR-34a repression of SIRT1 regulates apoptosis. *Proc Natl Acad Sci U S A*. 2008;105:13421–13426.
36. Schickel R, Park S-M, Murmann AE, Peter ME. miR-200c regulates induction of apoptosis through CD95 by targeting FAP-1. *Mol Cell*. 2010;38:908–915.
37. Otsubo T, Akiyama Y, Hashimoto Y, Shimada S, Goto K, Yuasa Y. MicroRNA-126 inhibits SOX2 expression and contributes to gastric carcinogenesis. *PLoS One*. 2011;6:e16617.

38. Hu Q, Du K, Mao X, Ning S. miR-197 is downregulated in cervical carcinogenesis and suppresses cell proliferation and invasion through targeting forkhead box M1. *Oncol Lett.* 2018;15:10063–10069.
39. Lu H, Buchan RJ, Cook SA. MicroRNA-223 regulates Glut4 expression and cardiomyocyte glucose metabolism. *Cardiovasc Res.* 2010;86:410–420.
40. Barde I, Rauwel B, Marin-Florez RM, et al. A KRAB/KAP1-miRNA cascade regulates erythropoiesis through stage-specific control of mitophagy. *Science.* 2013;340:350–353.
41. Kim J, Fiesel FC, Belmonte KC, et al. miR-27a and miR-27b regulate autophagic clearance of damaged mitochondria by targeting PTEN-induced putative kinase 1. *Mol Neurodegener.* 2016;11:55.
42. Cheng M, Liu L, Lao Y, et al. MicroRNA-181a suppresses parkin-mediated mitophagy and sensitizes neuroblastoma cells to mitochondrial uncoupler-induced apoptosis. *Oncotarget.* 2016;7:42274–42287.
43. Di Rita A, Maiorino T, Bruqi K, Volpicelli F, Bellenchi GC, Strappazzon F. miR-218 inhibits mitochondrial clearance by targeting PRKN E3 ubiquitin ligase. *Int J Mol Sci.* 2020;21(1):355.
44. Li W, Zhang X, Zhuang H, et al. MicroRNA-137 is a novel hypoxia-responsive microRNA that inhibits mitophagy via regulation of two mitophagy receptors FUNDC1 and NIX. *J Biol Chem.* 2014;289:10691–10701.
45. Li QQ, Zhang L, Wan HY, Liu M, Li X, Tang H. CREB1-driven expression of miR-320a promotes mitophagy by down-regulating VDAC1 expression during serum starvation in cervical cancer cells. *Oncotarget.* 2015;6:34924–34940.
46. Liu K, Yan L, Jiang X, et al. Inhibition of microRNA-124 protects against spinal cord ischemia-reperfusion injury partially through a mitophagy-dependent pathway. *J Thorac Cardiovasc Surg.* 2017;154:1498–1508.
47. Yang F, Li T, Dong Z, Mi R. MicroRNA-410 is involved in mitophagy after cardiac ischemia/reperfusion injury by targeting high-mobility group box 1 protein. *J Cell Biochem.* 2018;119:2427–2439.
48. Chen F, Feng L, Zheng YL, et al. 2, 2′, 4, 4′-tetrabromodiphenyl ether (BDE-47) induces mitochondrial dysfunction and related liver injury via eliciting miR-34a-5p-mediated mitophagy impairment. *Environ Pollut.* 2020;258:113693.
49. Zhang Z, Zhang T, Feng R, Huang H, Xia T, Sun C. circARF3 alleviates mitophagy-mediated inflammation by targeting miR-103/TRAF3 in mouse adipose tissue. *Mol Ther Nucleic Acids.* 2019;14:192–203.
50. Liu H, Huang H, Li R, et al. Mitophagy protects SH-SY5Y neuroblastoma cells against the TNFα-induced inflammatory injury: involvement of microRNA-145 and Bnip3. *Biomed Pharmacother.* 2019;109:957–968.
51. Liu D, Tan Q, Zhu J, et al. MicroRNA-33/33* inhibit the activation of MAVS through AMPK in antiviral innate immunity. *Cell Mol Immunol.* 2019;4:329–336.
52. Purohit PK, Edwards R, Tokatlidis K, Saini N. MiR-195 regulates mitochondrial function by targeting mitofusin-2 in breast cancer cells. *RNA Biol.* 2019;16:918–929.
53. Tsujimoto T, Mori T, Houri K, et al. miR-155 inhibits mitophagy through suppression of BAG5, a partner protein of PINK1. *Biochem Biophys Res Commun.* 2020;523:707–712.
54. Zhang S, Liu C, Zhang X. Mitochondrial damage mediated by miR-1 overexpression in cancer stem cells. *Mol Ther Nucleic Acids.* 2019;18:938–953.
55. Wu Q, Xu T, Liu Y, et al. miR-1224-5p mediates mitochondrial damage to affect silica-induced pulmonary fibrosis by targeting BECN1. *Int J Mol Sci.* 2017;18(11):2357.
56. Guo J, Yang Z, Yang X, Li T, Liu M, Tang H. miR-346 functions as a pro-survival factor under ER stress by activating mitophagy. *Cancer Lett.* 2018;413:69–81.
57. Markaki M, Palikaras K, Tavernarakis N. Novel insights into the anti-aging role of mitophagy. *Int Rev Cell Mol Biol.* 2018;340:169–208.
58. Enciu AM, Popescu BO, Gheorghisan-Galateanu A. MicroRNAs in brain development and degeneration. *Mol Biol Rep.* 2012;39:2243–2252.

59. Tekirdag KA, Korkmaz G, Ozturk DG, Agami R, Gozuacik D. MIR181A regulates starvation- and rapamycin-induced autophagy through targeting of ATG5. *Autophagy*. 2013;9:374–385.
60. Indrieri A, Carrella S, Romano A, et al. miR-181a/b downregulation exerts a protective action on mitochondrial disease models. *EMBO Mol Med*. 2019;11(5):e8734.
61. Amin ND, Bai G, Klug JR, et al. Loss of motoneuron-specific microRNA-218 causes systemic neuromuscular failure. *Science*. 2015;350:1525–1529.
62. Anderson CM, Macleod KF. Autophagy and cancer cell metabolism. *Int Rev Cell Mol Biol*. 2019;347:145–190.
63. Strappazzon F, Cecconi F. AMBRA1-induced mitophagy: a new mechanism to cope with cancer? *Mol Cell Oncol*. 2015;2(2):e975647.
64. Zhou J, Li G, Zheng Y, et al. A novel autophagy/mitophagy inhibitor liensinine sensitizes breast cancer cells to chemotherapy through DNM1L-mediated mitochondrial fission. *Autophagy*. 2015;11:1259–1279.
65. Mauro-Lizcano M, Esteban-Martinez L, Seco E, et al. New method to assess mitophagy flux by flow cytometry. *Autophagy*. 2015;11:833–843.
66. Yan C, Luo L, Guo CY, et al. Doxorubicin-induced mitophagy contributes to drug resistance in cancer stem cells from HCT8 human colorectal cancer cells. *Cancer Lett*. 2017;388:34–42.
67. Liu B, Tian Y, Li F, et al. Tumor-suppressing roles of miR-214 and miR-218 in breast cancer. *Oncol Rep*. 2016;35:3178–3184.
68. Song L, Huang Q, Chen K, et al. miR-218 inhibits the invasive ability of glioma cells by direct downregulation of IKK-beta. *Biochem Biophys Res Commun*. 2010;402:135–140.
69. Setty M, Helmy K, Khan AA, et al. Inferring transcriptional and microRNA-mediated regulatory programs in glioblastoma. *Mol Syst Biol*. 2012;8:605.
70. Skalsky RL, Cullen BR. Reduced expression of brain-enriched microRNAs in glioblastomas permits targeted regulation of a cell death gene. *PLoS One*. 2011;6:e24248.
71. Neault M, Couteau F, Bonneau É, De Guire V, Mallette FA. Molecular regulation of cellular senescence by MicroRNAs: implications in cancer and age-related diseases. *Int Rev Cell Mol Biol*. 2017;334:27–98.
72. Munk R, Panda AC, Grammatikakis I, Gorospe M, Abdelmohsen K. Senescence-associated MicroRNAs. *Int Rev Cell Mol Biol*. 2017;334:177–205.
73. Castoldi F, Vacchelli E, Zitvogel L, Maiuri MC, Pietrocola F, Kroemer G. Systemic autophagy in the therapeutic response to anthracycline-based chemotherapy. *Onco Targets Ther*. 2018;8(1):e1498285.
74. Lee SWL, Paoletti C, Campisi M, et al. MicroRNA delivery through nanoparticles. *J Control Release*. 2019;313:80–95.
75. Chakraborty C, Wen Z-H, Agoramoorthy G, Lin C-S. Therapeutic microRNA delivery strategies with special emphasis on cancer therapy and tumorigenesis: current trends and future challenges. *Curr Drug Metab*. 2016;17:469–477.

CHAPTER FOUR

Crosstalk between autophagy and apoptosis: Mechanisms and therapeutic implications

Andrew Thorburn*
Department of Pharmacology, University of Colorado School of Medicine, Aurora, CO, United States
*Corresponding author: e-mail address: andrew.thorburn@cuanschutz.edu

Contents

Abstract

The cellular recycling process of macroautophagy, which is the mechanism by which cellular material is delivered to lysosomes via double membraned vesicles called autophagosomes, is intimately connected to programmed cell death pathways, especially apoptosis. In this article, I discuss some underlying mechanisms and their implications for improving cancer therapy and propose that the approaches that have been taken to understand the autophagy-apoptosis connection to enhance cancer drug action can serve as a model for the kinds of information that should be developed to understand how autophagy controls other biological processes as well.

1. Autophagy in mammalian cells

The term "autophagy" refers to three mechanistically-distinct ways that cellular material is delivered to the lysosome. Although the other forms of autophagy (chaperone mediated autophagy and microautophagy) are important in biology, most work focuses on the type of autophagy known

Progress in Molecular Biology and Translational Science, Volume 172
ISSN 1877-1173
https://doi.org/10.1016/bs.pmbts.2020.04.023

as macroautophagy and this article will focus on this pathway and from hereon use "autophagy" to indicate only macroautophagy.

Autophagy is defined by the formation of double membrane vesicles called autophagosomes, which engulf cellular material including soluble proteins, organelles and parts of organelles.[1] Engulfed molecules can be passively captured simply because they are present nearby during the formation of the autophagosome or specifically targeted for engulfment; in both cases, they are then degraded following the fusion of the autophagosomes with lysosomes. This fusion event results in degradation of the inner autophagosomal membrane leading to exposure of the engulfed material to acidic lysosomal hydrolases that degrade the material into its component macromolecular precursors, which are used to make new macromolecules and to fuel metabolic pathways. The autophagy pathway is complicated with multiple distinct steps driven by specific complexes of proteins that are encoded by a class of genes called *ATG* genes that intersect with numerous cellular signaling pathways, especially the mTOR pathway, and other trafficking processes. This review focuses on how the relationship between autophagy and another process (in this case apoptosis) can be understood and manipulated. For details on the autophagy process itself, the reader is directed to a recent review of the autophagy pathway.[1]

2. Autophagy regulation of apoptosis

For many years, autophagy has been thought to affect apoptosis sensitivity.[2–4] This can involve autophagy stimulating apoptosis. For example, this occurs in some but not all cells treated with Fas ligand when a negative regulator of Fas signaling was degraded by autophagy[5] or when the inhibitor of apoptosis dBruce is degraded by autophagy during developmental cell death.[6] More commonly however, autophagy inhibits apoptosis. This is the basis for the most common type of clinical study whereby attempts are made to interfere with autophagy for clinical benefit. Somewhat surprisingly, although autophagy manipulation may be useful in many diseases,[7] the majority of clinical trials where deliberate attempts are being made to manipulate autophagy are in cancer therapy and all of them are trying to inhibit autophagy.[8] Such studies aim to combine an autophagy inhibitor with another drug in order to improve treatment by enhancing tumor cell killing by the other agent. Invariably, these clinical studies[9,10] use chloroquine (CQ) or hydroxychloroquine (HCQ). CQ and HCQ are inhibitors of the lysosome that prevent autophagosome-lysosome fusion that have the distinct advantages of being FDA approved, relatively inexpensive and available for immediate clinical use though other, more effective agents are also being developed.[11] Since autophagy is a way that cellular material is degraded, it is somewhat surprisingly that specific mechanisms by which this chemosensitization occurs have been sparse, i.e., understanding what it is whose degradation was blocked by treatment with a drug like CQ to make apoptosis more efficient. Additionally, drugs like CQ can affect chemosensitivity by autophagy-independent mechanisms.[12] It is therefore important to determine if one can elucidate a real autophagy-dependent mechanism by which the CQ is working. Recent work related to autophagy, cell death and cancer treatment has started to identify this mechanism and, perhaps more important, provide a framework for understanding what we should try to demonstrate as we aim to explain how autophagy has its biological effects.

3. What experimental data are needed to understand how autophagy controls a different cellular process?

Autophagy is controlled by the *ATG* genes, and the most rigorous way to demonstrate that autophagy is affecting another cellular process is by inhibiting those genes and testing if this elicits a given biological effect. Thus if knockout of an ATG enhances apoptosis after treatment with another agent like a cancer drug, one might conclude that autophagy was inhibiting the drug-induced apoptosis. Importantly however, the known ATG proteins are also involved in other autophagy-independent processes.[13] This means that it is possible to mistakenly conclude that autophagy was important in controlling a given process when in fact the biological effect was manifested by a different process that also used the ATG protein that was targeted. A good example of this problem arises when studying effects of mutations of the machinery (ATG7, ATG5, ATG12, ATG3 etc.[1]) responsible for protein-protein and protein-lipid conjugation steps leading to phosphatidylethanolamine (PE)-conjugation to members the LC3/GABARAP family of proteins. PE-conjugated LC3 (usually the LC3B isoform) is the most commonly used marker of autophagy. Because the PE-conjugated form of LC3 is easily measured and this is a critical step in the autophagy pathway, targeting of conjugation components (very often ATG5 or ATG7) has been frequently used to infer that autophagy was important for a given biological process.

There are two important and often under-appreciated caveats to such a conclusion. First, these molecules are also involved in a distinct but related trafficking process called LC3 Associated Phagocytosis (LAP),[14] thus even if one found that LC3 conjugation was disrupted and that this prevented the effect one was looking for, this might have been due to LAP rather than autophagy being important. In fact, the problem is even greater because these proteins are also involved in entirely different processes as well, e.g., ATG7 has been reported to be involved in regulating the activity of the TP53 transcription factor.[15] The second major caveat is that even without the protein conjugation machinery, mammalian cells are known to form double membrane vesicles that fuse with lysosomes albeit in a less efficient and slower manner than with conventional autophagy and with delayed degradation of the inner autophagosome membrane.[16] While it is presently unclear whether these "LC3-independent" autophagosomes deliver cargo to be degraded, it seems likely that at least non-specific substrates might be degraded in this way. The best understood mechanisms by which specific cargos are targeted to autophagosomes involve recognition of LC3s on the inner membrane of the autophagosome by adaptor proteins such as SQSTMI1/p62, so the inefficient autophagosomes seen in conjugation-deficient cells are likely to be at least somewhat deficient in selective cargo delivery.

Importantly, this means that if one sees an effect when you target a molecule like ATG5 or ATG7, then this evidence, on its own, is not sufficient to say that it was actually autophagy that was responsible for the effect. And, if one fails to see an effect with the same targeting, this evidence, on its own, is also not sufficient to say that autophagy was not involved. This raises an important question: What evidence would allow a valid conclusion that autophagy is involved in a given process? Because autophagy is, by definition, the degradation of cellular material via autophagosome-mediated delivery to the lysosome, it is critical to connect a specific degradation event to the biological pathway process that is being studied, and to show that this explains the effect of autophagy manipulation. Thus, in order to reliably conclude that autophagy itself was responsible for a particular biological effect one should, at a minimum, demonstrate the following:

1. Interference of autophagy using multiple methods (i.e., targeting different ATGs and using pharmacological inhibitors of the process) elicits a particular biological effect.
2. One can identify a substrate that is delivered to the lysosome in an ATG-dependent manner that controls the biological effect that is being studied.

3. One can interfere with *only* the specific activity of the autophagy target molecule identified in step 2 and demonstrate that this is sufficient to abolish the effect of autophagy manipulation on the process that is under investigation.

As discussed below, recent work on how autophagy enhances apoptosis sensitivity in response to anti-cancer drugs provides an example of these criteria being met. And, this work also provides an example of how developing this level of mechanistic understanding allows one to propose strategies to improve cancer therapy by altering the molecular basis for the therapeutic window that means it is possible to treat cancer patients at all and what possible clinical tests one might use to exploit the mechanism during patient treatment.

4. Autophagy regulation of cancer drug-induced apoptosis

Building on some previous studies, which found that basal autophagy controls the efficiency of apoptosis,[5,17,18] we recently described a molecular mechanism that we propose explains autophagy's ability to control apoptosis sensitivity of tumor cells and to enhance the action of a specific anti-cancer agent, the MDM2 inhibitor Nutlin.[19] Autophagy does this through PUMA/BBC3, a BH3 domain protein that sensitizes cells to undergo Mitochondrial Outer Membrane Permeabilization (MOMP)[20] and subsequent apoptosis.[21,22] MOMP is considered to be the key rate limiting step in the apoptosis pathway and although recent studies show that MOMP regulation is more complicated than initially thought,[23] in most cases, a cell where there has been widespread MOMP is destined to undergo apoptotic cell death. Importantly MOMP is controlled by the interplay of 3 different flavors of BCL-family proteins. Anti-apoptotic proteins such as BCL2 inhibit MOMP, pro-apoptotic proteins such as BAX and BAK are the molecules that form the pores that mediate MOMP. The BH3 only proteins includes PUMA, these molecules are sensors that activate apoptosis in response to different stimuli through one of two mechanisms, either inhibiting the inhibitors like BCL2 or activating BAX and BAK.[23] When autophagy is high, PUMA protein levels are low and cells are relatively resistant to apoptosis. Conversely, when autophagy is inhibited, PUMA is high and cells are sensitized to apoptosis.[17] Surprisingly, the underlying mechanism is not that autophagy degrades PUMA protein. Instead, basal autophagy turns over a transcription factor, FOXO3a, causing autophagy inhibitors to increase FOXO3-driven expression of the *PUMA* gene. Moreover, although

FOXO3 regulates other genes including pro-apoptotic genes like *BIM*, and other transcription factors (e.g., p53) control *PUMA*, mutation of a single FOXO3 binding site in the human genome (in an intron in the *PUMA* gene) completely blocked autophagy inhibitor enhancement of apoptosis by an anticancer drug, Nutlin.[19]

This example meets the three criteria described above. Importantly, the mechanism (autophagy regulation of a transcription factor) allowed a very specific way to test if this mechanism alone was responsible for the biological effect. The biological effect is enhancement of Nutlin-induced apoptosis by autophagy inhibition. This was demonstrated using both genetic inhibition of molecules such as ATG7 and pharmacological inhibitors including CQ,[19] a result that was known from other studies too.[24] In the cells that were studied, this enhancement was quite specific and neither CQ alone nor Nutlin alone had any significant ability to induce apoptosis but the combination did. The relevant target degraded by autophagy is FOXO3, which was shown to be delivered to lysosomes in an ATG-dependent manner and this activity was also blocked by pharmacological autophagy inhibition.[19] Finally, and most important, because the putative target whose autophagic degradation is a transcription factor and the hypothesis being tested was that preventing its degradation by inhibiting autophagy enhances apoptosis through a specific gene (*PUMA/BBC3*), it was possible to show that this activity alone explained the biological effect by mutating just the FOXO binding site in the endogenous *PUMA* gene. This was done by demonstrating that the ability to enhance Nutlin-mediated apoptosis with CQ (or ATG knockdown) was lost when the single FOXO binding site in the PUMA gene was deleted. Importantly, this mutation did not affect constitutive apoptosis by other agents in the presence of normal autophagy but only blocked the stimulation of apoptosis by autophagy inhibitors. A caveat is that the genetic evidence that this is driven by autophagy was targeting of molecules involved in LC3 conjugation and, as noted above, LC3 conjugation occurs in other processes such as LAP. However, in this case LAP is unlikely to be responsible since it involves lysosomal delivery of extracellular material following phagocytosis and the relevant target here is an intracellular protein.

By identifying the cargo whose degradation was critical for the autophagy mediated effect (resistance to apoptosis by higher levels of basal autophagy and thus enhanced apoptosis when that autophagy is blocked) and showing it worked through a specific mechanism (activating basal expression of one specific gene), this body of evidence suggests a model where what autophagy is doing is setting the apoptotic threshold by controlling the level of just one of the pro-apoptotic molecules in the BCL2 family.[25] If one thinks of that threshold as a cliff, the question becomes: Does the anti-cancer drug have sufficient force to push the cell over the edge of the cliff, i.e., to go over the apoptosis threshold. If a cell is far enough back from the edge, the drug will be unable to promote a strong enough apoptotic signal to push the cell over its threshold and the cell will survive. What autophagy inhibition is doing is to move the cell closer to the threshold, i.e., closer to the cliff edge. This provides an explanation for why autophagy inhibition itself needn't kill the cell and neither can the other agent but the combination of being close to the edge (because of autophagy inhibition) then getting an extra pro-apoptotic push (e.g., because of Nutlin treatment) was sufficient to induce apoptosis.

5. Implications for cancer therapy

Interestingly, the model described above suggests that by inhibiting autophagy in order to improve cancer therapy, what we are doing is exploiting the biological basis for why many cancer therapies work at all.[26,27] In the current era of cancer pharmacology, we often focus on "targeted therapeutics" that inhibit specific, often mutated cancer-driving molecules and agents that alter the immune response to cancer. However, conventional cytotoxic chemotherapy drugs have a long history of use to treat cancer [28] and these drugs still represent a mainstay of cancer treatment that is used to treat millions of people. Indeed, cytotoxic agents that ultimately target DNA or microtubules have cured millions of people who would otherwise have died of cancer and this number is undoubtedly much larger than have been cured by kinase inhibitors or other targeted agents, which with very high frequency lead to acquired resistance even when a tumor was initially responsive to treatment.[29] This raises an important question. How can we have a therapeutic index making it possible to treat a person with a fatal cancer that allows the patient to experience many years of being cancer-free when the drug target was a molecule like DNA that is just as essential in all the patient's normal cells as it is in their cancer cells? On the face of it, when targeting a molecule like DNA or microtubules one might expect no therapeutic index.

The mechanism that underlies the therapeutic index for conventional chemotherapy is that cancer cells tend to be closer to their apoptotic threshold than normal cells.[26] Parenthetically, it should be noted there is widespread confusion about the famous Hanahan and Weinberg hallmark of resisting apoptosis/cell death,[30,31] which is often cited as evidence that cancer cells are more resistant to apoptosis than their normal counterparts. In fact, we've known for over two decades that the opposite is true. Cancer cells are only just avoiding cell death; i.e. they are closer to their apoptotic threshold than normal cells.[32] Indeed, as demonstrated by Letai and colleagues (from whom the cliff metaphor was "borrowed") a cancer cell's proximity to the threshold can be measured using a technique called BH3 profiling. Importantly, this proximity as measured by BH3 profiling can predict clinical responses in patients[33,34] as well as normal tissue toxicities or lack thereof.[35]

If, as proposed here, autophagy controls proximity to the apoptotic threshold through the mechanism described above, this has several interesting implications. First, it suggests an explanation for why the strategy that we are adopting in dozens of clinical trials to try to inhibit autophagy in order to enhance the activity of other anti-cancer drugs seems to work with so many different kinds of drugs.[9] Since autophagy inhibition has simply moved the cancer cells closer to the cliff edge, autophagy inhibition would co-operate with any drug that could provide an extra push for that particular cancer. We have hints that this may be the case where CQ or HCQ treatment has elicited a beneficial clinical response to kinase inhibitors that one would expect to work based on mutational characteristics of the tumor but which were otherwise ineffective.[36–38] Second, this model implies that we may see enhanced normal tissue toxicities due to the other drug when we combine that agent with an autophagy inhibitor. This is because systemic autophagy inhibition might move normal cells closer to their apoptotic threshold too. Indeed, in the clinical studies that have been done there have been remarkably few instances of toxicities caused by CQ or HCQ, and the ones that have been seen, e.g., see Ref. 39 have been an enhancement of toxicities from the other drug. This is not what one would have expected for purely autophagy-dependent toxicity since what is seen in animals where there is acute inactivation of essential autophagy genes is neurodegeneration, disrupted metabolism and increased infection[40]; similar toxicities have not been reported amongst the hundred's of patients who have been treated with CQ or HCQ along with other anti-cancer agents. In a worst case scenario however, this enhancement of the other drug's toxicity might mean that there was no real enhancement of the therapeutic index but instead just the same difference between normal and cancer cells achieved with a lower dose of the other drug. One potential solution to this problem is to focus on tumor types and cancer cells that are inherently more reliant on autophagy than others.[41–44] These tumors may move even closer to the "edge of the cliff" with a combination treatment because of their inherent reliance on autophagy in the absence of other death stimuli. Interestingly, cases where CQ and HCQ in combination with kinase inhibitors has provided evidence of clinical benefit in people[36–38] are in tumor types (BRAF mutant brain tumors and pancreas tumors) that are thought to be autophagy-dependent. A third important implication of this model is that it may be possible to apply dynamic BH3 profiling[45] following treatment with an autophagy inhibitor as a way to identify individuals who are most likely to benefit from autophagy inhibition therapy. In this case one would be looking to treat the patients whose tumor cells move furthest towards the cliff edge in response to the autophagy inhibition.

In summary, by developing a rigorous mechanistic understanding of how autophagy regulates apoptosis and the apoptotic threshold we may be able to better refine future pre-clinical and clinical studies to more rapidly capitalize on the strategy of using autophagy inhibitors to enhance the action of other anti-cancer drugs. And, the mechanism outlined here as it relates to apoptosis and cancer therapy, might also serve as a model for how we should attempt to understand the myriad of other biological functions of autophagy. The idea would be that in order to conclude that autophagy is responsible for a given biological effect, one aims to not only show that multiple autophagy regulators alter the biological process that is being studied (in practice this will mean inhibiting several different ATGs including ones that discriminate between true autophagy and related processes like LAP). And, one should also identify the important molecules whose degradation is responsible for why autophagy has these effects and show that altering just specific functions of those molecules is sufficient to prevent autophagy manipulation from affecting the process in question.

Acknowledgment

Work in our laboratory is supported by grants from the National Cancer Institute (RO1 CA150925, RO1 CA190170).

References

1. Dikic I, Elazar Z. Mechanism and medical implications of mammalian autophagy. *Nat Rev Mol Cell Biol*. 2018;19:349–364.
2. Doherty J, Baehrecke EH. Life, death and autophagy. *Nat Cell Biol*. 2018;20:1110–1117.
3. Fitzwalter BE, Thorburn A. Recent insights into cell death and autophagy. *FEBS J*. 2015;282:4279–4288.

4. Mariño G, Niso-Santano M, Baehrecke EH, Kroemer G. Self-consumption: the interplay of autophagy and apoptosis. *Nat Rev Mol Cell Biol.* 2014;15:81–94.
5. Gump JM, Staskiewicz L, Morgan MJ, Bamberg A, Riches DWH, Thorburn A. Autophagy variation within a cell population determines cell fate through selective degradation of Fap-1. *Nat Cell Biol.* 2014;16:47–54.
6. Nezis IP, Shravage BV, Sagona AP, et al. Autophagic degradation of dBruce controls DNA fragmentation in nurse cells during late Drosophila melanogaster oogenesis. *J Cell Biol.* 2010;190:523–531.
7. Levine B, Kroemer G. Biological functions of autophagy genes: a disease perspective. *Cell.* 2019;176:11–42.
8. Towers CG, Thorburn A. Therapeutic targeting of autophagy. *EBioMedicine.* 2016;14:15–23.
9. Levy JMM, Towers CG, Thorburn A. Targeting autophagy in cancer. *Nat Rev Cancer.* 2017;17:528–542.
10. Onorati AV, Dyczynski M, Ojha R, Amaravadi RK. Targeting autophagy in cancer. *Cancer.* 2018;124:3307–3318.
11. Rebecca VW, Nicastri MC, McLaughlin N, et al. A unified approach to targeting the lysosome's degradative and growth signaling roles. *Cancer Discov.* 2017;7:1266–1283.
12. Maycotte P, Aryal S, Cummings CT, Thorburn J, Morgan MJ, Thorburn A. Chloroquine sensitizes breast cancer cells to chemotherapy independent of autophagy. *Autophagy.* 2012;8:200–212.
13. Galluzzi L, Green DR. Autophagy-independent functions of the autophagy machinery. *Cell.* 2019;177:1682–1699.
14. Heckmann BL, Green DR. LC3-associated phagocytosis at a glance. *J Cell Sci.* 2019;132: jcs222984.
15. Lee IH, Kawai Y, Fergusson MM, et al. Atg7 modulates p53 activity to regulate cell cycle and survival during metabolic stress. *Science.* 2012;336:225–228.
16. Tsuboyama K, Koyama-Honda I, Sakamaki Y, Koike M, Morishita H, Mizushima N. The ATG conjugation systems are important for degradation of the inner autophagosomal membrane. *Science.* 2016;354:1036–1041.
17. Thorburn J, Andrysik Z, Staskiewicz L, et al. Autophagy controls the kinetics and extent of mitochondrial apoptosis by regulating PUMA levels. *Cell Rep.* 2014;7:45–52.
18. Yonekawa T, Gamez G, Kim J, et al. RIP1 negatively regulates basal autophagic flux through TFEB to control sensitivity to apoptosis. *EMBO Rep.* 2015;16:700–708.
19. Fitzwalter BE, Towers CG, Sullivan K, et al. Autophagy inhibition mediates apoptosis sensitization in cancer therapy by relieving FOXO3a turnover. *Dev Cell.* 2018;44:555–565.
20. Chipuk JE, Green DR. PUMA cooperates with direct activator proteins to promote mitochondrial outer membrane permeabilization and apoptosis. *Cell Cycle.* 2009;8: 2692–2696.
21. Bouillet P, Strasser A. BH3-only proteins—evolutionarily conserved proapoptotic Bcl-2 family members essential for initiating programmed cell death. *J Cell Sci.* 2002;115:1567–1574.
22. Letai A, Bassik MC, Walensky LD, Sorcinelli MD, Weiler S, Korsmeyer SJ. Distinct BH3 domains either sensitize or activate mitochondrial apoptosis, serving as prototype cancer therapeutics. *Cancer Cell.* 2002;2:183–192.
23. Bock FJ, Tait SWG. Mitochondria as multifaceted regulators of cell death. *Nat Rev Mol Cell Biol.* 2019;17:608–616.
24. Sullivan KD, Palaniappan VV, Espinosa JM. ATM regulates cell fate choice upon p53 activation by modulating mitochondrial turnover and ROS levels. *Cell Cycle.* 2015;14:56–63.

25. Fitzwalter BE, Thorburn A. FOXO3 links autophagy to apoptosis. *Autophagy*. 2018;14:1467–1468.
26. Letai A. Cell death and cancer therapy: don't forget to kill the cancer cell!. *Clin Cancer Res*. 2015;21:5015–5020.
27. Sarosiek KA, Ni Chonghaile T, Letai A. Mitochondria: gatekeepers of response to chemotherapy. *Trends Cell Biol*. 2013;23:612–619.
28. Devita VT, Chu E. A history of cancer chemotherapy. *Cancer Res*. 2008;68:8643–8653.
29. Konieczkowski DJ, Johannessen CM, Garraway LA. A convergence-based framework for cancer drug resistance. *Cancer Cell*. 2018;33:801–815.
30. Hanahan D, Weinberg RA. The hallmarks of cancer. *Cell*. 2000;100:57–70.
31. Hanahan D, Weinberg RA. Hallmarks of cancer: the next generation. *Cell*. 2011;144:646–674.
32. Green DR, Evan GI. A matter of life and death. *Cancer Cell*. 2002;1:19–30.
33. Ni Chonghaile T, Sarosiek KA, Vo T-T, et al. Pretreatment mitochondrial priming correlates with clinical response to cytotoxic chemotherapy. *Science*. 2011;334:1129–1133.
34. Vo T-T, Ryan J, Carrasco R, et al. Relative mitochondrial priming of myeloblasts and normal HSCs determines chemotherapeutic success in AML. *Cell*. 2012;151:344–355.
35. Sarosiek KA, Fraser C, Muthalagu N, et al. Developmental regulation of mitochondrial apoptosis by c-Myc governs age- and tissue-specific sensitivity to cancer therapeutics. *Cancer Cell*. 2017;31:142–156.
36. Kinsey CG, Camolotto SA, Boespflug AM, et al. Protective autophagy elicited by RAF→MEK→ERK inhibition suggests a treatment strategy for RAS-driven cancers. *Nat Med*. 2019;25:620–627.
37. Levy JMM, Thompson JC, Griesinger AM, et al. Autophagy inhibition improves chemosensitivity in BRAFV600E brain tumors. *Cancer Discov*. 2014;4:773–780.
38. Mulcahy Levy JM, Zahedi S, Griesinger AM, et al. Autophagy inhibition overcomes multiple mechanisms of resistance to BRAF inhibition in brain tumors. *Elife*. 2017;6: e19671.
39. Barnard RA, Wittenburg LA, Amaravadi RK, Gustafson DL, Thorburn A, Thamm DH. Phase I clinical trial and pharmacodynamic evaluation of combination hydroxychloroquine and doxorubicin treatment in pet dogs treated for spontaneously occurring lymphoma. *Autophagy*. 2014;10:1415–1425.
40. Karsli-Uzunbas G, Guo JY, Price S, et al. Autophagy is required for glucose homeostasis and lung tumor maintenance. *Cancer Discov*. 2014;4:914–927.
41. Guo JY, Chen H-Y, Mathew R, et al. Activated Ras requires autophagy to maintain oxidative metabolism and tumorigenesis. *Genes Dev*. 2011;25:460–470.
42. Maycotte P, Gearheart CM, Barnard R, et al. STAT3-mediated autophagy dependence identifies subtypes of breast cancer where autophagy inhibition can be efficacious. *Cancer Res*. 2014;74:2579–2590.
43. Towers CG, Fitzwalter BE, Regan D, et al. Cancer cells upregulate NRF2 signaling to adapt to autophagy inhibition. *Dev Cell*. 2019;50:690–703.
44. Yang S, Wang X, Contino G, et al. Pancreatic cancers require autophagy for tumor growth. *Genes Dev*. 2011;25:717–729.
45. Montero J, Sarosiek KA, Deangelo JD, et al. Drug-induced death signaling strategy rapidly predicts cancer response to chemotherapy. *Cell*. 2015;160:977–989.

CHAPTER FIVE

Autophagy in immunity

Christian Münz*
Viral Immunobiology, Institute of Experimental Immunology, University of Zürich, Zürich, Switzerland
*Corresponding author: e-mail address: christian.muenz@uzh.ch

Contents

Abstract

The molecular machinery of macroautophagy consists of Atg proteins and supports cytoplasmic constituent degradation in lysosomes as its canonical function, phagosome maturation and exocytosis. These different biological processes contribute to cell intrinsic, innate and adaptive immunity. For the respective immune responses, Atg proteins mediate direct pathogen degradation, inflammation restriction, antigen presentation on MHC molecules and survival of memory lymphocyte populations. During adaptive immunity MHC class II presentation of antigens is supported and MHC class I presentation restricted by the macroautophagy machinery. Considering these various functions might allow us to predict the outcome of interventions that manipulate the machinery of Atg proteins as immunotherapies for the benefit of human health.

1. ATG product supported cellular pathways

Yoshinori Ohsumi received in 2017 the Nobel Prize in Physiology or Medicine for his discovery of the molecular machinery of macroautophagy.[1] This machinery of autophagy related (atg) gene products modifies membranes and recruits substrates and trafficking molecules to them.[2,3] In its canonical function this leads to the formation of a double membrane surrounded vesicle, the autophagosome, delivering its cytoplasmic cargo and

Progress in Molecular Biology and Translational Science, Volume 172
ISSN 1877-1173
https://doi.org/10.1016/bs.pmbts.2020.03.005

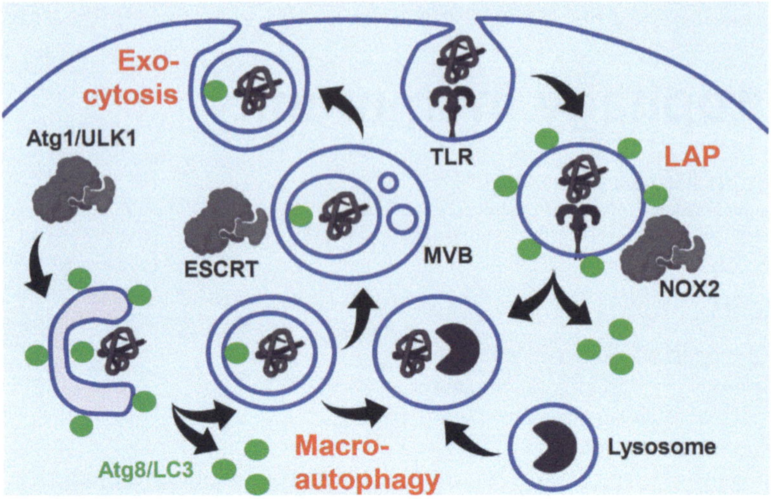

Fig. 1 Atg supported pathways. The canonical function of Atg proteins, including membrane modification by Atg8/LC3 proteins, is macroautophagy during which Atg proteins form double membrane engulfed vesicles, so-called autophagosomes, around cytoplasmic cargo for its delivery to lysosomal degradation (lower left). This pathway depends on the Atg1/ULK1 kinase complex. Without lysosomal fusion this cytoplasmic cargo can be exocytosed, involving at least parts of the multivesicular body (MVB) machinery, including ESCRT complexes (upper middle). The Atg proteins can also influence phagosome maturation during LC3-associated phagocytosis (LAP) and this pathway depends on NADPH oxidase 2 (NOX2; right).

the inner membrane for lysosomal degradation (Fig. 1). Five protein groups are involved in this process: the Atg1/ULK1 kinase complex, the Atg6/Beclin-1 containing phosphatidylinositol 3 (PI3) kinase complex, the Atg8/LC3 lipidation complex, autophagic cargo receptors and the fusion machinery with lysosomes.[4]

The first of these complexes containing Atg1/ULK1, FIP200 and ATG13 (Atg1/ULK1 complex) integrates metabolic cues that signal the demand for autophagic turnover to generate energy and building blocks for macromolecules. Accordingly, the Atg1/ULK1 complex is inhibited by phosphorylation by the mammalian target of rapamycin (mTOR) and activated by phosphorylation by the AMP kinase (AMPK). Atg1/ULK1 then transfers its activation status by phosphorylating the Atg6/Beclin-1 complex containing VPS34, VPS15 and during autophagosome generation Atg14. The VPS34 mediated PI3 kinase activity marks membranes with a PI3P mark that then recruits via WIPI1 and 2 the Atg8/LC3 lipidation complex containing Atg16L1, Atg5 and Atg12. This complex is assembled by ubiquitin like conjugation of Atg12 to Atg5 with the E1-like enzyme Atg7 and the E2-like enzyme Atg10, followed by association with Atg16L1. This Atg8/LC3 lipidation complex conjugates the ubiquitin like molecule Atg8 in a E3-like reaction to phosphatidylethanolamine in the phagophore, the developing autophagosome. Prior to this conjugation Atg8 and its six mammalian orthologues (MAP1LC3A, MAP1LC3B, MAP1LC3C, GABARAP, GABARAP-L1, GABARAP-L2) are processed by the protease Atg4 and then activated by the E1-like enzyme Atg7 and the E2-like enzyme Atg3. These Atg8/LC3 molecules mediate membrane fusion events to grow the autophagosome, likely from Atg9 associated membrane vesicles, and recruitment of its cargo. Substrate recruitment is achieved by cargo receptors that bind cytoplasmic constituents and contain a LC3-interacting region (LIR). Cytoplasmic cargo can be ubiquitinated protein aggregates, organelles or pathogens, cytosol exposed glycans during rupture of endosomes or integral organelle proteins that are only exposed to the cytosol or recruited if the organelle is damaged and needs to be discarded. Some examples of such cargo receptors are the ubiquitin binding proteins p62/sequestosome 1, NBR1, NDP52, optineurin and TAX1BP1,[5] glycan recognizing galectin 8[6] and the mitochondrial protein NIX.[5] Once the autophagosome is completed the Atg8/LC3 proteins are mostly recycled from the outer autophagosome membrane by Atg4 mediated cleavage, while some residual Atg8/LC3 lipidation supports autophagosome transport along microtubules and the recruitment of the fusion machinery with lysosomes.[7,8] This fusion machinery containing syntaxin 17, YKT6, the HOPS complex and Rab7[8–10] then allows fusion with lysosomes for the degradation of autophagosome cargo and of the inner autophagosome membrane including the lipidated Atg8/LC3 molecules at this membrane and the substrate adaptor proteins. Therefore, LC3B and p62/sequestosome 1 turnover is often used to monitor macroautophagy.

In studies by the Ohsumi lab it became however clear that Atg8/LC3 lipidation is not exclusive to autophagosome membranes.[11] In Atg4 deficient yeast that expresses the protease processed form of Atg8, Atg8 conjugation to many more membranes was observed, including endosomes. Similar conjugation to phagosomes was observed in higher eukaryotic cells and the process was coined LC3-associated phagocytosis or LAP[12–17] (Fig. 1). This process does not require the Atg1/ULK1 complex. Instead PI3 phosphorylation on phagosome membranes that have engulfed cargo which engages microbe associated molecular pattern (MAMP) receptor TLR2, the glycan receptor Dectin-1, the apoptotic material receptor

TIM4 or the immune complex receptor FcγRIIa[12,15,18,19] is achieved by a Atg6/Beclin-1 complex that in mouse phagocytes seems to contain UVRAG and RUBICON.[14] The resulting PI3P facilitates NADPH oxidase recruitment to the respective phagosomes and its reactive oxygen species (ROS) production is required for LAP.[13,14,20] The Atg8/LC3 lipidation complex might be recruited to LAP phagosomes independent of PI3P via the WD40 domain of Atg16L1[16] and the resulting Atg8/LC3 conjugation to phagosome membranes facilitates endocytosed bacteria degradation,[12] endocytosed cargo delivery to TLR containing endosomes[21] and antigen presentation on major histocompatibility complex (MHC) class II molecules to $CD4^+$ T cells.[13,16,18]

Finally, a third process in addition to macroautophagy and endocytosis that utilizes Atg proteins is exocytosis (Fig. 1). This was originally noted for acyl-CoA binding protein and cytokines like IL-1β that leave cells not via the canonical secretory pathway from the endoplasmic reticulum (ER) via the Golgi apparatus to the cell membrane.[22–25] It requires the formation of a compartment for unconventional protein secretion (CUPS) at the interphase of ER and Golgi, involving PI3 phosphorylation, Atg8/LC3 lipidation, the Golgi-associated protein GRASP65 and machinery for multivesicular body (MVB) formation including ESCRTI and III complexes.[26–28] This machinery might also support exosome release[29] and get high-jacked by viruses for release from cells.[30–34]

Thus the Atg machinery that was originally characterized during macroautophagy also contributes to phagocytosis and exocytosis. Its role in immunity will be discussed in the next sections.

2. Atg proteins in cell intrinsic immunity

At least two of these cellular pathways that are supported by Atg proteins, macroautophagy and LAP, can be used by cells for the degradation of invading microorganisms.[35,36] This constitutes a means of cell intrinsic immunity by which cells can rid themselves of pathogens in an autonomous fashion (Fig. 2). Atg protein supported cell intrinsic immunity targets all stages of microbial entry, including endosomes, escape of viruses and bacteria into the cytosol and replication in the cytosol. When still in endosomes, *Listeria monocytogenes* is delivered by LAP for lysosomal degradation.[37] NOX2 mediated ROS formation is required for *Listeria* clearance after phagocytosis, as is the Atg8/LC3 lipidation machinery, but not the Atg1/ULK1 complex. Other pathogens also recruit NOX2 to phagosomes after

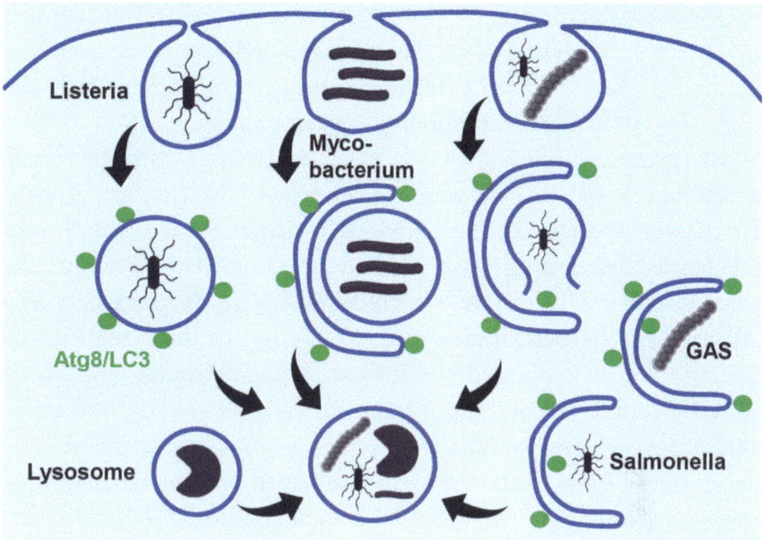

Fig. 2 Cell intrinsic pathogen restriction by Atg proteins. The Atg machinery can restrict pathogens like *Listeria monocytogenes* with LAP (left). Alternatively, it has been described to eliminate *Mycobacterium tuberculosis* by macroautophagy of pathogen containing phagosomes (middle). Furthermore, it can also target *Salmonella typhimurium* and group A *Streptococcus* in damaged endosomes or in the cytosol by macroautophagy (right).

engulfment, including *Salmonella typhimurium,*[20] but many others block LAP, including *Mycobacterium tuberculosis* and *Aspergillus fumigatus*. In the absence of its virulence factors CpsA, NOX2 and LC3 are recruited to *Mycobacterium tuberculosis* containing phagosomes.[38] CpsA deficient *Mycobacterium tuberculosis* is then efficiency degraded by LAP. Similarly, *Aspergillus fumigatus* blocks NOX2 recruitment and conjugation of LC3 to phagosomes by melanin mediated Ca^{2+} sequestration,[39] even so other fungi like *Candida albicans* and *Saccharomyces cerevisiae* can efficiently stimulate LAP and get degraded by it.[12,13] In addition to the direct modification of the phagosomal membrane, in some instances macroautophagy can also engulf endosomes. This has been reported for *Mycobacterium tuberculosis* containing endosomes after TLR mediated activation of the DNA-damage regulated autophagy modulator 1 (DRAM1).[40] Indeed, *Mycobacterium tuberculosis* was one of the first pathogens for which restriction by the Atg machinery was shown.[41] However, in most instances this requires damage to the pathogen containing endosome.[6] This damage is sensed by cytosolic exposure of glycans which are usually confined to the extracellular space or lumen of endosomes or secretory vesicles. Cytosolic galectins like galectin 3 and 8 detect these glycans.[42,43] The damaged endosomes are then targeted to autophagosomes either via galectin 8 binding to NDP52 which attaches to LC3C in autophagic membranes[44] or via ubiquitination of luminal endosomal proteins and Atg16L1.[42] Such endosomal damage is induced by *Salmonella, Listeria, Shigella* and Adenovirus.[43,45] However, Adenovirus inhibits macroautophagy of the damaged endosomes via a PPxY motif in its viral capsid.[45] Finally, also bacteria and viruses in the cytosol get targeted by macroautophagy.[46–49] These include group A *Streptococci, Salmonella,* Sindbis virus and herpes simplex virus. Targeting of these requires usually ubiquitination of the pathogen by for example the ubiquitin ligases LUBAC and SMURF1, followed by recruitment to phagosomes by one or several LC3 binding cargo receptors like p62/sequestosome 1 and optineurin.[47,49–51] Many pathogens have, however, also developed escape mechanisms from macroautophagy in the cytosol. These include the actin polymerization capacity of *Listeria monocytogenes*, walling itself of against autophagosomes.[52] As other examples *Shigella* inhibits Atg5 via its VirG protein[53] and *Legionella pneumophila* cleaves LC3 with its protease RavZ.[54] Some pathogens even use autophagic membranes for their own benefit. Along these lines several picorna- and herpesviruses acquire envelopes that contain lipidated Atg8/LC3 proteins.[30–34] These pathogens block autophagic membrane fusion with lysosomes and divert them to exocytosis, while other members of the same virus families such as herpes simplex and Kaposi sarcoma associated herpesvirus block autophagosome formation by inhibiting Atg6/Beclin-1.[55,56] Thus, Atg proteins support with macroautophagy and LAP powerful cell intrinsic restriction mechanisms against infectious agents, but successful pathogens have developed escape mechanism to evade them or even use parts of them like autophagic membrane generation for their own benefit.

3. Regulation of innate immunity by Atg proteins

In addition to cell intrinsic clearance of pathogens Atg proteins also influence innate MAMP sensing, paracrine immune activation and innate leucocyte populations that react rapidly as a first line of defense against infections.[57] Sensing of DNA and RNA as MAMPs by TLRs requires their transport into endosomal compartments, where the respective TLR7, TLR8 and TLR9 receptors reside.[58] In contrast to cytosolic sensing of these MAMPs, endosomal sensing is primarily restricted to immune cells, such as classical and plasmacytoid dendritic cells (cDCs and pDCs), B cells and macrophages. Macroautophagy has been reported to deliver cytosolic RNA during viral infections to TLR7 containing vesicular compartments of pDCs.[59] In addition to support of endogenous nucleic acid sensing by TLRs exogenous nucleic acid delivery to TLR containing endosomes benefits from the Atg machinery.[60] Furthermore, LC3-associated phagosomes have been reported to deliver DNA containing immune complexes to TLR9 containing endosomes for efficient MAMP sensing in pDCs.[21] Thus, endosomal TLR sensing benefits from the Atg machinery in pDCs and B cells.

In contrast to these beneficial roles of Atg proteins in delivering endogenous and exogenous nucleic acid based MAMPs for TLR sensing by macroautophagy and LAP, cytosolic MAMP mediated immune activation is rather restricted by Atg proteins. A prominent feature of Atg deficiency in myeloid cells is hyperactivity of inflammasomes.[61] These multiprotein complexes support caspase 1 and 11 mediated activation of cytokines such as IL-1β and IL-18, which are expressed as precursors in response to MAMP sensing by for example TLRs.[62] At least in part this increased inflammasome activation during Atg deficiency is due to diminished removal of ligands for these cytosolic MAMP sensors. Along these lines it has been shown that DNA that leaks from damaged mitochondria which are not efficiently removed by macroautophagy activates the NLRP3 inflammasome.[63,64] Also inflammasome activation during virus infection is increased in the absence of macroautophagy.[65] Furthermore, LAP removes substrates for inflammasome activation. It has been shown that mice with deficiency in Atgs involved in LAP formation, but not those exclusively required for macroautophagy develop a lupus erythematosus-like disease due to inefficient clearance of apoptotic cell debris by LAP.[66] Moreover, LAP deficiency in microglia exacerbates amyloid plaque induced inflammation in the brain of a mouse model of Alzheimer's disease.[67] Thus, Atg proteins restrict inflammasome activation, primarily by degrading MAMPs that are recognized by these immune sensors.

Furthermore, Atg proteins also negatively regulate cytosolic nucleic acid sensing for type I interferon (IFNα/β) production. Both cytosolic DNA recognition by the cyclic GMP-AMP (cGAMP) synthase (cGAS) and cytosolic RNA detection by retinoic acid-inducible gene I (RIG-I) are restricted by Atg proteins.[57] In addition to the already discussed autophagic clearance of MAMPs for these sensors, Atg6/Beclin-1 binds and inhibits cGAS, preventing its activation of the stimulator of interferon genes (STING) and IFNα/β production.[68] Moreover, cGAMP activates ULK1 to inhibit

STING by phosphorylation.[69] Finally, Atg9 prevents perinuclear STING translocation for efficient IFNα/β production.[70] Similar to inflammasome activation, moreover, LAP restricts substrate availability for STING dependent type I IFN production.[71] In the absence of efficient apoptotic cell debris clearance in the tumor microenvironment macrophages acquire proinflammatory features due to this increased IFNα/β production which increases anti-tumor immunity. Cytosolic RNA sensing by RIG-I is inhibited by the Atg8/LC3 lipidation complex via disruption of RIG-I interaction and mitochondrial localization with MAVS which is essential for its activation of the type I IFN pathway.[72–76] The Atg5-Atg12 conjugate has been reported to disrupt RIG-I interaction with MAVS. In addition, the Atg16L1/Atg5-Atg12 complex has been suggested to displace the RIG-I/MAVS complex from mitochondria. Thus, Atg proteins inhibit cytosolic MAMP sensing for both inflammasome and type I IFN activation, while they seem to support MAMP delivery to endosomes for TLR mediated innate immune activation.

In addition to a role of mitochondrial degradation by macroautophagy in cytosolic inflammasome activation, accumulation of damaged mitochondria has also been suggested to compromise innate immune cell function. Along these lines it has been shown that the innate lymphocyte populations natural killer (NK) cells, invariant NKT cells and innate lymphoid cells type 1, 2 and 3 (ILC1, ILC2 and ILC3) require Atg proteins for their survival.[77–80] In the absence of Atg proteins damaged mitochondria and their ROS production compromises thymic development of NKT cells, expanded NK cell subset maintenance and ILCs during homeostatic proliferation in for example lymphodepleted hosts. On the other hand internalization of CD1d molecules that present glycolipids to NKT cells is enhanced by Atg proteins, compromising NKT cell responses[81] Furthermore, lipid droplet degradation by macroautophagy to generate pyruvate for mitochondrial respiration was found to be necessary for the development of functional neutrophils, an innate leucocyte population that is important for restriction of extracellular pathogens.[82] Thus, innate lymphocyte survival and the associated innate immune functions are compromised in the absence of Atg proteins, probably mainly due to canonical macroautophagy functions in the metabolism of these cells.

4. Influence of Atg proteins on adaptive immunity

In addition to the role of Atg proteins in the rapid responses of cell intrinsic and innate immunity to infection, they also play a role in the slower, but more tailored adaptive immune responses by T and B cells, which also guarantee long-lived memory of infections. Atg proteins mainly influence antigen presentation on MHC molecules (Fig. 3) and adaptive memory lymphocyte survival.[4,57]

T cells recognize their antigens as peptides presented on MHC molecules on the cell surface.[83] A contribution of Atg proteins to this antigen presentation has initially been described for MHC class II restricted intracellular antigens of bacterial, viral and self-protein origin to helper $CD4^+$ T cells.[84–86] It was then noted that macroautophagy inhibition compromised the presentation of a subset of MHC class II ligands processed from cytosolic and nuclear proteins,[87] including the Atg8/LC3 orthologues LC3B, GABARAP and GABARAP-L2 as well as the adaptor protein TAX1BP1.[87,88] This self-protein processing for MHC class II presentation by macroautophagy is important for thymic selection of $CD4^+$ T cells.[89,90] Furthermore, some pathogen derived proteins are presented via this pathway and their immune evasion strategies target it.[91–96] Finally, coupling of antigens to LC3B enhances their MHC class II presentation by targeting

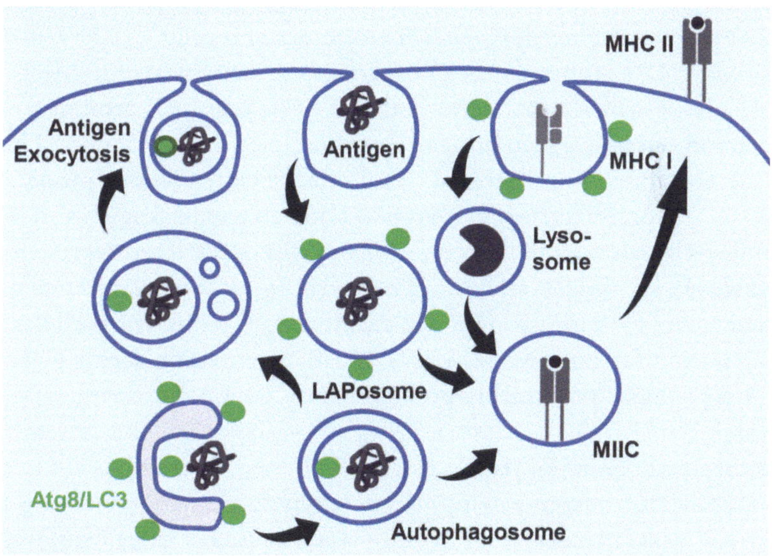

Fig. 3 Role of Atg proteins in antigen presentation on MHC molecules. Both macroautophagy of intracellular antigens and LAP of extracellular antigens support processing for MHC class II presentation (right). The Atg machinery, however, also supports antigen release, which has been shown to be efficiently cross-presented on MHC class I molecules of neighboring antigen presenting cells (left). In contrast to these beneficial functions for adaptive immune responses, Atg proteins also aid in the internalization of MHC class I molecules and thereby dampen $CD8^+$ T cell responses (upper right).

to autophagosomes.[89,97–101] In contrast to this beneficial role of Atg proteins for MHC class II presentation of intracellular antigens, MHC class I presentation of these to cytotoxic $CD8^+$ T cells is compromised in the presence of the Atg machinery.[102,103] The Atg8/LC3 lipidation complex facilitates MHC class I internalization and thereby removes MHC class I bound peptide complexes from the cell surface, thereby diminishing anti-viral $CD8^+$ T cell responses.[102] Thus, Atg proteins support MHC class II, but inhibit MHC class I presentation of intracellular antigens.

It was however already noted in the first in vivo experiments that loss of the Atg8/LC3 lipidation machinery also compromised extracellular antigen presentation on MHC class II molecules.[104] Further studies determined that LAP supports endocytosed antigen presentation on MHC class II molecules to $CD4^+$ T cells for fungal and self-proteins.[13,16,18,105] Accordingly, MHC class II presentation of extracellular autoantigen in a mouse model of CNS autoimmunity was compromised in the absence of the Atg8/LC3 lipidation machinery.[106] Furthermore, bacterial outer membrane vesicles are not efficiently processed for MHC class II presentation to regulatory $CD4^+$ T cells without LAP, resulting in inflammatory bowel disease symptoms in mice.[107] Moreover, LAP seems to mediate extracellular antigen presentation on MHC class II molecules of B cells.[108] These findings suggest that LAP supports extracellular antigen presentation onto MHC class II molecules. Furthermore, some evidence exists that LAP promotes cross-presentation of extracellular antigen for MHC class I presentation. This has been demonstrated for bacterial, viral, fungal and soluble model proteins.[109–112] Moreover, Atg proteins play also a role in packaging and releasing antigens by tumor and virus infected cells for optimal cross-presentation on MHC class I molecules.[113,114] The resulting extracellular vesicles have even been used for vaccination in mouse models.[115] Thus, extracellular antigen processing for both MHC class I and II presentation seems to benefit from Atg proteins, presumably predominantly via LAP and Atg supported exocytosis.

As for innate lymphocytes, as discussed above, Atg proteins are in addition essential for adaptive lymphocyte survival, primarily for long-lived memory T cells, memory B cells and plasma cells. For T cell survival and memory T cell maintenance clearance of malfunctioning mitochondria and ER by macroautophagy has been found to be crucial.[116–119] Without macroautophagy of these organelles memory $CD8^+$ or $CD4^+$ T cell responses to viral infections or model antigens could not be maintained.[120–122] The same organelles and their quality control by macroautophagy influence memory B and plasma cell maintenance. Macroautophagy of ER is particularly important to sustain antibody producing plasma cells.[123,124] This also extends to plasma cell maintenance during a systemic lupus erythematosus-like autoimmune disease in mice.[125] Damaged mitochondrial clearance in autophagosomes is required for memory B cell responses to viral infections.[126,127] Thus, healthy organelle homeostasis of primarily mitochondria and ER is required to maintain adaptive lymphocyte responses and establish T and B cell memory.

5. Considerations regarding the net outcome of autophagy regulation on immune responses

As discussed above Atg proteins influence many aspects of cell intrinsic, innate and adaptive immunity. Therefore, it is often difficult to estimate how therapeutic interference with this pathway might influence immune responses. This becomes especially important at the current time when stimulation of macroautophagy is considered to ameliorate pathologies that are associated with aging.[128] In this respect, stimulation of the autophagy machinery might have beneficial effects, rendering immune cells more potent in clearing intracellular and extracellular pathogens, curbing immunopathological inflammation, presenting intracellular and extracellular antigens more efficiently on MHC class II molecules and establishing immunological memory more easily. However, immunity to tumors and virus infections require often inflammation and $CD8^+$ T cell responses and the increased inflammation in the absence of macroautophagy and MHC class I stabilization without Atg protein supported endocytosis have been described to protect from influenza A virus infection and tumors.[65,71,102] Therefore, stimulation of the macroautophagy machinery has many beneficial effects on immunity but might compromise it in situations in which a robust inflammatory environment and cytotoxic $CD8^+$ T cell responses are needed.

How should one stimulate the Atg machinery to harness beneficial effects for immune responses? Many immune functions are similarly regulated by both macroautophagy and LAP. These include pathogen clearance,[37,43] hyperinflammation[64,66] and antigen processing for MHC class II presentation to helper $CD4^+$ T cells.[13,85] Therefore, treatments that could harness both pathways and their respective Atg machineries could be considered for immunotherapies. Such treatments could include stimulation of the Atg6/Beclin-1 complexes with peptides or small molecules that disrupt the interaction of Atg6/Beclin-1 with Bcl-2, and thereby increase the availability for the VPS34 complexes.[129,130] Another possibility would be to target the master transcription factor of both lysosomal biogenesis and the Atg machinery, TFEB.[131] Indeed some natural compounds, including curcumin and its derivatives have been reported to stimulate macroautophagic flux in this fashion.[132] In contrast to stimulation of macroautophagy and LAP, specific inhibitors that would compromise both pathways and thereby increase inflammation in tumor and infectious disease settings have still not been identified. The extrapolation of findings with lysosomal acidification inhibitors like chloroquine to contributions of LAP and macroautophagy in the respective disease settings seems to be difficult due to the low specificity of the respective compounds for Atg assisted pathways.[133] Thus, future more specific interventions will have to demonstrate if we can manipulate Atg supported immune mechanisms for therapeutic benefit despite the pleiotropic functions of LAP, macroautophagy and Atg supported exocytosis in immunity.

Acknowledgments

Research in the author's laboratory is supported by Cancer Research Switzerland (KFS-4091-02-2017), KFSP-Precision^{MS} of the University of Zürich, the Vontobel Foundation, the Baugarten Foundation, the Sobek Foundation, the Swiss Vaccine Research Institute, the Swiss Multiple Sclerosis Society, Roche, Novartis, ReiThera and the Swiss National Science Foundation (310030B_182827 and CRSII5_180323).

References

1. Tsukada M, Ohsumi Y. Isolation and characterization of autophagy-defective mutants of Saccharomyces cerevisiae. *FEBS Lett.* 1993;333(1–2):169–174.
2. Mizushima N, Yoshimori T, Ohsumi Y. The role of Atg proteins in autophagosome formation. *Annu Rev Cell Dev Biol.* 2011;27:107–132.
3. Cadwell K, Debnath J. Beyond self-eating: the control of nonautophagic functions and signaling pathways by autophagy-related proteins. *J Cell Biol.* 2018;217(3):813–822.
4. Levine B, Kroemer G. Biological functions of autophagy genes: a disease perspective. *Cell.* 2019;176(1–2):11–42.
5. Rogov V, Dotsch V, Johansen T, Kirkin V. Interactions between autophagy receptors and ubiquitin-like proteins form the molecular basis for selective autophagy. *Mol Cell.* 2014;53(2):167–178.
6. Randow F, Münz C. Autophagy in the regulation of pathogen replication and adaptive immunity. *Trends Immunol.* 2012;33(10):475–487.
7. Verlhac P, Gregoire IP, Azocar O, et al. Autophagy receptor NDP52 regulates pathogen-containing autophagosome maturation. *Cell Host Microbe.* 2015;17(4):515–525.
8. McEwan DG, Popovic D, Gubas A, et al. PLEKHM1 regulates autophagosome-lysosome fusion through HOPS complex and LC3/GABARAP proteins. *Mol Cell.* 2015;57(1):39–54.

9. Itakura E, Kishi-Itakura C, Mizushima N. The hairpin-type tail-anchored SNARE syntaxin 17 targets to autophagosomes for fusion with endosomes/lysosomes. *Cell*. 2012;151(6):1256–1269.
10. Matsui T, Jiang P, Nakano S, Sakamaki Y, Yamamoto H, Mizushima N. Autophagosomal YKT6 is required for fusion with lysosomes independently of syntaxin 17. *J Cell Biol*. 2018;217(8):2633–2645.
11. Nakatogawa H, Ishii J, Asai E, Ohsumi Y. Atg4 recycles inappropriately lipidated Atg8 to promote autophagosome biogenesis. *Autophagy*. 2012;8(2):177–186.
12. Sanjuan MA, Dillon CP, Tait SW, et al. Toll-like receptor signalling in macrophages links the autophagy pathway to phagocytosis. *Nature*. 2007;450(7173):1253–1257.
13. Romao S, Gasser N, Becker AC, et al. Essential autophagy proteins stabilize pathogen containing phagosomes for prolonged MHC class II antigen processing. *J Cell Biol*. 2013;203(5):757–766.
14. Martinez J, Malireddi RK, Lu Q, et al. Molecular characterization of LC3-associated phagocytosis reveals distinct roles for Rubicon, NOX2 and autophagy proteins. *Nat Cell Biol*. 2015;17(7):893–906.
15. Martinez J, Almendinger J, Oberst A, et al. Microtubule-associated protein 1 light chain 3 alpha (LC3)-associated phagocytosis is required for the efficient clearance of dead cells. *Proc Natl Acad Sci U S A*. 2011;108(42):17396–17401.
16. Fletcher K, Ulferts R, Jacquin E, et al. The WD40 domain of ATG16L1 is required for its non-canonical role in lipidation of LC3 at single membranes. *EMBO J*. 2018;37: e97840.
17. Florey O, Kim SE, Sandoval CP, Haynes CM, Overholtzer M. Autophagy machinery mediates macroendocytic processing and entotic cell death by targeting single membranes. *Nat Cell Biol*. 2011;13(11):1335–1343.
18. Ma J, Becker C, Lowell CA, Underhill DM. Dectin-1-triggered recruitment of light chain 3 protein to phagosomes facilitates major histocompatibility complex class II presentation of fungal-derived antigens. *J Biol Chem*. 2012;287(41):34149–34156.
19. Cemma M, Grinstein S, Brumell JH. Autophagy proteins are not universally required for phagosome maturation. *Autophagy*. 2016;12(9):1440–1446.
20. Huang J, Canadien V, Lam GY, et al. Activation of antibacterial autophagy by NADPH oxidases. *Proc Natl Acad Sci U S A*. 2009;106(15):6226–6231.
21. Henault J, Martinez J, Riggs JM, et al. Noncanonical autophagy is required for type I interferon secretion in response to DNA-immune complexes. *Immunity*. 2012;37(6):986–997.
22. Manjithaya R, Anjard C, Loomis WF, Subramani S. Unconventional secretion of Pichia pastoris Acb1 is dependent on GRASP protein, peroxisomal functions, and autophagosome formation. *J Cell Biol*. 2010;188(4):537–546.
23. Duran JM, Anjard C, Stefan C, Loomis WF, Malhotra V. Unconventional secretion of Acb1 is mediated by autophagosomes. *J Cell Biol*. 2010;188(4):527–536.
24. Dupont N, Jiang S, Pilli M, Ornatowski W, Bhattacharya D, Deretic V. Autophagy-based unconventional secretory pathway for extracellular delivery of IL-1beta. *EMBO J*. 2011;30(23):4701–4711.
25. Zhang M, Kenny S, Ge L, Xu K, Schekman R. Translocation of interleukin-1beta into a vesicle intermediate in autophagy-mediated secretion. *Elife*. 2015;4:e11205.
26. Bruns C, McCaffery JM, Curwin AJ, Duran JM, Malhotra V. Biogenesis of a novel compartment for autophagosome-mediated unconventional protein secretion. *J Cell Biol*. 2011;195(6):979–992.
27. Cruz-Garcia D, Curwin AJ, Popoff JF, Bruns C, Duran JM, Malhotra V. Remodeling of secretory compartments creates CUPS during nutrient starvation. *J Cell Biol*. 2014;207(6):695–703.
28. Curwin AJ, Brouwers N, Alonso YAM, et al. ESCRT-III drives the final stages of CUPS maturation for unconventional protein secretion. *Elife*. 2016;5:e16299.

29. Pallet N, Sirois I, Bell C, et al. A comprehensive characterization of membrane vesicles released by autophagic human endothelial cells. *Proteomics*. 2013;13(7):1108–1120.
30. Robinson SM, Tsueng G, Sin J, et al. Coxsackievirus B exits the host cell in shed microvesicles displaying autophagosomal markers. *PLoS Pathog*. 2014;10(4):e1004045.
31. Chen YH, Du W, Hagemeijer MC, et al. Phosphatidylserine vesicles enable efficient en bloc transmission of enteroviruses. *Cell*. 2015;160(4):619–630.
32. Nowag H, Guhl B, Thriene K, et al. Macroautophagy proteins assist Epstein Barr virus production and get incorporated into the virus particles. *EBioMedicine*. 2014;1(2–3): 116–125.
33. Buckingham EM, Jarosinski KW, Jackson W, Carpenter JE, Grose C. Exocytosis of varicella-zoster virions involves a convergence of endosomal and autophagy pathways. *J Virol*. 2016;90:8673–8685.
34. Taisne C, Lussignol M, Hernandez E, Moris A, Mouna L, Esclatine A. Human cytomegalovirus hijacks the autophagic machinery and LC3 homologs in order to optimize cytoplasmic envelopment of mature infectious particles. *Sci Rep*. 2019;9(1):4560.
35. Herb M, Gluschko A, Schramm M. LC3-associated phagocytosis—the highway to hell for phagocytosed microbes. *Semin Cell Dev Biol*. 2019;29:S1084–9521.
36. Hu W, Chan H, Lu L, et al. Autophagy in intracellular bacterial infection. *Semin Cell Dev Biol*. 2019;S1084–9521.
37. Gluschko A, Herb M, Wiegmann K, et al. The beta2 integrin mac-1 induces protective LC3-associated phagocytosis of Listeria monocytogenes. *Cell Host Microbe*. 2018;23(3): 324–337 e5.
38. Koster S, Upadhyay S, Chandra P, et al. Mycobacterium tuberculosis is protected from NADPH oxidase and LC3-associated phagocytosis by the LCP protein CpsA. *Proc Natl Acad Sci U S A*. 2017;114(41):E8711–E8720.
39. Akoumianaki T, Kyrmizi I, Valsecchi I, et al. Aspergillus cell wall melanin blocks LC3-associated phagocytosis to promote pathogenicity. *Cell Host Microbe*. 2016; 19(1):79–90.
40. Meijer AH, van der Vaart M. DRAM1 promotes the targeting of mycobacteria to selective autophagy. *Autophagy*. 2014;10(12):2389–2391.
41. Gutierrez MG, Master SS, Singh SB, Taylor GA, Colombo MI, Deretic V. Autophagy is a defense mechanism inhibiting BCG and mycobacterium tuberculosis survival in infected macrophages. *Cell*. 2004;119(6):753–766.
42. Fujita N, Morita E, Itoh T, et al. Recruitment of the autophagic machinery to endosomes during infection is mediated by ubiquitin. *J Cell Biol*. 2013;203(1):115–128.
43. Thurston TL, Wandel MP, von Muhlinen N, Foeglein A, Randow F. Galectin 8 targets damaged vesicles for autophagy to defend cells against bacterial invasion. *Nature*. 2012;482(7385):414–418.
44. von Muhlinen N, Akutsu M, Ravenhill BJ, et al. LC3C, bound selectively by a noncanonical LIR motif in NDP52, is required for antibacterial autophagy. *Mol Cell*. 2012;48(3):329–342.
45. Montespan C, Marvin SA, Austin S, et al. Multi-layered control of Galectin-8 mediated autophagy during adenovirus cell entry through a conserved PPxY motif in the viral capsid. *PLoS Pathog*. 2017;13(2):e1006217.
46. Nakagawa I, Amano A, Mizushima N, et al. Autophagy defends cells against invading group A Streptococcus. *Science*. 2004;306(5698):1037–1040.
47. Wild P, Farhan H, McEwan DG, et al. Phosphorylation of the autophagy receptor optineurin restricts Salmonella growth. *Science*. 2011;333:228–233.
48. Orvedahl A, MacPherson S, Sumpter Jr. R, Talloczy Z, Zou Z, Levine B. Autophagy protects against Sindbis virus infection of the central nervous system. *Cell Host Microbe*. 2010;7(2):115–127.

49. Orvedahl A, Sumpter Jr. R, Xiao G, et al. Image-based genome-wide siRNA screen identifies selective autophagy factors. *Nature*. 2011;480(7375):113–117.
50. Thurston TL, Ryzhakov G, Bloor S, von Muhlinen N, Randow F. The TBK1 adaptor and autophagy receptor NDP52 restricts the proliferation of ubiquitin-coated bacteria. *Nat Immunol*. 2009;10(11):1215–1221.
51. Noad J, von der Malsburg A, Pathe C, Michel MA, Komander D, Randow F. LUBAC-synthesized linear ubiquitin chains restrict cytosol-invading bacteria by activating autophagy and NF-kappaB. *Nat Microbiol*. 2017;2:17063.
52. Yoshikawa Y, Ogawa M, Hain T, et al. Listeria monocytogenes ActA-mediated escape from autophagic recognition. *Nat Cell Biol*. 2009;11(10):1233–1240.
53. Ogawa M, Yoshimori T, Suzuki T, Sagara H, Mizushima N, Sasakawa C. Escape of intracellular Shigella from autophagy. *Science*. 2005;307(5710):727–731.
54. Choy A, Dancourt J, Mugo B, et al. The Legionella effector RavZ inhibits host autophagy through irreversible Atg8 deconjugation. *Science*. 2012;338:1072–1076.
55. Orvedahl A, Alexander D, Talloczy Z, et al. HSV-1 ICP34.5 confers neurovirulence by targeting the Beclin 1 autophagy protein. *Cell Host Microbe*. 2007;1:23–35.
56. Pattingre S, Tassa A, Qu X, et al. Bcl-2 antiapoptotic proteins inhibit Beclin 1-dependent autophagy. *Cell*. 2005;122(6):927–939.
57. Matsuzawa-Ishimoto Y, Hwang S, Cadwell K. Autophagy and inflammation. *Annu Rev Immunol*. 2018;36:73–101.
58. Tan X, Sun L, Chen J, Chen ZJ. Detection of microbial infections through innate immune sensing of nucleic acids. *Annu Rev Microbiol*. 2018;72:447–478.
59. Lee HK, Lund JM, Ramanathan B, Mizushima N, Iwasaki A. Autophagy-dependent viral recognition by plasmacytoid dendritic cells. *Science*. 2007;315(5817):1398–1401.
60. Chaturvedi A, Dorward D, Pierce SK. The B cell receptor governs the subcellular location of toll-like receptor 9 leading to hyperresponses to DNA-containing antigens. *Immunity*. 2008;28(6):799–809.
61. Saitoh T, Fujita N, Jang MH, et al. Loss of the autophagy protein Atg16L1 enhances endotoxin-induced IL-1beta production. *Nature*. 2008;456:264–268.
62. Schroder K, Tschopp J. The inflammasomes. *Cell*. 2010;140(6):821–832.
63. Zhou R, Yazdi AS, Menu P, Tschopp J. A role for mitochondria in NLRP3 inflammasome activation. *Nature*. 2011;469(7329):221–225.
64. Zhong Z, Umemura A, Sanchez-Lopez E, et al. NF-kappaB restricts Inflammasome activation via elimination of damaged mitochondria. *Cell*. 2016;164(5):896–910.
65. Lu Q, Yokoyama CC, Williams JW, et al. Homeostatic control of innate lung inflammation by vici syndrome gene Epg5 and additional autophagy genes promotes influenza pathogenesis. *Cell Host Microbe*. 2016;19(1):102–113.
66. Martinez J, Cunha LD, Park S, et al. Noncanonical autophagy inhibits the autoinflammatory, lupus-like response to dying cells. *Nature*. 2016;533(7601):115–119.
67. Heckmann BL, Teubner BJW, Tummers B, et al. LC3-associated endocytosis facilitates beta-amyloid clearance and mitigates neurodegeneration in murine Alzheimer's disease. *Cell*. 2019;178(3):536–551 e14.
68. Liang Q, Seo GJ, Choi YJ, et al. Crosstalk between the cGAS DNA sensor and Beclin-1 autophagy protein shapes innate antimicrobial immune responses. *Cell Host Microbe*. 2014;15(2):228–238.
69. Konno H, Konno K, Barber GN. Cyclic dinucleotides trigger ULK1 (ATG1) phosphorylation of STING to prevent sustained innate immune signaling. *Cell*. 2013; 155(3):688–698.
70. Saitoh T, Fujita N, Hayashi T, et al. Atg9a controls dsDNA-driven dynamic translocation of STING and the innate immune response. *Proc Natl Acad Sci U S A*. 2009;106(49): 20842–20846.

71. Cunha LD, Yang M, Carter R, et al. LC3-associated phagocytosis in myeloid cells promotes tumor immune tolerance. *Cell.* 2018;175(2):429–441 e16.
72. Jounai N, Takeshita F, Kobiyama K, et al. The Atg5 Atg12 conjugate associates with innate antiviral immune responses. *Proc Natl Acad Sci U S A.* 2007;104(35):14050–14055.
73. Tal MC, Sasai M, Lee HK, Yordy B, Shadel GS, Iwasaki A. Absence of autophagy results in reactive oxygen species-dependent amplification of RLR signaling. *Proc Natl Acad Sci U S A.* 2009;106(8):2770–2775.
74. Lei Y, Wen H, Yu Y, et al. The mitochondrial proteins NLRX1 and TUFM form a complex that regulates type I interferon and autophagy. *Immunity.* 2012;36(6):933–946.
75. Zhao Y, Sun X, Nie X, et al. COX5B regulates MAVS-mediated antiviral signaling through interaction with ATG5 and repressing ROS production. *PLoS Pathog.* 2012;8(12):e1003086.
76. Mathew R, Khor S, Hackett SR, Rabinowitz JD, Perlman DH, White E. Functional role of autophagy-mediated proteome remodeling in cell survival signaling and innate immunity. *Mol Cell.* 2014;55(6):916–930.
77. Salio M, Puleston DJ, Mathan TS, et al. Essential role for autophagy during invariant NKT cell development. *Proc Natl Acad Sci U S A.* 2014;111(52):E5678–E5687.
78. Pei B, Zhao M, Miller BC, et al. Invariant NKT cells require autophagy to coordinate proliferation and survival signals during differentiation. *J Immunol.* 2015;194(12): 5872–5884.
79. O'Sullivan TE, Johnson LR, Kang HH, Sun JC. BNIP3- and BNIP3L-mediated mitophagy promotes the generation of natural killer cell memory. *Immunity.* 2015; 43(2):331–342.
80. O'Sullivan TE, Geary CD, Weizman OE, et al. Atg5 is essential for the development and survival of innate lymphocytes. *Cell Rep.* 2016;15(9):1910–1919.
81. Keller CW, Loi M, Ewert S, et al. The autophagy machinery restrains iNKT cell activation through CD1D1 internalization. *Autophagy.* 2017;13(6):1025–1036.
82. Riffelmacher T, Clarke A, Richter FC, et al. Autophagy-dependent generation of free fatty acids is critical for normal neutrophil differentiation. *Immunity.* 2017;47(3): 466–480 e5.
83. Cresswell P. A personal retrospective on the mechanisms of antigen processing. *Immunogenetics.* 2019;71(3):141–160.
84. Brazil MI, Weiss S, Stockinger B. Excessive degradation of intracellular protein in macrophages prevents presentation in the context of major histocompatibility complex class II molecules. *Eur J Immunol.* 1997;27(6):1506–1514.
85. Paludan C, Schmid D, Landthaler M, et al. Endogenous MHC class II processing of a viral nuclear antigen after autophagy. *Science.* 2005;307(5709):593–596.
86. Nimmerjahn F, Milosevic S, Behrends U, et al. Major histocompatibility complex class II-restricted presentation of a cytosolic antigen by autophagy. *Eur J Immunol.* 2003; 33(5):1250–1259.
87. Dengjel J, Schoor O, Fischer R, et al. Autophagy promotes MHC class II presentation of peptides from intracellular source proteins. *Proc Natl Acad Sci U S A.* 2005;102: 7922–7927.
88. Suri A, Walters JJ, Rohrs HW, Gross ML, Unanue ER. First signature of islet β-cell-derived naturally processed peptides selected by diabetogenic class II MHC molecules. *J Immunol.* 2008;180(6):3849–3856.
89. Aichinger M, Wu C, Nedjic J, Klein L. Macroautophagy substrates are loaded onto MHC class II of medullary thymic epithelial cells for central tolerance. *J Exp Med.* 2013;210(2):287–300.
90. Nedjic J, Aichinger M, Emmerich J, Mizushima N, Klein L. Autophagy in thymic epithelium shapes the T-cell repertoire and is essential for tolerance. *Nature.* 2008; 455(7211):396–400.

91. Jagannath C, Lindsey DR, Dhandayuthapani S, Xu Y, Hunter Jr. RL, Eissa NT. Autophagy enhances the efficacy of BCG vaccine by increasing peptide presentation in mouse dendritic cells. *Nat Med.* 2009;15(3):267–276.
92. Saini NK, Baena A, Ng TW, et al. Suppression of autophagy and antigen presentation by mycobacterium tuberculosis PE_PGRS47. *Nat Microbiol.* 2016;1(9):16133.
93. Leung CS, Haigh TA, Mackay LK, Rickinson AB, Taylor GS. Nuclear location of an endogenously expressed antigen, EBNA1, restricts access to macroautophagy and the range of CD4 epitope display. *Proc Natl Acad Sci U S A.* 2010;107(5): 2165–2170.
94. Blanchet FP, Moris A, Nikolic DS, et al. Human immunodeficiency virus-1 inhibition of immunoamphisomes in dendritic cells impairs early innate and adaptive immune responses. *Immunity.* 2010;32(5):654–669.
95. Leib DA, Alexander DE, Cox D, Yin J, Ferguson TA. Interaction of ICP34.5 with Beclin 1 modulates herpes simplex virus type 1 pathogenesis through control of $CD4^+$ T-cell responses. *J Virol.* 2009;83(23):12164–12171.
96. Gobeil PA, Leib DA. Herpes simplex virus gamma34.5 interferes with autophagosome maturation and antigen presentation in dendritic cells. *MBio.* 2012;3(5):e00267-12.
97. Schmid D, Pypaert M, Münz C. MHC class II antigen loading compartments continuously receive input from autophagosomes. *Immunity.* 2007;26:79–92.
98. Fonteneau JF, Brilot F, Münz C, Gannage M. The tumor antigen NY-ESO-1 mediates direct recognition of melanoma cells by $CD4^+$ T cells after intercellular antigen transfer. *J Immunol.* 2016;196(1):64–71.
99. Comber JD, Robinson TM, Siciliano NA, Snook AE, Eisenlohr LC. Functional macroautophagy induction by influenza A virus without a contribution to MHC-class II restricted presentation. *J Virol.* 2011;85:6453–6463.
100. Coulon PG, Richetta C, Rouers A, et al. HIV-infected dendritic cells present endogenous MHC class II-restricted antigens to HIV-specific $CD4^+$ T cells. *J Immunol.* 2016;197(2):517–532.
101. Jin Y, Sun C, Feng L, et al. Regulation of SIV antigen-specific $CD4^+$ T cellular immunity via autophagosome-mediated MHC II molecule-targeting antigen presentation in mice. *PLoS One.* 2014;9(3):e93143.
102. Loi M, Muller A, Steinbach K, et al. Macroautophagy proteins control MHC class I levels on dendritic cells and shape anti-viral $CD8^+$ T cell responses. *Cell Rep.* 2016;15(5):1076–1087.
103. Hubbard-Lucey VM, Shono Y, Maurer K, et al. Autophagy gene Atg16l1 prevents lethal T cell alloreactivity mediated by dendritic cells. *Immunity.* 2014;41(4):579–591.
104. Lee HK, Mattei LM, Steinberg BE, et al. In vivo requirement for Atg5 in antigen presentation by dendritic cells. *Immunity.* 2010;32(2):227–239.
105. Keller CW, Loi M, Ligeon LA, Gannage M, Lunemann JD, Münz C. Endocytosis regulation by autophagy proteins in MHC restricted antigen presentation. *Curr Opin Immunol.* 2018;52:68–73.
106. Keller CW, Sina C, Kotur MB, et al. ATG-dependent phagocytosis in dendritic cells drives myelin-specific CD4(+) T cell pathogenicity during CNS inflammation. *Proc Natl Acad Sci U S A.* 2017;114(52):E11228–E11237.
107. Chu H, Khosravi A, Kusumawardhani IP, et al. Gene-microbiota interactions contribute to the pathogenesis of inflammatory bowel disease. *Science.* 2016;352(6289): 1116–1120.
108. Arbogast F, Arnold J, Hammann P, et al. ATG5 is required for B cell polarization and presentation of particulate antigens. *Autophagy.* 2018;15:1–15.
109. De Luca A, Iannitti RG, Bozza S, et al. $CD4^+$ T cell vaccination overcomes defective cross-presentation of fungal antigens in a mouse model of chronic granulomatous disease. *J Clin Invest.* 2012;122(5):1816–1831.

110. Fiegl D, Kagebein D, Liebler-Tenorio EM, et al. Amphisomal route of MHC class I cross-presentation in bacteria-infected dendritic cells. *J Immunol*. 2013;190(6): 2791–2806.
111. Dasari V, Rehan S, Tey SK, Smyth MJ, Smith C, Khanna R. Autophagy and proteasome interconnect to coordinate cross-presentation through MHC class I pathway in B cells. *Immunol Cell Biol*. 2016;94:964–974.
112. Mintern JD, Macri C, Chin WJ, et al. Differential use of autophagy by primary dendritic cells specialized in cross-presentation. *Autophagy*. 2015;11(6):906–917.
113. Uhl M, Kepp O, Jusforgues-Saklani H, Vicencio JM, Kroemer G, Albert ML. Autophagy within the antigen donor cell facilitates efficient antigen cross-priming of virus-specific $CD8^+$ T cells. *Cell Death Differ*. 2009;16(7):991–1005.
114. Li Y, Wang LX, Yang G, Hao F, Urba WJ, Hu HM. Efficient cross-presentation depends on autophagy in tumor cells. *Cancer Res*. 2008;68(17):6889–6895.
115. Yu G, Li Y, Cui Z, et al. Combinational immunotherapy with Allo-DRibble vaccines and anti-OX40 co-stimulation leads to generation of cross-reactive effector T cells and tumor regression. *Sci Rep*. 2016;6:37558.
116. Pua HH, Dzhagalov I, Chuck M, Mizushima N, He YW. A critical role for the autophagy gene Atg5 in T cell survival and proliferation. *J Exp Med*. 2007;204(1):25–31.
117. Pua HH, Guo J, Komatsu M, He YW. Autophagy is essential for mitochondrial clearance in mature T lymphocytes. *J Immunol*. 2009;182(7):4046–4055.
118. Stephenson LM, Miller BC, Ng A, et al. Identification of Atg5-dependent transcriptional changes and increases in mitochondrial mass in Atg5-deficient T lymphocytes. *Autophagy*. 2009;5(5):625–635.
119. Jia W, Pua HH, Li QJ, He YW. Autophagy regulates endoplasmic reticulum homeostasis and calcium mobilization in T lymphocytes. *J Immunol*. 2011;186(3):1564–1574.
120. Xu X, Araki K, Li S, et al. Autophagy is essential for effector $CD8^+$ T cell survival and memory formation. *Nat Immunol*. 2014;15(12):1152–1161.
121. Puleston DJ, Zhang H, Powell TJ, et al. Autophagy is a critical regulator of memory $CD8^+$ T cell formation. *Elife*. 2014;3:e03706.
122. Murera D, Arbogast F, Arnold J, Bouis D, Muller S, Gros F. CD4 T cell autophagy is integral to memory maintenance. *Sci Rep*. 2018;8(1):5951.
123. Pengo N, Scolari M, Oliva L, et al. Plasma cells require autophagy for sustainable immunoglobulin production. *Nat Immunol*. 2013;14(3):298–305.
124. Conway KL, Kuballa P, Khor B, et al. ATG5 regulates plasma cell differentiation. *Autophagy*. 2013;9(4):528–537.
125. Arnold J, Murera D, Arbogast F, Fauny JD, Muller S, Gros F. Autophagy is dispensable for B-cell development but essential for humoral autoimmune responses. *Cell Death Differ*. 2016;23(5):853–864.
126. Chen M, Hong MJ, Sun H, et al. Essential role for autophagy in the maintenance of immunological memory against influenza infection. *Nat Med*. 2014;20(5):503–510.
127. Chen M, Kodali S, Jang A, Kuai L, Wang J. Requirement for autophagy in the long-term persistence but not initial formation of memory B cells. *J Immunol*. 2015; 194(6):2607–2615.
128. Madeo F, Carmona-Gutierrez D, Hofer SJ, Kroemer G. Caloric restriction mimetics against age-associated disease: targets, mechanisms, and therapeutic potential. *Cell Metab*. 2019;29(3):592–610.
129. Liu Y, Shoji-Kawata S, Sumpter Jr. RM, et al. Autosis is a Na^+,K^+-ATPase-regulated form of cell death triggered by autophagy-inducing peptides, starvation, and hypoxia-ischemia. *Proc Natl Acad Sci U S A*. 2013;110(51):20364–20371.
130. Chiang WC, Wei Y, Kuo YC, et al. High-throughput screens to identify autophagy inducers that function by disrupting Beclin 1/Bcl-2 binding. *ACS Chem Biol*. 2018;13(8):2247–2260.

131. Settembre C, Di Malta C, Polito VA, et al. TFEB links autophagy to lysosomal biogenesis. *Science*. 2011;332(6036):1429–1433.
132. Deng S, Shanmugam MK, Kumar AP, Yap CT, Sethi G, Bishayee A. Targeting autophagy using natural compounds for cancer prevention and therapy. *Cancer*. 2019;125(8):1228–1246.
133. Levy JMM, Towers CG, Thorburn A. Targeting autophagy in cancer. *Nat Rev Cancer*. 2017;17(9):528–542.

CHAPTER SIX

Autophagy in cardiovascular health and disease

Mahmoud Abdellatif[a,*], Senka Ljubojevic-Holzer[a,b], Frank Madeo[b,c], Simon Sedej[a,b,d,*]

[a]Department of Cardiology, Medical University of Graz, Graz, Austria
[b]BioTechMed Graz, Graz, Austria
[c]Institute of Molecular Biosciences, NAWI Graz, University of Graz, Graz, Austria
[d]Institute of Physiology, Faculty of Medicine, University of Maribor, Maribor, Slovenia
*Corresponding authors: e-mail address: mahmoud.abdellatif@medunigraz.at; simon.sedej@medunigraz.at

Contents

Abstract

Autophagy is a cellular housekeeping and quality control mechanism that is essential for homeostasis and survival. By virtue of this role, any perturbations to the flow of this process in cardiac or vascular cells can elicit harmful effects on the cardiovascular system, and subsequently affect whole organismal health. In this chapter, we summarize the preclinical evidence supporting the role of autophagy in sustaining cardiovascular health during homeostasis and disease. Furthermore, we discuss how autophagy activation by dietary, genetic and pharmaceutical interventions can be exploited to counteract common cardiovascular disorders, including atherosclerosis, coronary artery disease, diabetic cardiomyopathy, arrhythmia, chemotherapy-induced cardiotoxicity and heart failure.

Progress in Molecular Biology and Translational Science, Volume 172
ISSN 1877-1173
https://doi.org/10.1016/bs.pmbts.2020.04.022

1. Introduction

Macroautophagy (herein referred to as autophagy) is a homeostatic and survival mechanism required for quality control in almost all eukaryotic cells, including those comprising the cardiovascular system.[1] In particular, long-lived cardiac myocytes critically depend on the housekeeping function of autophagy as they are terminally-differentiated cells.[2] Therefore, the heart, and subsequently the whole circulatory system, is particularly vulnerable to autophagic perturbations. In this chapter, we focus on available evidence that shaped the current understanding of the role of autophagy in cardiovascular health maintenance, and discuss how autophagy can be exploited to counteract pertinent cardiovascular diseases.

2. Autophagy in cardiovascular health

The fundamental role of autophagy in cardiovascular health can be easily inferred from the consequences of modifying its activity in cardiac or vascular cell types. In this regard, several experimental studies employed genetic, dietary and pharmacological interventions to activate or attenuate autophagy, showing that compromised autophagy causes a significant deterioration in cardiovascular health; meanwhile its activation coincides with improved cardiac and vascular functions.

2.1 Efficient autophagy is essential for cardiovascular homeostasis

Autophagy is indispensable for survival as mice with global deletion of an essential autophagy gene die during embryogenesis or soon after birth.[3] However, conditional knock-out of autophagy genes in cardiac or vascular cell types does not affect perinatal viability and, thus, allows for studying the physiological role of autophagy in an organ-, tissue- and even cell-specific manner.

In the heart, eliminating the autophagy-related gene 5 (*Atg5*) specifically in the cardiomyocytes causes sarcoplasmic accumulation of cellular waste in the form of membranous sheaths and cisternae, which is associated with myofibrillar disarray, mitochondrial dysfunction, and an early decline in cardiac function. Ultimately, *Atg5* deficiency in cardiac myocytes leads to dilated heart failure, characterized by reduced ejection fraction and maladaptive remodeling (*e.g.*, hypertrophy and fibrosis), as well as premature mortality by 10 months of age.[4] Similarly, global ablation of lysosome-associated membrane protein-2 (LAMP-2), which is essential for lysosomal fusion with the autophagosome for subsequent degradation of its contents, leads to early postnatal mortality by 20–40 days of age. Although *Lamp2*-deficient mice do not exhibit any obvious cardiac abnormalities at young age, the animals that survive to older age develop a severe cardiomyopathy, characterized by excessive hypertrophy and contractile dysfunction, which is associated with a substantial accumulation of undigested autophagosomes in the cardiomyocytes.[5] Of note, aggravating the accumulation of autophagosomes in young *Lamp2*-deficient mice by alternate-day fasting accelerates the decline in cardiac function,[6] suggesting a causal role of blocked autophagic flux in the observed cardiomyopathy when they reach old age. Together, these findings demonstrate that the integrity of autophagy-lysosome machinery is indispensable for long-term myocardial homeostasis at baseline and also in response to fasting. In line with this notion, humans harboring a mutated *Lamp2* suffer from cardiac and also skeletal myopathies due to the glycogen storage syndrome, known as Danon disease.[7]

Along similar lines, vascular homeostasis is massively compromised if any of its cell types lack a non-redundant autophagy gene. For instance, mice deprived of autophagy-related gene 7 (*Atg7*) in their vascular smooth muscle cells show abnormal vascular contractility and calcium homeostasis.[8] Consistently, endothelial-specific knock-out of *Atg7* or *Atg5* leads to increased bleeding tendency due to diminished endothelial supply of von Willebrand factor, which is a haemostatic binding protein required for platelet adhesion.[9]

It is important to note that autophagy inhibition during embryogenesis, even if limited to a single cell type, might affect the development of relevant organs.[10] That being said, further efforts are warranted to comprehensively characterize all autophagy-related genes during mammalian cardiovascular morphogenesis. Until then, it cannot be entirely excluded that the observed cardiovascular abnormalities in these constitutive knock-out models have, at least in part, a developmental origin. For this purpose, postnatal or preferably late-life inhibition of autophagy could be a better approach to study the role of autophagy in maintaining cardiovascular health, without potential interference with the cardiovascular system development. Another valid alternative is the partial attenuation of autophagy through manipulating the molecular pathways regulating this housekeeping process. Mammalian target of rapamycin (mTOR) signaling is one such pathway, which can be genetically manipulated in various ways. For instance, knock-out of glycogen synthase kinase-3a (*Gsk3a*) effectively upregulates mTOR signaling and, thus, attenuates autophagy in mice. Interestingly, *Gsk3a* knock-out mice exhibit signs of accelerated cardiac aging already during their mid-life. In detail, accumulation of autophagosomes in *Gsk3a*-deficient cardiomyocytes is associated with clearly disrupted mitochondria and sarcomeres, leading to myocardial hypertrophy, fibrosis and cardiac dysfunction. These cardiac abnormalities further exacerbate with aging, culminating in heart failure and shortened lifespan.[11] Conversely, reactivation of autophagy in *Gsk3a* knock-out mice by the mTOR inhibitor everolimus restores cardiac function, excluding other pleotropic side effects of *Gsk3a* deletion to be responsible for the observed phenotype.[11] Another strategy for diminishing autophagy is through knocking out the mTOR regulator tuberous sclerosis complex 2 (TSC2). Mice lacking TSC2 exhibit mTOR-dependent autophagy inhibition, which leads to premature deterioration of cardiac contractility and hypertrophy, an effect that was effectively mitigated by the autophagy activator trehalose.[12]

In the vasculature, partial inhibition of autophagy is also sufficient to cause a significant endothelial dysfunction. For instance, attenuating autophagy by deleting its transcriptional regulator Kruppel-like transcription factor-4 (KLF4) in the endothelium leads to a stunted *in vivo* response to the vasodilator acetylcholine, indicative of endothelial dysfunction.[13] Along similar lines, reducing endothelial autophagic activity by ablating AMP-activated protein kinase alpha subunit (PRKAA) clearly impairs vascular distensibility. However, autophagy re-activation by rapamycin improves this functional deficit, which denotes a cause and effect relationship between intact autophagy and vascular functionality.[14] Taken together, these findings provide convincing evidence that autophagy plays an essential role in maintaining the homeostasis of the whole circulatory system, including the heart and its vascular tree.

2.2 Stimulating autophagy promotes cardiovascular health

Building on the premise that autophagic activity proportionally correlates with cardiovascular health, various interventions have been tested to examine whether reinforcing autophagy can promote cardiovascular resilience, especially in the context of aging when the cardiovascular system and autophagy itself progressively decline.[15]

2.2.1 Genetic activation of autophagy

Several transgenic mouse models with increased autophagy have been shown to maintain a healthier cardiovascular system. For instance, mice harboring a cardiomyocyte-specific dominant negative phosphoinositide 3-kinase (dnPI3K) mutation live longer and show attenuated signs of cardiac aging, including improved cardiac functional reserve and attenuated fibrotic and hypertrophic remodeling. At the cardiomyocyte level, dnPI3K mice also have a rejuvenated phenotype, denoted by compressed markers of senescence, oxidative stress and inflammation, as also lower lipofuscin accumulation.[16] Similarly, stimulating autophagy by Akt2 deletion exerts various cardioprotective effects in aged mice, including improved contractile function and cardiomyocyte calcium homeostasis, resulting in extended lifespan.[17] Interestingly, these effects seem to be primarily driven by autophagy induction as these cardiomyocyte improvements were abolished by the autophagy inhibitor 3-methyladenine.[17]

Consistently, activating mitophagy, a specific form of autophagy targeting dysfunctional and damaged mitochondria, promotes cardiovascular health. Specifically, genetic overexpression of the mitophagy mediator Parkin in the cardiomyocytes improves mitochondrial structure and function, and also reduces senescence and oxidative stress, altogether resulting in improved cardiac functional reserve in old Parkin-overexpressing mice.[18] Peripherally, KPL4 overexpression in the endothelium significantly improves endothelial vasodilatory function, an effect that is abolished by the autophagy inhibitor chloroquine.[13] Taken together, genetic activation of autophagy in the heart and vessels seems to effectively extend the healthspan (*i.e.*, disease-free period) of the cardiovascular system.

2.2.2 Dietary activation of autophagy

Caloric restriction, defined as reduced daily intake of calories without incurring malnutrition, effectively stimulates autophagy across species. Experimental testing of caloric restriction, ranging from 15% to 40% reduction in caloric intake, proved efficient in improving different aspects of cardiovascular health. At the heart, caloric restriction has been shown to improve hypertrophy and fibrosis as well as systolic and diastolic dysfunctions in aged animals.[19,20] Caloric restriction does so by conferring multiple improvements on the cardiomyocyte itself, including enhanced mitochondrial fitness and limited oxidative stress, apoptotic cell death, senescence and inflammation.[19] At the vasculature, caloric restriction reduces arterial stiffness by improving endothelial cell function and attenuating collagen deposition, elastin remodeling and oxidative stress.[21] Interestingly, these cardiovascular beneficial effects of caloric restriction are evident even if applied for a short duration at a later life stage,[22–24] and not necessarily throughout life.[20,21] In support of this concept, short-term periodic fasting, which is also a pro-autophagic dietary regimen, has been recently shown in humans to improve various markers of cardiovascular health.[25]

It is important to note that, unlike aging, it is still not clear whether dietary restriction applied during early thriving phases of life is beneficial. Surprisingly, young mice exposed to caloric restriction during their early adulthood showed a reduced cardiac performance,[19] which further deteriorates if autophagy is deactivated by AMP-activated protein kinase (AMPK) deletion.[26] Although this suggests that autophagy is essential for the cardioprotective effects of dietary restriction, future studies are warranted to elucidate the extent to which the degree and duration of caloric restriction affects its salutary effects when applied at young age. Most importantly, unlike earlier work in the field, future studies testing these conditions will need to rigorously assess autophagic flux following established guidelines.[27]

2.2.3 Pharmacological activation of autophagy

Various natural and pharmaceutical compounds with intrinsic pro-autophagic action have been shown to improve cardiovascular health. One such compound is the natural polyamine spermidine, which exerts pleiotropic cardioprotective effects in mice and rats. In particular, spermidine reduces, or even prevents signs of cardiac decline, including maladaptive hypertrophy, myocardial stiffness and diastolic abnormalities. These beneficial effects are primarily driven by improvements at the cellular and molecular levels, culminating in overall rejuvenation of cardiomyocyte structure and function.[28] At the vasculature, spermidine improves blood pressure control and reverses age-related arterial stiffening, endothelial dysfunction and ventricular-vascular uncoupling.[29,30] Additionally, spermidine feeding lowers vascular accumulation of collagen, advanced glycation end-products and oxidative stress.[30] Most importantly, autophagy is causally implicated in these cardiovascular protective effects of spermidine as (i) cardiomyocyte-specific *Atg5*-deficient mice lack the cardiac benefits of spermidine supplementation,[28] and (ii) the aortic rings of spermidine-fed mice showed no functional advantages over their non-treated controls upon incubation with the autophagy inhibitor chloroquine.[30]

In addition to spermidine, other autophagy-promoting compounds have been shown to confer beneficial effects on the cardiovascular system.

For instance, the mTOR inhibitor and autophagy inducer rapamycin attenuates cardiac hypertrophy and functional decline in aged mice. It also limits the levels of pro-inflammatory cytokines—systemically and locally at the heart.[31] Recently, it has been shown that a decline in cardiac hypertrophy is detectable as early as 2 weeks of rapamycin administration. Interestingly, the pro-autophagic action of rapamycin seems to be limited to these first 2 weeks of treatment, which, despite being correlative, may suggest a link between autophagy and the cardioprotective effect of rapamycin.[32]

Furthermore, the natural autophagy inducer trehalose has been shown to restore nitric oxide (NO) bioavailability and improve aged vessels function in mice. Trehalose does so by reducing vascular oxidative stress and pro-inflammatory cytokines, meanwhile increasing endothelial NO synthase expression.[33] Importantly, trehalose has been shown to exert similar beneficial effects in humans. When supplemented to the elderly, trehalose improves vasodilatory responses in resistance arteries by ameliorating the bioavailability and smooth muscle sensitivity of NO.[34] Interestingly, the vascular protective effects of trehalose are abolished in cultured endothelial cells co-incubated with the autophagy inhibitor 3-methyladenine, indicating that the vascular effects of trehalose are essentially autophagy-dependent.[33]

Similarly, autophagy induction through the activation of sirtuins by various compounds (*e.g.*, resveratrol, SRT1720 and NAD^+ precursors) has been shown to exert health-promoting effects on the heart and vasculature. For instance, resveratrol delays the functional and molecular signatures of aging in the heart.[35] Consistently, resveratrol improves vascular function in aged mice through the reduction of endothelial apoptosis, oxidative stress and inflammation. Moreover, resveratrol reduces vascular stiffness and arterial wall inflammation in rhesus monkeys fed a diet that was abnormally high in fat and sucrose content.[36] That being said, although sirtuins activation is known to induce autophagy,[37] further research is warranted to delineate to what extent the beneficial cardiovascular effects of sirtuins activation depend on autophagy.

Collectively, orally administered activators of autophagy, also known as caloric restriction mimetics,[38] hold a promise—at least in preclinical testing—to substitute caloric restriction as a more feasible and safe strategy to attenuate, retard or maybe even completely avert the decline in cardiovascular health associated with aging and possibly other risk factors.

3. Autophagy in cardiovascular disease

Recent years have witnessed an ample number of experimental studies demonstrating that repeated or continuous stimulation of autophagy could be also beneficial against specific cardiovascular diseases.[39] To this end, we will present currently prevailing views and highlight recent advances in understanding the involvement of autophagy in a subset of common and life-threatening cardiovascular disorders.

3.1 Diabetic cardiomyopathy

Diabetic cardiomyopathy affects more than one third of diabetic patients and its prevalence is projected to grow due to current obesity pandemics.[40] Cardiac abnormalities associated with this disease comprise myocardial hypertrophy, fibrosis and diastolic dysfunction.[40] Mechanistically, the diabetic heart is in a state of "metabolic rigidity" due to excessive utilization of fatty acids, which subsequently leads to increased oxidative stress burden in the cardiomyocytes. Mitochondrial dysfunction ensues leading to reduced cardiac performance, and eventually increased risk of heart failure in these patients.

Research over the past years has revealed that cardiac autophagic flux is altered in animal models of diet-induced type 2 diabetes. Although the majority of studies suggest that high-fat diet (HFD) inhibits autophagy,[41,42] a clear consensus as to whether autophagy is protective or detrimental in the diabetic heart has not been accomplished until very recently. In a longitudinal study conducted by Sadoshima and colleagues, HFD feeding to mice was found to inhibit cardiac autophagic flux, despite initially activating it between 3 and 6 weeks of feeding. On the contrary, mitophagy was reported to progressively increase even after 2 months of HFD administration.[43] The authors proposed that the activation of autophagy and mitophagy helps protect the heart during diabetic cardiomyopathy by counteracting excessive accumulation of dysfunctional mitochondria and cellular waste. In support of this notion, cardiomyocyte-specific deletion of *Atg7* aggravated cardiac dysfunction in HFD-fed mice. In line with this finding, the absence of Parkin induced a similar adverse cardiac phenotype. Most importantly, the authors showed that systemic delivery of the autophagy inducer TAT (trans-activator of transcription)-Beclin1 peptide to wild-type mice fed a HFD promotes autophagy and mitophagy, improves mitochondrial respiration capacity and diastolic dysfunction.[43] Taken together, these findings support the concept that autophagy and mitophagy are adaptive mechanisms that can be exploited to protect from or treat diabetic cardiomyopathy.

3.2 Atherosclerosis

Atherosclerosis is a chronic inflammatory disease characterized by the formation of lipid-containing plaques in the arterial walls under the influence of various risk factors (*e.g.*, old age, hypertension, diabetes mellitus, smoking or high cholesterol levels).[44] Multiple lines of evidence suggest that autophagy becomes dysfunctional during the development of atherosclerotic plaques in almost all cell types involved, including macrophages, vascular smooth muscle cells and endothelial cells. Although defective autophagy in all these cell types contributes to plaque progression, the underlying mechanisms are different.[45] For instance, aberrant autophagy in macrophages promotes plaque instability through increased apoptosis and necrosis.[46] On the contrary, autophagy-deficient vascular smooth muscle cells contribute to accelerated plaque formation *via* senescence induction[47]; meanwhile endothelial cells lacking essential autophagy-related genes exhibit both apoptosis and senescence.[48] This indicates a cell type-dependent mechanistic role of autophagy in the development and progression of atherosclerosis, which can be harnessed to develop suitable autophagy-dependent treatments for atherosclerosis.

It is well accepted that defective autophagy in atherosclerosis does not originate from disrupted autophagosomal formation, but rather from interrupted lysosomal-mediated degradation of its cargo. Indeed, Sergin et al. found a decline in the autophagy-lysosome system of macrophages in both human and mouse atherosclerotic plaques.[49] Furthermore, severe oxidative stress and cholesterol crystals in advanced atherosclerotic plaques have been shown to damage the lysosomal membrane, leading to cytosolic leakage of potent acidic hydrolases.[50,51] From a therapeutic point of view, reversing the dysfunction of autophagy in plaque macrophages by trehalose or by genetically overexpressing the transcription factor EB (TFEB) effectively alleviates the atherosclerotic lesions.[49] Mechanistically, re-activated autophagy exerts this beneficial effect through attenuating the apoptotic and pro-inflammatory signals of plaque macrophages.[49] Along similar line, mitophagy activation by overexpressing PINK1 or Parkin has been shown to protect vascular smooth muscle cells exposed to atherogenic lipids, at least *in vitro*.[52] These findings collectively demonstrate that targeting autophagy or mitophagy activation may reduce atherosclerotic plaque burden and, thus, confer protection against atherosclerotic cardiovascular disease.

3.3 Coronary artery disease

Coronary artery disease (CAD), which is typically associated with atherosclerosis and may lead to myocardial infarction, is a leading cause of morbidity and mortality worldwide. The myocardial injury associated with CAD consists of ischemic and reperfusion phases. Both ischemia and reperfusion injuries trigger cardiomyocyte stress and activate autophagy in cardiomyocytes. However, they differ in terms of oxygen and nutrients availability, and also levels of reactive oxygen species (ROS) accumulation.

During ischemia, decreased oxygen and nutrient supplies reduce intracellular ATP production, causing AMPK activation and mTOR inhibition, both of which stimulate autophagy to sustain cellular energy levels and preserve cardiomyocyte viability.[53] Hence, the activation of autophagy by a variety of autophagy inducers including trehalose and spermidine has been shown to limit myocardial damage and improve cardiac outcomes of ischemic injury in rodents.[54,55] Conversely, inhibition of autophagy by pharmacological (*e.g.*, 3-methyladenine or bafilomycin A1)[56,57] or genetic means (*e.g.*, mTOR activation or AMPK inhibition)[53,58] aggravates myocardial injury and facilitates adverse ischemic cardiac remodeling in mice subjected to myocardial infarction. Consistent with this notion, Parkin knockout mice, which have reduced mitophagy, are more sensitive to ischemic heart injury as they show reduced survival and develop larger infarcts than their wildtype counterparts.[59]

During reperfusion, despite the replenishment of oxygen and nutrients to the ischemic myocardium, there is a harmful accumulation of ROS.[60] This excessive oxidative stress causes an exaggerated activation of autophagy during reperfusion, which exacerbates the reperfusion injury by igniting cardiomyocyte death.[53,61] Hence, autophagy plays a dichotomous role in promoting cell survival during ischemia, but provoking cell death during reperfusion. This context-dependent effect of autophagy calls for attention during the development of autophagy-based therapies for coronary artery disease, particularly to avoid potential adverse effect during myocardial reperfusion. In this regard, it has been recently demonstrated that the detrimental effect of autophagy during reperfusion could be due to a surge in mitochondrial permeability upon reperfusion. Increased mitochondrial permeability boosts autophagy/mitophagy leading to excessive mitochondrial sequestration, but without sufficient autophagosomal clearance (by lysosomes) due to reduced LAMP-2 levels.[61,62] Interestingly, pharmacological inhibition of mitochondrial permeability transition pore (mPTP) *per se* has shown beneficial effects against myocardial reperfusion injury, at least preclinically.[61,63] Hence, it is speculated that blocking the mPTP opening during myocardial reperfusion may render autophagy activators beneficial also during reperfusion through cleaning the aftermath of damaged intracellular constituents, meanwhile exaggerated mitochondrial clearance is precluded by controlling mPTP opening.[64]

3.4 Heart failure

In the presence of other risk factors, the typical end point of all previously discussed cardiovascular disorders is heart failure. Regardless of its form, heart failure with reduced or preserved ejection fraction (HFrEF and HFpEF, respectively), is a devastating condition with poor life quality and survival rates. Although HFrEF and HFpEF affect a comparable fraction of patients, available data related to autophagy is disproportionally HFrEF-centric. Initial preclinical studies on pressure overload-induced HFrEF suggested that cardiac autophagy is an adaptive response that regulates cardiomyocyte homeostasis and, thus, protects the heart against hemodynamic stress. At odds with this notion, cardiomyocyte-specific *Atg5* knockout mice have shown high vulnerability to develop HFrEF in response to pressure overload.[65] However, in another study that was published around the same time,[66] cardiac autophagy was shown to be counterintuitively activated during pressure overload-induced HFrEF, despite concurring hypertrophic growth. Nevertheless, this controversy has been recently resolved by a longitudinal study, which revealed that autophagy and mitophagy are only temporarily activated upon pressure overload, but later on both decline significantly, thereby facilitating the development of HFrEF. In support of this notion, reactivating autophagy by TAT-Beclin 1 peptide attenuates HFrEF and associated mitochondrial dysfunction in mice subjected to pressure overload.[67] Furthermore, reduced mitochondrial autophagy due to *Drp1* haploinsufficiency exacerbates HFrEF development, meanwhile the beneficial effect of TAT-Beclin 1 was abrogated in mice lacking either autophagy or mitophagy due to mutations in *Beclin 1* or *Drp1*, respectively.[67]

As for HFpEF, available evidence for the role of autophagy is rather indirect and scarce. For instance, the autophagy inducer spermidine has been shown to protect hypertensive *Dahl* salt-sensitive rats from a HFpEF-like cardiac phenotype, characterized by diastolic dysfunction, lung congestion and preserved ejection fraction.[28] More importantly, caloric restriction, which is one of the strongest autophagy promoters, has been demonstrated to exert beneficial effects in HFpEF.[68] Specifically, reduced caloric intake by 350–400 kcal per day was sufficient to improve cardiopulmonary functional capacity in HFpEF patients within a duration of 20 weeks.[68] However, despite the well-known abilities of caloric restriction and spermidine to induce autophagy, it is still largely unknown whether their effects on HFpEF are necessarily autophagy-dependent. Nonetheless, autophagy activation seems to hold great promise as a potential therapy for heart failure, be it with reduced or preserved ejection fraction.

3.5 Other cardiovascular conditions

Other cardiovascular disorders where altered activity of autophagy and/or mitophagy has been reported, include but are not limited to, cardiac arrhythmias and cardiotoxicity induced by chemotherapeutic agents. However, unlike the previously described cardiovascular diseases, it is still not completely clear whether autophagy is beneficial or detrimental in these conditions.

3.5.1 Cardiac arrhythmia

Cardiac arrhythmia, also known as irregular heartbeat, is a group of conditions in which the heart does not follow a normal sinus rhythm, and instead beats at an abnormal pace (*i.e.*, too fast or too slow) or rhythm (*i.e.*, irregularly). Levels of autophagy markers have been examined in some of these conditions. For instance, ventricular tachycardia induced by ischemia-reperfusion in isolated mouse hearts has been shown to be associated with increased markers of autophagy, such as LC3B-II/LC3B-I ratio and Beclin-1.[69] Similarly, LC3B-II levels are elevated in humans and dogs suffering from atrial fibrillation.[70] On the contrary, another study demonstrated reduced levels of LC3II and LC3B-II/LC3B-I ratio in the atrial tissue of patients with postoperative atrial fibrillation, despite increased accumulation of electron microscopy-detected autophagosomes.[71] A possible explanation of this discrepancy is decreased autophagosome degradation due to impaired autophagic flux. However, none of the aforementioned studies examined autophagic flux in these arrhythmic conditions, which precludes the possibility to draw a clear conclusion on whether the increase in autophagy markers is due to increased or interrupted autophagic flow. That being said, these studies clearly demonstrate that autophagy is altered in cardiac arrhythmic conditions, and, thus, could be of interest as potential therapeutic targets.[72] Indeed, some of the currently used antiarrhythmic drugs reportedly induce autophagy, at least *in vitro*.[72] However, to establish a causal relationship between autophagy and cardiac arrhythmia or antiarrhythmic drugs, genetically modified models with increased or decreased autophagy activity should be employed.

To this end, future studies are required to (i) assess changes in autophagic flux in cardiac arrhythmias, (ii) elucidate whether altered autophagy (activated or interrupted) is adaptive or maladaptive, and most importantly (iii) decipher the mechanistic links between autophagy and cardiac arrhythmias.

3.5.2 Chemotherapy-induced cardiotoxicity

Many chemotherapeutic agents, including anthracyclines—effective and widely used chemotherapeutic agents in a variety of cancers—adversely affect cardiovascular health. Such chemotherapy-induced cardiotoxicity occurs in an acute or a chronic manner and manifests in various forms, including disturbed blood pressure control (*e.g.*, hypo- or hypertension), cardiac arrhythmias, left ventricular dysfunction, and ultimately congestive heart failure. Early studies assigned anthracycline-induced cardiotoxicity to the excessive production of ROS. Nonetheless, later findings showed that antioxidants do not prevent anthracycline-induced cardiotoxicity, which led to the hypothesis that other mechanisms may be involved, including autophagy. To this end, whether autophagy protects the myocardium from chemotherapy-induced injury or contributes to its underlying mechanisms has been a matter of debate. In this regard, several studies reported elevated autophagy-related markers in response to the anthracycline doxorubicin, meanwhile autophagy inhibition has been suggested to prevent its associated cardiotoxicity.[73,74] On the contrary, other studies examined autophagy flux and reported that anthracyclines inhibit autophagosomal clearance by impairing autophagosome-lysosome fusion, and reactivating autophagic flux rescues the myocardium from the injury provoked by doxorubicin.[75–78] A plausible explanation for these disparate findings could be the lack of a standardized approach to evaluate *in vivo* autophagic flux—which allows for the differentiation between enhanced autophagosome formation *vs.* blocked clearance of the autophagosome[27]—when some of these studies were carried out. Alternatively, it is possible that the effect of anthracyclines on autophagy depends on the dose and/or duration of treatment (*e.g.*, acute high dose *vs.* chronic low dose), which could in turn affect the magnitude and direction of autophagy modulation, and ultimately the resulting cardiac outcomes. Although extensive investigation is still required to conclusively clarify the feasibility of targeting cardiac autophagy to prevent anthracycline-induced chemotherapy, the emerging view in this field seems to support a positive role of autophagy in maintaining cardiac health during chemotherapy.[79] Regardless, further studies will still need to clarify this conundrum. In addition, future work should assess the role of other forms of autophagy, such as mitophagy and chaperone-mediated autophagy not only in cardiomyocytes, but also other cardiac cell types (*e.g.*, endothelial cells, fibroblasts and cardiac progenitor cells), which have been largely neglected so far.

4. Concluding remarks

Any perturbations in autophagy that affect the formation of autophagosomes or their subsequent degradation by lysosomes lead to impaired function and homeostasis of long-lived cardiomyocytes and other cells in the cardiovascular system. Hence, the cytoprotective function of autophagy is essential to sustain cardiovascular health and protect from disease (Fig. 1). With that in mind, future experimental studies will need to determine the long-term impact, including the benefits and potential side effects, of autophagy activators in order to define the most suitable compounds for future clinical testing. Research efforts will also need to accurately determine the duration and timing of autophagy-targeting interventions as some cardiovascular diseases show a course-dependent modulation of autophagy.[80] It is also worth mentioning that the majority of cardiovascular patients are elderly and so they typically suffer from multiple comorbidities. This issue necessitates a special consideration because autophagy activation may not be equally recommended in different pathologies. Therefore, a personalized approach combined with the development of small molecules with specific pro-autophagic actions at the cardiovascular system represents a tantalizing therapeutic prospect. If successfully translated to humans, autophagy activation may then become a standard practice to treat cardiovascular disorders.

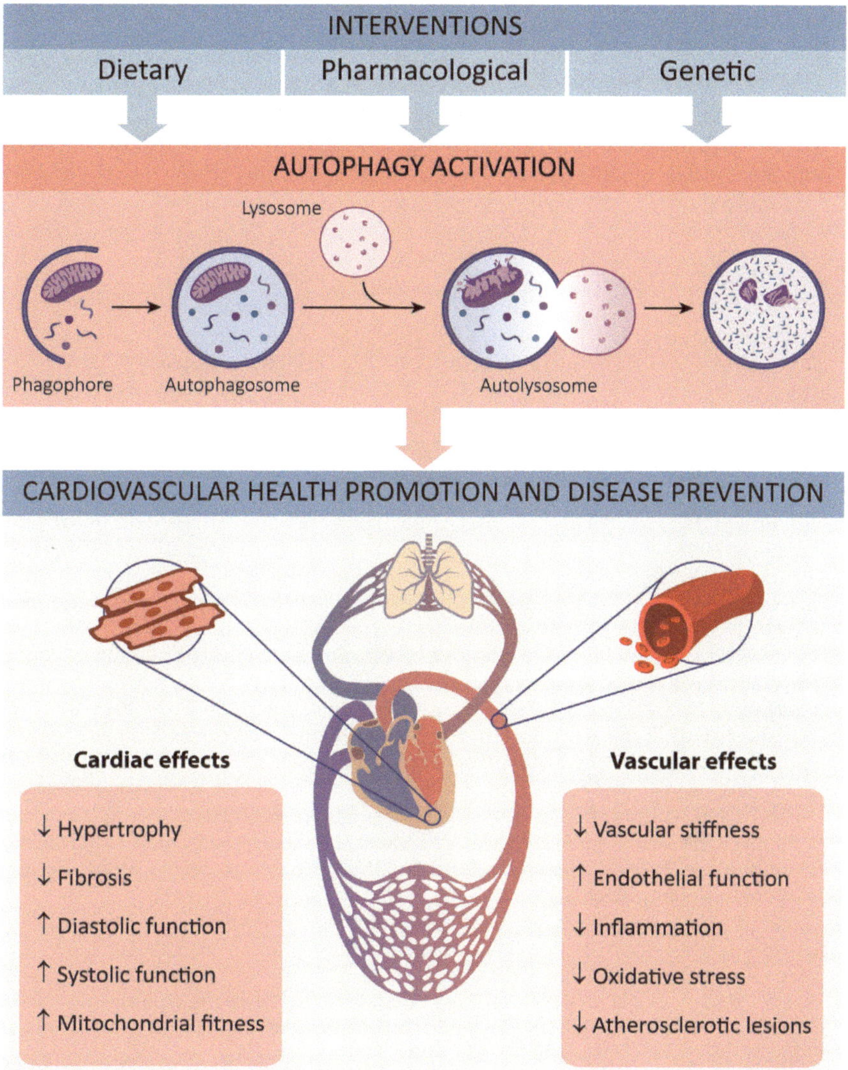

Fig. 1 Promoting cardiovascular health by pro-autophagic interventions. During autophagy, damaged, dysfunctional, or long-lived organelles and protein aggregates are sequestrated within a double-membraned vesicle, called autophagosome, which subsequently fuses with a lysosome for cargo degradation by acidic hydrolases. Stimulating this process by dietary, pharmacological or genetic interventions exerts pleiotropic effect on the heart and vasculature at the molecular, cellular and whole organ levels, which, in turn, promote the whole cardiovascular system health and protect it from disease, at least in preclinical testing.

Acknowledgments

The authors are grateful to support from the Austrian Science Fund FWF (grants I3301 ERA-CVD [MINOTAUR] to S.S.; V530 to S.L.H.; SFB LIPOTOX F3007 & F3012, W1226, P29203, P29262, P27893, P31727 to F.M.), the Austrian Society of Cardiology, (to M.A.), the European Society of Cardiology (to M.A.), the Austrian Federal Ministry of Education, Science and Research (to F.M.), and the BioTechMed Graz (to F.M., S.L.H. and S.S).

References

1. Galluzzi L, Pietrocola F, Levine B, Kroemer G. Metabolic control of autophagy. *Cell.* 2014;159(6):1263–1276. https://doi.org/10.1016/j.cell.2014.11.006.
2. Bergmann O, Bhardwaj RD, Bernard S, et al. Evidence for cardiomyocyte renewal in humans. *Science.* 2009;324(5923):98–102. https://doi.org/10.1126/science.1164680.
3. Mizushima N, Levine B. Autophagy in mammalian development and differentiation. *Nat Cell Biol.* 2010;12(9):823–830.
4. Taneike M, Yamaguchi O, Nakai A, et al. Inhibition of autophagy in the heart induces age-related cardiomyopathy. *Autophagy.* 2010;6(5):600–606. https://doi.org/10.4161/auto.6.5.11947.
5. Tanaka Y, Guhde G, Suter A, et al. Accumulation of autophagic vacuoles and cardiomyopathy in LAMP-2-deficient mice. *Nature.* 2000;406(6798):902–906. https://doi.org/10.1038/35022595.
6. Godar RJ, Ma X, Liu H, et al. Repetitive stimulation of autophagy-lysosome machinery by intermittent fasting preconditions the myocardium to ischemia-reperfusion injury. *Autophagy.* 2015;11(9):1537–1560. https://doi.org/10.1080/15548627.2015.1063768.
7. Nishino I, Fu J, Tanji K, et al. Primary LAMP-2 deficiency causes X-linked vacuolar cardiomyopathy and myopathy (Danon disease). *Nature.* 2000;406(6798):906–910. https://doi.org/10.1038/35022604.
8. Michiels CF, Fransen P, De Munck DG, De Meyer GRY, Martinet W. Defective autophagy in vascular smooth muscle cells alters contractility and Ca(2)(+) homeostasis in mice. *Am J Physiol Heart Circ Physiol.* 2015;308(6):H557–H567. https://doi.org/10.1152/ajpheart.00659.2014.
9. Torisu T, Torisu K, Lee IH, et al. Autophagy regulates endothelial cell processing, maturation and secretion of von Willebrand factor. *Nat Med.* 2013;19(10):1281–1287. https://doi.org/10.1038/nm.3288.
10. Lee E, Koo Y, Ng A, et al. Autophagy is essential for cardiac morphogenesis during vertebrate development. *Autophagy.* 2014;10(4):572–587. https://doi.org/10.4161/auto.27649.
11. Zhou J, Freeman TA, Ahmad F, et al. GSK-3alpha is a central regulator of age-related pathologies in mice. *J Clin Invest.* 2013;123(4):1821–1832. https://doi.org/10.1172/JCI64398.
12. Taneike M, Nishida K, Omiya S, et al. mTOR hyperactivation by ablation of tuberous sclerosis complex 2 in the mouse heart induces cardiac dysfunction with the increased number of small mitochondria mediated through the down-regulation of autophagy. *PLoS One.* 2016;11(3): e0152628. https://doi.org/10.1371/journal.pone.0152628.
13. Hsieh PN, Zhou G, Yuan Y, et al. A conserved KLF-autophagy pathway modulates nematode lifespan and mammalian age-associated vascular dysfunction. *Nat Commun.* 2017;8(1):914. https://doi.org/10.1038/s41467-017-00899-5.
14. Wang Q, Wu S, Zhu H, et al. Deletion of PRKAA triggers mitochondrial fission by inhibiting the autophagy-dependent degradation of DNM1L. *Autophagy.* 2017;13(2):404–422. https://doi.org/10.1080/15548627.2016.1263776.

15. Abdellatif M, Sedej S, Carmona-Gutierrez D, Madeo F, Kroemer G. Autophagy in cardiovascular aging. *Circ Res*. 2018;123(7):803–824.
16. Inuzuka Y, Okuda J, Kawashima T, et al. Suppression of phosphoinositide 3-kinase prevents cardiac aging in mice. *Circulation*. 2009;120(17):1695–1703. https://doi.org/10.1161/circulationaha.109.871137.
17. Ren J, Yang L, Zhu L, et al. Akt2 ablation prolongs life span and improves myocardial contractile function with adaptive cardiac remodeling: role of Sirt1-mediated autophagy regulation. *Aging Cell*. 2017;16(5):976–987. https://doi.org/10.1111/acel.12616.
18. Hoshino A, Mita Y, Okawa Y, et al. Cytosolic p53 inhibits Parkin-mediated mitophagy and promotes mitochondrial dysfunction in the mouse heart. *Nat Commun*. 2013;4:2308. https://doi.org/10.1038/ncomms3308.
19. Sheng Y, Lv S, Huang M, et al. Opposing effects on cardiac function by calorie restriction in different-aged mice. *Aging Cell*. 2017;16(5):1155–1167. https://doi.org/10.1111/acel.12652.
20. Shinmura K, Tamaki K, Sano M, et al. Impact of long-term caloric restriction on cardiac senescence: caloric restriction ameliorates cardiac diastolic dysfunction associated with aging. *J Mol Cell Cardiol*. 2011;50(1):117–127. https://doi.org/10.1016/j.yjmcc.2010.10.018.
21. Donato AJ, Walker AE, Magerko KA, et al. Life-long caloric restriction reduces oxidative stress and preserves nitric oxide bioavailability and function in arteries of old mice. *Aging Cell*. 2013;12(5):772–783. https://doi.org/10.1111/acel.12103.
22. Dai D-F, Karunadharma PP, Chiao YA, et al. Altered proteome turnover and remodeling by short-term caloric restriction or rapamycin rejuvenate the aging heart. *Aging Cell*. 2014;13(3):529–539. https://doi.org/10.1111/acel.12203.
23. Yan L, Gao S, Ho D, et al. Calorie restriction can reverse, as well as prevent, aging cardiomyopathy. *Age (Dordr)*. 2013;35(6):2177–2182. https://doi.org/10.1007/s11357-012-9508-5.
24. Rippe C, Lesniewski L, Connell M, LaRocca T, Donato A, Seals D. Short-term calorie restriction reverses vascular endothelial dysfunction in old mice by increasing nitric oxide and reducing oxidative stress. *Aging Cell*. 2010;9(3):304–312. https://doi.org/10.1111/j.1474-9726.2010.00557.x.
25. Stekovic S, Hofer SJ, Tripolt N, et al. Alternate day fasting improves physiological and molecular markers of aging in healthy, non-obese humans. *Cell Metab*. 2019;30(3):462–476.e5. https://doi.org/10.1016/j.cmet.2019.07.016.
26. Zheng Q, Zhao K, Han X, et al. Inhibition of AMPK accentuates prolonged caloric restriction-induced change in cardiac contractile function through disruption of compensatory autophagy. *Biochim Biophys Acta*. 2015;1852(2):332–342. https://doi.org/10.1016/j.bbadis.2014.04.023.
27. Klionsky DJ, Abdelmohsen K, Abe A, et al. Guidelines for the use and interpretation of assays for monitoring autophagy (3rd edition). *Autophagy*. 2016;12(1):1–222. https://doi.org/10.1080/15548627.2015.1100356.
28. Eisenberg T, Abdellatif M, Schroeder S, et al. Cardioprotection and lifespan extension by the natural polyamine spermidine. *Nat Med*. 2016;22(12):1428–1438. https://doi.org/10.1038/nm.4222.
29. Eisenberg T, Abdellatif M, Zimmermann A, et al. Dietary spermidine for lowering high blood pressure. *Autophagy*. 2017;13(4):767–769. https://doi.org/10.1080/15548627.2017.1280225.
30. LaRocca TJ, Gioscia-Ryan RA, Hearon CMJ, Seals DR. The autophagy enhancer spermidine reverses arterial aging. *Mech Ageing Dev*. 2013;134(7–8):314–320. https://doi.org/10.1016/j.mad.2013.04.004.
31. Flynn JM, O'Leary MN, Zambataro CA, et al. Late-life rapamycin treatment reverses age-related heart dysfunction. *Aging Cell*. 2013;12(5):851–862. https://doi.org/10.1111/acel.12109.

32. Chiao YA, Kolwicz SC, Basisty N, et al. Rapamycin transiently induces mitochondrial remodeling to reprogram energy metabolism in old hearts. *Aging (Albany NY)*. 2016;8(2):314–327. https://doi.org/10.18632/aging.100881.
33. LaRocca TJ, Henson GD, Thorburn A, Sindler AL, Pierce GL, Seals DR. Translational evidence that impaired autophagy contributes to arterial ageing. *J Physiol*. 2012;590(14):3305–3316. https://doi.org/10.1113/jphysiol.2012.229690.
34. Kaplon RE, Hill SD, Bispham NZ, et al. Oral trehalose supplementation improves resistance artery endothelial function in healthy middle-aged and older adults. *Aging (Albany NY)*. 2016;8(6):1167–1183. https://doi.org/10.18632/aging.100962.
35. Barger JL, Kayo T, Vann JM, et al. A low dose of dietary resveratrol partially mimics caloric restriction and retards aging parameters in mice. *PLoS One*. 2008;3(6): e2264. https://doi.org/10.1371/journal.pone.0002264.
36. Mattison JA, Wang M, Bernier M, et al. Resveratrol prevents high fat/sucrose diet-induced central arterial wall inflammation and stiffening in nonhuman primates. *Cell Metab*. 2014;20(1):183–190. https://doi.org/10.1016/j.cmet.2014.04.018.
37. Ou X, Lee MR, Huang X, Messina-Graham S, Broxmeyer HE. SIRT1 positively regulates autophagy and mitochondria function in embryonic stem cells under oxidative stress. *Stem Cells*. 2014;32(5):1183–1194. https://doi.org/10.1002/stem.1641.
38. Madeo F, Carmona-Gutierrez D, Hofer SJ, Kroemer G. Caloric restriction mimetics against age-associated disease: targets, mechanisms, and therapeutic potential. *Cell Metab*. 2019;29(3):592–610. https://doi.org/10.1016/j.cmet.2019.01.018.
39. Bravo-San Pedro JM, Kroemer G, Galluzzi L. Autophagy and mitophagy in cardiovascular disease. *Circ Res*. 2017;120(11):1812–1824. https://doi.org/10.1161/circresaha.117.311082.
40. Jia G, Hill MA, Sowers JR. Diabetic cardiomyopathy: an update of mechanisms contributing to this clinical entity. *Circ Res*. 2018;122(4):624–638. https://doi.org/10.1161/circresaha.117.311586.
41. Hsu H-C, Chen C-Y, Lee B-C, Chen M-F. High-fat diet induces cardiomyocyte apoptosis via the inhibition of autophagy. *Eur J Nutr*. 2016;55(7):2245–2254. https://doi.org/10.1007/s00394-015-1034-7.
42. Che Y, Wang Z-P, Yuan Y, et al. Role of autophagy in a model of obesity: a longterm high fat diet induces cardiac dysfunction. *Mol Med Rep*. 2018;18(3):3251–3261. https://doi.org/10.3892/mmr.2018.9301.
43. Tong M, Saito T, Zhai P, et al. Mitophagy is essential for maintaining cardiac function during high fat diet-induced diabetic cardiomyopathy. *Circ Res*. 2019;124(9):1360–1371. https://doi.org/10.1161/circresaha.118.314607.
44. Libby P, Buring JE, Badimon L, et al. Atherosclerosis. *Nat Rev Dis Prim*. 2019;5(1):56. https://doi.org/10.1038/s41572-019-0106-z.
45. Grootaert MOJ, Roth L, Schrijvers DM, De Meyer GRY, Martinet W. Defective autophagy in atherosclerosis: to die or to senesce? *Oxid Med Cell Longev*. 2018;2018:7687083. https://doi.org/10.1155/2018/7687083.
46. Ackermann K, Bonaterra GA, Kinscherf R, Schwarz A. Growth differentiation factor-15 regulates oxLDL-induced lipid homeostasis and autophagy in human macrophages. *Atherosclerosis*. 2019;281:128–136. https://doi.org/10.1016/j.atherosclerosis.2018.12.009.
47. Osonoi Y, Mita T, Azuma K, et al. Defective autophagy in vascular smooth muscle cells enhances cell death and atherosclerosis. *Autophagy*. 2018;14(11):1991–2006. https://doi.org/10.1080/15548627.2018.1501132.
48. Kheloufi M, Vion A-C, Hammoutene A, et al. Endothelial autophagic flux hampers atherosclerotic lesion development. *Autophagy*. 2018;14(1):173–175. https://doi.org/10.1080/15548627.2017.1395114.

49. Sergin I, Evans TD, Zhang X, et al. Exploiting macrophage autophagy-lysosomal biogenesis as a therapy for atherosclerosis. *Nat Commun*. 2017;8:15750. https://doi.org/10.1038/ncomms15750.
50. Kiffin R, Bandyopadhyay U, Cuervo AM. Oxidative stress and autophagy. *Antioxid Redox Signal*. 2006;8(1–2):152–162. https://doi.org/10.1089/ars.2006.8.152.
51. Duewell P, Kono H, Rayner KJ, et al. NLRP3 inflammasomes are required for atherogenesis and activated by cholesterol crystals. *Nature*. 2010;464(7293):1357–1361. https://doi.org/10.1038/nature08938.
52. Swiader A, Nahapetyan H, Faccini J, et al. Mitophagy acts as a safeguard mechanism against human vascular smooth muscle cell apoptosis induced by atherogenic lipids. *Oncotarget*. 2016;7(20):28821–28835. https://doi.org/10.18632/oncotarget.8936.
53. Matsui Y, Takagi H, Qu X, et al. Distinct roles of autophagy in the heart during ischemia and reperfusion: roles of AMP-activated protein kinase and Beclin 1 in mediating autophagy. *Circ Res*. 2007;100(6):914–922. https://doi.org/10.1161/01.RES.0000261924.76669.36.
54. Sciarretta S, Yee D, Nagarajan N, et al. Trehalose-induced activation of autophagy improves cardiac remodeling after myocardial infarction. *J Am Coll Cardiol*. 2018;71(18):1999–2010. https://doi.org/10.1016/j.jacc.2018.02.066.
55. Yan J, Yan J-Y, Wang Y-X, et al. Spermidine-enhanced autophagic flux improves cardiac dysfunction following myocardial infarction by targeting the AMPK/mTOR signalling pathway. *Br J Pharmacol*. 2019;176(17):3126–3142. https://doi.org/10.1111/bph.14706.
56. Wu X, He L, Chen F, et al. Impaired autophagy contributes to adverse cardiac remodeling in acute myocardial infarction. *PLoS One*. 2014;9(11): e112891. https://doi.org/10.1371/journal.pone.0112891.
57. Kanamori H, Takemura G, Goto K, et al. Autophagy limits acute myocardial infarction induced by permanent coronary artery occlusion. *Am J Physiol Heart Circ Physiol*. 2011;300(6):H2261–H2271. https://doi.org/10.1152/ajpheart.01056.2010.
58. Sciarretta S, Zhai P, Shao D, et al. Rheb is a critical regulator of autophagy during myocardial ischemia: pathophysiological implications in obesity and metabolic syndrome. *Circulation*. 2012;125(9):1134–1146. https://doi.org/10.1161/circulationaha.111.078212.
59. Kubli DA, Zhang X, Lee Y, et al. Parkin protein deficiency exacerbates cardiac injury and reduces survival following myocardial infarction. *J Biol Chem*. 2013;288(2):915–926. https://doi.org/10.1074/jbc.M112.411363.
60. Ma X, Liu H, Murphy JT, et al. Regulation of the transcription factor EB-PGC1alpha axis by beclin-1 controls mitochondrial quality and cardiomyocyte death under stress. *Mol Cell Biol*. 2015;35(6):956–976. https://doi.org/10.1128/MCB.01091-14.
61. Ma X, Liu H, Foyil SR, et al. Impaired autophagosome clearance contributes to cardiomyocyte death in ischemia/reperfusion injury. *Circulation*. 2012;125(25):3170–3181. https://doi.org/10.1161/circulationaha.111.041814.
62. Zhou B, Kreuzer J, Kumsta C, et al. Mitochondrial permeability uncouples elevated autophagy and lifespan extension. *Cell*. 2019;177(2):299–314.e16. https://doi.org/10.1016/j.cell.2019.02.013.
63. Gomez L, Thibault H, Gharib A, et al. Inhibition of mitochondrial permeability transition improves functional recovery and reduces mortality following acute myocardial infarction in mice. *Am J Physiol Heart Circ Physiol*. 2007;293(3):H1654–H1661. https://doi.org/10.1152/ajpheart.01378.2006.
64. Abdellatif M. Autophagy promotes longevity-except in the presence of "leaky" mitochondria. *Cardiovasc Res*. 2019;115(12):e118–e120. https://doi.org/10.1093/cvr/cvz224.

65. Nakai A, Yamaguchi O, Takeda T, et al. The role of autophagy in cardiomyocytes in the basal state and in response to hemodynamic stress. *Nat Med.* 2007;13(5):619–624. https://doi.org/10.1038/nm1574.
66. Zhu H, Tannous P, Johnstone JL, et al. Cardiac autophagy is a maladaptive response to hemodynamic stress. *J Clin Invest.* 2007;117(7):1782–1793. https://doi.org/10.1172/JCI27523.
67. Shirakabe A, Zhai P, Ikeda Y, et al. Drp1-dependent mitochondrial autophagy plays a protective role against pressure overload-induced mitochondrial dysfunction and heart failure. *Circulation.* 2016;133(13):1249–1263. https://doi.org/10.1161/circulationaha.115.020502.
68. Kitzman DW, Brubaker P, Morgan T, et al. Effect of caloric restriction or aerobic exercise training on peak oxygen consumption and quality of life in obese older patients with heart failure with preserved ejection fraction: a randomized clinical trial. *JAMA.* 2016;315(1):36–46. https://doi.org/10.1001/jama.2015.17346.
69. Meyer G, Czompa A, Reboul C, et al. The cellular autophagy markers Beclin-1 and LC3B-II are increased during reperfusion in fibrillated mouse hearts. *Curr Pharm Des.* 2013;19(39):6912–6918. https://doi.org/10.2174/138161281939131127122510.
70. Yuan Y, Zhao J, Yan S, et al. Autophagy: a potential novel mechanistic contributor to atrial fibrillation. *Int J Cardiol.* 2014;172(2):492–494. https://doi.org/10.1016/j.ijcard.2014.01.027.
71. Garcia L, Verdejo HE, Kuzmicic J, et al. Impaired cardiac autophagy in patients developing postoperative atrial fibrillation. *J Thorac Cardiovasc Surg.* 2012;143(2):451–459. https://doi.org/10.1016/j.jtcvs.2011.07.056.
72. van Bavel JJA, Vos MA, van der Heyden MAG. Cardiac arrhythmias and antiarrhythmic drugs: an autophagic perspective. *Front Physiol.* 2018;9:127. https://doi.org/10.3389/fphys.2018.00127.
73. Kobayashi S, Volden P, Timm D, Mao K, Xu X, Liang Q. Transcription factor GATA4 inhibits doxorubicin-induced autophagy and cardiomyocyte death. *J Biol Chem.* 2010;285(1):793–804. https://doi.org/10.1074/jbc.M109.070037.
74. Wang X, Wang X-L, Chen H-L, et al. Ghrelin inhibits doxorubicin cardiotoxicity by inhibiting excessive autophagy through AMPK and p38-MAPK. *Biochem Pharmacol.* 2014;88(3):334–350. https://doi.org/10.1016/j.bcp.2014.01.040.
75. Kawaguchi T, Takemura G, Kanamori H, et al. Prior starvation mitigates acute doxorubicin cardiotoxicity through restoration of autophagy in affected cardiomyocytes. *Cardiovasc Res.* 2012;96(3):456–465. https://doi.org/10.1093/cvr/cvs282.
76. Sishi BJN, Loos B, van Rooyen J, Engelbrecht A-M. Autophagy upregulation promotes survival and attenuates doxorubicin-induced cardiotoxicity. *Biochem Pharmacol.* 2013;85(1):124–134. https://doi.org/10.1016/j.bcp.2012.10.005.
77. Li M, Sala V, De Santis MC, et al. Phosphoinositide 3-kinase gamma inhibition protects from anthracycline cardiotoxicity and reduces tumor growth. *Circulation.* 2018;138(7):696–711. https://doi.org/10.1161/circulationaha.117.030352.
78. Li DL, Wang ZV, Ding G, et al. Doxorubicin blocks cardiomyocyte autophagic flux by inhibiting lysosome acidification. *Circulation.* 2016;133(17):1668–1687. https://doi.org/10.1161/circulationaha.115.017443.
79. Bartlett JJ, Trivedi PC, Pulinilkunnil T. Autophagic dysregulation in doxorubicin cardiomyopathy. *J Mol Cell Cardiol.* 2017;104:1–8. https://doi.org/10.1016/j.yjmcc.2017.01.007.
80. Schiattarella GG, Hill JA. Therapeutic targeting of autophagy in cardiovascular disease. *J Mol Cell Cardiol.* 2016;95:86–93. https://doi.org/10.1016/j.yjmcc.2015.11.019.

CHAPTER SEVEN

Autophagy in kidney disease: Advances and therapeutic potential

Divya Bhatia, Mary E. Choi*
Division of Nephrology and Hypertension, Joan and Sanford I. Weill Department of Medicine, NewYork-Presbyterian Hospital, Weill Cornell Medicine, New York, NY, United States
*Corresponding author: e-mail address: mechoi@med.cornell.edu

Contents

Abstract

Autophagy is a highly conserved intracellular catabolic process for the degradation of cytoplasmic components that has recently gained increasing attention for its importance in kidney diseases. It is indispensable for the maintenance of kidney homeostasis both in physiological and pathological conditions. Investigations utilizing various kidney cell-specific conditional autophagy-related gene knockouts have facilitated the advancement in understanding of the role of autophagy in the kidney. Recent findings are raising the possibility that defective autophagy exerts a critical role in different pathological conditions of the kidney. An emerging body of evidence reveals that

Progress in Molecular Biology and Translational Science, Volume 172
ISSN 1877-1173
https://doi.org/10.1016/bs.pmbts.2020.01.008

autophagy exhibits cytoprotective functions in both glomerular and tubular compartments of the kidney, suggesting the upregulation of autophagy as an attractive therapeutic strategy. However, there is also accumulating evidence that autophagy could be deleterious, which presents a formidable challenge in developing therapeutic strategies targeting autophagy. Here, we review the recent advances in research on the role of autophagy during different pathological conditions, including acute kidney injury (AKI), focusing on sepsis, ischemia-reperfusion injury, cisplatin, and heavy metal-induced AKI. We also discuss the role of autophagy in chronic kidney disease (CKD) focusing on the pathogenesis of tubulointerstitial fibrosis, podocytopathies including focal segmental glomerulosclerosis, diabetic nephropathy, IgA nephropathy, membranous nephropathy, HIV-associated nephropathy, and polycystic kidney disease.

1. Introduction

Autophagy plays a critical role in maintaining the morphology and viability of various cells in the kidney and their homeostasis during both healthy and diseased conditions.[1,2] The autophagy-dependent recycling of damaged or superfluous organelles and elimination of oxidative stress is particularly crucial for an organ like the kidney with high metabolic requirements involved in functions such as electrolyte reabsorption, fluid exchange, and removal of waste.[3] Autophagy is a highly dynamic self-degradative process that facilitates the lysosomal-dependent enzymatic digestion of cytosolic components. Autophagosome is a double-membraned vacuole that sequesters dysfunctional cytosolic materials and fusing with the lysosome to execute the process of cargo degradation.[4,5] The formation of autophagosome containing damaged organelles was first reported in renal proximal convoluted tubules in mice with drug-induced nephrotoxicity in 1969.[6] Based on the mode of delivery of the cargo to the lysosome, there are three distinct forms of autophagy: macroautophagy, microautophagy, and chaperone-mediated autophagy (CMA). Macroautophagy, most commonly referred as autophagy, is a widely studied major catabolic pathway. The process is regulated by a set of evolutionarily conserved autophagy-related (ATG) genes that orchestrate and support the formation of autophagosome.[7] The ATG genes were initially recognized in yeast and later in mammals.[5,8]

In microautophagy, the lysosomal membrane invaginates around the cytosolic substrate for degradation.[9] In CMA, the heat shock cognate protein of 70 or 90 kDa (hsc70 or hsc90) identifies proteins with amino acid recognition sequence KFERQ. The lysosome, through a lysosome-associated membrane protein (LAMP)-2A, recognizes and eliminates the hsc and substrate complex.[10] An increase in the levels of ammonia associated with metabolic acidosis and hypokalemia in chronic kidney disease (CKD) may downregulate CMA.[11] Also, in the streptozotocin-induced diabetic rat model, levels of LAMP-2A in the renal cortex were abundant in lysosomal fractions, and a decrease in CMA was associated with an increase in protein aggregates and renal hypertrophy.[12]

Selective autophagy contributes to maintaining cellular homeostasis by targeted degradation of dysfunctional organelles, including mitochondria (mitophagy), endoplasmic reticulum (ERphagy), peroxisomes (pexophagy), aggregates of proteins (aggrephagy), glycogen (glycophagy), lipids (lipophagy), defective ribosomes (ribophagy) or nuclei (nucleophagy) and the removal of microorganisms (xenophagy). Mitochondrial dysfunction has been implicated in the pathogenesis of kidney disease.[13] The emerging role of mitochondrial autophagy, or mitophagy, in kidney diseases has recently gained much attention and is a topic of recent investigations and reviewed elsewhere.[14–19]

Here, we focus our review on the recent advances in understanding the role of autophagy in acute kidney injury (AKI) and CKD. The seminal studies by Professor Yoshinori Ohsumi (the 2016 Nobel Laureate in Physiology or Medicine) on the molecular mechanism of autophagy have provided the basis to decipher its role for the clinical applications.[19] The kidney cell (podocyte, endothelial cell, proximal, and distal tubular epithelial cell)-specific conditional knockout models of Atg genes have enabled a better understanding of the functions of autophagy in kidney diseases.[20–27] In this chapter, we will review the role of autophagy in the renal glomerular and tubular compartments during various pathological conditions of the kidney. Growing evidence suggests that maladaptive autophagy is directly linked to various pathological conditions of the kidney and is implicated in AKI and CKD.[28–30]

2. Autophagy in acute kidney injury

Sepsis, renal ischemia, drug, and heavy metal-induced nephrotoxicity are among the common causes of AKI. In response to these stresses, the autophagy pathway is activated.[31–34] Autophagy exerts cytoprotective functions *via* mechanisms that help cellular adaptation against oxidative stress through recycling of damaged cellular components.[28] However, autophagy can also be deleterious in certain conditions of AKI. Below, we discuss the current understanding of the role of autophagy in various forms of AKI (Fig. 1).

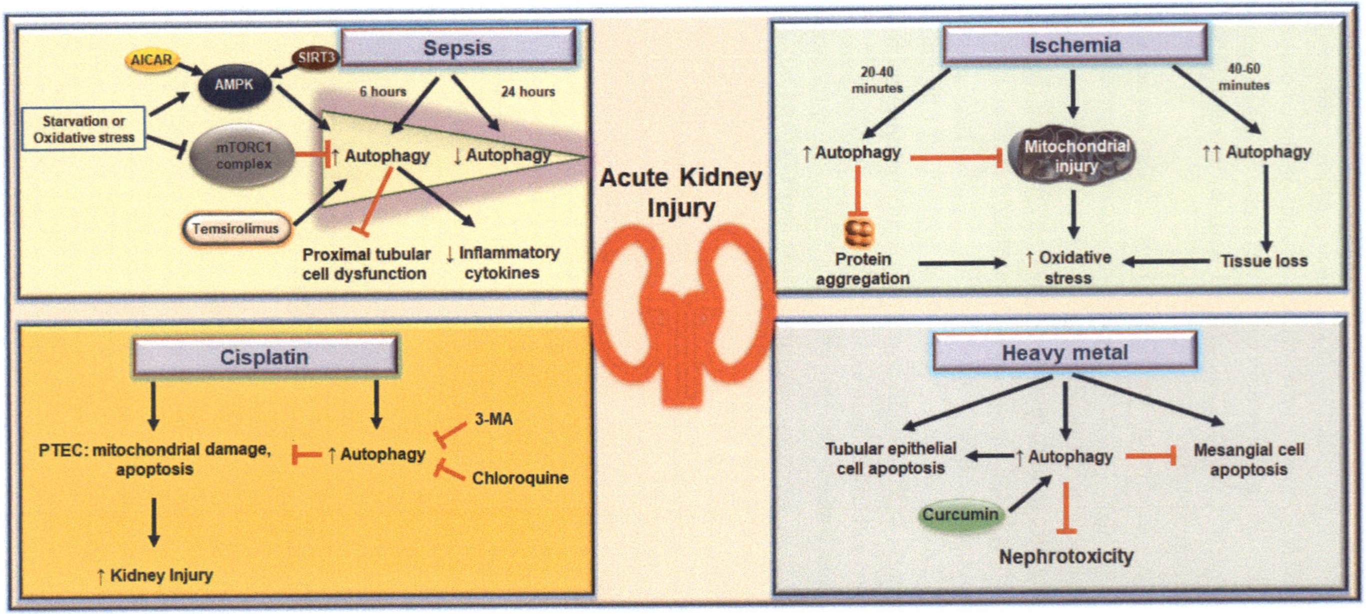

Fig. 1 Autophagy in acute kidney injury. Autophagy exerts important functions in acute kidney injury (AKI). Schema showing the role of autophagy during sepsis, ischemia-reperfusion injury (IRI), cisplatin, and heavy-metal-induced AKI. The model-specific functions of autophagy are discussed in the chapter. Autophagy has been shown to play protective roles against kidney injury. In contrast, persistent activation of autophagy may contribute to tissue loss and renal damage. AICAR, 5-aminoimidazole-4-carboxamide ribonucleotide; SIRT3, Sirtuin-3; AMPK, 5′ Adenosine monophosphate-activated protein kinase; mTORC1, mechanistic target of rapamycin complex 1; PTEC, proximal tubular epithelial cells; 3-MA, 3-methyladenine.

2.1 Sepsis-induced AKI

Sepsis is frequently associated with multiorgan dysfunction that leads to increased mortality and is the most common cause of AKI in critically ill patients.[35] Studies suggest that adaptive or maladaptive responses of renal tubular epithelial cells, inflammation, and vascular dysfunction are responsible for sepsis-induced AKI. Autophagy represents an important mechanism for the maintenance of cellular homeostasis during pathologic stress conditions in the kidney, including sepsis. In cecal ligation and puncture (CLP)-induced model of sepsis in rats, autophagy initially increases at 4 h and then starts declining after a period of 9 h, and the studies by Hsiao et al. suggest that the decline of autophagy contributes to proximal tubular dysfunction during sepsis.[36] Similarly, in the murine model of sepsis induced by CLP, Sunahara et al. reported that the number of autophagosomes, the expression of the lipidated form of microtubule-associated protein 1 light chain 3 (LC3-II), and its colocalization with lysosomes initially increased at 6 h and then started declining after a period of 24 h after CLP, indicating decrease in the autophagic flux.[37] In both the CLP and lipopolysaccharide (LPS) models, 5-aminoimidazole-4-carboxamide ribonucleotide (AICAR)-mediated induction of the 5′ Adenosine monophosphate-activated protein kinase (AMPK) pathway and activation of autophagy were associated with reduced circulating inflammatory cytokines, activation of endothelial cells, neutrophils adhesion and improved renal function.[38] The 3-methyladenine (3-MA)-induced decrease in LC3-II and Beclin 1 expression and suppression of autophagy were associated with worse kidney injury after sepsis.[32] The proximal tubular epithelial cells are more susceptible to common renal insults.[30,39] Howell et al. reported that the activation of autophagy by temsirolimus treatment prevented proximal tubular dysfunction and endotoxin-induced renal injury.[40] Sirtuin-3 (SIRT3), an inhibitor of nucleotide-binding oligomerization domain (NOD)-like receptor protein 3 (NLRP3) inflammasome, has been shown to activate autophagy and reduce CLP-induced kidney injury.[41] Zhang et al. reported that the activation of autophagy in the LPS-induced sepsis model involves calcium/calmodulin-dependent protein kinase IV (CaMKIV)-mediated activation of the mechanistic target of rapamycin complex 1 (mTORC1) pathway to exert cytoprotective effects to mitigate AKI.[42] Interestingly, CaMKIV prevented ubiquitin-mediated proteasomal degradation of mTOR and thereby permitting mTOR expression and activity, which were required for LPS-induced autophagy. These findings are contrary to the current paradigm that mTORC1 activation prototypically suppresses autophagy. Further studies to clarify the role of mTORC1 pathway in activation of autophagy in sepsis-induced AKI are warranted.

2.2 Ischemia-reperfusion injury-induced AKI

Renal ischemia-reperfusion injury (IRI) frequently follows a sudden temporary impairment of blood flow and reoxygenation of tissue after a period of hypoxia or ischemia. IRI-induced AKI is a common cause of kidney injury in patients undergoing kidney transplantation.[43] Several *in vivo* and *in vitro* studies have observed the activation of autophagy after IRI.[24,26,27,44–48] The protective effects of autophagy against IRI-induced AKI have also been reported.[24,27,49] The increase in autophagy in proximal tubular epithelial cells after IRI was confirmed using autophagy reporter mouse model expressing red fluorescent protein (RFP)-enhanced green fluorescent protein (EGFP)-tagged LC3 transgene under the control of a CAG promoter (CAG-RFP-EGFP-LC3).[50] The autolysosome formation was detected through an acidic environment-induced increase in the RFP signal, and a decrease in the EGFP fluorescence. The *Map1lc3-GFP* reporter mice, in which LC3 was fused to EGFP at the N-terminus, allowed identification of LC3-II through the presence of the GFP-positive puncta formation.[27] The activation of autophagy after a brief duration of ischemia (20–40 min) was cytoprotective, as it prevented stress-induced mitochondrial dysfunction and kidney injury.[29,51,52] However, prolonged ischemia (40–60 min) caused overactivation of autophagy, that contributed to cell death, tissue loss, and renal function impairment.[51,53,54] The proximal tubule-specific conditional deletion of *Atg5* resulted in worse tubular atrophy, renal damage, and reduced renal function after IRI-induced AKI.[24,55] Similarly, both the proximal tubular epithelial and distal tubular epithelial cell-specific *Atg5* conditional knockout and proximal tubular epithelial cell-specific *Atg7* conditional knockout mice showed higher susceptibility to ischemia-related mitochondrial stress, more numbers of aggregated proteins, apoptosis and reduced renal function after IRI.[26,55] Xie et al. demonstrated the cytoprotective functions of ischemia preconditioning-induced activation of autophagy against apoptosis and renal injury using the IRI model.[56] Beclin 1 knockdown by small interfering RNA (siRNA) or 3-MA-mediated suppression of autophagy in proximal tubular epithelial cells exacerbated oxidative stress-induced cell death.[26,57] However, Tan et al. found that inhibition of excessive autophagy using fibroblast growth factor 10 (FGF-10) exerted a cytoprotective role against IRI-mediated inflammation and kidney injury.[58] The bafilomycin A1-dependent suppression of autophagy and reduced expression of LC3 and ATG5 proteins were associated with a reduction in apoptosis and caspase-3/7 activity.[46] The *ATG7* knockdown by siRNA or 3-MA-mediated autophagy inhibition in human embryonic kidney (HEK)293 cells protected against oxidative stress-induced cell death.[59] Altogether, these studies provide evidence that the activation of autophagy can confer cytoprotective effects by promoting cell survival and mitigate renal damage and preserve renal function depending on the condition and duration of ischemic injury. On the other hand, prolonged stress leads to excessive autophagy that may be detrimental and contributes to tissue loss.

2.3 Cisplatin-induced AKI

Cisplatin, a chemotherapeutic agent, is a known nephrotoxin and can cause AKI. The cisplatin treatment induces a time-dependent increase in the LC3 expression in proximal tubular epithelial cells.[23] The inhibition of autophagy through either treatment with 3-MA or knockdown of *Becn1* by short-hairpin RNA or siRNA, or *Atg5* by siRNA resulted in severe cisplatin-induced mitochondrial damage, higher caspase activation and apoptosis in proximal tubular epithelial cells.[60,61] The conditional deletion of *Atg5* in proximal tubular epithelial cells enhanced tubular injury, oxidative stress, and worsened renal function after cisplatin treatment.[23] Similarly, proximal tubule-specific *Atg7* knock out mice displayed higher sensitivity to cisplatin-induced renal dysfunction, increased oxidative stress, apoptosis, and tubular damage.[55] The treatment with apoptotic inhibitor zVAD-fmk (a lysomotropic agent) suppressed cisplatin injury-induced cathepsin B and calpain activation, inhibited autophagic flux, and contributed to worsening renal function.[62] The treatment with chloroquine (autophagy inhibitor) also aggravated cisplatin-induced renal tubular injury.[55] These studies suggest that autophagy exerts protective functions against cisplatin-induced AKI.

2.4 Heavy metal-induced AKI

Heavy metals, including cadmium, arsenic, lead, and mercury are known to be nephrotoxic and can induce AKI.[63] Exposure to heavy metal is associated with impaired autophagic flux.[64,65] The activation of the cell death pathway or cell survival is entirely dependent upon the exposure/condition. Autophagy exhibits paradoxical functions. For example, under starvation conditions, autophagy prevented cell death, while inhibition of autophagy exaggerated apoptosis.[66] However, exposure to heavy metal such as cadmium or arsenic resulted in autophagy-dependent tubular epithelial cell death.[65,67] Wang et al. observed that cadmium-induced nephrotoxicity resulted in the activation of autophagy, that was associated with enhanced apoptosis and production of reactive oxygen species (ROS).[68] The pharmacological inhibition of autophagy in cadmium-treated mesangial cells improved their viability, suggesting that autophagy promotes cell death induced by heavy metal exposure.[67] However, *Atg16l1* knockdown in rat mesangial cells showed increased cleavage of pro-caspase-3 after cadmium treatment, suggesting a role of autophagy in inhibiting the initiation of apoptosis.[69] Autophagy-mediated cell death after severe oxidative stress is a type II programmed cell death, that is characterized by extensive degradation of the endoplasmic reticulum (ER), Golgi apparatus and polyribosomes.[70] In type I programmed cell death or apoptosis, there is a cytoplasmic condensation, while the organelles are preserved without degradation. Autophagy-dependent activation of apoptosis may serve as a protective mechanism for the proximal tubular epithelial cells to cope against heavy metal-induced nephrotoxicity. The treatment with curcumin has been shown to restore autophagic flux and reduce heavy metal-induced AKI.[71]

Taken together, the studies discussed above provide evidence that autophagy plays a dual role in the pathogenesis of AKI, exerting either cytoprotective effects or detrimental effects in a context-specific manner. These findings underscore the need for further investigations to clarify the seemingly opposing functions of autophagy in AKI. This is particularly important as it presents unique challenges in developing therapeutic strategies targeting autophagy in the treatment of AKI.

3. Autophagy in chronic kidney disease

Autophagy has been implicated to play a significant role in the pathogenesis of CKD.[2] Development of renal fibrosis is a hallmark of CKD and represents the common pathway in progressive kidney disease that accompanies a gradual decline in renal function and progression to end-stage kidney failure.[72] Here, we summarize the current knowledge of the role of autophagy in CKD focusing on its role in the pathogenesis of kidney fibrosis and the various forms of CKD including podocytopathies such as focal segmental glomerulosclerosis (FSGS), diabetic nephropathy (DN), IgA nephropathy (IgAN), human immunodeficiency virus (HIV)-associated nephropathy (HIVAN), and polycystic kidney disease (PKD) (Fig. 2).

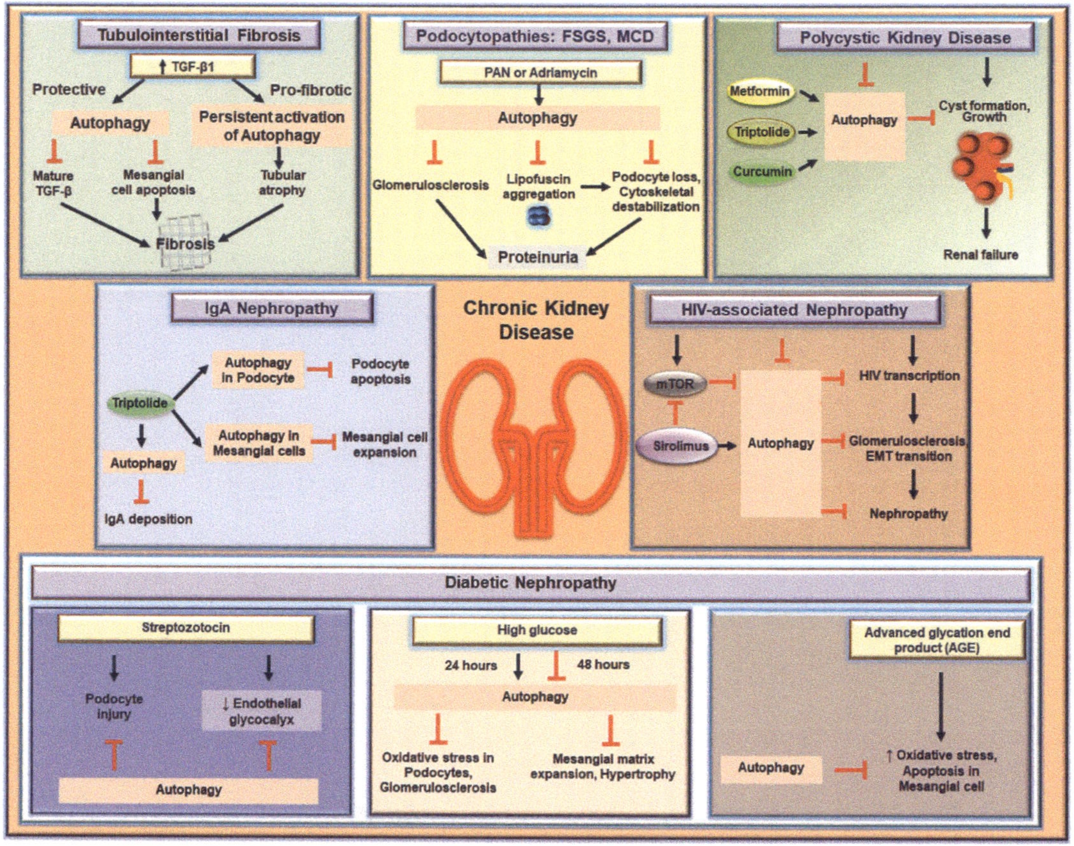

Fig. 2 Autophagy in chronic kidney disease. Autophagy plays a critical role in the pathogenesis of chronic kidney diseases (CKD). Schematic representation showing the functions of autophagy in different diseases, including tubulointerstitial fibrosis, podocytopathies: focal segmental glomerulosclerosis (FSGS), minimal change disease (MCD), diabetic nephropathy, IgA nephropathy, HIV-associated nephropathy, and polycystic kidney disease. Autophagy has been shown to exert cytoprotective functions against fibrosis. However, persistent activation of autophagy may promote tubular atrophy and extracellular matrix accumulation. PTEC, proximal tubular epithelial cells; EC, endothelial cells.

3.1 Autophagy in tubulointerstitial fibrosis

The major pathologic manifestation and outcome of progression of CKD is tubulointerstitial fibrosis (TIF).[73] The presence of fibrosis is characterized by the infiltration of inflammatory cells, increased number of fibroblasts, accumulation of extracellular matrix (ECM), and tubular apoptosis and atrophy.[74] Increased expression of autophagy markers, including LC3 and Beclin 1 is observed in the obstructed renal tubules after unilateral ureteral obstruction (UUO), which is a well-established model of kidney fibrosis.[75] The findings of the role of autophagy during fibrosis are heterogeneous. Autophagy has been shown to exert both profibrotic and anti-fibrotic functions after UUO.[17,76]

Evidence supporting a renoprotective role of autophagy is demonstrated by numerous lines of investigation, including 3-MA-dependent inhibition of autophagy worsening tubular cell apoptosis and TIF induced by UUO.[76] Recently, Nam et al. reported that the conditional deletion of *Atg7* in the distal tubule also resulted in higher mitochondrial damage, p62/Sequestosome 1 (SQSTM1) accumulation, NLRP3 inflammasome activation, apoptosis, and severe TIF after UUO.[77] In addition, kidney epithelium-specific *Atg5* deleted mice developed tubulointerstitial damage by 4 months of age and died by 6 months with evidence of tubulointerstitial inflammation, apoptosis, necrosis, and TIF.[22] Transforming growth factor-beta 1 (TGF-β1), a pleiotropic profibrogenic cytokine, acts as an inducer as well as an inhibitor of autophagy. We have previously reported that TGF-β1 promotes collagen production and activates autophagy in mesangial cells, and activated autophagy, in turn, favors intracellular degradation of type I collagen.[78] Our group has also observed that the activation of autophagy in mesangial cells by TGF-β1 treatment prevented serum starvation-induced apoptosis *via* downregulation of cyclin-dependent kinase inhibitor p27 and upregulation of cell cycle regulators, cyclin D1 and E.[79] Recently, Shi et al. also reported similar observations that the activation of autophagy in mesangial cells was linked to a lower percentage of β-galactosidase-positive senescent cells, and aged human and rodent kidneys displayed downregulation of autophagy proteins associated with an aging-related phenotype of glomerulosclerosis and TIF.[80] We have also demonstrated that autophagy, through degrading mature TGF-β protein, suppresses TIF in the UUO model.[81] In addition, the deficiency of autophagic proteins resulted in increased renal expression of TGF-β1 and collagen, enhanced tubular epithelial cell apoptosis, and severe TIF after UUO, as confirmed in $LC3^{-/-}$ or Beclin 1 haploinsufficient ($Becn1^{+/-}$) mice.[78]

Moreover, reduced expression of Beclin 1 by genetic deletion or siRNA knockdown in mesangial cells resulted in enhanced type I collagen stimulation by TGF-β1.[78] We also demonstrated that TGF-β1 induces autophagy in mesangial cells *via* the TGF-β-activated kinase 1 (TAK1)/mitogen-activated protein kinase 3 (MKK3)-p38 signaling pathway.[78] Interestingly, TGF-β1 (*via* activation of mTOR through the phosphoinositide 3-kinase (*PI3K*)/AKT pathway) negatively regulates autophagy in HEK293 cells.[82] Our study also confirmed that treatment with low-dose of carbon monoxide (CO), shown to protect against UUO-induced renal fibrosis, or CO-releasing molecule 2 (CORM-2) suppressed collagen expression induced by TGF-β1 stimulation in mesangial cells through activation of autophagy.[78,83]

While a substantial body of evidence, as discussed above, supports the concept that autophagy is protective against kidney fibrosis, there are studies demonstrating contrasting findings. In cultured human renal proximal tubular epithelial cells, TGF-β1 treatment induced activation of autophagy through the generation of ROS to promote apoptosis.[84] Persistent autophagic activity in kidney proximal tubules after UUO contributes to tubular atrophy and interstitial fibrosis, whereas proximal tubule-specific deletion of *Atg7* suppressed tubulointerstitial fibrosis in the UUO model.[17] The increase in the mitochondrial related oxidative stress, renal tubular injury, and apoptosis after UUO were associated with increased autophagic activity.[85] Furthermore, 3-MA-mediated inhibition of autophagy suppressed the infiltration of inflammatory cells and attenuated TIF in a rat model of hyperuricemic nephropathy.[86] The reasons for the heterogeneous findings remain unclear, and they highlight the complexity of the role of autophagy in the pathogenesis of renal fibrosis and CKD. The multifunctional nature of both TGF-β1 actions and autophagy may be context and cell type-dependent, and further rigorous investigations are clearly warranted.

3.2 Autophagy in podocytopathies

Podocytes exhibit a high basal level of autophagy, as reported in mouse podocytes, GFP-LC3 transgenic mice, and human kidney biopsy.[20,87,88] Autophagic activity in podocytes plays an essential role in preventing the progression of podocytopathies such as minimal change disease (MCD) and FSGS. MCD and primary FSGS are glomerular diseases associated with nephrotic syndrome and may lead to chronic renal failure and, ultimately end-stage renal disease (ESRD).[89,90] A prominent pathological feature of

MCD is podocyte foot process effacement in the absence of visible alterations by light microscopy, whereas FSGS is characterized by the presence of focal and segmental sclerotic glomerular lesions and podocyte loss, and decrease in the number of podocytes results in poor renal prognosis.[89,90] Podocyte-specific deletion of *Mtor* gene (*Mtor* pod-KO) inhibited autophagic flux, and the mice displayed heavy proteinuria as early as 3 weeks of age, associated with pronounced foot process effacement and progressive podocyte loss that resulted in ESRD by 5 weeks post-birth.[91] The increase in oxidative stress due to absence of autophagy contributed to loss of podocytes, and reduction in podocyte number and/or function, and was associated with development of glomerulosclerosis and the pathogenesis of pathological conditions, including FSGS.[92,93] The administration of puromycin aminonucleoside (PAN) or adriamycin (ADR) induces podocyte injury and proteinuria, resembling human MCD or FSGS. Podocyte-targeted depletion of *Atg5* resulted in increased apoptosis and glomerulosclerosis in aged mice, and was associated with aggregation of lipofuscin, oxidized and ubiquitinated proteins, and proteinuria in both PAN- and ADR-induced models of nephrosis.[20,21] The autophagic activity was higher during the recovery phase after PAN-induced nephrosis.[88] Moreover, mice with *Atg7* deletion in the podocytes displayed increased proteinuria, podocyte cytoskeleton destabilization, foot process effacement, and podocyte loss 1 day after unilateral nephrectomy compared to control wild-type mice.[94] The podocyte-specific deletion of prorenin receptor (PRR), a receptor for renin and prorenin, and an accessory subunit of the vacuolar proton pump H(+)-ATPase, resulted in defective autophagic flux, and within 2 weeks after birth, mice developed albuminuria, podocyte foot process effacement and increased podocyte apoptosis, and died between 2 and 3 weeks of age.[95,96] The association of an increase in accumulation of autophagosomes and p62/SQSTM1 in the glomeruli of podocyte-specific PRR knockout mice with a reduction in renal function highlights the importance of autophagy-mediated clearance and protein turnover in podocyte quality control.[96] Taken together, these studies demonstrate that autophagy in podocytes plays an essential role in maintaining podocyte function and survival.

The B-cell lymphoma 2 (Bcl-2) homology 3 (BH3)-only protein carrying one BH3 domain is implicated in the activation of autophagy.[97] Apolipoprotein L1 (APOL1) is a BH3-only lipid-binding protein, and disease-associated risk variants in *APOL1* gene have been identified that exert cytotoxic effects, at least in part, through modulation of autophagic flux.[98]

Wild-type and risk variant APOL1 differentially localize to intracellular lipid droplets and the ER, respectively, and translocation of risk variant APOL1 from the ER to lipid droplets results in a reduction of autophagic flux and cytotoxicity.[98] The genetic variants in two alleles of *APOL1* gene are associated with an increased risk of non-diabetic kidney disease in African Americans.[99,100] APOL1 variants also increase risk for FSGS in African Americans.[101–103] Regulating autophagic flux through modulation of APOL1 risk variants by recruitment to lipid droplets represents an intriguing potential therapeutic intervention in African American patients with APOL1-associated kidney disease.

In addition, the podocytes from patients with MCD displayed higher expression of Beclin 1 than patients with FSGS.[104] The decrease in the podocyte autophagic activity was associated with higher apoptosis, glomerulosclerosis, and worsening of podocytopathy.[105] The Trehalose-induced activation of autophagy protected against PAN-induced podocyte injury in cultured human podocytes.[106] These studies further confirm that the autophagic activity exerts critical functions in preventing the progression of podocytopathies.

3.3 Autophagy in diabetic nephropathy

DN is the leading cause of CKD and ESRD, which affects patients with type 1 (15–40%) and type 2 (5–20%) diabetes.[107] The progressive loss of renal function in DN is associated with increased podocyte and tubular injury, cytoskeletal destabilization, foot process effacement, glomerular and tubular hypertrophy, and mesangial cell expansion.[108,109] High glucose-induced activation of autophagy in podocytes showed protective effects against oxidative stress.[110,111] However, prolonged exposure to glucose suppressed autophagy in podocytes.[112] Similarly, in streptozotocin-induced diabetic mice, decreased autophagy in podocytes was associated with progression of nephropathy.[111] The podocyte-specific deletion of *Atg5* gene resulted in severe proteinuria and higher podocyte injury in a high-fat diet-induced diabetic model.[113] Also, the kidney biopsy samples from patients with DN, who had massive proteinuria, show a higher expression of p62, suggesting insufficient autophagic activity.[113]

In addition, hyperactivation of mTORC1 is implicated in the progression of DN.[114,115] Rapamycin (mTORC1 inhibitor) has been shown to restore autophagy in high glucose-treated podocytes[111] and attenuate glomerulosclerosis, hypertrophy, and fibrotic response.[116]

Thus, up-regulating autophagy by the use of rapamycin may be beneficial for treatment of DN. However, adverse effects of rapamycin, such as proteinuria in non-diabetic kidney disease have raised concerns about its safety and efficacy.[117] Hence, the jury is still out on whether inhibitors of mTORC1 are suitable therapeutic options in patients with DN, and a target that bears a better potential for more effective treatment with fewer side effects still needs to be identified.

The protective functions of autophagy against DN have also been reported in other cell types. For example, the ursolic acid-mediated activation of autophagy in mesangial cells attenuated high glucose-induced hypertrophy, proliferation, and fibrotic response.[118] The advanced glycation end product (AGE)-dependent damage is also one of the chief causes of glomerular injury in diabetic kidney disease.[119] Xu and colleagues observed that autophagy prevented the AGE-mediated apoptosis and oxidative stress in mesangial cells.[120]

In addition, defective clearance of autophagosomes during diabetic conditions also resulted in phenotypical changes in endothelial cells.[21] The *Atg5* endothelial and hematopoietic cell-specific conditional knockout mice displayed enhanced characteristic features of DN, including enlarged glomerular capillary loops, and mesangial matrix expansion.[16] Autophagy in endothelial cells displayed cytoprotective functions against oxidative stress and glomerular injury.[121] Furthermore, endothelial cell-specific conditional deletion of *Atg5* worsened renal function in a diabetic model.[21] The decrease in the endothelial glycocalyx is commonly observed during progression of DN.[122] However, suppression of autophagy did not affect endothelial glycocalyx.[21] Autophagy in endothelial cells has been shown to combat metabolic stress and protect against glomerulosclerosis during diabetic conditions.[21]

In proximal and distal tubular epithelial cells, autophagy was downregulated in a streptozotocin-induced diabetic model.[123] The reduction of autophagy in tubular epithelial cells was accompanied by activated mTOR and downregulated AMPK pathway, suggesting mTOR mediated suppression of autophagy.[123] Sakai et al. recently reported differential autophagic activity in proximal tubular epithelial cells in type I and type 2 diabetic models, which was regulated by insulin through the mTOR pathway. In type 2 diabetic model of kidney disease, autophagy was repressed in proximal tubular epithelial cells even under starvation conditions.[124] However, in streptozotocin-induced diabetic mice, autophagy in the proximal tubular epithelial cells was induced even under fed conditions and was further enhanced after chloroquine administration.[124] Thus, the activation or suppression of autophagy is dependent on the specific pathological condition, and it may vary from the glomerular to the tubular compartment. Overall, in DN, autophagy has been shown to exert renoprotective functions in both the glomerular and tubular cells and prevent the progression of diabetic kidney disease.

3.4 Autophagy in IgA nephropathy

The IgAN involves the deposition of IgA in the kidney, resulting in local inflammation-induced kidney damage that may lead to ESRD.[125] Autophagy in patients with IgAN was first observed in the podocytes through electron microscopy and found to be increased compared to healthy subjects.[126,127] Sato et al. reported two different types of autophagy (I and II) in the kidney biopsy samples from patients with IgAN.[127] The higher number of type II associated autophagosomes was associated with a better prognosis than type I autophagy,[127] and the latter rarely displays formation of autophagic vacuoles or classical autophagic activity.[127] The LC3-II puncta formation in the podocytes decreased after treatment with IgA conditioned media.[126] The triptolide-induced activation of autophagy in IgA conditioned media-treated podocytes protected against apoptotic cell death.[126] Triptolide was also helpful in reducing mesangial cell proliferation and mitigating mesangial expansion and deposition of IgA in a rat model of IgAN.[128] These studies indicate that autophagy may be beneficial in reducing the deposition of IgA and preventing the inflammation-related renal damage in IgA nephropathy.

3.5 Autophagy in membranous nephropathy

Membranous nephropathy (MN) is one of the leading causes of nephrotic syndrome in nondiabetic adult patients.[129] Proteinuria in MN is facilitated by activation of the complement system and membrane attack complex (MAC) (C5b-9)-mediated podocyte injury.[130] The MAC or C5b-9 inhibits podocyte autophagy through blocking the fusion of lysosome with autophagosome.[131] The inhibition of autophagy has been shown to exaggerate proteinuria.[96] Hartleben et al. reported an increase in LC3 puncta formation in podocytes of renal allograft biopsies from patients with MN.[20] Another study observed not only increases in LC3 puncta but also a higher accumulation of p62/SQSTM1 in podocytes of patients with MN, suggesting that the autophagic pathway is blocked.[132] Based on these limited published findings, inhibition of autophagy may contribute to podocyte injury in MN. Further research to characterize the functions of ATG proteins in MN is required.

3.6 Autophagy in HIV-associated nephropathy

HIV infection has been associated with AKI and CKD. Pathologically, chronic form of HIVAN is similar to collapsing FSGS and also involves tubular atrophy and tubulointerstitial inflammation, with an outcome of TIF.[103,133] Autophagic activity plays an essential role in the pathogenesis of HIVAN. The treatment with rapamycin (autophagy activator) was able to attenuate HIV-induced nephropathy.[134] Sirolimus also helped in suppressing the transcription of HIV and inhibiting epithelial to mesenchymal transition (EMT) as confirmed through *in vivo* and *in vitro* studies.[135] Kumar et al. showed that the mTOR pathway during HIVAN was highly active in both the glomerular and tubular compartments of the kidney.[136] Rapamycin treatment was beneficial in ameliorating glomerulosclerosis, proteinuria, and improving renal function.[134,136] The genetic variants in *APOL1* gene, known to be associated with increased risk of FSGS and non-diabetic kidney disease among African Americans and shown to modulate autophagic flux, also increase risk of HIVAN.[102,103,137] The *nef* gene plays a significant role in the pathogenesis of HIVAN.[138,139] This gene has been shown to mimic the function of Bafilomycin A1 and inhibit the process of autophagy,[139] suggesting that the *nef* gene may favor the progression of HIVAN by suppressing autophagic activity.

3.7 Autophagy in polycystic kidney disease

PKD is characterized by the formation of clusters of cysts in the renal parenchyma, and eventual loss of renal function and progression to ESRD. Autosomal dominant PKD (ADPKD) is more common than autosomal recessive PKD (ARPKD).[140] ADPKD results from mutations in the polycystin proteins: polycystin1 or PC1 (encoded by *PKD1* or *PKD2* genes) and polycystin-2 or PC-2, which are involved in cell growth and survival. Several pieces of evidence suggest that the activation of autophagy may exert protective functions in ADPKD.[141–151] Treatment with various autophagy inducers has been shown to inhibit the signaling pathways that favor cyst formation.[142–144] For example, autophagy activators including metformin,[142] curcumin,[143] triptolide,[144] and inhibitors that act against various pathways that suppress autophagy such as mTOR,[145] tyrosine kinase,[146] and caspase,[147] have been shown to act against cyst formation.[140] Treatment with metformin (AMPK inhibitor and autophagy activator) slows the process of cyst formation.[148] In *Pkd1* deleted mice, curcumin averted cyst formation through targeting STAT3.[143] The role of STAT3, which negatively regulates the autophagy pathway[149] has also been implicated in the animal model and human ADPKD.[150] In the animal model of ADPKD, triptolide (autophagy inducer) successfully thwarted cell proliferation and cystogenesis by favoring the release of calcium ions.[151] Treatment with mTOR inhibitors has also been shown to prevent the progression of PKD and renal failure.[152] In Cy/+ rat, that carries *Anks6* (*Pkdr1*) mutation, administration of everolimus (a rapamycin analog) successfully normalized the blood pressure, kidney volume, and improved renal function.[152] However, in patients with PKD, rapamycin improved[125] I-iothalamate glomerular filtration rate[153] and everolimus reduced the increase in total kidney volume[154] but therapies with both failed to prevent the progressive loss of renal function. The failure of an increase in autophagic flux in the kidney of Cy/+ mice after bafilomycin treatment[155] and lack of autophagic response in *Pkd1* knockdown and *Pkd1* knockout cells[141] suggest that autophagy is impaired during PKD. The activation of autophagy in PKD may have therapeutic potential against cyst formation and growth.

4. Potential therapeutic approaches and clinical applications

Autophagy activation may be a potential therapeutic approach for treating patients with different kidney diseases. The currently available activators of autophagy like mTOR inhibitors (rapamycin, sirolimus, and everolimus) were successfully used in pre-clinical studies using various animal models[152,156] but to date failed to show complete protection in patients with kidney diseases.[153,154] While designing therapeutic targets, it is important to consider that both mTOR dependent and mTOR independent mechanisms of autophagy exist.[157,158] Most of the studies in rodent models of kidney diseases are focused on studying the effect of global or conditional deletion of *Becn1* (haploinsufficient) or *Atg5* or *Atg7* genes.[22,60,77,78,113] However, autophagic activity and autophagosome formation have been reported even in the absence of these genes.[159–161] Using the search terms "kidney" and "autophagy", ClinicalTrials.gov showed only five clinical trials, out of which four studied the effect of inhibition of autophagy on renal carcinoma. One out of five trials (phase 4) focused on determining the effects of low protein diet and keto acids on autophagy and improvement of muscle wasting in patients with CKD, the results of which are not available on the ClinicalTrials.gov. Currently, there is an unmet need to develop a more complete understanding of the mechanism and functions of ATG genes during different pathological states of the kidney. We need to obtain a clearer knowledge of the step(s) of the autophagy pathway (autophagosome formation and/or lysosomal-mediated clearance) that is/are affected during different pathological conditions. Moreover, we need to better understand the response in a cell type and context-specific manner to the particular injury or defect(s) and its consequences that pertain to disease progression. In manipulating the autophagic pathway, the dose of the drug also needs to be carefully designed, given that persistent and overactivation of the autophagy pathway may be deleterious. Transient activation of autophagy may be potentially beneficial in reducing aggregated cellular components, eliminating oxidative stress, and preventing the progression of kidney diseases.

5. Concluding remarks

In this chapter, we have highlighted the role of autophagy and ATG genes in different cells of the kidney and reviewed their functions during various pathological states. The insight gained from the various cell-specific conditional knockout mouse models, summarized in Table 1, have improved our understanding of the role of ATG genes.

Table 1 Disease phenotype after kidney cell-specific deletion of autophagy-related genes using different models of kidney diseases.

Autophagy-related gene/targeted cell	Model/kidney disease	Disease phenotype	Reference
Atg5/podocyte	LPS or PAN/ nephrosis	Albuminuria, loss of podocytes, foot process effacement	20
	STZ/DN	Severe podocytopathy, glomerulosclerosis, GBM thickening	21
	High-fat diet/ DN	Massive proteinuria, glomerulopathy, fibrosis	113

Table 1 Disease phenotype after kidney cell-specific deletion of autophagy-related genes using different models of kidney diseases.—Cont'd

Autophagy-related gene/targeted cell	Model/kidney disease	Disease phenotype	Reference
Atg7/podocytes	Unliteral nephrectomy/progressive kidney injury	Severe proteinuria, foot process effacement, and decrease in the podocyte number	94
Atg5/proximal tubular epithelial cells	Cisplatin/AKI	Tubular injury, apoptosis, and atrophy, cast formation, loss of brush border, and increase in mitochondrial-derived oxidative stress	23
	IRI/AKI	Massive tubular vacuolation, tubular injury, elevated BUN and serum creatinine	24
	FFA-albumin overload/DN	Proteinuria, apoptosis of tubular cells, and tubulointerstitial damage	25
Atg7/proximal tubular epithelial cells	LPS/AKI	Severe LPS-mediated tissue damage	31
	Cisplatin or IRI/AKI	Loss of renal function, tubular apoptosis, and damage to renal cortex and outer medulla	55
Atg5/all tubular segments	Ischemia-reperfusion injury/AKI	Renal oxidative stress and tubular injury	27
Atg5/endothelial cells	STZ/DN	Microalbuminuria, endothelial lesions, and fibrotic response	21

LPS, lipopolysaccharide; PAN, puromycin aminonucleoside; STZ, streptozotocin; DN, diabetic nephropathy; GBM, glomerular basement membrane; AKI, acute kidney injury; IRI, ischemia-reperfusion injury; BUN, blood urea nitrogen; FFA, free fatty acid.

However, the functions of autophagy are more intricate and may vary during different stages of the disease, and autophagy appears to be activated during AKI and suppressed during CKD. Strategies to enhance autophagy may have significant implications to prevent the progression of kidney diseases.

Acknowledgments

M.E.C. is supported by National Institutes of Health (NIH) R01 HL133801, R01 HL060234, R01 HL055330, and R01 HL132198.

Disclosure statement

The spouse of M.E.C. is a cofounder and shareholder, and serves on the Scientific Advisory Board of Proterris, Inc.

References

1. Melk A, Baisantry A, Schmitt R. The yin and yang of autophagy in acute kidney injury. *Autophagy*. 2016;12(3):596–597.
2. Lin TA, Wu VC, Wang CY. Autophagy in chronic kidney diseases. *Cell*. 2019;8(1): E61.
3. Hoenig MP, Zeidel ML. Homeostasis, the milieu intérieur, and the wisdom of the nephron. *Clin J Am Soc Nephrol*. 2014;9(7):1272–1281.
4. Choi AMK, Ryter SW, Levine B. Autophagy in human health and disease. *N Engl J Med*. 2013;368(7):651–662.
5. Mizushima N, Komatsu M. Autophagy: renovation of cells and tissues. *Cell*. 2011;147(4):728–741.
6. Ericsson JLE. Studies on induced cellular autophagy: I electron microscopy of cells with in vivo labeled lysosomes. *Exp Cell Res*. 1969;55(1):95–106.
7. Klionsky DJ, Cregg JM, Dunn WA, et al. A unified nomenclature for yeast autophagy-related genes. *Dev Cell*. 2003;5(4):539–545.
8. Ravikumar B, Sarkar S, Davies JE, et al. Regulation of mammalian autophagy in physiology and pathophysiology. *Physiol Rev*. 2010;90(4):1383–1435.
9. Li WW, Li J, Bao JK. Microautophagy: lesser-known self-eating. *Cell Mol Life Sci*. 2012;69(7):1125–1136.
10. Cuervo AM, Wong E. Chaperone-mediated autophagy: roles in disease and aging. *Cell Res*. 2014;24(1):92–104.
11. Franch HA. Chaperone-mediated autophagy in the kidney: the road more traveled. *Semin Nephrol*. 2014;34(1):72–83.
12. Sooparb S, Price SR, Shaoguang J, Franch HA. Suppression of chaperone-mediated autophagy in the renal cortex during acute diabetes mellitus. *Kidney Int*. 2004;65(6): 2135–2144.
13. Eirin A, Lerman A, Lerman LO. The emerging role of mitochondrial targeting in kidney disease. *Handb Exp Pharmacol*. 2017;240:229–250.
14. Bhatia D, Chung KP, Nakahira K, et al. Mitophagy dependent macrophage reprogramming protects against kidney fibrosis. *JCI Insight*. 2019;4(23): e132826.
15. Lin Q, Li S, Jiang N, et al. PINK1-parkin pathway of mitophagy protects against contrast-induced acute kidney injury via decreasing mitochondrial ROS and NLRP3 inflammasome activation. *Redox Biol*. 2019;26: 101254.
16. Namba T, Takabatake Y, Kimura T, et al. Autophagic clearance of mitochondria in the kidney copes with metabolic acidosis. *J Am Soc Nephrol*. 2014;25(10):2254–2266.
17. Livingston MJ, Ding HF, Huang S, et al. Persistent activation of autophagy in kidney tubular cells promotes renal interstitial fibrosis during unilateral ureteral obstruction. *Autophagy*. 2016;12(6):976–998.
18. Tang C, Han H, Yan M, et al. PINK1-PRKN/PARK2 pathway of mitophagy is activated to protect against renal ischemia-reperfusion injury. *Autophagy*. 2018; 14(5):880–897.

19. Ryter SW, Bhatia D, Choi ME. Autophagy: a lysosome-dependent process with implications in cellular redox homeostasis and human disease. *Antioxid Redox Signal.* 2018;30(1):138–159.
20. Hartleben B, Godel M, Meyer-Schwesinger C, et al. Autophagy influences glomerular disease susceptibility and maintains podocyte homeostasis in aging mice. *J Clin Invest.* 2010;120(4):1084–1096.
21. Lenoir O, Jasiek M, Henique C, et al. Endothelial cell and podocyte autophagy synergistically protect from diabetes-induced glomerulosclerosis. *Autophagy.* 2015;11(7):1130–1145.
22. Kawakami T, Gomez IG, Ren S, et al. Deficient autophagy results in mitochondrial dysfunction and FSGS. *J Am Soc Nephrol.* 2015;26(5):1040–1052.
23. Takahashi A, Kimura T, Takabatake Y, et al. Autophagy guards against cisplatin-induced acute kidney injury. *Am J Pathol.* 2012;180(2):517–525.
24. Kimura T, Takabatake Y, Takahashi A, et al. Autophagy protects the proximal tubule from degeneration and acute ischemic injury. *J Am Soc Nephrol.* 2011; 22(5):902–913.
25. Yamahara K, Kume S, Koya D, et al. Obesity-mediated autophagy insufficiency exacerbates proteinuria-induced tubulointerstitial lesions. *J Am Soc Nephrol.* 2013; 24(11):1769–1781.
26. Jiang M, Liu K, Luo J, Dong Z. Autophagy is a renoprotective mechanism during in vitro hypoxia and in vivo ischemia-reperfusion injury. *Am J Pathol.* 2010; 176(3):1181–1192.
27. Liu S, Hartleben B, Kretz O, et al. Autophagy plays a critical role in kidney tubule maintenance, aging and ischemia-reperfusion injury. *Autophagy.* 2012;8(5):826–837.
28. Kaushal GP, Shah SV. Autophagy in acute kidney injury. *Kidney Int.* 2016;89(4): 779–791.
29. De Rechter S, Decuypere JP, Ivanova E, et al. Autophagy in renal diseases. *Pediatr Nephrol.* 2016;31(5):737–752.
30. Wang Z, Choi ME. Autophagy in kidney health and disease. *Antioxid Redox Signal.* 2014;20(3):519–537.
31. Mei S, Livingston M, Hao J, et al. Autophagy is activated to protect against endotoxic acute kidney injury. *Sci Rep.* 2016;6: 22171.
32. Wu Y, Wang L, Meng L, et al. Biological effects of autophagy in mice with sepsis-induced acute kidney injury. *Exp Ther Med.* 2019;17(1):316–322.
33. Shi M, Flores B, Gillings N, et al. αKlotho mitigates progression of AKI to CKD through activation of autophagy. *J Am Soc Nephrol.* 2016;27(8):2331–2345.
34. Liu W, Dai N, Wang Y, et al. Role of autophagy in cadmium-induced apoptosis of primary rat osteoblasts. *Sci Rep.* 2016;6: 20404.
35. Zarbock A, Gomez H, Kellum JA. Sepsis-induced acute kidney injury revisited: pathophysiology, prevention and future therapies. *Curr Opin Crit Care.* 2014;20(6): 588–595.
36. Hsiao HW, Tsai KL, Wang LF, et al. The decline of autophagy contributes to proximal tubular dysfunction during sepsis. *Shock.* 2012;37(3):289–296.
37. Sunahara S, Watanabe E, Hatano M, et al. Influence of autophagy on acute kidney injury in a murine cecal ligation and puncture sepsis model. *Sci Rep.* 2018;8(1): 1050.
38. Escobar DA, Botero-Quintero AM, Kautza BC, et al. Adenosine monophosphate-activated protein kinase activation protects against sepsis-induced organ injury and inflammation. *J Surg Res.* 2015;194(1):262–272.
39. Barnett LMA, Cummings BS. Nephrotoxicity and renal pathophysiology: a contemporary perspective. *Toxicol Sci.* 2018;164(2):379–390.
40. Howell GM, Gomez H, Collage RD, et al. Augmenting autophagy to treat acute kidney injury during endotoxemia in mice. *PLoS One.* 2013;8(7): e69520.

41. Zhao W, Zhang L, Chen R, et al. SIRT3 protects against acute kidney injury via AMPK/mTOR-regulated autophagy. *Front Physiol.* 2018;9:1526.
42. Zhang X, Howell GM, Guo L, et al. CaMKIV-dependent preservation of mTOR expression is required for autophagy during lipopolysaccharide-induced inflammation and acute kidney injury. *J Immunol.* 2014;193(5):2405–2415.
43. Malek M, Nematbakhsh M. Renal ischemia/reperfusion injury; from pathophysiology to treatment. *J Renal Inj Prev.* 2015;4(2):20–27.
44. Ling H, Chen H, Wei M, et al. The effect of autophagy on inflammation cytokines in renal ischemia/reperfusion injury. *Inflammation.* 2016;39(1):347–356.
45. Funk JA, Schnellmann RG. Persistent disruption of mitochondrial homeostasis after acute kidney injury. *Am J Physiol Renal Physiol.* 2012;302(7):F853–F864.
46. Turkmen K, Martin J, Akcay A, et al. Apoptosis and autophagy in cold preservation ischemia. *Transplantation.* 2011;91(11):1192–1197.
47. Isaka Y, Suzuki C, Abe T, et al. Bcl-2 protects tubular epithelial cells from ischemia/reperfusion injury by dual mechanisms. *Transplant Proc.* 2009;41(1):52–54.
48. Chien CT, Shyue SK, Lai MK. Bcl-xL augmentation potentially reduces ischemia/reperfusion-induced proximal and distal tubular apoptosis and autophagy. *Transplantation.* 2017;84(9):1183–1190.
49. He L, Livingston MJ, Dong Z. Autophagy in acute kidney injury and repair. *Nephron Clin Pract.* 2014;127(1–4):56–60.
50. Li L, Wang ZV, Hill JA, Lin F. New autophagy reporter mice reveal dynamics of proximal tubular autophagy. *J Am Soc Nephrol.* 2014;25(2):305–315.
51. Decuypere JP, Pirenne J, Jochmans I. Autophagy in renal ischemia-reperfusion injury: friend or foe? *Am J Transplant.* 2014;14(6):1464–1465.
52. Havasi A, Dong Z. Autophagy and tubular cell death in the kidney. *Semin Nephrol.* 2016;36(3):174–188.
53. Chen Y, Klionsky DJ. The regulation of autophagy—unanswered questions. *J Cell Sci.* 2011;124(2):161–170.
54. Takabatake Y, Kimura T, Takahashi A, Isaka Y. Autophagy and the kidney: health and disease. *Nephrol Dial Transplant.* 2014;29(9):1639–1647.
55. Jiang M, Wei Q, Dong G, et al. Autophagy in proximal tubules protects against acute kidney injury. *Kidney Int.* 2012;82(12):1271–1283.
56. Xie Y, Jiang D, Xiao J, et al. Ischemic preconditioning attenuates ischemia/reperfusion-induced kidney injury by activating autophagy via the SGK1 signaling pathway. *Cell Death Dis.* 2018;9(3): 338.
57. Zhang YL, Zhang J, Cui LY, Yang S. Autophagy activation attenuates renal ischemia-reperfusion injury in rats. *Exp Biol Med (Maywood).* 2015;240(12): 1590–1598.
58. Tan X, Zhu H, Tao Q, et al. FGF10 protects against renal ischemia/reperfusion injury by regulating autophagy and inflammatory signaling. *Front Genet.* 2018;9:556.
59. Suzuki C, Isaka Y, Takabatake Y, et al. Participation of autophagy in renal ischemia/reperfusion injury. *Biochem Biophys Res Commun.* 2008;368(1):100–106.
60. Yang C, Kaushal V, Shah SV, Kaushal GP. Autophagy is associated with apoptosis in cisplatin injury to renal tubular epithelial cells. *Am J Physiol Renal Physiol.* 2008;294(4): F777–F787.
61. Periyasamy-Thandavan S, Jiang M, Wei Q, et al. Autophagy is cytoprotective during cisplatin injury of renal proximal tubular cells. *Kidney Int.* 2008;74(5):631–640.
62. Herzog C, Yang C, Holmes A, Kaushal GP. z-VAD-fmk prevents cisplatin-induced cleavage of autophagy proteins but impairs autophagic flux and worsens renal function. *Am J Physiol Renal Physiol.* 2012;303(8):F1239–F1250.
63. Petejova N, Martinek A, Zadrazil J, Teplan V. Acute toxic kidney injury. *Ren Fail.* 2019;41(1):576–594.

64. Liu F, Wang XY, Zhou XP, et al. Cadmium disrupts autophagic flux by inhibiting cytosolic Ca2+-dependent autophagosome-lysosome fusion in primary rat proximal tubular cells. *Toxicology*. 2017;383:13–23.
65. Kimura A, Ishida Y, Nosaka M, et al. Exaggerated arsenic nephrotoxicity in female mice through estrogen-dependent impairments in the autophagic flux. *Toxicology*. 2016;339:918.
66. Debnath J, Baehrecke EH, Kroemer G. Does autophagy contribute to cell death? *Autophagy*. 2005;1(12):66–74.
67. Wang SH, Shih YL, Ko WC, Wei YH, Shih CM. Cadmium-induced autophagy and apoptosis are mediated by a calcium signaling pathway. *Cell Mol Life Sci*. 2008;65(22): 3640–3652.
68. Wang SH, Shih YL, Kuo TC, Ko W, Shih CM. Cadmium toxicity toward autophagy through ROS-activated GSK-3beta in mesangial cells. *Toxicol Sci*. 2009; 108(1):124–131.
69. Fujishiro H, Liu Y, Ahmadi B, Templeton DM. Protective effect of cadmium-induced autophagy in rat renal mesangial cells. *Arch Toxicol*. 2018;92(2):619–631.
70. Bursch W, Hochegger K, Torok L, et al. Autophagic and apoptotic types of programmed cell death exhibit different fates of cytoskeletal filaments. *J Cell Sci*. 2000;113(7):1180–1198.
71. Lira-León A, Aparicio-Trejo OE, Reyes-Fermín LM, Pedraza-Chaverri J. Role of autophagy on heavy metal-induced renal damage and the protective effects of curcumin in autophagy and kidney preservation. *Medicina (Kaunas)*. 2019;55(7): E360.
72. Choi ME, Ding Y, Kim SI. TGF-β signaling via TAK1 pathway: role in kidney fibrosis. *Semin Nephrol*. 2012;32:244–252.
73. Lee SY, Kim SI, Choi ME. Therapeutic targets for treating fibrotic kidney diseases. *Transl Res*. 2015;165(4):512–530.
74. Zeisberg M, Neilson EG. Mechanisms of tubulointerstitial fibrosis. *J Am Soc Nephrol*. 2010;21(11):1819–1834.
75. Li L, Zepeda-Orozco V, Black R, et al. Autophagy is a component of epithelial cell fate in obstructive uropathy. *Am J Pathol*. 2010;176(4):1767–1778.
76. Kim WY, Nam SA, Song HC, et al. The role of autophagy in unilateral ureteral obstruction rat model. *Nephrology (Carlton)*. 2012;17(2):148–159.
77. Nam SA, Kim WY, Kim JW, et al. Autophagy attenuates tubulointerstitial fibrosis through regulating transforming growth factor-β and NLRP3 inflammasome signaling pathway. *Cell Death Dis*. 2019;10(2):78–87.
78. Kim SI, Na HJ, Ding Y, et al. Autophagy promotes intracellular degradation of type I collagen induced by transforming growth factor (TGF)-β1. *J Biol Chem*. 2012;287(15):11677–11688.
79. Ding Y, Kim JK, Kim SI, et al. TGF-β1 protects against mesangial cell apoptosis via induction of autophagy. *J Biol Chem*. 2010;285(48):37909–37919.
80. Shi M, Yang S, Zhu X, et al. The RAGE/STAT5/autophagy axis regulates senescence in mesangial cells. *Cell Signal*. 2019;62: 109334.
81. Ding Y, Kim SI, Lee SY, et al. Autophagy regulates TGF-β expression and suppresses kidney fibrosis induced by unilateral ureteral obstruction. *J Am Soc Nephrol*. 2014;2 5(12):2835–2846.
82. Huang C, Lin MZ, Cheng D, et al. KCa3.1 mediates dysfunction of tubular autophagy in diabetic kidneys via PI3k/Akt/mTOR signaling pathways. *Sci Rep*. 2016;6: 23884.
83. Wang L, Lee JY, Kwak JH, et al. Protective effects of low-dose carbon monoxide against renal fibrosis induced by unilateral ureteral obstruction. *Am J Physiol Renal Physiol*. 2008;294(3):F508–F517.
84. Xu Y, Yang S, Huang J, et al. Tgfbeta1 induces autophagy and promotes apoptosis in renal tubular epithelial cells. *Int J Mol Med*. 2012;29(5):781–790.

85. Xu Y, Ruan S, Wu X, et al. Autophagy and apoptosis in tubular cells following unilateral ureteral obstruction are associated with mitochondrial oxidative stress. *Int J Mol Med.* 2013;31(3):628–636.
86. Bao J, Shi Y, Tao M, et al. Pharmacological inhibition of autophagy by 3-MA attenuates hyperuricemic nephropathy. *Clin Sci.* 2018;132(21):2299–2322.
87. Asanuma K, Tanida I, Shirato I, et al. MAP-LC3, a promising autophagosomal marker, is processed during the differentiation and recovery of podocytes from PAN nephrosis. *FASEB J.* 2003;17(9):1165–1167.
88. Sato S, Kitamura H, Adachi A, et al. Two types of autophagy in the podocytes in renal biopsy specimens: ultrastructural study. *J Submicrosc Cytol Pathol.* 2006;38(2-3): 167–174.
89. Vivarelli M, Massella L, Ruggiero B, Emma F. Minimal change disease. *Clin J Am Soc Nephrol.* 2017;12(2):332–345.
90. Rosenberg AZ, Kopp JB. Focal segmental glomerulosclerosis. *Clin J Am Soc Nephrol.* 2017;12(3):502–517.
91. Cina DP, Onay T, Paltoo A, et al. Inhibition of MTOR disrupts autophagic flux in podocytes. *J Am Soc Nephrol.* 2012;23(3):412–420.
92. Wiggins RC. The spectrum of podocytopathies: a unifying view of glomerular diseases. *Kidney Int.* 2007;71(12):1205–1214.
93. White KE, Bilous RW, Marshall SM, et al. Podocyte number in normotensive type 1 diabetic patients with albuminuria. *Diabetes.* 2002;51(10):3083–3089.
94. Oliva Trejo JA, Asanuma K, Kim EH, et al. Transient increase in proteinuria, polyubiquitylated proteins and ER stress markers in podocyte-specific autophagy-deficient mice following unilateral nephrectomy. *Biochem Biophys Res Commun.* 2014; 446(4):1190–1196.
95. Oshima Y, Kinouchi K, Ichihara A, et al. Prorenin receptor is essential for normal podocyte structure and function. *J Am Soc Nephrol.* 2011;22(12):2203–2212.
96. Riediger F, Quack I, Qadri F, et al. Prorenin receptor is essential for podocyte autophagy and survival. *J Am Soc Nephrol.* 2011;22(12):2193–2202.
97. Maiuri MC, Criollo A, Tasdemir E, et al. BH3-only proteins and BH3 mimetics induce autophagy by competitively disrupting the interaction between Beclin 1 and Bcl-2/ Bcl-X(L). *Autophagy.* 2007;3(4):374–376.
98. Chun J, Zhang JY, Wilkins MS, et al. Recruitment of APOL1 kidney disease risk variants to lipid droplets attenuates cell toxicity. *Proc Natl Acad Sci USA.* 2019;116(9):3712–3721.
99. Freedman BI, Kopp JB, Langefeld CD, et al. The apolipoprotein L1 (APOL1) gene and nondiabetic nephropathy in African Americans. *J Am Soc Nephrol.* 2010;21:1422–1426.
100. Genovese G, Friedman DJ, Ross MD, et al. Association of trypanolytic ApoL1 variants with kidney disease in African Americans. *Science.* 2010;329:841–845.
101. Madhavan SM, O'Toole JF, Konieczkowski M, Ganesan S, Bruggeman LA, Sedor JR. APOL1 localization in normal kidney and nondiabetic kidney disease. *J Am Soc Nephrol.* 2011;22:2119–2128.
102. Kopp JB, Nelson GW, Sampath K, et al. APOL1 genetic variants in focal segmental glomerulosclerosis and HIV-associated nephropathy. *J Am Soc Nephrol.* 2011;22: 2129–2137.
103. Papeta N, Kiryluk K, Patel A, et al. APOL1 variants increase risk for FSGS and HIVAN but not IgA nephropathy. *J Am Soc Nephrol.* 2011;22:1991–1996.
104. Cybulsky AV. The intersecting roles of endoplasmic reticulum stress, ubiquitin-proteasome system, and autophagy in the pathogenesis of proteinuric kidney disease. *Kidney Int.* 2013;84(1):25–33.
105. Zeng C, Fan Y, Wu J, et al. Podocyte autophagic activity plays a protective role in renal injury and delays the progression of podocytopathies. *J Pathol.* 2014;234(2):203–213.

106. Kang YL, Saleem MA, Chan KW, Yung BY, Law HK. Trehalose, an mTOR independent autophagy inducer, alleviates human podocyte injury after puromycin aminonucleoside treatment. *PLoS One*. 2014;9(11): e113520.
107. Gross JL, de Azevedo MJ, Silveiro SP, et al. Diabetic nephropathy: diagnosis, prevention, and treatment. *Diabetes Care*. 2005;28(1):164–176.
108. Stitt-Cavanagh E, MacLeod L, Kennedy C. The podocyte in diabetic kidney disease. *ScientificWorldJournal*. 2009;9:1127–1139.
109. Najafian B, Alpers CE, Fogo AB. Pathology of human diabetic nephropathy. *Contrib Nephrol*. 2011;170:36–47.
110. Ma T, Zhu J, Chen X, et al. High glucose induces autophagy in podocytes. *Exp Cell Res*. 2013;319(6):779–789.
111. Fang L, Zhou Y, Cao H, et al. Autophagy attenuates diabetic glomerular damage through protection of hyperglycemia-induced podocyte injury. *PLoS One*. 2013; 8(4): e60546.
112. Zhang MZ, Wang Y, Paueksakon P, Harris RC. Epidermal growth factor receptor inhibition slows progression of diabetic nephropathy in association with a decrease in endoplasmic reticulum stress and an increase in autophagy. *Diabetes*. 2014;63(6): 2063–2072.
113. Tagawa A, Yasuda M, Kume S, et al. Impaired podocyte autophagy exacerbates proteinuria in diabetic nephropathy. *Diabetes*. 2016;65(3):755–767.
114. Godel M, Hartleben B, Herbach N, et al. Role of mTOR in podocyte function and diabetic nephropathy in humans and mice. *J Clin Invest*. 2011;121(6):2197–2209.
115. Ding Y, Choi ME. Autophagy in diabetic nephropathy. *J Endocrinol*. 2015;224(1): R15–R30.
116. Bonegio RG, Fuhro R, Wang Z, et al. Rapamycin ameliorates proteinuria-associated tubulointerstitial inflammation and fibrosis in experimental membranous nephropathy. *J Am Soc Nephrol*. 2005;16(7):2063–2072.
117. Huber TB, Walz G, Kuehn EW. mTOR and rapamycin in the kidney: signaling and therapeutic implications beyond immunosuppression. *Kidney Int*. 2011;79(5):502–511.
118. Lu X, Fan Q, Xu L, et al. Ursolic acid attenuates diabetic mesangial cell injury through the up-regulation of autophagy via miRNA-21/PTEN/Akt/mTOR suppression. *PLoS One*. 2015;10(2): e0117400.
119. Saulnier PJ, Wheelock KM, Howell S, et al. Advanced glycation end products predict loss of renal function and correlate with lesions of diabetic kidney disease in American Indians with type 2 diabetes. *Diabetes*. 2016;65(12):3744–3753.
120. Xu L, Fan Q, Wang X, Zhao X, Wang L. Inhibition of autophagy increased AGE/ROS-mediated apoptosis in mesangial cells. *Cell Death Dis*. 2016;7(11):e2445.
121. Matsuda J, Namba T, Takabatake Y, et al. Antioxidant role of autophagy in maintaining the integrity of glomerular capillaries. *Autophagy*. 2018;14(1):53–65.
122. Boels MG, Avramut MC, Koudijs A, et al. Atrasentan reduces albuminuria by restoring the glomerular endothelial glycocalyx barrier in diabetic nephropathy. *Diabetes*. 2016;65(8):2429–2439.
123. Barbosa Junior AA, Zhou H, Hultenschmidt D, et al. Inhibition of cellular autophagy in proximal tubular cells of the kidney in streptozotocin-diabetic and uninephrectomized rats. *Virchows Arch B Cell Pathol Incl Mol Pathol*. 1992;61(6):359–366.
124. Sakai S, Yamamoto T, Takabatake Y, et al. Proximal tubule autophagy differs in type 1 and 2 diabetes. *J Am Soc Nephrol*. 2019;30(6):929–945.
125. Rodrigues JC, Haas M, Reich HN. IgA nephropathy. *Clin J Am Soc Nephrol*. 2017;12(4):677–686.
126. Liang S, Jin J, Gong J, et al. How many podocyte autophagosomes are there in immunoglobulin A nephropathy and idiopathic membranous nephropathy? *Int Urol Nephrol*. 2016;48(12):2109–2114.

127. Sato S, Yanagihara T, Ghazizadeh M, et al. Correlation of autophagy type in podocytes with histopathological diagnosis of IgA nephropathy. *Pathobiology*. 2009;76(5): 221–226.
128. Liang S, Jin J, Shen X, et al. Triptolide protects podocytes via autophagy in immunoglobulin A nephropathy. *Exp Ther Med*. 2018;16(3):2275–2280.
129. He L, Peng X, Liu G, et al. Anti-inflammatory effects of triptolide on IgA nephropathy in rats. *Immunopharmacol Immunotoxicol*. 2015;37(5):421–427.
130. Cattran DC, Brenchley PE. Membranous nephropathy: integrating basic science into improved clinical management. *Kidney Int*. 2017;91(3):566–574.
131. Luo W, Olaru F, Miner JH, et al. Alternative pathway is essential for glomerular complement activation and proteinuria in a mouse model of membranous nephropathy. *Front Immunol*. 2018;9:1433–1444.
132. Liu WJ, Li ZH, Chen XC, et al. Blockage of the lysosome-dependent autophagic pathway contributes to complement membrane attack complex-induced podocyte injury in idiopathic membranous nephropathy. *Sci Rep*. 2017;7(1): 8643.
133. Waheed S, Sakr A, Chheda ND, et al. Outcomes of renal transplantation in HIV-1 associated nephropathy. *PLoS One*. 2015;10(6): e0129702.
134. Rai P, Plagov A, Kumar D, et al. Rapamycin-induced modulation of HIV gene transcription attenuates progression of HIVAN. *Exp Mol Pathol*. 2013;94(1):255–261.
135. Yadav A, Kumar D, Salhan D, et al. Sirolimus modulates HIVAN phenotype through inhibition of epithelial mesenchymal transition. *Exp Mol Pathol*. 2012;93(1):173–181.
136. Kumar D, Konkimalla S, Yadav A, et al. HIV-associated nephropathy: role of mammalian target of rapamycin pathway. *Am J Pathol*. 2010;177(2):813–821.
137. Kopp JB, Heymann J, Winkler CA. APOL1 renal risk variants: fertile soil for HIV-associated nephropathy. *Semin Nephrol*. 2017;37(6):514–519.
138. Medapalli RK, He JC, Klotman PE. HIV-associated nephropathy: pathogenesis. *Curr Opin Nephrol Hypertens*. 2011;20(3):306–311.
139. Saribas AS, Khalili K, Sariyer IK. Dysregulation of autophagy by HIV-1 Nef in human astrocytes. *Cell Cycle*. 2015;14(18):2899–2904.
140. Bergmann C, Guay-Woodford LM, Harris PC, et al. Polycystic kidney disease. *Nat Rev Dis Primers*. 2018;4(1): 50.
141. Ravichandran K, Edelstein CL. Polycystic kidney disease: a case of suppressed autophagy? *Semin Nephrol*. 2014;34(1):27–33.
142. Pisani A, Riccio E, Bruzzese D, et al. Metformin in autosomal dominant polycystic kidney disease: experimental hypothesis or clinical fact? *BMC Nephrol*. 2018;19(1): 282.
143. Leonhard WN, van der WA, Novalic Z, et al. Curcumin inhibits cystogenesis by simultaneous interference of multiple signaling pathways: in vivo evidence from a Pkd1-deletion model. *Am J Physiol Renal Physiol*. 2011;300(5):F1193–F1202.
144. Leuenroth SJ, Bencivenga N, Chahboune H, Hyder F, Crews CM. Triptolide reduces cyst formation in a neonatal to adult transition Pkd1 model of ADPKD. *Nephrol Dial Transplant*. 2010;25(7):2187–2194.
145. Torres VE, Boletta A, Chapman A, et al. Prospects for mTOR inhibitor use in patients with polycystic kidney disease and hamartomatous diseases. *Clin J Am Soc Nephrol*. 2010;5(7):1312–1329.
146. Sweeney WE, Chen Y, Nakanishi K, Frost P, Avner ED. Treatment of polycystic kidney disease with a novel tyrosine kinase inhibitor. *Kidney Int*. 2000;57(1):33–40.
147. Tao Y, Kim J, Stanley M, et al. Pathways of caspase-mediated apoptosis in autosomal-dominant polycystic kidney disease (ADPKD). *Kidney Int*. 2005;67(3):909–919.
148. Takiar V, Nishio S, Seo-Mayer P, et al. Activating AMP-activated protein kinase (AMPK) slows renal cystogenesis. *Proc Natl Acad Sci USA*. 2011;108(6):2462–2467.
149. Pietrocola F, Izzo V, Niso-Santano M, et al. Regulation of autophagy by stress-responsive transcription factors. *Semin Cancer Biol*. 2013;23(5):310–322.

150. Talbot JJ, Shillingford JM, Vasanth S, et al. Polycystin-1 regulates STAT activity by a dual mechanism. *Proc Natl Acad Sci USA*. 2011;108(19):7985–7990.
151. Leuenroth SJ, Okuhara D, Shotwell JD, et al. Triptolide is a traditional Chinese medicine-derived inhibitor of polycystic kidney disease. *Proc Natl Acad Sci USA*. 2007;104(11):4389–4394.
152. Zafar I, Belibi FA, He Z, Edelstein CL. Long-term rapamycin therapy in the Han: SPRD rat model of polycystic kidney disease (PKD). *Nephrol Dial Transplant*. 2009;24(8):2349–2353.
153. Braun WE, Schold JD, Stephany BR, Spirko RA, Herts BR. Low-dose rapamycin (sirolimus) effects in autosomal dominant polycystic kidney disease: an open-label randomized controlled pilot study. *Clin J Am Soc Nephrol*. 2014;9(5):881–888.
154. Walz W, Budde K, Mannaa M, et al. Everolimus in patients with autosomal dominant polycystic kidney disease. *N Engl J Med*. 2010;363(9):830–840.
155. Belibi F, Zafar I, Ravichandran K, et al. Hypoxia-inducible factor-1α (HIF-1α) and autophagy in polycystic kidney disease (PKD). *Am J Physiol Renal Physiol*. 2011; 300(5):F1235–F1243.
156. Kajiwara M, Masuda S. Role of mTOR inhibitors in kidney disease. *Int J Mol Sci*. 2016;17(6): E975.
157. Sarkar S. Regulation of autophagy by mTOR-dependent and mTOR-independent pathways: autophagy dysfunction in neurodegenerative diseases and therapeutic application of autophagy enhancers. *Biochem Soc Trans*. 2013;41(5):1103–1130.
158. Ahumada-Castro U, Silva-Pavez E, Lovy A, et al. MTOR-independent autophagy induced by interrupted endoplasmic reticulum-mitochondrial Ca2+ communication: a dead end in cancer cells. *Autophagy*. 2019;15(2):358–361.
159. Scarlatti F, Maffei R, Beau I, Codogno P, Ghidoni R. Role of non-canonical Beclin 1-independent autophagy in cell death induced by resveratrol in human breast cancer cells. *Cell Death Differ*. 2008;15(8):1318–1329.
160. Nishida Y, Arakawa S, Fujitani K, et al. Discovery of Atg5/Atg7-independent alternative macroautophagy. *Nature*. 2009;461(7264):654–658.
161. Arakawa S, Honda S, Yamaguchi H, Shimizu S. Molecular mechanisms and physiological roles of Atg5/Atg7-independent alternative autophagy. *Proc Jpn Acad Ser B Phys Biol Sci*. 2017;93(6):378–385.

CHAPTER EIGHT

Autophagy in chronic lung disease

Alexandra C. Racanelli[a,b], **Augustine M.K. Choi**[a,b], **Mary E. Choi**[b,c,*]
[a]Division of Pulmonary and Critical Care Medicine, Joan and Sanford I. Weill Department of Medicine, Weill Cornell Medicine, New York, NY, United States
[b]NewYork-Presbyterian Hospital, Weill Cornell Medicine, New York, NY, United States
[c]Division of Nephrology and Hypertension, Joan and Sanford I. Weill Department of Medicine, Weill Cornell Medicine, New York, NY, United States
*Corresponding author: e-mail address: mechoi@med.cornell.edu

Contents

Abstract

The development of chronic lung disease occurs as a consequence of multiple cellular events that involve an initial insult which often leads to the development of chronic inflammation, and the dysregulation of cellular proliferation and cell death mechanisms. Multiple cell types in the lung are key to the respiratory and protective/barrier functions necessary to manage the chronic exposures to environmental, mechanical, and oxidative stressors. Autophagy is essential to lung development and homeostasis, as well as the prevention and development of disease. The cellular process involves the collection and removal of unwanted organelles and proteins through lysosomal degradation. In recent years, investigations have addressed the roles of autophagy and selective autophagy in numerous chronic lung diseases. Here, we highlight recent advances on the role of autophagy in the pathogenesis of asthma, chronic obstructive pulmonary disease and emphysema, pulmonary arterial hypertension, and idiopathic pulmonary fibrosis.

Progress in Molecular Biology and Translational Science, Volume 172
ISSN 1877-1173
https://doi.org/10.1016/bs.pmbts.2020.02.001

1. Introduction

Macroautophagy (autophagy) is a highly conserved cellular process that drives the degradation and removal of unwanted proteins and organelles. This occurs under basal conditions to maintain homeostasis and is also stimulated during periods of cellular stress (*e.g.*, nutrient deprivation, hypoxia, growth factor depletion, and infection).[1] Activated autophagic machinery recover nutrients through non-specific breakdown of proteins and organelles by the formation of autophagosomes that are then guided toward lysosomal degradation. In contrast, a process known as selective autophagy, drives the removal of distinct cytosolic structures such as mitochondria (mitophagy), protein aggregates (aggrephagy), and bacteria (xenophagy).[1] Beyond its maintenance role, autophagy is key to normal and pathological conditions and has been described as fundamental to the severity of several human diseases.[2]

During nutrient deprivation, mechanistic target of rapamycin complex 1 (mTORC1), a negative regulatory of autophagy, is inhibited by the presence of abundant levels of adenosine monophosphate (AMP), through the activity of 5′ AMP-activated protein kinase (AMPK).[3] Inhibition of mTORC1 leads to the activation and subcellular localization of the Unc-51-like kinase 1 (ULK1) complex to the endoplasmic reticulum. ULK1 complex then triggers the nucleation phase of autophagy, a critical step in phagophore formation, through phosphorylation of components of the class III phosphoinositide 3-kinase (PI3KC3) complex I that includes key autophagic proteins Beclin-1, ATG14 (autophagy-related 14), VPS34 (vacuolar protein sorting 34), p115, and AMBRA1 (autophagy and Beclin-1 regulator 1) (Fig. 1). Activation of Beclin-1 and associated proteins enhances phosphatidylinositol 3-phosphate (PI3P) activity leading to the recruitment of WD repeat domain phosphoinositide interacting proteins and zinc finger protein FYVE domain-containing protein-1 to the developing autophagosome *via* interaction with PI3P-binding domains.[3] Phagophore expansion occurs with further recruitment of autophagy-related proteins ATG5, ATG12, ATG3, ATG16L1, and, eventually, microtubule-associated protein 1 light chain 3 (LC3) to the phosphatidylethanolamine located in the lipid membrane. LC3 recruitment facilitates further sequestration of ubiquitinated material into the developing autophagosome. Sealing of the autophagosomal membrane gives rise to the double membraned mature autophagosome which eventually fuses with the lysosome to allow degradation of the autophagic cargo (Fig. 1).[3]

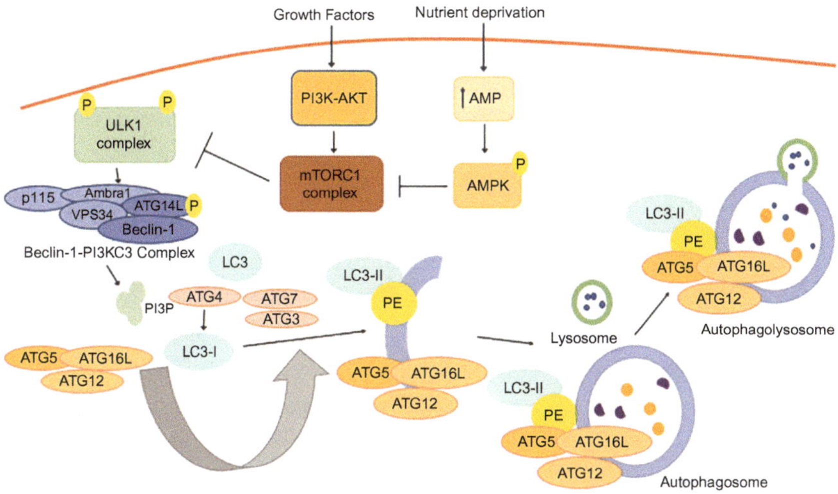

Fig. 1 Activation of autophagy and autophagosomal formation. In a cellular state of nutrient deprivation, activation of 5′ AMP-activated protein kinase (AMPK) occurs when cellular levels of adenosine monophosphate (AMP) accumulate. This leads to the inhibition of mTORC1 which in turn supports the activation Unc-51-like kinase 1 (ULK1) complex. The ULK1 complex facilitates the initiation of the nucleation phase of autophagy through phosphorylation of the Beclin-1-PI3KC3 complex which includes Beclin-1, ATG14 (autophagy-related 14, VPS34 (vacuolar protein sorting 34), p115, and AMBRA1 (autophagy and Beclin-1 regulator 1). Activation of these proteins enhances phosphatidylinositol 3-phosphate (PI3P) activity promoting autophagosomal membrane formation. Autophagosomal elongation is stimulated by conjugation of ATG5 to ATG12 and ATG16L. Through the function of the autophagy related proteins (ATG4, ATG12, ATG16L, ATG4, ATG3, and ATG7) the microtubule-associated protein 1 light chain 3 (LC3) is modified with the cellular lipid phosphatidylethnanolamine (PE) and converted to the conjugated form LC3-II. LC3-II recruitment facilitates further sequestration of ubiquitinated material into the developing autophagosome. As a final stage, the autophagosome fuses with the lysosome to allow degradation of autophagic cargo.

In recent years, alterations in the autophagy pathway have been implicated in an increasing number of human diseases, including cancer, neurodegenerative disease, cardiovascular disease, and inflammatory disease.[2] While a complete understanding of the role played in either prevention or progression of certain diseases remains to be fully understood, several lines of research have highlighted changes in autophagic cellular functions as key to the pathogenesis of numerous chronic lung diseases. In this review, we aim to highlight the importance of autophagy as both a friend and foe in the pathogenesis of several chronic lung diseases.

2. Autophagy and the lung

The primary function of the lungs is to facilitate the delivery of oxygen gas to the arterial system, allowing perfusion of distal tissues, and the elimination of waste gases from the body. The respiratory system is composed of nasal passages, conducting airways (trachea and bronchi), bronchioles, and the alveoli-capillaries network, making up the primary ventilatory unit.[4] The right and left bronchi expand off the trachea and contain cartilage and submucosal mucus secreting glands and neuroendocrine cells (that release serotonin, calcitonin, and gastrin releasing peptide) within their walls. They branch further to become bronchioles, small airways that lack both glands and cartilage, that extend to form even smaller (<2 mm diameter) terminal bronchioles. Respiratory bronchioles stem from terminal bronchioles which form into alveolar ducts then sacs.[4] The alveolar wall serves as the site of gas exchange and consists of capillary endothelium, basement membrane and surrounding interstitial tissue, alveolar epithelium (type I and type III), and alveolar macrophages.

The components of the lung are directly exposed to the outside environment and serve as the first lines of defense to mitigate the challenges presented by chronic exposure to environmental (bacteria, virus, toxins), mechanical, and oxidative stress.[5] Antioxidant defenses, innate and adaptable immune responses, and several mechanisms of cell death play key roles in the maintenance of normal lung function. A growing body of evidence supports the paradigm that autophagy activated by environmental and cellular stresses represents a critical intracellular process for the maintenance of lung homeostasis. The cytoprotective functions of autophagy facilitate the removal of damaged organelles and protein aggregates that occur as a consequence of acute and/or chronic stressors, aging, and genetic or epigenetic predispositions. When these cellular processes are disrupted, pathological states along the components of the respiratory tree ensue. Significant roles for autophagy have been defined across various types of chronic lung disease (obstructive lung diseases, pulmonary vascular disease, and restrictive lung disease) and whether it is protective or harmful in certain disease states remains to be understood.

3. Obstructive lung diseases

3.1 Asthma

Asthma is an obstructive disease of the airways typified by chronic inflammation, hyperresponsiveness (driven by muscular bronchoconstriction), mucus overproduction, and airway remodeling. As a clinical entity, asthma is a debilitating disease that effects over 300 million people worldwide and patients suffer from recurrent episodes of wheezing, breathlessness, chest tightness, and cough.[6] Symptoms are triggered by diverse environmental exposures, infections, stress, and exercise, where excessive type 2 helper T (Th2) cell activation drives cytokine and chemokine production leading to airway narrowing and mucus production. Over time, abnormal Th1, Th2, and Th17 responses lead to chronic persistent inflammation and airways remodel with bronchial smooth muscle hypertrophy and increased vascularity, collagen deposition, and presence of mucus glands.[6–8] The mechanism driving the acute and chronic phases of asthma are complex, but many studies have established cross-talk between intracellular autophagic events and the pathogenesis of asthma (Table 1).

Initial *in vivo* and *in vitro* studies directly correlated the degree of interleukin-13 (IL-13) response, a cytokine central to the Th2 response observed in asthma (mucus hypersecretion and ROS generation), with the level of autophagic activation observed in airway epithelial cells.[9,25] Specifically, mice deficient in autophagy through the deletion of *Atg16l1* had a blunted IL-13 response, evidenced by less mucin MUC5AC production following cytokine release when compared to wild-type control animals.[9] *In vitro* studies using human airway epithelial cells, where genetic (*ATG5 and ATG14)* or pharmacological (3-methyladenine or bafilomycin A_1) inhibition of autophagy also revealed a dampened IL-13 response.[9] Additional studies also linked activation of autophagy to secretion of IL-18 in response to *Alternaria alternata*, an outdoor allergen that causes allergic airway disease.[26] Autophagic events in bronchial epithelial cells are now positioned downstream of transforming growth factor-beta 3 (TGF-ß3) signaling which drives MUC5AC hyperexpression and mucus production and inhibition of autophagy appears to halt the TGF-ß3 effects.[10] In airway B cells, through functions of IL-4, autophagy is upregulated and when deficient in autophagic proteins the asthma phenotype is perturbed both *in vivo* and *in vitro.*[11] Of high clinical relevance are the findings from several studies in the murine ovalbumin model of asthma where autophagy inhibition, *via* 3-methyladenine treatment or *Atg5* shRNA knockdown, showed a reduction in airway hyperresponsiveness, eosinophilia, and inflammation.[27,28]

Human studies also demonstrate the relevance of autophagy to the pathogenesis of asthma, as autophagosomes are increased in bronchial fibroblasts and epithelial cells from patients with moderately severe asthma.[12] Additionally, genetic association identified single nucleotide polymorphisms

Table 1 Murine and human studies highlighting the deleterious role of autophagy in asthma, COPD, and PAH.

Disease	Autophagy deleterious
Obstructive	
Asthma	**Murine studies** – IL-13 response and mucus hypersecretion in **airway epithelial cells** reduced in *Atg16ll* knockout mice[9] – Autophagy inhibition blocks TGF-ß3 signaling and mucus production in **bronchial epithelial cells[10]** – Autophagy blockade in **airway B cells** reduces asthma phenotype *in vivo* and in vitro[11] **Human data** – **Bronchial fibroblasts** and **epithelial cells** from severe asthmatics with increase autophagosomes[12] – SNPs in *SQSTM1, MAP1LC3B, BECN1,* and *ATG5* were found in severe asthmatics with reduced lung function[12] – Elevated levels of autophagy in **neutrophils** and **eosinophils** isolated from severe asthmatics[13,14]
Chronic obstructive pulmonary disease (COPD)	**Murine studies** – Reduction in autophagy dampened inflammation, cellular dysfunction and apoptosis in chronic cigarette smoke exposure[15,16] – Loss of *Atg5* in **bronchial epithelial cells** reduced airway inflammation to particulate matter[17] – Global loss of *Beclin-1* and *Lc3b* blocked cigarette smoke induced emphysema[15] – Cellular dysfunction limited in autophagy impaired mice[18] – Inhibition of mitophagy *via* loss of *Pink1* −/− reduced response to cigarette smoke[19] – Cigarette smoke exposed **neutrophils** have increased autophagy and compromised ability to ingest pathogens[20] **Human data** – Increased autophagosomes in human tissue[15] – Polymorphisms in *EGR-1* and *ATG16L1* identified as risk factors in COPD tissue[15,21]
Vascular	
Pulmonary arterial hypertension (PAH)	**Murine studies** – Loss of *Ucp2* and increased mitophagy worsened hypoxia-induced pulmonary hypertension[22] – Reduction in autophagy was linked to efficacy of liraglutide in monocrotaline model[23] – Autophagy activation linked to progression of right ventricular remodeling and fibrosis[24]

(SNPs) in genes encoding for key autophagic proteins ULK1, sequestosome 1 (p62/*SQSTM1*), microtubule-associated protein 1 light chain 3 beta (*MAP1LC3B*), Beclin-1 (*BECN1)*, and ATG5 that correlated with a reduction in lung function.[12] Likewise, a variant in *ATG5* has also been identified in childhood asthma.[29] Sputum and peripheral blood studies revealed elevated levels of autophagy in sputum granulocytes and peripheral blood eosinophils and neutrophils isolated from patients with severe asthma when compared to non-severe asthmatics and healthy controls.[13,14] In these studies, persistent activation of autophagy was found when key cytokines IL-5 and IL-1β were elevated.[13] Additionally, in severe asthmatics, ATG5 protein expression was positively associated with expression of collagen type V alpha I (COL5A1), suggesting a role for autophagy in airway remodeling and fibrosis observed in progressive asthma.[30]

While the notion that autophagy is pathogenic and may contribute to the asthma phenotype is evidenced in both animal and human studies, there are suggestions that the "pro-asthma" autophagic function represents a cell type-specific response. For example, in certain studies a loss of autophagy-related proteins in bronchial epithelial cells, and hence, a reduction in autophagy, revealed increased airway hyper-responsiveness to anticholinergic exposure and oxidative stress-mediated apoptosis.[31,32] Similarly, a loss of autophagy in Cd11c+immune cells led to an augmentation of neutrophilic airway inflammation observed in the murine asthma model.[33] Furthermore, simvastatin-induced autophagy provided cytoprotection in murine bronchial smooth muscle cells by suppression of IL-4, IL-5, and IL-13, resulting in decreased hypertrophy and fibrosis when compared to controls.[34]

Several efficacious asthma therapies (*e.g.*, monoclonal anti-IL5 antibody, anti-nerve growth antibody, astragalin) have been shown to downregulate autophagy *in vivo*.[35–37] Moreover, direct inhibition of autophagy through intranasal administration of chloroquine was also found to thwart airway inflammation, hyper-responsiveness, and remodeling.[38] Additionally, ketamine, an advanced therapy for refractory asthma, was found to block cellular autophagic activity in diseased airways, providing a potential explanation for how this agent may assist in the treatment patients with symptoms of severe refractory asthma.[39,40] Taken together, the therapeutic targeting of autophagy may represent a feasible treatment in asthma, but it should be approached with caution as autophagy seems to play a protective role in airway inflammation and hyperreactivity.

3.2 COPD and emphysema

COPD is the third leading cause of death in the United States and there is currently no curative treatment.[41] COPD represents a chronic inflammatory disease state, often following long-term exposure to cigarette smoke, where involvement of both the proximal airways as well as the distal lung tissue, including the terminal alveolar subunits, drives the symptoms of chronic bronchitis and emphysema. Airway epithelial cells are primarily targeted by exposure to cigarette smoke and the unchecked chronic inflammation and oxidative stress that ensue drives the pathogenesis of COPD with the involvement of multiple forms of regulated cell death (*i.e.*, apoptosis, necroptosis, and autophagy).[5] Alveolar macrophages derived from circulating monocytes also participate in the pathogenesis of disease through the release of chemokines that lead to the recruitment of CD8 lymphocytes, monocytes, and neutrophils to the lung.[42] A complete understanding of the molecular mechanisms underpinning the pathogenesis of COPD are not fully understood, but several lines of investigation have shown that dysregulated autophagy contributes to the pathogenesis of COPD (Table 1).

Lung tissue from COPD patients or from mice chronically exposed to cigarette smoke revealed increase expression of autophagic proteins. Increased autophagy was observed in human lung tissue across all GOLD stages, making this cellular process a marker of both early and late disease states.[15] The initial suggestions of the importance of autophagy in COPD progression came from studies examining upstream regulators of autophagy (Toll-like receptor 4 (TLR4) and Early growth response-1 (EGR-1)), where dampening of the autophagic response limited the inflammation, cellular dysfunction, and apoptosis observed in response to chronic cigarette smoke exposure both *in vitro* and *in vivo*.[15,16] Loss of bronchial epithelial-specific *Atg5* was also found to reduce airway inflammation in response to a murine model of particulate matter.[17] Gene expression arrays have revealed activators of autophagy (*e.g.*, *EGR-1*) to be upregulated in COPD tissue and several polymorphisms were identified in *EGR-1* and *ATG16L1* as significant risk factors for the development of COPD.[21,43] Furthermore, global loss of key autophagic proteins (*Beclin-1, Lc3b*) blocked cigarette smoke-induced epithelial cell death and the development of airspace enlargement.[15] Likewise, autophagy impaired mice were also resistant to mucociliary clearance disruption in the airways of subchronic cigarette smoke exposure.[18] Interestingly, studies have also shown that autophagy activation drives systemic complications of COPD, namely chronic kidney disease, when loss of autophagic proteins prevented the development of kidney abnormalities in a murine model of COPD.[44]

A link between activation of selective autophagy of mitochondria, namely mitophagy, and other regulated forms of cell death (*e.g.*, apoptosis and necroptosis) as drivers of the COPD phenotype stressed the importance of these processes in normal lung homeostasis and the pathogenesis of COPD[19,45–49] Inhibition of mitophagy through the genetic deletion of PTEN-induced kinase 1 (*Pink1−/−*) reduced the development of airspace enlargement, mucociliary clearance disruption, and mitochondrial dysfunction commonly observed in epithelial cells exposed to cigarette smoke.[19] Genetic or pharmacological manipulation of autophagy activation in *in vitro* models linked activation of autophagy with chronic inflammation and mucus production.[17,50,51] More recent studies showed enhanced autophagic degradation of ferritin, a process known as ferritinophagy, in airway epithelial cells with enhanced ferroptosis, an additional form of regulated cell death mediated by phospholipid peroxidation in association with free iron, in the pathogenesis of COPD utilizing both *in vitro* and *in vivo* models.[52]

While there are accumulating data that activation of autophagy and selective autophagy in epithelial cells are key to the pathogenesis of COPD, there are also several lines of evidence that support the notion that it is insufficient autophagy or mitophagy that drives the pathogenesis of COPD[53–57] (Table 2). Initial studies using human bronchial epithelial cells found that cigarette smoke-induced cellular senescence was blocked by the pharmacological activation of autophagy using the mTOR inhibitor, Torin-1.[56] Additionally, epithelial cells isolated from COPD patients had a blunted observable activation of autophagy in response to cigarette smoke exposure when compared to epithelial cells obtained from non-smokers or non-COPD smokers.[56]

Parkin, a key regulator of mitophagy, is downregulated in human COPD tissue.[53] Genetic deletion of *Parkin (Prkn−/−)* in a murine model of COPD revealed worsening airway thickening and airspace enlargement.[54] *In vitro* experiments where Parkin was overexpressed in epithelial cells resulted in dampening of mitochondrial reactive oxygen species and cellular senescence observed in response to cigarette smoke extract.[54] These investigations suggest that increased levels of Parkin enhance mitophagy and perturb the development of COPD both *in vitro* and *in vivo*.

Cigarette smoke is known to induce proteostasis that drives an accumulation of ubiquitinated proteins (aggresomes) within epithelial cells and potentiates chronic inflammation.[58] Furthermore, decreased proteosomal

Table 2 Murine and human studies highlighting the protective role of autophagy in COPD, IPF, and PAH.

Disease	Autophagy protective
Obstructive	
Chronic obstructive pulmonary disease (COPD)	**Murine studies** – deletion of *Parkin* (reduction in mitophagy) revealed worsening airway thickening and airspace enlargement[54] – pharmacological activation of autophagy blocked the development of aggresomes in COPD mouse model[58,59] – Cigarette smoke-exposed **alveolar macrophages** also show increased aggresomes and impaired autophagy[60,61] **Human data** – **Bronchial epithelial** cells isolated from COPD patients have a blunted observable activation of autophagy in response to cigarette smoke exposure *in vitro*[56,57] – Parkin, a key regulator of mitophagy, is downregulated in human COPD tissue[53] – Lung homogenates from COPD tissue have increased p62, suggesting a defective autophagy[18]
Restrictive	
Idiopathic pulmonary fibrosis (IPF)	**Murine studies** – Autophagy decreased in **airway epithelial cells** and **fibroblasts** in fibrosis murine model[62–66] – Induction of autophagy in bleomycin fibrosis model was found following neutralization of pro-inflammatory cytokine Il-17A, the effect was dependent on autophagy induction[67] – Loss of *Atg4, Pink1, Park2* response to bleomycin induced fibrosis was exaggerated[64–66,68] **Human data** – Autophagy decreased in human tissue (reduction in LC3 and p62 levels[67] – Beclin 1 is downregulated in fibroblasts isolated from IPF patients[69] – Nintedanib and pirfenidone treatment results in activation of autophagy through mitophagy and Beclin-1 and Atg7 activation[70,71]
Vascular	
Pulmonary arterial hypertension (PAH)	**Murine studies** – Upregulation of autophagy in response to experimental PAH (hypoxia)[72,73] – Loss of *Beclin-1, Egr-1, Lc3b* worsened PAH phenotype in experimental models[72,73] – Rapamycin treatment perturbed the development of PAH in Sugen treated rats[74] **Human data** – Increase autophagosomes and autophagic proteins in pulmonary arterial hypertension tissue[72,75]

function has been shown to accelerate cigarette smoke-induced airspace enlargement[58,59] in animal models. Lung homogenates from COPD patients show an increase in ubiquitinated proteins and p62/SQSTM1, a key adaptor protein responsible for delivering ubiquitinated substrates to LC3 and autophagosome formation.[18] Defective autophagy was considered the cause of cigarette smoke-induced formation of aggresomes when pharmacological activation of autophagy blocked this phenotype in the COPD mouse model.[60] In those studies, the investigators surmised that the accumulation of p62 and ubiquitinated proteins observed in severe COPD-emphysema lungs indicated the pathogenic role of insufficient autophagic clearance in COPD.[60]

While the relevance of autophagy to the inflammatory response and impaired mucociliary clearance observed in cigarette smoke exposed airway epithelial cells has been realized, the upregulation of autophagy in alveolar macrophages and neutrophils has also been observed.[20,61] Interestingly, impaired aggresome clearance and p62 accumulation are also key features of cigarette smoke-exposed alveolar macrophages, suggesting that impaired autophagy impacts homeostasis in these cells as well.[61] In those same studies, cigarette smoke exposed alveolar macrophages were found to be defective in xenophagy, a selective autophagy responsible for the clearance of bacteria, suggesting an explanation for the increased infection rates observed in smokers with COPD.[61] Further analysis revealed that alveolar macrophages from lung tissue or bronchoalveolar lavage fluid from patients with COPD are defective in phagocytosis and the ability to regulate mitochondrial reactive oxygen species production. However, how this relates to autophagy remains to be elucidated. Extensive studies in neutrophils are lacking, but initial investigation uncovered that cigarette smoke-exposed neutrophils have increased autophagy associated with a compromised ability to ingest the respiratory pathogen, *Staphylococcus aureus*.[20] Further studies are required to fully understand the contribution of the cigarette smoke-induced autophagy observed in alveolar macrophages and neutrophils to the pathogenesis of COPD.

Accumulating evidence argues for an essential role of autophagy in multiple cell types in the pathogenesis of COPD. It appears that both overactive and insufficient autophagy can drive the inflammation, cell death, and cellular dysfunction observed in COPD. Therapeutic options for the treatment of COPD continue to remain rather limited, and the potential of targeting autophagy as a treatment for COPD warrants further exploration. Pharmacological activation of autophagy has been used in murine models, but the clinical relevance of the observed protection remains unclear since therapy was administered prophylactically during the course of cigarette smoke exposure.[18,19,60,76,77] Autophagy activation through mTOR inhibition, using rapamycin, has also been tested and the results are inconclusive.[78] While administration of rapamycin to cigarette smoke-exposed mice appeared to reduce alveolar inflammation, the drug promoted cellular inflammation and apoptosis in room air-exposed mice. These findings underscore the complexity of the role of targeting autophagy as a therapeutic modality in COPD.

4. Pulmonary vascular disease

Pulmonary arterial hypertension (PAH) is a severely debilitating and often fatal disease typified by vascular remodeling and rise in pulmonary artery pressure associated with progression to right ventricular failure. Pulmonary hypertension affects more than 100 million people worldwide and the current treatments fail to halt or reverse ongoing vascular remodeling.[79] The distal arterioles are the targets of the disease and current tenet supports the notion that endothelial cell injury and apoptosis serve as an initial trigger, and vascular cells, namely endothelial cells and smooth muscle cells, progress to develop a pro-proliferative/anti-apoptotic phenotype that leads to unchecked vascular remodeling.

The importance of the initial endothelial injury and apoptotic cell death in the development of PAH is highlighted in several murine studies, where treatment with the vascular endothelial growth factor (VEGF) inhibitor SU5416, a strong inducer of endothelial apoptosis, resulted in PAH with features similar to that observed in human disease.[80] However, following the initial phase both endothelial and smooth muscle cells are highly proliferative in nature leading to vessel wall thickening and the formation of plexiform lesions. In this case, signaling driven by growth factors namely VEGF, platelet-derived growth factor (PDGF), and fibroblast growth factor (FGF) maintain cell cycle progression in vascular cells.[81,82] Pro-survival signaling pathways (AKT/ERK/NF-kB/STAT3) are upregulated in endothelial and smooth muscle cells that enhance persistent vascular disarray.[83] Although the molecular mechanisms involved in the pathogenesis of PAH have been extensively studied, new therapeutic targets that can halt and reverse progression remain lacking.

In recent years, several studies have questioned the role of autophagy in driving the pathogenesis of PAH and the conclusions remain controversial

(Tables 1 and 2). Human PAH lung tissue samples show increase in autophagic markers, where LC3B and autophagosome numbers are elevated when compared to control lungs.[72] Both *in vivo* and *in vitro* studies showed up-regulation of the same autophagic markers in vascular cells in response to the hypoxia-induced experimental PAH models.[72,75] Additionally, genetic deletion of *Lc3b* or *Egr-1,* a known regulator of LC3B, led to a rise in right ventricular systolic pressure and loss of LC3B in the vascular cells led to increased reactive oxygen species production and cellular proliferation *in vitro*.[72] Additional studies revealed a loss of *Beclin-1* enhanced PAH phenotype with increased angiogenesis and oxidative stress in pulmonary artery endothelial cells from fetal lambs with persistent pulmonary hypertension.[73] In the PAH rat model, using a VEGF inhibitor (SU5416), investigators highlighted the need for increased autophagy in endothelial cells as a mechanism to halt progression of pulmonary hypertension when treatment with rapamycin, an indirect activator of autophagy, perturbed the development of PAH in SU5416 treated rats.[74] In a separate study, investigators tested the ability of dihydroartemisinin (DHA), a compound with known anti-inflammatory and anti-tumor activities, to prevent the development of both hypoxia and monocrotaline-induced pulmonary hypertension.[84,85] *In vitro* studies clarified that this protection appears to be linked to the activation of endothelial cell autophagy in response to DHA.[86]

While the data supporting a protective role of autophagy in the pathogenesis of PAH is robust, there are several studies that identify cases where autophagy may be pathological. In the case of HIV-related PAH, severe and exaggerated pulmonary vascular disease in patients with concomitant opioid usage was observed.[87] *In vitro* studies linked the presence of HIV virus and opioid dosing with an induction of key autophagic markers and autophagosome production in human arterial endothelial cells. Investigators linked the aggressive nature of this HIV-related PAH to the observed impact on cellular autophagy.[87] In recent years, roles for mitochondrial and metabolic dysfunction in the pathogenesis of PAH has been established.[88] The abnormalities identified in PAH include, but are not limited to, alterations in the pentose phosphate pathway, glutaminolysis, fatty acid oxidation.[88] A link between excessive mitophagy, a selective form of autophagy, mitochondrial dysfunction, and a pathogenic endothelial cell phenotype in response to hypoxia was observed in an *in vitro* experimental model of pulmonary hypertension.[89] Additionally, loss of the key mitochondrial protein, uncoupling protein 2, revealed increased mitophagy and worsening hypoxia-induced pulmonary hypertension.[22] Given the relevance of metabolic derangements to the pathogenesis of PAH, metabolic therapies are being tested in pre-clinical models of pulmonary hypertension. Liraglutide, a glucagon-like peptide-1 receptor (GLP-1), is widely used to treat diabetes, and was found to prevent and reverse monocrotaline-induced PAH.[23] Follow up studies found that the observed effect appears to be related to decrease in mitochondrial fusion involving dynamin-related protein-1 (Drp1) and a reduction in cellular autophagic activity in the smooth muscle cells.[90] Notably, recent studies have also linked autophagy activation to the progression of right ventricular remodeling and fibrosis.[24]

Further studies are needed to clarify the conditions where autophagy protects against or augments the pathogenesis of PAH. Murine models utilizing cell-specific genetic deletion strategies will likely assist in clarifying these conditions in experimental models of PAH. Pre-clinical animal studies examining the impact of pharmacological activation of autophagy on the severity of the observed PAH phenotype have been inconclusive.[91,92] Clinically, everolimus, an mTOR inhibitor and activator of autophagy, has been used in cases of severe PAH secondary to chronic thromboembolic disease and there was measurable improvement in exercise tolerance and a decrease in pulmonary vascular resistance.[93] Currently, there are Phase I clinical trials underway testing the utility of mTOR inhibitors, thus, autophagy activators, in the treatment of patients with severe PAH.

5. Restrictive lung disease

Idiopathic pulmonary fibrosis (IPF) is a chronic, progressive, and fatal fibroproliferative disorder characterized by loss of alveolar structure, accumulation of myofibroblasts, and extensive remodeling of lung parenchyma. It is the most common and severe form of restrictive lung disease with over three million people worldwide who are afflicted with IPF with limited available effective therapies. Current dogma supports the idea that endogenous or exogenous injury to the alveolar epithelium drives cell-specific apoptosis and activation of fibroblasts/myofibroblasts that leads to fibroblast deposition and extensive lung remodeling. TGF-ß1 is a pro-fibrotic cytokine that is increased in IPF lung tissues and has been shown to enhance production of type I collagen from myofibroblasts, as well as, induce apoptosis of epithelial cells in culture. In murine models of fibrosis, overexpression of TGF-ß1 causes alveolar epithelial apoptosis and fibrosis.

In recent years, several investigations have focused on the role of autophagy in the pathogenesis of IPF (Table 2). These data support the current paradigm that autophagy represents a protective mechanism in normal wound healing and limiting fibroblast activation and collagen deposition, while promoting autophagy associated cell death with resolution of the injury. In contrast, in the case of IPF, suppressed or defective autophagy has been shown to facilitate epithelial cell dysfunction and myofibroblast activation and production.

Autophagic pathways are decreased in lung epithelial cells and fibroblasts from IPF patients and in murine models of IPF.[62–65,67] Human tissues of patients with IPF demonstrated evidence of decreased autophagic activity *via* reduction in LC3 and p62 expression.[67] Initial murine studies using bleomycin model of lung fibrosis found that neutralizing the pro-inflammatory cytokine, IL-17A, led to induction of autophagy and resolution of acute inflammation, attenuation of pulmonary fibrosis, and increased cellular survival.[94] When autophagy was blocked by the administration of 3-methyladenine the protective effects of IL-17A blockade did not persist.[94] In addition to IL-17-mediated inhibition of autophagy, bioactive lipids such as sphingosine 1-phosphate and lysophosphatidic acid are also known to inhibit autophagy and to promote fibrogenesis.[95] Beclin-1, a key autophagic protein, is downregulated in fibroblasts isolated from patients with IPF.[69] Similarly, human fibroblasts from IPF donors displayed reduction in autophagic proteins and the unfolded protein response in parallel with increased collagen production *in vitro*.[96] Additionally, when autophagy is inhibited in epithelial cells the epithelial-to-mesenchymal transition is enhanced and fibrosis ensues.[97]

In multiple *in vivo* studies, where genetic deletions led to autophagy deficient (*Atg4−/−*) or mitophagy deficient (*Pink1−/−, Prkn−/−*) mice, the fibrotic phenotype of the lung in response to bleomycin was exaggerated.[64–66,68] Interestingly, mice deficient in *matrix metalloproteinase 19 (Mmp19−/−*) have exaggerated fibrotic response to bleomycin which was linked to autophagic suppression *via* downregulation of ATG4C, a cysteine peptidase necessary for autophagosomal production.[98] Furthermore, genome-wide association studies identified the autophagy gene Cep55 as having a polymorphism significantly associated with susceptibility to bleomycin-induced lung fibrosis in mice.[99] Other studies focused on epithelial cells have shown that *Atg4b*-deficient mice have a greater inflammatory response 7 days post bleomycin and augmented apoptosis of alveolar and bronchial epithelial cells leading to more extensive fibrosis and collagen accumulation.[68] More recently, using 3D lung organoids, investigators showed that inhibition of vimentin intermediate filament assembly promoted autophagic clearance of type I collagen.[100] Aging studies revealed that mice with loss of autophagic proteins (*Lc3b*−/− and *Atg4b*−/−) were more susceptible to bleomycin-induced lung fibrosis and linked cathepsin A, a binding partner to LC3B, and ER stress to increased apoptosis of epithelial cells in culture.[101,102]

The role of mTOR activation in dampening the autophagic response in IPF is supported by evidence that fibroblasts harvested from IPF patients were found to have aberrant PTEN/AKT/mTOR pathway that permits unchecked fibroblast proliferation through the suppression of autophagy.[62] Accordingly, inhibition of autophagy *in vitro* led to increase in the development of myofibroblast phenotype *in vitro*, whereas, autophagy activation through mTOR inhibition with rapamycin reversed these effects.[62] Similarly, conditional knockdown of the tuberous sclerosis-1 (*Tsc1*) gene in epithelial cells caused mice to be more susceptible to bleomycin-induced lung fibrosis, a condition that was reversible when rapamycin or chloroquine was administered to stimulate autophagy.[103]

Targeting autophagy as a therapy for patients with IPF may have great potential and pre-clinical and clinical studies are underway. Immunomodulators including IL-17A neutralizing antibodies, mir449A, and PDGF4B inhibitors are known activators of autophagy and show promise in pre-clinical models.[65,94,104] Additionally, rapamycin, through the activation of autophagy, appears to blunt the development of pulmonary fibrosis in murine models.[66,103] Berberine, an isoquinolone alkaloid, known to prevent activation of TGF-ß1 signaling, has been shown to ameliorate IPF in the bleomycin model.[105] Further studies revealed that berberine leads to inhibition of mTOR and stimulation of autophagy. Interestingly both nintedanib and pirfenidone, two recently FDA approved therapies offered to patients with IPF, show anti-fibrotic activity through the induction of both non-selective autophagy through Beclin-1 and ATG7 and mitophagy, respectively.[70,71] Further studies are required to evaluate the benefits of targeting autophagy in IPF and whether or not these targeting strategies should be cell type-specific.

6. Concluding remarks

The cellular components of the lung are challenged by oxidative and environmental stressors that require appropriate cellular responses to maintain homeostasis. Whether autophagy functions as a protector or promoter of a given lung disease is a complex question that needs further investigations to better understand the physiologic functions of autophagy in lung disease.

The multiple cell types, and spatial and temporal heterogeneity of lung tissue injury mediating the development of lung diseases such as asthma, COPD, PAH, and IPF make it even more challenging to interpret confidently the results from studies using global knockout mouse models. Further clarity will likely evolve with future studies utilizing cell-specific autophagy gene targeted mice in a given disease model. Additionally, further defining phenotypes of the heterogenous disease states of asthma and COPD will likely uncover new roles of autophagy and selective autophagy in the pathogenesis of these diseases. While pre-clinical models have advanced our understanding of the functional roles of autophagic pathways in multiple chronic lung diseases, further investigations in pre-clinical and, importantly, clinical studies are needed to translate the findings to potential therapeutic targets.

Acknowledgments

M.E.C. and A.M.K.C. are supported by National Institutes of Health (NIH) R01 HL132198, R01 HL133801, R01 HL055330, and R01 HL060234. A.C.R. is supported by NIH T32HL134629.

Disclosure statement

The spouse of M.E.C. is a cofounder and shareholder, and serves on the Scientific Advisory Board of Proterris, Inc. A.M.K.C. is a cofounder, stockholder and serves on the Scientific Advisory Board for Proterris, which develops therapeutic uses for carbon monoxide (CO). A.M.K.C. also has a use patent on CO.

References

1. Levine B, Kroemer G. Biological functions of autophagy genes: a disease perspective. *Cell.* 2019;176(1–2):11–42.
2. Dikic I, Elazar Z. Mechanism and medical implications of mammalian autophagy. *Nat Rev Mol Cell Biol.* 2018;19(6):349–364.
3. Yin Z, Pascual C, Klionsky DJ. Autophagy: machinery and regulation. *Microb Cell.* 2016;3(12):588–596.
4. Klaassen CD, Watkins JB, Casarett LJ. *Casarett & Doull's Essentials of Toxicology.* 2nd ed. New York: McGraw-Hill Medical; 2010.
5. Sauler M, Bazan IS, Lee PJ. Cell death in the lung: the apoptosis-necroptosis axis. *Annu Rev Physiol.* 2019;81:375–402.
6. Mims JW. Asthma: definitions and pathophysiology. *Int Forum Allergy Rhinol.* 2015;-5(suppl 1):S2–S6.
7. Jyothula SS, Eissa NT. Autophagy and role in asthma. *Curr Opin Pulm Med.* 2013;19(1):30–35.
8. Lambrecht BN, Hammad H. The immunology of asthma. *Nat Immunol.* 2015; 16(1):45–56.
9. Dickinson JD, Alevy Y, Malvin NP, et al. IL13 activates autophagy to regulate secretion in airway epithelial cells. *Autophagy.* 2016;12(2):397–409.

10. Zhang Y, Tang H, Yuan X, et al. TGF-beta3 promotes MUC5AC hyper-expression by modulating autophagy pathway in airway epithelium. *EBioMedicine*. 2018;33: 242–252.
11. Xia F, Deng C, Jiang Y, et al. IL4 (interleukin 4) induces autophagy in B cells leading to exacerbated asthma. *Autophagy*. 2018;14(3):450–464.
12. Poon AH, Chouiali F, Tse SM, et al. Genetic and histologic evidence for autophagy in asthma pathogenesis. *J Allergy Clin Immunol*. 2012;129(2):569–571.
13. Ban GY, Pham DL, Trinh TH, et al. Autophagy mechanisms in sputum and peripheral blood cells of patients with severe asthma: a new therapeutic target. *Clin Exp Allergy*. 2016;46(1):48–59.
14. Pham DL, Ban GY, Kim SH, et al. Neutrophil autophagy and extracellular DNA traps contribute to airway inflammation in severe asthma. *Clin Exp Allergy*. 2017;47(1): 57–70.
15. Chen ZH, Kim HP, Sciurba FC, et al. Egr-1 regulates autophagy in cigarette smoke-induced chronic obstructive pulmonary disease. *PLoS One*. 2008;3(10), e3316.
16. An CH, Wang XM, Lam HC, et al. TLR4 deficiency promotes autophagy during cigarette smoke-induced pulmonary emphysema. *Am J Physiol Lung Cell Mol Physiol*. 2012;303(9):L748–L757.
17. Wu YF, Li ZY, Dong LL, et al. Inactivation of MTOR promotes autophagy-mediated epithelial injury in particulate matter-induced airway inflammation. *Autophagy*. 2020;16(3):435–450.
18. Lam HC, Cloonan SM, Bhashyam AR, et al. Histone deacetylase 6-mediated selective autophagy regulates COPD-associated cilia dysfunction. *J Clin Invest*. 2013;123(12): 5212–5230.
19. Mizumura K, Cloonan SM, Nakahira K, et al. Mitophagy-dependent necroptosis contributes to the pathogenesis of COPD. *J Clin Invest*. 2014;124(9):3987–4003.
20. Guzik K, Skret J, Smagur J, et al. Cigarette smoke-exposed neutrophils die unconventionally but are rapidly phagocytosed by macrophages. *Cell Death Dis*. 2011;2, e131.
21. Ning W, Li CJ, Kaminski N, et al. Comprehensive gene expression profiles reveal pathways related to the pathogenesis of chronic obstructive pulmonary disease. *Proc Natl Acad Sci U S A*. 2004;101(41):14895–14900.
22. Haslip M, Dostanic I, Huang Y, et al. Endothelial uncoupling protein 2 regulates mitophagy and pulmonary hypertension during intermittent hypoxia. *Arterioscler Thromb Vasc Biol*. 2015;35(5):1166–1178.
23. Lee MY, Tsai KB, Hsu JH, Shin SJ, Wu JR, Yeh JL. Liraglutide prevents and reverses monocrotaline-induced pulmonary arterial hypertension by suppressing ET-1 and enhancing eNOS/sGC/PKG pathways. *Sci Rep*. 2016;6:31788.
24. Ibrahim YF, Shults NV, Rybka V, Suzuki YJ. Docetaxel reverses pulmonary vascular remodeling by decreasing autophagy and resolves right ventricular fibrosis. *J Pharmacol Exp Ther*. 2017;363(1):20–34.
25. Zeki AA, Yeganeh B, Kenyon NJ, Post M, Ghavami S. Autophagy in airway diseases: a new frontier in human asthma? *Allergy*. 2016;71(1):5–14.
26. Murai H, Okazaki S, Hayashi H, et al. Alternaria extract activates autophagy that induces IL-18 release from airway epithelial cells. *Biochem Biophys Res Commun*. 2015;464(4):969–974.
27. Silveira JS, Antunes GL, Kaiber DB, et al. Autophagy induces eosinophil extracellular traps formation and allergic airway inflammation in a murine asthma model. *J Cell Physiol*. 2020;235(1):267–280.
28. Liu JN, Suh DH, Trinh HK, Chwae YJ, Park HS, Shin YS. The role of autophagy in allergic inflammation: a new target for severe asthma. *Exp Mol Med*. 2016;48(7), e243.
29. Martin LJ, Gupta J, Jyothula SS, et al. Functional variant in the autophagy-related 5 gene promotor is associated with childhood asthma. *PLoS One*. 2012;7(4), e33454.

30. Poon AH, Choy DF, Chouiali F, et al. Increased autophagy-related 5 gene expression is associated with collagen expression in the airways of refractory asthmatics. *Front Immunol.* 2017;8:355.
31. Inoue D, Kubo H, Taguchi K, et al. Inducible disruption of autophagy in the lung causes airway hyper-responsiveness. *Biochem Biophys Res Commun.* 2011;405(1):13–18.
32. Tsai YG, Wen YS, Wang JY, et al. Complement regulatory protein CD46 induces autophagy against oxidative stress-mediated apoptosis in normal and asthmatic airway epithelium. *Sci Rep.* 2018;8(1):12973.
33. Suzuki Y, Maazi H, Sankaranarayanan I, et al. Lack of autophagy induces steroid-resistant airway inflammation. *J Allergy Clin Immunol.* 2016;137(5):1382–1389.e1389.
34. Gu W, Cui R, Ding T, et al. Simvastatin alleviates airway inflammation and remodelling through up-regulation of autophagy in mouse models of asthma. *Respirology.* 2017;22(3):533–541.
35. Liu H, Mi S, Li Z, Hua F, Hu ZW. Interleukin 17A inhibits autophagy through activation of PIK3CA to interrupt the GSK3B-mediated degradation of BCL2 in lung epithelial cells. *Autophagy.* 2013;9(5):730–742.
36. Wang ZG, Li H, Huang Y, et al. Nerve growth factor-induced Akt/mTOR activation protects the ischemic heart via restoring autophagic flux and attenuating ubiquinated protein accumulation. *Oncotarget.* 2017;8(3):5400–5413.
37. Cho IH, Choi YJ, Gong JH, Shin D, Kang MK, Kang YH. Astragalin inhibits autophagy-associated airway epithelial fibrosis. *Respir Res.* 2015;16:51.
38. McAlinden KD, Deshpande DA, Ghavami S, et al. Autophagy activation in asthma airways remodeling. *Am J Respir Cell Mol Biol.* 2019;60(5):541–553.
39. Zou H, Wang LX, Wang M, et al. MTOR-mediated autophagy is involved in the protective effect of ketamine on allergic airway inflammation. *J Immunol Res.* 2019; 2019:5879714.
40. Goyal S, Agrawal A. Ketamine in status asthmaticus: a review. *Indian J Crit Care Med.* 2013;17(3):154–161.
41. Burney PG, Patel J, Newson R, Minelli C, Naghavi M. Global and regional trends in COPD mortality, 1990-2010. *Eur Respir J.* 2015;45(5):1239–1247.
42. Brusselle GG, Joos GF, Bracke KR. New insights into the immunology of chronic obstructive pulmonary disease. *Lancet.* 2011;378(9795):1015–1026.
43. Chen CZ, Ou CY, Wang RH, et al. Association of Egr-1 and autophagy-related gene polymorphism in men with chronic obstructive pulmonary disease. *J Formos Med Assoc.* 2015;114(8):750–755.
44. Pabon MA, Patino E, Bhatia D, et al. Beclin-1 regulates cigarette smoke-induced kidney injury in a murine model of chronic obstructive pulmonary disease. *JCI Insight.* 2018;3(18), e99592.
45. Wang G, Zhou H, Strulovici-Barel Y, et al. Role of OSGIN1 in mediating smoking-induced autophagy in the human airway epithelium. *Autophagy.* 2017;13(7):1205–1220.
46. Hou HH, Cheng SL, Chung KP, et al. Elastase induces lung epithelial cell autophagy through placental growth factor: a new insight of emphysema pathogenesis. *Autophagy.* 2014;10(9):1509–1521.
47. Wen W, Yu G, Liu W, et al. Silencing FUNDC1 alleviates chronic obstructive pulmonary disease by inhibiting mitochondrial autophagy and bronchial epithelium cell apoptosis under hypoxic environment. *J Cell Biochem.* 2019;120(10):17602–17615.
48. Zhang M, Shi R, Zhang Y, et al. Nix/BNIP3L-dependent mitophagy accounts for airway epithelial cell injury induced by cigarette smoke. *J Cell Physiol.* 2019;234(8):14210–14220.
49. Han D, Wu X, Liu L, Shu W, Huang Z. Sodium tanshinone IIA sulfonate protects ARPE-19 cells against oxidative stress by inhibiting autophagy and apoptosis. *Sci Rep.* 2018;8(1):15137.

50. Li D, Hu J, Wang T, et al. Silymarin attenuates cigarette smoke extract-induced inflammation via simultaneous inhibition of autophagy and ERK/p38 MAPK pathway in human bronchial epithelial cells. *Sci Rep*. 2016;6:37751.
51. Zhou JS, Zhao Y, Zhou HB, et al. Autophagy plays an essential role in cigarette smoke-induced expression of MUC5AC in airway epithelium. *Am J Physiol Lung Cell Mol Physiol*. 2016;310(11):L1042–L1052.
52. Yoshida M, Minagawa S, Araya J, et al. Involvement of cigarette smoke-induced epithelial cell ferroptosis in COPD pathogenesis. *Nat Commun*. 2019;10(1):3145.
53. Ito S, Araya J, Kurita Y, et al. PARK2-mediated mitophagy is involved in regulation of HBEC senescence in COPD pathogenesis. *Autophagy*. 2015;11(3):547–559.
54. Araya J, Tsubouchi K, Sato N, et al. PRKN-regulated mitophagy and cellular senescence during COPD pathogenesis. *Autophagy*. 2019;15(3):510–526.
55. Roscioli E, Tran HB, Jersmann H, et al. The uncoupling of autophagy and zinc homeostasis in airway epithelial cells as a fundamental contributor to COPD. *Am J Physiol Lung Cell Mol Physiol*. 2017;313(3):L453–L465.
56. Fujii S, Hara H, Araya J, et al. Insufficient autophagy promotes bronchial epithelial cell senescence in chronic obstructive pulmonary disease. *Oncoimmunology*. 2012;1(5):630–641.
57. Zhu L, Barrett EC, Xu Y, Liu Z, Manoharan A, Chen Y. Regulation of cigarette smoke (CS)-induced autophagy by Nrf2. *PLoS One*. 2013;8(4), e55695.
58. van Rijt SH, Keller IE, John G, et al. Acute cigarette smoke exposure impairs proteasome function in the lung. *Am J Physiol Lung Cell Mol Physiol*. 2012;303(9):L814–L823.
59. Yamada Y, Tomaru U, Ishizu A, et al. Decreased proteasomal function accelerates cigarette smoke-induced pulmonary emphysema in mice. *Lab Invest*. 2015;95(6):625–634.
60. Tran I, Ji C, Ni I, Min T, Tang D, Vij N. Role of cigarette smoke-induced aggresome formation in chronic obstructive pulmonary disease-emphysema pathogenesis. *Am J Respir Cell Mol Biol*. 2015;53(2):159–173.
61. Monick MM, Powers LS, Walters K, et al. Identification of an autophagy defect in smokers' alveolar macrophages. *J Immunol*. 2010;185(9):5425–5435.
62. Nho RS, Hergert P. IPF fibroblasts are desensitized to type I collagen matrix-induced cell death by suppressing low autophagy via aberrant Akt/mTOR kinases. *PLoS One*. 2014;9(4), e94616.
63. Araya J, Kojima J, Takasaka N, et al. Insufficient autophagy in idiopathic pulmonary fibrosis. *Am J Physiol Lung Cell Mol Physiol*. 2013;304(1):L56–L69.
64. Sosulski ML, Gongora R, Danchuk S, Dong C, Luo F, Sanchez CG. Deregulation of selective autophagy during aging and pulmonary fibrosis: the role of TGFbeta1. *Aging Cell*. 2015;14(5):774–783.
65. Kobayashi K, Araya J, Minagawa S, et al. Involvement of PARK2-mediated mitophagy in idiopathic pulmonary fibrosis pathogenesis. *J Immunol*. 2016;197(2):504–516.
66. Patel AS, Song JW, Chu SG, et al. Epithelial cell mitochondrial dysfunction and PINK1 are induced by transforming growth factor-beta1 in pulmonary fibrosis. *PLoS One*. 2015;10(3), e0121246.
67. Patel AS, Lin L, Geyer A, et al. Autophagy in idiopathic pulmonary fibrosis. *PLoS One*. 2012;7(7), e41394.
68. Cabrera S, Maciel M, Herrera I, et al. Essential role for the ATG4B protease and autophagy in bleomycin-induced pulmonary fibrosis. *Autophagy*. 2015;11(4):670–684.
69. Ricci A, Cherubini E, Scozzi D, et al. Decreased expression of autophagic beclin 1 protein in idiopathic pulmonary fibrosis fibroblasts. *J Cell Physiol*. 2013;228(7):1516–1524.
70. Rangarajan S, Kurundkar A, Kurundkar D, et al. Novel mechanisms for the antifibrotic action of nintedanib. *Am J Respir Cell Mol Biol*. 2016;54(1):51–59.

71. Kurita Y, Araya J, Minagawa S, et al. Pirfenidone inhibits myofibroblast differentiation and lung fibrosis development during insufficient mitophagy. *Respir Res.* 2017; 18(1):114.
72. Lee SJ, Smith A, Guo L, et al. Autophagic protein LC3B confers resistance against hypoxia-induced pulmonary hypertension. *Am J Respir Crit Care Med.* 2011;183(5): 649–658.
73. Lee SJ, Kim HP, Jin Y, Choi AM, Ryter SW. Beclin 1 deficiency is associated with increased hypoxia-induced angiogenesis. *Autophagy.* 2011;7(8):829–839.
74. Kato F, Sakao S, Takeuchi T, et al. Endothelial cell-related autophagic pathways in Sugen/hypoxia-exposed pulmonary arterial hypertensive rats. *Am J Physiol Lung Cell Mol Physiol.* 2017;313(5):L899–L915.
75. Teng RJ, Du J, Welak S, et al. Cross talk between NADPH oxidase and autophagy in pulmonary artery endothelial cells with intrauterine persistent pulmonary hypertension. *Am J Physiol Lung Cell Mol Physiol.* 2012;302(7):L651–L663.
76. Vij N, Chandramani-Shivalingappa P, Van Westphal C, Hole R, Bodas M. Cigarette smoke-induced autophagy impairment accelerates lung aging, COPD-emphysema exacerbations and pathogenesis. *Am J Physiol Cell Physiol.* 2018;314(1):C73–C87.
77. Li Y, Yu G, Yuan S, et al. Cigarette smoke-induced pulmonary inflammation and autophagy are attenuated in Ephx2-deficient mice. *Inflammation.* 2017;40(2):497–510.
78. Yoshida T, Mett I, Bhunia AK, et al. Rtp801, a suppressor of mTOR signaling, is an essential mediator of cigarette smoke-induced pulmonary injury and emphysema. *Nat Med.* 2010;16(7):767–773.
79. Farber HW, Miller DP, Poms AD, et al. Five-year outcomes of patients enrolled in the REVEAL registry. *Chest.* 2015;148(4):1043–1054.
80. Sakao S, Taraseviciene-Stewart L, Lee JD, Wood K, Cool CD, Voelkel NF. Initial apoptosis is followed by increased proliferation of apoptosis-resistant endothelial cells. *FASEB J.* 2005;19(9):1178–1180.
81. Tu L, Dewachter L, Gore B, et al. Autocrine fibroblast growth factor-2 signaling contributes to altered endothelial phenotype in pulmonary hypertension. *Am J Respir Cell Mol Biol.* 2011;45(2):311–322.
82. Izikki M, Guignabert C, Fadel E, et al. Endothelial-derived FGF2 contributes to the progression of pulmonary hypertension in humans and rodents. *J Clin Invest.* 2009;119(3):512–523.
83. Prins KW, Thenappan T, Weir EK, Kalra R, Pritzker M, Archer SL. Repurposing medications for treatment of pulmonary arterial hypertension: what's old is new again. *J Am Heart Assoc.* 2019;8(1), e011343.
84. Yan J, Chen R, Liu P, Gu Y. Docosahexaenoic acid inhibits development of hypoxic pulmonary hypertension: in vitro and in vivo studies. *Int J Cardiol.* 2013;168(4): 4111–4116.
85. Chen R, Zhong W, Shao C, et al. Docosahexaenoic acid inhibits monocrotaline-induced pulmonary hypertension via attenuating endoplasmic reticulum stress and inflammation. *Am J Physiol Lung Cell Mol Physiol.* 2018;314(2):L243–L255.
86. Yu H, Liu J, Dong Y, et al. Anti-hypoxic effect of dihydroartemisinin on pulmonary artery endothelial cells. *Biochem Biophys Res Commun.* 2018;506(4):840–846.
87. Dalvi P, Sharma H, Chinnappan M, et al. Enhanced autophagy in pulmonary endothelial cells on exposure to HIV-tat and morphine: role in HIV-related pulmonary arterial hypertension. *Autophagy.* 2016;12(12):2420–2438.
88. Marshall JD, Bazan I, Zhang Y, Fares WH, Lee PJ. Mitochondrial dysfunction and pulmonary hypertension: cause, effect, or both. *Am J Physiol Lung Cell Mol Physiol.* 2018;314(5):L782–L796.
89. Ryan J, Dasgupta A, Huston J, Chen KH, Archer SL. Mitochondrial dynamics in pulmonary arterial hypertension. *J Mol Med (Berl).* 2015;93(3):229–242.

90. Wu YC, Wang WT, Lee SS, et al. Glucagon-like peptide-1 receptor agonist attenuates autophagy to ameliorate pulmonary arterial hypertension through Drp1/NOX- and Atg-5/Atg-7/beclin-1/LC3beta pathways. *Int J Mol Sci*. 2019;20(14), e3435.
91. Long L, Yang X, Southwood M, et al. Chloroquine prevents progression of experimental pulmonary hypertension via inhibition of autophagy and lysosomal bone morphogenetic protein type II receptor degradation. *Circ Res*. 2013;112(8):1159–1170.
92. Spiekerkoetter E, Tian X, Cai J, et al. FK506 activates BMPR2, rescues endothelial dysfunction, and reverses pulmonary hypertension. *J Clin Invest*. 2013;123(8): 3600–3613.
93. Seyfarth HJ, Hammerschmidt S, Halank M, Neuhaus P, Wirtz HR. Everolimus in patients with severe pulmonary hypertension: a safety and efficacy pilot trial. *Pulm Circ*. 2013;3(3):632–638.
94. Mi S, Li Z, Yang HZ, et al. Blocking IL-17A promotes the resolution of pulmonary inflammation and fibrosis via TGF-beta1-dependent and -independent mechanisms. *J Immunol*. 2011;187(6):3003–3014.
95. Huang LS, Berdyshev EV, Tran JT, et al. Sphingosine-1-phosphate lyase is an endogenous suppressor of pulmonary fibrosis: role of S1P signalling and autophagy. *Thorax*. 2015;70(12):1138–1148.
96. Ghavami S, Yeganeh B, Zeki AA, et al. Autophagy and the unfolded protein response promote profibrotic effects of TGF-beta1 in human lung fibroblasts. *Am J Physiol Lung Cell Mol Physiol*. 2018;314(3):L493–L504.
97. Hill C, Li J, Liu D, et al. Autophagy inhibition-mediated epithelial-mesenchymal transition augments local myofibroblast differentiation in pulmonary fibrosis. *Cell Death Dis*. 2019;10(8):591.
98. Jara P, Calyeca J, Romero Y, et al. Matrix metalloproteinase (MMP)-19-deficient fibroblasts display a profibrotic phenotype. *Am J Physiol Lung Cell Mol Physiol*. 2015;308(6):L511–L522.
99. Paun A, Lemay AM, Tomko TG, Haston CK. Association analysis reveals genetic variation altering bleomycin-induced pulmonary fibrosis in mice. *Am J Respir Cell Mol Biol*. 2013;48(3):330–336.
100. Surolia R, Li FJ, Wang Z, et al. Vimentin intermediate filament assembly regulates fibroblast invasion in fibrogenic lung injury. *JCI Insight*. 2019;4(7), e123253.
101. Kesireddy VS, Chillappagari S, Ahuja S, et al. Susceptibility of microtubule-associated protein 1 light chain 3beta (MAP1LC3B/LC3B) knockout mice to lung injury and fibrosis. *FASEB J*. 2019;33(11):12392–12408.
102. Maciel M, Hernandez-Barrientos D, Herrera I, Selman M, Pardo A, Cabrera S. Impaired autophagic activity and ATG4B deficiency are associated with increased endoplasmic reticulum stress-induced lung injury. *Aging (Albany NY)*. 2018;10(8): 2098–2112.
103. Gui YS, Wang L, Tian X, et al. mTOR Overactivation and compromised autophagy in the pathogenesis of pulmonary fibrosis. *PLoS One*. 2015;10(9), e0138625.
104. Han R, Ji X, Rong R, et al. MiR-449a regulates autophagy to inhibit silica-induced pulmonary fibrosis through targeting Bcl2. *J Mol Med (Berl)*. 2016;94(11):1267–1279.
105. Guan C, Qiao S, Lv Q, et al. Orally administered berberine ameliorates bleomycin-induced pulmonary fibrosis in mice through promoting activation of PPAR-gamma and subsequent expression of HGF in colons. *Toxicol Appl Pharmacol*. 2018;343:1–15.

CHAPTER NINE

Autophagy in motor neuron diseases

Natalia Rodríguez-Muela[a,b,c,*]

[a]Deutsche Zentrum für Neurodegenerative Erkrankungen e.V, DZNE/German Center for Neurodegenerative Diseases, Dresden, Germany
[b]Technische Universität Dresden, CRTD/Center for Regenerative Therapies Dresden, Dresden, Germany
[c]Max Planck Institute for Molecular Cell Biology and Genetics, Dresden, Germany
*Corresponding author: e-mail address: natalia.rodriguez-muela@dzne.de

Contents

Abstract

Motor neuron diseases (MNDs) are a wide group of neurodegenerative disorders characterized by the degeneration of a specific neuronal type located in the central nervous system, the motor neuron (MN). There are two main types of MNs, spinal and cortical MNs and depending on the type of MND, one or both types are affected. Cortical MNs innervate spinal MNs and these control a variety of cellular targets, being skeletal muscle their main one which is also affected in MNDs. A correct functionality of autophagy is necessary for the survival of all cellular types and it is particularly crucial for neurons, given their postmitotic and highly specialized nature. Numerous studies have identified alterations of autophagy activity in multiple MNDs. The scientific community has been particularly prolific in reporting the role that autophagy plays in the most common adult MND, amyotrophic lateral sclerosis, although many studies have started to identify physiological and pathological functions of this catabolic system in other MNDs, such as spinal muscular atrophy and spinal and bulbar muscular atrophy. The degradation of selective cargo by autophagy and how this process is altered upon the presence of MND-causing mutations is currently also a matter of intense investigation, particularly

Progress in Molecular Biology and Translational Science, Volume 172
ISSN 1877-1173
https://doi.org/10.1016/bs.pmbts.2020.03.009

regarding the selective autophagic clearance of mitochondria. Thorough reviews on this field have been recently published. This chapter will cover the current knowledge on the functionality of autophagy and lysosomal homeostasis in the main MNDs and other autophagy-related topics in the MND field that have risen special interest in the research community.

1. Autophagy in amyotrophic lateral sclerosis (ALS)

ALS is a late onset, rapidly progressing neurodegenerative disease, characterized by the loss of upper (cortical) and lower (spinal) MNs followed by severe muscle denervation and wasting. Like in most neurodegenerative diseases, ALS presents a complex etiology where only about 10% of the cases are inherited and 90% sporadic. Today, more than 30 genes have been involved in the pathology, being mutations in the *C9ORF72* gene the most common among the familial cases. Despite more than 25 years have passed since the discovery of the first ALS-linked gene, *SOD1*, no consensus on the main toxicity of mutant SOD1 or any other mutant ALS-related proteins has arisen. Nonetheless, accumulation of ubiquitinated SOD1, FUS or TDP-43 aggregates when the respective genes are mutated (and aggregates mainly composed of the latter protein in most sporadic ALS patients), highlights protein aggregation as a main disease mechanism in ALS, which suggests a shared decline of autophagy activity, the catabolic system in charge of clearing these species.

Indeed, the belief that autophagy dysfunction exerts a causative role in the pathogenesis of ALS is supported by the fact that many of the ALS-linked familial genes, which in some cases have been also found mutated in sporadic cases, are genes involved in autophagy regulation and genes related to autophagy-associated pathways (Table 1). However, it remains unclear whether enhancing or reducing autophagy could constitute a potential therapeutic avenue, even in the monogenic forms of the disease caused by highly penetrant gene mutations.

1.1 Toxic gain of function and loss of function mutations in ALS-associated proteins and their impact on autophagy activity

Accumulation of evidence through years of research indicates that most ALS-linked mutations result in the pathology by both the loss of function of the wild-type protein and the toxic gain of function that the resulting

Table 1 Summary of ALS-linked proteins involved in autophagy.

ALS-linked protein	OMIM	Gene	Autophagy-related function (references)	Autophagy-related binding partners	References
SOD1	105400	Superoxide dismutase 1	• Aggresome formation • Mutants impair AP maturation/transport by disrupting dynein-dynactin complex • Mutants induce autophagy via undefined mechanisms	OPTN	1, 2, 45, 47–49, 58, 62
C9ORF72	105550	Chromosome 9 open reading frame 72	• Endosomal trafficking regulation • AP maturation • Lysosomal degradation of autophagy and lipid metabolism regulator CARM1 • Degradation of endocytosed material and lysosomal homeostasis maintenance • Positively regulates mTORC1 signaling	SQSTM1, UBQLN2, ULK1, SMCR8 HNRNPA1, HNRNPA2B1	7, 12–19, 22–24, 165, 167, 168

Continued

Table 1 Summary of ALS-linked proteins involved in autophagy.—cont'd

ALS-linked protein	OMIM	Gene	Autophagy-related function (references)	Autophagy-related binding partners	References
TDP-43	612069	TAR DNA-binding protein 43	• Positively regulates mTORC1 signaling; regulates AP-lysosome fusion • Endosome trafficking and maturation • Autophagy initiation • Loss of nuclear function caused by mutation results in reduced expression of autophagy proteins • Mutations associate with SGs	ATXN2, UBQLN2, SQSTM1, VCP, FUS, HNRNPA1, HNRNPA2B1	28–30, 32, 55, 65–67, 162, 173
FUS	608030	Fused in sarcoma/translated in liposarcoma	• AP maturation/AP-lysosome fusion • Mutations associate with SGs	TDP-43, SMN, VCP, SQSTM1	32, 34, 161, 176, 177, 187
TBK1	616439	TANK-binding kinase 1	• Activation of autophagy adaptors • Autophagy initiation via phosphorylation of SMCR8 (c9orf72 binding partner) and phagophore elongation • AP maturation via phosphorylation of microtubule-binding protein CEP170	OPTN, SQSTM1	35, 36, 62, 63

UBQLN2	300857	Ubiquilin-2	• Recruits APs to polyubiquitinated aggregates • Mutants hamper the endosomal pathway	TDP-43, HNRNPA1	39, 40, 44, 187
OPTN	613435	Optineurin	• Autophagy receptor • AP formation and maturation	TBK1, SQSTM1, SDO1	41, 42, 188
VCP	613954	Valosin-containing protein	• Mutants reduce mTOR activity aberrantly inducing autophagy • Mutants impair AP maturation and fusion • Mutants repress TDP-43-containing RNP complex disassembly	FUS, TDP-43, HNRNPA1	43, 154–160, 184
SQSTM1/p62	616437	Sequestosome-1	• Autophagy receptor • Inclusion body formation • Mutants inhibit cargo recognition	TBK1, TDP-43, OPTN, FUS, HNRNPA2B1	37, 68, 70, 71

mutated protein exerts. The most common toxic effect of these diseased associated proteins is the disruption of the proteostasis control. Multiple studies have reported alterations in several steps of the autophagy machinery induced by mutated ALS-linked proteins, starting with the studies that focused on the accumulation of mutant SOD1.[1] SOD1 is a Cu, Zn-superoxide dismutase located both in the cytoplasm and at the mitochondrial intermembrane space responsible for catalyzing the conversion of superoxide to hydrogen peroxide and dioxygen. How mutant SOD1 disrupts autophagy is just recently starting to be clarified. Using the transgenic SOD1-G93A ALS mouse model, Xie *et al.* demonstrated that the presence of mutated SOD1 was accompanied by lysosomal and endolysosomal deficits in spinal MNs and impaired autophagy degradation at early phases of the disease.[2] Their results showed undegraded autophagosomes (APs) containing damaged mitochondria that accumulated in the MN axons, incapable of being transported by the dynein-driven retrograde transport. Interestingly they could revert this transport defect by expressing snapin, a dynein-adaptor, which rescued the autophagy-lysosomal deficit and enhanced mitochondrial turnover leading to improved MN survival and ameliorated disease phenotype. These findings greatly contributed to the molecular understanding of the pathological function that mutant SOD1 exerts on AP/endolysosomal trafficking. *SOD1* was the first gene to be linked to familial ALS, however only 10% of all familial cases are caused by mutations in this gene. Much later *C9ORF72* was identified as the most common genetic cause of ALS,[3,4] being responsible for about 40% of the familial cases and, interestingly, its mutation can also be found in 8% of the sporadic patients. While hundreds of different mutations have been described for *SOD1*, the pathogenic mutation in *C9ORF72* constitutes a GGGGCC repeat expansion in the first intronic region of the gene (up to 1,000 GGGGCC repeats) that give rise to dipeptide repeat proteins produced by repeat-associated non-ATG (RAN) translation.[3,4] The molecular mechanism that governs the transcription and translation of these non-ATG repeats is currently under intense investigation,[5] however we know from Almeida *et al.*'s work in MNs derived from *C9ORF72* patient iPSCs that these RAN translation products lead to elevated p62 levels and increased sensitivity to cellular stress induced by autophagy inhibitors.[6] This was one of the first studies that identified autophagy defects in the presence of *C9ORF72* repeat expansions. In order to fully understand the physiopathology of *C9ORF72*-induced ALS it will be necessary to first elucidate all the functions that the protein is involved in. Despite there is still much to be unraveled, work from Julie Atkin's group significantly moved this field forward by revealing that wild-type C9ORF72 protein regulates endosomal trafficking by activating the Rab complex involved in endosomal transport (Rab1, Rab5, Rab7 and Rab11).[7] Rab proteins are essential for a large variety of vesicular trafficking events. They observed that *C9ORF72* knock-down resulted in defects in the cellular internalization of different proteins. These observations led them to suggest that C9ORF72 functions as a guanine exchange factor (GEF), as previously predicted by two different bioinformatics studies.[8,9] The fact that an altered LC3 II to LC3 I ratio was also displayed when C9ORF72 was deficient and that C9ORF72 colocalized with LC3 positive vesicles and migrated along neurites together with lysosomes represented the first indication that C9ORF72 also regulated autophagy. Another key finding of this study was the detection of C9ORF72 in extracellular vesicles, which may open a new path for monitoring autophagy activity by studying exosome-contained proteins.[7,10] After Atkin's findings several groups simultaneously reported multiple roles of C9ORF72 in regulating autophagy activity. C9ORF72, via its interaction with SMCR8, WDR41 and TBK1, controls the initiation of autophagy by regulating the Rab1a-dependent trafficking of ATG1/ULK1 complex to the phagophore which leads to AP formation.[11–14] Webster *et al.* further showed that *C9ORF72* knock-down in mouse primary neurons and *C9ORF72* ALS patient-derived neurons displayed accumulation of p62 puncta, TDP-43 positive inclusions and defective autophagy induction, while overexpressing *C9ORF72* activated autophagy leading to increased AP numbers in cell lines.[11] Besides this function in the early stages of autophagy pathway, several studies have also reported the involvement of C9ORF72 in the maturation of autophagosomes. Charlet-Berguerand's and Chen's groups discovered that the GEF function of the C9ORF72/SMCR8 complex also regulates RAB8a and RAB39b activity promoting autophagosome maturation and controlling autophagy flux.[12,15–17]

Charlet-Berguerand's results also pointed out that the neuronal degeneration induced by C9ORF72 deficiency was intensified upon combination with poly-glutamine expansion mutations in *Ataxin-2* gene, which have also been identified in familial ALS patients. Besides C9ORF72 participation in membrane movement and vesicle fusion events, C9ORF72 has also been involved in regulating mTORC1 signaling since it has been found localized to lysosomes upon amino acid deficiency and CRISPR/Cas9-mediated *C9orf72* KO MEFs present unresponsive mTORC1 to changes in amino acid availability and enlarged lysosomes.[18] This potential mTORC1 regulation by C9ORF72 correlates with the observed alterations in autophagy and lysosomal gene expression associated with changes in C9ORF72 levels.[12,13,19] One of the well-known effectors of the genetic regulation of autophagy and lysosomal activity is TFEB,[20,21] whose contribution to this C9ORF72-mTORC1-lysosome axis has already been proposed.[22] In this study, using MEF and HEK293T cells, the authors showed that loss of C9ORF72 led to decreased mTORC1 activity, which up-regulated TFEB function and resulted in enhanced autophagy flux. It is still to be determined how exactly C9ORF72 exerts this function.

Other studies have continued to deepen our understanding of the functions of wild-type C9ORF72 in relationship to autophagy. It has been recently shown that C9ORF72 is required for targeting the epigenetic regulator of autophagy and lipid metabolism CARM1 for lysosomal degradation.[23] In C9ORF72 absence, CARM1 levels are abnormally high leading to increased autophagy-mediated lipid degradation and enhanced *de novo* fatty acid synthesis, especially under nutrient starvation. Latter studies have confirmed C9ORF72 role in the degradation of endocytosed material and in the lysosomal homeostasis maintenance[16,24] exposing the possibility that altered immune responses and the accumulation of undegraded material often detected in C9ORF72 ALS patients could be underlined by impaired lysosomal degradation and possibly by altered lysosomal exocytosis.[16] It is evident from all these studies that the C9ORF72 regulation of autophagy is notably complex and it is not only limited to the regulation of the AP formation or degradation, but it also coordinates the endosomal and lysosomal systems. The multiple roles of C9ORF72 in autophagy regulation is nicely compiled in a recent review.[25] It is tempting to believe that other C9ORF72 functions in proteostasis regulation are still to be discovered and without them our understanding of C9ORF72-related pathology will not be complete and therefore the development of efficacious therapies will continue to be a major challenge.

C9ORF72 is not the only ALS-associated protein that has been linked to lysosomal homeostasis. In the recent years a new role for the multifunctional TDP-43 protein in controlling lysosomal biology has also been discovered. TDP-43 is a DNA/RNA binding protein originally described as a transcriptional repressor that also functions as a pre-mRNA splicing and translational regulator. Mainly located in the nucleus, it also shuffles to the cytoplasm where it was initially described to regulate mRNA stability, transport and local translation in neuronal projections.[26] In an ALS context, mutated TDP-43 (occurring in 1–5% of familial ALS patients) aberrantly accumulates in the cytoplasm depleting therefore the nucleus from its physiological nuclear functions. Cytoplasmic TDP-43 positive inclusions are pathological hallmarks of all ALS forms, sporadic or familial regardless of the gene mutated, except of *SOD1* mutations, where protein inclusions are mainly SOD1 positive.[27] Rare TDP-43 mutations cause ALS, however abnormal levels of the wild-type protein can alter the balance of nuclear/cytoplasmic TDP-43 and lead to disease. Overexpression of *TDP-43* results in cytoplasmic TDP-43 aggregates in all model systems studied, from yeast to mice models and human neurons. Interestingly, in the last years it has been shown that loss of TDP-43 impairs mTORC1 signaling resulting in an abnormally increased autophagy-lysosomal activity that is however dysfunctional. Xia *et al.* showed that TDP-43 binds to Raptor mRNA, a mTORC1 complex adaptor, stabilizing it,[28] therefore TDP-43 loss results in decreased Raptor levels, mTORC1 inactivation and this leads to TFEB nuclear translocation. Consequently, autophagy and lysosomal genes are highly expressed; however, and unexpectedly, the authors showed that the autophagosome-lysosome fusion was impaired due to the downregulation of dynactin 1, a mediator of this process, leading to the accumulation of APs and the overloading of the autophagy machinery. Treatment of TDP-43 mutant flies with rapamycin worsened the neurodegenerative phenotype whereas activating mTORC1 signaling by phosphatidic acid (PA) ameliorated it. Aberrant levels of cytoplasmic TDP-43 have been also directly linked to cytotoxicity induced by alterations in the endolysosomal pathway. Recent studies have shown that the endolysosomal system, independent of autophagy, contributes to the clearance of TDP-43 aggregates.[29,30] Indeed, using yeast, Leibiger *et al.* reported that the endolysosomal pathway is more relevant for the clearance of TDP-43 aggregates than autophagy. They reported that TDP-43 overexpression results in endosome/vacuole fragmentation, impaired vacuolar proteolytic function—possibly due to the deficient transport of vacuolar/lysosomal protease Pep4/Cathepsin D by the endosomal system– and autophagy overactivation, likely as a mechanism to compensate for the blockage of the endolysosomal system upon TDP-43 cytoplasmic accumulation. However, the TDP-43-induced autophagy the authors found in yeast seems to increase cytotoxicity since deleting *Atg1* reduced cell death.[30] Therefore, in agreement with the previous study both loss of TDP-43 and its abnormal accumulation in the cytoplasm when mutated, increase lysosomal and autophagy activities through possibly multiple mechanisms, that are however dysfunctional and therefore contribute to the degeneration. It is still unknown whether different TDP-43 species follow, and possibly impair, a preferable degradation pathway. Further studies are also necessary to unravel whether enhancing the activity of the endosomal pathway would overcome the autophagy impairment caused by aggregated TDP-43 and lead to neuronal survival.

Another ALS-related protein that has been recently reported to impair autophagy function is fused in sarcoma/translocated in liposarcoma (FUS/TLS) or FUS. FUS is another RNA-binding protein involved in transcriptional activation and for the last years it has been in the spotlight since it was found mutated in 1–5% of familial ALS patients. Most *FUS*-ALS causing mutations clustered in its nuclear-localization domain[31,32] and result in the appearance of FUS cytoplasmic inclusions. Therefore, mutated FUS toxicity is due to both its cytoplasmic accumulation and the disruption of its RNA processing functions caused by its nuclear loss. It has been shown that FUS regulates its own levels, and it has been suggested that mutations in non-coding 3′UTR region of *FUS* lead to its accumulation and this to neurodegeneration.[33] Don Cleveland's group recently generated a mouse ALS model by overexpressing wild-type *FUS* or ALS-linked mutant *FUS* broadly in CNS under the control of a mouse prion promoter and found that both scenarios led to motor neuron degeneration and markedly reduced lifespan.[34] The overexpression of either form of FUS (wild-type or mutated) in Neuro2A cells led to inhibition of autophagy flux and accumulation of p62.[34] However, the molecular mechanism through which the accumulation of cytoplasmic FUS or the presence of mutated FUS exerts this negative effect on autophagy is still unresolved.

Mutations in other more recently identified ALS-linked genes are also starting to be associated with autophagy dysregulation. *TBK1* mutations resulting in its haploinsufficiency causes ALS and has been recently reported to result in ubiquitin-positive cytoplasmic inclusions indicative of deficient autophagy.[35–37] Similar studies focusing on mutations in the ESCRT-III subunit CHMP2B,[38] ubiquilin-2 (UBQLN2),[39,40] OPTN[41,42] and VCP[43] are also starting to arise.[44] These findings have been extremely instructive for exposing the contribution of autophagy dysregulation induced by ALS-linked mutations to ALS pathology; however, in most cases we are still lacking the underlying molecular mechanisms. The upcoming studies will have to unravel the molecular players mediating such dysregulation in order to fully understand the contribution of autophagy to the disease biology and therefore to eventually develop efficacious therapies targeting this degradative pathway.

1.2 Cytoprotective role of autophagy in ALS

Historically conflicting data has been reported to tackle the role of autophagy in ALS onset and progression. A pioneer report pointing out an autophagy dysregulation in ALS was published by Morimoto *et al.*[45] Analyzing the spinal cord MNs of the mutated SOD1-G93A transgenic ALS mouse model they observed increased levels of lipidated LC3 and decreased levels of phosphorylated mTOR at Ser2448[45]—a site specifically phosphorylated via the PI3 kinase/Akt signaling pathway. These data suggested an increased autophagy activity in this ALS mouse model. Later studies have confirmed autophagy induction in the same mouse.[1,46] However, most of those studies were performed at a steady-state level analyzing the amounts of autophagy-related proteins in mouse spinal cords by using immunohistochemical, western blot or electron microscopy approaches. It was therefore plausible that the accumulation of LC3 II and APs in the spinal MNs of these ALS mice reflected a blockage of the autophagy machinery at the degradation level rather than an overactivated autophagy. One of the studies which brought up this possibility was published in 2011. The same authors that a few years earlier had proposed an autophagy impairment in the MNs of the SOD1-G93A mouse model, now showed that p62 was also increased in these mouse spinal cords suggesting that autophagy could be blocked and that the increased presence of undegraded APs that several groups had discovered was actually an indication of such blockage.[47] Further supporting this new theory, the authors found that inducing autophagy activity with rapamycin worsened the MNs degeneration, led to mitochondrial dysfunction, shortened the ALS mice life span and did not reduce the accumulation of SOD1 positive aggregates. The activity and functions that autophagy exerts in ALS are highly complex and therefore the observed results may lead to conflicting interpretations. Such complexity can be exemplified by the fact that that three years after this last study was published, the same group reported that whereas inducing mTOR-depending autophagy had a deleterious effect in the SOD1-G93A mouse model, inducing mTOR-independent autophagy with trehalose had completely opposite effects: reduced accumulation of ubiquitinated proteins and SOD1 and p62 positive aggregates, improved autophagic flux in the MNs, reduced skeletal muscle denervation and spinal cord MN loss and even extended mice life span.[48] This report validated the data published before by one of the pioneers in the study of autophagy activity in ALS pathology, Claudio Hetz. In 2009 he published that genetical upregulation of autophagy by deleting XBP-1—a transcription factor that regulates genes involved in the unfolded-protein-response (UPR)—ameliorated the neurodegeneration phenotype in a mutant SOD1-G86R ALS mouse model by reducing mutant SOD1 aggregates in the spinal cord and delaying the onset of the disease.[49] Hetz's group published in 2013 that induction of autophagy with trehalose ameliorated ALS disease phenotype in the same mutant SOD1 transgenic mice.[50] The exact molecular mechanism by which trehalose induces autophagy is still a mystery, and therefore it is still difficult to reconcile how and why two different forms of activating the same process, autophagy, could lead to drastically opposite results. A recent study by Poletti's group provided interesting insights in this regard. Using NSC34 cells they showed that trehalose causes lysosomal membrane permeabilization leading to phosphatase PPP3/calcineurin activation and dephosphorylation and subsequent nuclear translocation of TFEB.[51] These results indicate that at least one of the mechanisms by which trehalose induces autophagy is via TFEB-mediated upregulation of autophagic and lysosomal genes and proposed the intriguing hypothesis that mild lysosomal damage may induce cytoprotective autophagy in a neurodegenerative context. Despite in the recent years multiple groups have described cytoprotective effects of autophagy in a variety of cellular and animal ALS models,[52–57] it is still under debate whether autophagy plays a cytoprotective or detrimental role in ALS. What seems to be increasingly accepted is that the outcome will depend on the mutation and cellular context. It is conceivable that the contribution of autophagy to ALS pathophysiology will depend on the cell type—or even on subpopulations of the same cellular type, disease stage and the functionality of other autophagy-related cellular processes. The field is already realizing about the need for such systematic and specific exploration of autophagy activity in ALS. Maniatis' group recently reported that, using an SOD1-G93A ALS mouse model crossed with a conditional neuronal *Atg7* KO mouse, in early stages of the disease autophagy is essential to maintain neuromuscular junction functionality by degrading ubiquitinated aggregates. However, once the disease has progressed autophagy may exert a pathogenic role by activating a toxic response from glial cells leading to inflammation and MN degeneration. Indeed, autophagy deficient ALS SOD1-G93A mice survived longer than the mice that had intact *Atg7*.[58]

This dual role of autophagy in ALS has also been recently reported associated to the more recently identified ALS gene *TBK1*.[59,60] TBK1 is a serine/threonine protein kinase that belongs to the IKK-kinase family of kinases that are involved in innate immunity signaling pathways. TBK1 regulates cell proliferation, apoptosis, autophagy and mitophagy, specifically through the phosphorylation of autophagy adaptors such as p62, OPTN and NDP52.[35,61] Brenner *et al.* reported that *Tbk1* hemizygosis in SOD1-G93A ALS mice led to impaired autophagy in MNs and muscle denervation at early disease stages. However, at a late stage, *Tbk1* deficiency rescued microglial-induced neuroinflammation, slowed down disease progression and extended survival.[62] These data further indicate that autophagy may exert different, and even opposite, functions depending on the disease stage. An interesting report exposing possibly a new function of TBK1 in autophagy regulation was recently published by Boeckers' group. They showed that iPSC-derived MNs from ALS patients carrying *TBK1* mutations present immature phagophores that fail to properly elongate, accompanied by cytosolic p62 positive aggregates.[63] The authors suggest that, although inducing autophagy activity could constitute an evident approach to promote the clearance of those aggregates, it would not lead to the expected outcome in all ALS paradigms but instead exacerbate the autophagy overload when one of the phases of the autophagy machinery is not working properly. Therefore, understanding the precise molecular mechanism/s that go awry upon each specific disease mutation would be necessary to determine what the best approach is to modulate autophagy activity and preserve cellular proteostasis.

Both autophagy and the endolysosomal pathways have been involved in the clearance of TDP-43 aggregates.[54,64–67] Most studies published to date have proposed enhancing autophagy as a cytoprotective strategy to clear TDP-43 protein aggregates and promote neuronal survival. Using a mouse model for the study of TDP-43 proteinopathies Wang *et al.* reported that rapamycin treatment reduced TDP-43 cytoplasmic inclusions and neuronal loss and ameliorated the learning/memory deficits observed in these mice. Motor dysfunction was also improved with other autophagy inducers such as spermidine, carbamazepine and tamoxifen.[64] These results were validated by Barmada *et al.* inducing autophagy with fluphenazine (FPZ), methotrimeprazine (MTM) and 10-(4′-(N-diethylamino) butyl)-2-chlorophenoxazine (NCP). Additionally, using an image-based single-cell measurement approach on rat primary neurons and human iPSC-derived neurons harboring two different TDP-43 mutants, they showed improved clearance of mutant TDP-43 aggregates.[65]

As mentioned earlier, the endosomal and the autophagy pathways are connected by an intricate network. The involvement of the endocytic pathway to ALS pathology has gained attention in the last years. Liu *et al.* recently reported that TDP-43 colocalizes with endocytic proteins and that it is degraded by this pathway probably to a greater extent than by autophagy.[29] Using yeast and HEK293A cells they showed that blocking endocytosis increases TDP-43 aggregation and toxicity and it also worsened the disease phenotype in a TDP-43 ALS fly model. They even proposed that a dysfunction of endocytosis activity may play a causative role in the development of ALS. These results align with the findings by Leibiger *et al.* reporting that TDP-43 accumulation impairs endolysosomal function leading to an increase in its own levels and an enhanced autophagy activity that become deleterious.[30] Besides the TDP-43 pathological molecular mechanisms unraveled by this last study, we still know very little about the underlying toxicity mediated by this mutated and/or aberrantly accumulated protein. Nonetheless, new light is being constantly shed in this field. Recently, it was shown that mouse and fly models of TDP-43 deficiency display a downregulation of *Atg7*, p62 accumulation and motor deficits.[55] These studies suggest that many ALS-associated proteins exert a plethora of functions controlling lysosomal and autophagy activities, at transcriptional and posttranslational levels that are just beginning to be understood. Fig. 2 depicts the proteostasis-related processes involved in ALS pathology.

In 2007 Komatsu in Tanaka's group discovered the indispensable role of p62 on cytoplasmic inclusion formation when autophagy is deficient.[68] Mutations in the gene encoding this autophagy receptor have also been reported in familial ALS.[69,70] In agreement with this seminal study by Komatsu, recent research has shown that loss of p62 in mutant SOD1-H46R ALS mice exacerbates the disease phenotype. Conversely, overexpressing p62 in a double transgenic mouse SQSTM1;SOD1-H46R led to earlier disease onset and shorter lifespan with deposits of p62 and poly-ubiquitinated proteins in the mice spinal cord.[71] These results indicate that aberrantly high levels of p62 in a mutant SOD1 ALS context accelerate disease onset by compromising proteostasis, already altered by the presence of aggregates induced by the mutated p62.

Of note, most of the currently available autophagy modulators are remarkably unspecific, particularly the ones that induce autophagy by targeting mTOR, which entails that their off-targets effects may affect the disease pathology in an autophagy-independent manner. It is therefore of acute necessity to develop new efficient and specific compounds that selectively target key autophagy machinery regulators in order to fully understand the contribution of this catabolic pathway to the onset and progression of MN diseases and to prevent the appearance of studies reporting conflicting results.

In the last decade we have learnt a great deal about the functions that the lysosome-autophagy system displays in healthy neurons, including MNs, about how MND-causing mutations impact its functionality and about the multiple strategies neurons undertake to cope with the homeostasis challenges that these mutations unleash. However, a less explored yet intriguing aspect of autophagy contribution to ALS pathogenesis relates to the degeneration induced by non-cell autonomous factors. Spinal MNs constitute 5–10% of all cells in the spinal cord and neurons in general account just for about 50% of the cells in the brain. A much deeper focus has been addressed at understanding neuronal autophagy than on exploring how the abovementioned mutations—affecting ubiquitously expressed genes—impact autophagy in glial cells, and how these in return contribute to MND progression, and even onset, or how alterations in the biology of these cells can impact autophagy activity in the neighboring MNs, directly or indirectly through the secretion of certain factors. A detailed compilation of our current knowledge of glial autophagy in ALS can be found in Strohm and Behrends.[72]

2. Autophagy in spinal muscular atrophy (SMA)

Unlike ALS, SMA is a fully genetic, autosomal recessive, childhood onset neuromuscular disease and the leading genetic cause of infant mortality. Nonsense mutations or deletions of one single gene, *SMN1*, are responsible for almost the entirety of the cases, and as a result patients have low to very low levels of the *SMN1* coding protein SMN. SMN stands for Survival of Motor Neuron protein, which main functions are the assembly of the spliceosome machinery—required for mRNA splicing– and axonal mRNA transport, although other SMN functions are starting to be identified[73]: cytoskeleton dynamics,[74–76] endocytosis,[77–79] mitochondrial dysfunction,[80–82] ribosome biology[83] and autophagy.[84–86]

A genetic duplication phenomenon occurring during the evolutionary separation of human and chimpanzee lineages more than five million years ago resulted in the appearance of a second gene, called *SMN2*.[87] *SMN2* is identical to *SMN1* except for five point mutations, two of them being in coding regions and just one of them—a transcriptionally silent C-to-T transition at position 6 of exon 7—being largely responsible for the aberrant skipping of the exon 7 in the majority of the transcription events.[88,89] As a result, only a small fraction of *SMN2* transcripts retain exon 7, encoding therefore a full length, functional SMN protein.

The majority of the *SMN2* transcripts produce a truncated protein, called SMNdelta7, that is highly unstable and quickly degraded, although it retains some function.[90]

Unlike most neurodegenerative disease, SMA is not characterized by the presence of a mutated protein prone to aggregate and lead to cytotoxic undegradable material. Such is the case in Parkinson´s disease, where the presence of intracellular alpha-synuclein positive aggregates constitutes a hallmark of the pathology. Similarly, intracellular tangles of hyperphosphorylated Tau and both intracellular and extracellular deposits of beta-amyloid represent the pathological signature of Alzheimer´s disease. Intracellular aggregates of mutated huntingtin can be found in Huntington´s disease brains and, as described above, accumulation of ubiquitinated SOD1, FUS or TDP-43 cytoplasmic inclusions are present in most ALS patients. However, the presence of such misfolded or aberrant protein deposits in SMA-affected cells was not expected, and maybe for that reason the involvement of autophagy in SMA pathophysiology has been traditionally neglected and only recently had become a new research focus. The main discoveries regarding autophagy and endosomal system dysregulations in a SMN-deficiency context are illustrated in Fig. 1.

One of the first studies placing autophagy in the spotlight of SMA was published in 2011 by Soler´s group. They described an increased autophagy activity in *in vitro* mouse spinal cord MNs obtained from SMA mouse models, with higher amounts of APs and increased levels of lipidated LC3 and Beclin-1, compared to healthy control MNs.[91] They proposed an aberrant overactivation of autophagy as contributor to SMA pathology and its inhibition as a potential therapeutic target for the disease. However, these were steady-state observations that could very well be indicative of an increased autophagy activity but also of a blockage of the latest steps of the autophagic pathway. Further work from the group years later confirmed the enhanced numbers of APs and Beclin-1 levels in the spinal cord ventral horns of SMA mice *in vivo*, although immunostaining using specific MN markers allowing the identification of those alterations specifically in the affected MNs was not performed.[85,92] Their later experiments to analyze autophagy flux showed a reduction of SMN protein when inhibiting autophagy with bafilomycin-A1 and increased SMN when autophagy was induced with rapamycin. These results, though, would suggest the induction of autophagy as the preferred treatment to increase SMN protein levels, in opposition to the treatment proposed by their previous findings. They also showed that treating SMA mice with the PI3K inhibitor 3-MA

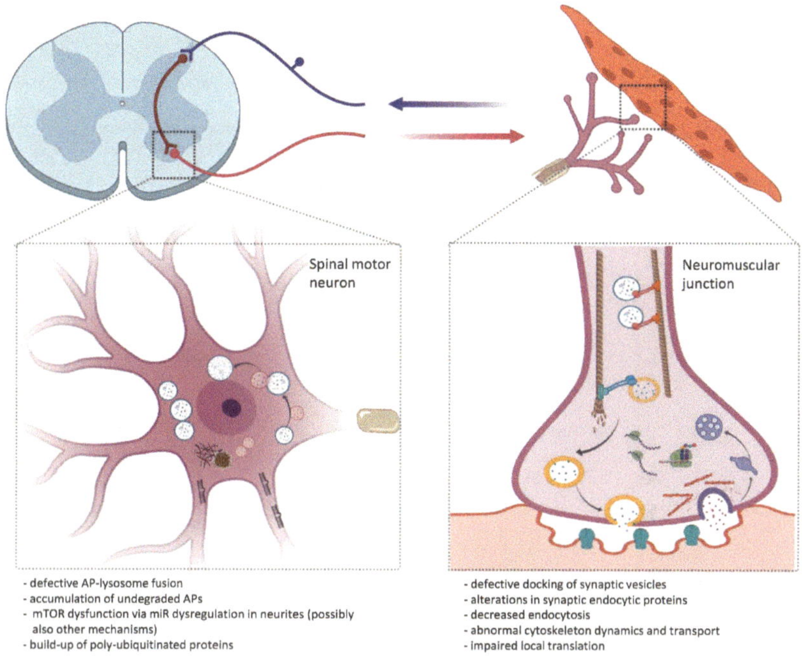

Fig. 1 Schematic representation of dysregulations in autophagy and endosomal pathways induced by SMN deficiency in the motor system of a SMA *in vivo* model. On the left, alterations of autophagy in the motor neurons of the ventral horn of the spinal cord are depicted. On the right, reported impairments of the endosomal system, vesicle trafficking and local translation in the axon terminal of a neuromuscular junction are indicated. (Figure created with BioRender).

delayed MN death and improved the disease phenotype, and therefore advocated that autophagy is involved in the MN death occurring in this SMA model.[92] These data appeared somehow counterintuitive given that if autophagy activation results in higher SMN protein levels and SMA cells have highly induced autophagy, SMA cells would not be expected to present a SMN deficiency. This controversy could be explained by the fact that autophagy is a very dynamic and complex process that may be playing a different role in different cells—even though carrying the same genetic mutation—or at different phases of the disease. Since the beginning of autophagy research there has been controversy about whether autophagy could constitute a mechanism of programmed cell death caused by excessive bulk degradation of cellular components and cytoplasmic vacuolization.[93] However, a high abundance of APs in a dying cell does not necessarily imply that autophagy is the cause of the death but maybe a cellular attempt to

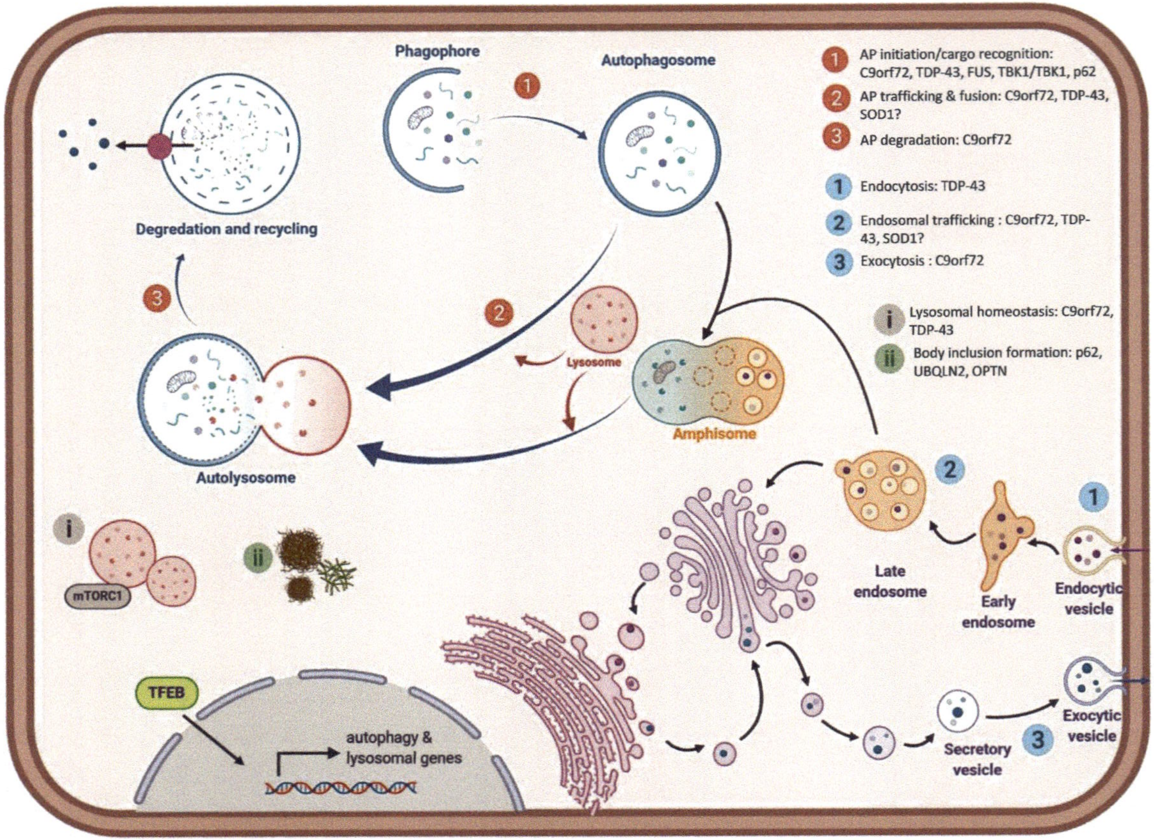

Fig. 2 Schematic representation of proteostasis-related processes implicated in ALS pathology. The different steps of autophagy and endo-lysosomal systems as well as TFEB-mediated lysosomal biogenesis and inclusion body formation where ALS-associated proteins have been reported to exert a modulatory effect are illustrated herein. (Figure created with BioRender).

prevent it.[94] Several examples of autophagy-dependent cell death—only involving autophagy machinery components and no other cell death pathway—have been reported in lower animal models for cellular clearance essential during development.[95–97] To date, the strongest evidence for a pro-death function of autophagy *in vivo* has been reported under hypoxia-ischemia and excitotoxicity conditions. In this case the authors, by deleting essential autophagy genes such as *Atg7* or *Beclin-1*, proved that the neuronal loss caused by the ischemic insult was reduced compared to control animals where autophagy was not impaired.[98,99] It would be very interesting therefore to validate the results obtained by Soler and Vercelli´s groups by genetically inhibiting autophagy in SMA animal models, or even better, by reducing APs numbers and basal autophagy to wild-type levels, to undoubtedly demonstrate whether the function of autophagy in SMA is pro-survival or pro-death as they propose. Another piece of evidence indicating autophagy impairment upon SMN reduction was provided by Androphy´s lab, when they showed attenuated autophagy flux in cell lines where SMN had been knocked-down.[84] However, why autophagy is dysfunctional in SMA, or what is the interplay between SMN protein and autophagy activity, were still completely ignored questions.

Trying to give answer to some of these queries, we have recently showed that the function that autophagy plays in SMA pathology seems to be dual. On the one hand we demonstrated that the key protein in SMA, SMN, is degraded by autophagy and not only by the proteasome, contrary to what was long-time accepted in the field.[86] Indeed, we found that not only the functional, full-length SMN but also the truncated form of the protein—SMNdelta7—and SMN binding partners essential for the assembly of the spliceosome machinery were targeted by the selective form of this degradative pathway. We demonstrated that the autophagy receptor p62 mediates such degradation. *p62* Knock-down led to marked increases in SMN protein in mouse and human MNs, which resulted in enhanced MN survival *in vitro* and *in vivo*, reduced muscle waste, restoration of the neuromuscular junction defects classically observed in SMA pathology and extended lifespan of SMA mouse and fly models. It is universally accepted that the most efficacious treatment for SMA patients is increasing the levels of functional SMN protein. Therefore, the discovery that selective autophagy regulates SMN degradation opened a new avenue for potential therapy development, which could be used in combination with other approaches aimed, for example, at correcting the defective *SMN2* splicing. On the second hand, we observed that autophagy is dysfunctional in SMA, confirming some of the results that other groups had shown before. Using multiple cell types—including mouse embryonic stem cell and iPSC-derived MNs and mouse spinal cords—we found that there is indeed an abnormal accumulation of APs and increased levels of polyubiquitinated proteins and p62 upon SMN deficiency, and an impaired autophagy flux measured with the mCherry-GFP-LC3 reporter and by LC3 lipidation upon inhibition of lysosomal activity.[86] These results indicated that the transport of APs to the lysosomes for degradation or the fusion with them is partially blocked when SMN protein is lacking. p62, present at aberrantly high levels in a SMA context could therefore bind SMN and sequester it inside APs that, even if there are not properly degraded because of the partial blockage in the latest steps of the autophagy machinery, would reduce the amount of functional SMN available in the cytoplasm and nucleus of the cells. This study suggested that restoring proper autophagy activity while preventing the selective target of SMN for autophagic degradation would solve two major issues in SMA: the altered autophagic function and the deficient levels of functional SMN. Further studies are needed to elucidate how and upon which circumstances p62 binds SMN, whether specific post-translational modifications of SMN are needed, if other autophagy receptors exert a similar control of SMN and the molecular underpinnings of SMN regulation of autophagy.

2.1 Endocytic pathway defects in SMA

Accumulation of evidence suggests that multiple steps of the endocytic pathway are essential for autophagy. Through endocytosis cells sense their environment and internalize nutrients and growth factors into early endosomes that turn into late endosomes or multivesicular bodies (MVBs). The autophagic and endocytic pathways intersect when endosomes or MVBs fuse with APs to form an amphisome, which subsequently fuses with a lysosome to form an autolysosome.[100] Recent studies are showing that biogenesis of APs needs membranes from endosomes[101,102] and that autophagy also controls the formation of early endosomes[103]; however, a detailed understanding of the molecular pathways regulating all interactions between both systems is still missing. Alterations in endocytosis have also been reported as part of SMA pathology, a research field pioneered by Anne Hart´s and Brunhilde Wirth´s groups. In 2016, Hart´s lab demonstrated that SMN deficiency in *C. elegans* causes defective docking of synaptic vesicles and alterations in synaptic endocytic proteins.[77] Soon after, Wirth´s lab, using ex vivo muscle explants from SMA mice after stimulation of their innervating MNs, found that SMN deficiency is accompanied by reduced levels of F-actin in the axon terminal, which leads to decreased endocytosis in the presynaptic terminal of NMJs.[78] They further showed that overexpression of the actin-binding protein Plastin3 (PLS3)—which they previously identified as a potent SMA disease modifier in humans and various animal models[104]—restores this endocytic defect. Increasing the levels of CORO1C—another actin-binding protein and PLS3 molecular partner—also elevated F-actin and rescued the axonal branching defects in a SMA zebrafish model. How the molecular machineries regulating autophagy and the endocytic pathway interact in a SMA context has not been explored yet.

Endocytic processes are not only essential for intracellular trafficking and degradation of molecules, but also for a healthy synaptic function, which has also been shown defective in SMA. Indeed, neuromuscular junction defects—such as accumulation of neurofilaments in the axon terminals, reduced endplate size, aberrant clustering of acetylcholine receptors in the muscle fibers among others—appear in SMA affected patients and in mouse models before the abnormalities in mRNA splicing.[105] In the last years it has become evident that protein quality control at the synapse is particularly important to maintain neuronal functionality.[106] Autophagic degradation of pre- and postsynaptic components has been reported as a main contributor to synaptic remodeling, which seems to be required to maintain synaptic function and plasticity during development and disease.[107] This autophagic function may be especially important for MNs since, given their long axons (the longest of all neuronal types), APs must travel large distances from the synapse to the neuronal soma where most of the lysosomes are located. An increasing number of studies are untangling the multiple roles that autophagy displays at the synapse. Thorough reviews on that matter are available[106,108]; however very little is known about autophagy regulation and functions at the NMJ. Alterations of components of the presynaptic autophagy machinery have not been described yet as part of SMA pathology. Nevertheless, since SMN mediates actin-dependent axonal transport of RNA and protein granules[109,110] required for synaptogenesis and neurite and axonal morphogenesis,[111] and SMN has also been involved in microtubule-dependent mitochondria trafficking,[81,112] it would not be unlikely that it also regulates AP transport at the synapse. Whether other molecular factors play a role in the axon terminals of MNs and/or at the postsynaptic level in the muscle fibers under basal conditions or upon the presence of MN causing mutations is still to be determined.

In the recent years a few studies in SMA mice have also linked mTOR activity dysregulation to SMA pathology. Two studies have suggested that mTOR is downregulated in SMA and that increasing its activity by depleting the mTOR negative regulator PTEN or by inhibiting the microRNA miR-183—both shown to be enhanced in the neurites of SMN-deficient MNs—can promote MN survival and enhance lifespan in SMA mice.[113–115] Many studies have been published on the crosstalk between autophagy and miRNA metabolism,[116–118] however little is known in the SMA context. It has been recently published that MNs from SMA mice display an enhanced autophagy function, induced by neuronal activity, which degrades DROSHA[119]—a critical enzyme for miRNA processing. These decreased DROSHA levels in SMA could therefore be one of the reasons behind the increased miR-183 in SMA MNs and downregulated mTOR function. However, another study reported that muscle biopsies from patients affected by a mild form of SMA, type III, showed over-expression of mTOR.[120] In agreement with the latest study, our results indicate an overactivation of mTOR upon SMN deficiency in all the different cell types studied. There is therefore disagreement in the field as to how SMN regulates mTOR signaling, however it is possible that the activity of mTOR in SMA differs between different tissues or even among cell populations within the same tissue. Further studies are needed to unravel the connection between different levels of SMN protein and mTOR signaling and autophagy activity, and whether various alterations in these two major homeostasis control networks in SMA are cell-type specific and even linked to distinct disease-stages.

3. Autophagy in spinal and bulbar muscular atrophy (SBMA)

There are at least nine neurodegenerative diseases caused by CAG expansions. These poly-glutamine diseases include Huntington's disease, several spinocerebellar ataxias and spinobulbar muscular atrophy or SBMA.[121] SBMA is an adult-onset, X-linked neuromuscular disorder caused by a CAG trinucleotide repeat in exon 1 of the androgen receptor (AR), which is activated by its ligand testosterone (or dihydrotestosterone).[122] Mutated AR misfolds and aggregates in the nucleus and cytoplasm of primarily MNs and muscle cells causing multiple disruptions in cellular function and leading to neurodegeneration.[123] One of the first studies exploring whether these cytosolic aggregates could be removed by autophagy was published in 2009 by Merry's lab. Using cultured MNs from a transgenic SBMA mouse model expressing the mutated AR, the authors showed that pharmacological activation of autophagy rescued MN death induced by the toxic AR aggregates, even when the mutated AR was modified to reside in the nucleus.[124] Other reports validating the cytoprotective role of autophagy in SBMA pathology have been published since then[125,126] although the mechanisms of action of those autophagy-activating molecules are still unclear. Similarly, the combinatorial treatment of the anti-androgen bicalutamide (that promotes cytoplasmic retention of polyQ-AR reducing therefore its nuclear translocation) with the autophagy inducer trehalose has been shown to reduce the polyQ-AR deposit formation in *in vitro* cultured MNs.[125] These studies evidenced that autophagy functions as a cytoprotective pathway to clear AR cytoplasmic aggregates. However, a detailed understanding of the underlying molecular mechanisms and whether a dysfunctional autophagy unable to perform such degradative role is part of the SBMA pathology was still to be investigated. As described earlier, p62 is not only an autophagy receptor but it also mediates inclusion formation, particularly when autophagy does not work properly.[68] In 2013, an elegant study showed that deleting p62 in a SBMA mouse model exacerbated the motor dysfunction by increasing the levels of monomeric mutant AR and mutant AR protein complexes as a result of autophagy impairment. On the contrary, p62 overexpression protected against mutant AR toxicity by inducing the formation of cytoprotective protein inclusions.[126] These results therefore not only supported the positive function of autophagy at clearing these aggregates, but also suggested that inclusion formation of mutant AR may represent a cellular attempt to cope with the presence of the aberrant receptor.

Chaperone-assisted selective autophagy, or CASA, is a less studied type of autophagy that has been strongly linked to SBMA pathophysiology. Around the year 2000, it was discovered that the BAG (Bcl-2-associated athanogene) protein family, HSC/HSP70 co-chaperones, are proteostasis modulators. The human BAG protein family contains six members, namely BAG1-6,[127,128] that interact with Bcl-2 to suppress apoptosis,[129] but they are also involved in other cellular processes, such as protein folding, stress response and autophagy. Except for BAG5 all members of the BAG proteins physically interact with HSP70 via their BAG domain, and all of them have an ubiquitin-like domain through which they bind ubiquitinated cargo to target it for degradation. The two BAG proteins more studied to date, and linked to neurodegenerative diseases, are BAG1 and BAG3. Robust work from Christian Behl's group showed that whereas BAG1 is involved in the removal of polyubiquitinated proteins by the proteasomal system, BAG3 is linked to the autophagy receptor p62 and involved in the turnover of polyubiquitinated proteins by autophagy, thus discovering a pathway for selective autophagy that is mediated by this multifunctional HSP70 co-chaperone.[130] They also found that under cellular stress, aging and neurodegenerative-prone conditions the ratio BAG3:BAG1 increases in neurons and mouse brain which enhances autophagy activity.[131] This chaperone-based system to degrade cellular components has been shown to be particularly important to maintain homeostasis in muscle cells, the targets of spinal MNs. Unlike chaperone-mediated autophagy (CMA), which is induced by stress, CASA functions under normal conditions. The involvement of BAG3 and its associated chaperons in the pathology of MNDs, particularly SBMA and ALS, and the crosstalk of this degradative network with autophagy has been extensively studied by Poletti's group. In 2015 his group showed upregulation of autophagic genes (*Beclin-1*, *Atg10*, *p62/SQSTM1*, *LC3*) and the gene encoding the BAG3 co-chaperone small heat shock protein *Hspb8* in the muscle of symptomatic SBMA mice. BAG3 and other HSPB8 interactors (HSPB2 and HSPB3) were also upregulated at the mRNA and protein level.[69] They had previously shown that HSPB8 was a key protein involved in the autophagic removal of not only polyQ-AR but also the mutated ALS-linked proteins SOD1 and TDP-43 and that its expression was highly increased upon blockage of the ubiquitin proteasome system to facilitate autophagy clearance of these pro-aggregation proteins.[46,132,133] They have further showed that the BAG3:BAG1 ratio was increased in SBMA muscle suggesting that the mutant polyQ-AR induces a potent autophagic response in muscle cells and that autophagy is the main degradation pathway for the misfolded AR over the proteasome.[132] They concluded that the levels of HSPB8-based protein quality control machinery could be used as muscle-specific biomarkers to assess SBMA progression and/or response to pharmacological treatments. Moreover, they have recently demonstrated that the activation of both pathways together, CASA and autophagy, through the overexpression of BAG3 and HSPB8 in combination with trehalose treatment leads to the complete clearance of polyQ-AR aggregates in skeletal muscle cells,[134] thus opening a promising combination approach to treat SBMA patients.

The chaperone network, in charge of controlling protein folding and turnover, comprises more than 180 chaperones and co-regulators. The levels of multiple of them are induced by stressful conditions, particularly proteotoxicity, and the regulation of chaperone HSPB8 has specifically linked to MND. Accumulation of misfolded proteins in ALS and SBMA MNs and muscle has been shown to induce HSPB8 transcription as an attempt by these cells to limit the formation of protein aggregates. Interestingly, HSPB8 is highly expressed in MNs (compared to other spinal cord cells) and skeletal muscle, its levels decrease with age[46] and mutations in HSPB8 cause Charcot-Marie-Tooth type 2L disease, hereditary distal motor neuropathy type II or distal myopathy.[135–137] HSPB8 also facilitates autophagy activity by binding BAG3, which functions as a scaffold protein binding dynein.[138] It has been proposed that the HSPB8-BAG3-HSP70 complex is transported via dynein through microtubules to the microtubule organization center (MTOC) where misfolded proteins are engulfed into nascent APs to be degraded via autophagy.[139,140] Great advance has been made in the characterization of the BAG1 and BAG3 protein complexes involved in the proteasomal and autophagic degradation of misfolded proteins, respectively.[138] Looking forward, a detailed understanding of the molecular mechanism that controls the *de novo* synthesis of BAG1, BAG3 and HSPB8 as well as the cues that control the crosstalk between BAG1-proteasome and BAG3-autophagy degradation upon proteostasis stress will be needed.

The vast majority of studies focused on the role of autophagy in SBMA pathology agree on its neuroprotective function as a mechanism to clear protein aggregates; however, some data has been also published suggesting that autophagy could exert a detrimental function at least when other proteostasis control mechanisms do not work properly. A study showed that skeletal muscle from SBMA patients or mouse models (AR113Q knock-in mice) displayed activated UPR mediated by CHOP and that CHOP deletion worsened the disease phenotype through the activation of autophagy, suggesting a deleterious function of autophagy in this case. This finding was validated in Beclin-1 haploinsufficient SBMA mice, where muscle wasting was decreased, and the lifespan extended compared to SBMA mice with normal Beclin-1 levels.[141] Additionally, the authors further showed that autophagy induction caused by CHOP deficiency worsened muscle atrophy and concluded that autophagy is aberrantly upregulated in SBMA skeletal muscle contributing to the pathology. However, the interplay among the UPR and autophagy ensuring protein homeostasis is intricate. It is possible that if the UPR is a natural response of SBMA skeletal muscle to cope with the polyQ-AR-induced proteostasis stress, the blockage of this response could also alter the normal autophagy functionality in the cells and aggravate the degeneration. In 2014, the same group described that TFEB nuclear localization and TFEB-target genes were upregulated in skeletal muscle from this SBMA mouse model and patient cells.[142] They further showed that these cells displayed an enhanced autophagy response upon stimulation compared to wild-type cells, while the nuclear localization of TFEB transcriptional antagonist ZKSCAN3 was reduced compared to healthy controls.[142] Interestingly, La Spada´s group found different results regarding TFEB's role in SBMA pathology. Using motor neuron-like cells and a different SBMA mouse model (YAC AR100), they showed that wild-type AR interacts with TFEB acting as a co-activator while mutated AR interferes with its transactivation. As a consequence, polyQ-AR reduces the turnover of long-term proteins and impairs autophagic flux.[143] They further proved that TFEB over-expression restored the autophagic flux defect in iPSC-derived MNs from SBMA patients, and suggested TFEB and compounds that favor autophagosome–lysosome fusion as a potential candidate targets for therapy development endeavors. These studies proposing opposite scenarios for the activity of TFEB and the lysosome-autophagy axis in SBMA pathogenesis, exposes that autophagy´s role in the pathology of SBMA is still debated. It is indeed possible that the autophagic response to the mutation in the AR differs between skeletal muscle and neuronal cells. It would be therefore interesting to unravel whether autophagy is differentially active and exerts distinct functions between these two cellular types and even within the same muscle type at different disease stages. In this regard, Maria Pennuto´s group recently reported a glycolytic-to-oxidative metabolic switch in SBMA affected muscle fibers during the early phases of the disease, accompanied by an enhancement of mTOR activity. However, at later stages, they found increased TFEB expression and autophagy and mTOR inactivation.[144] Similar observations in other neurodegenerative diseases are making evident that cellular type and disease stage dependency are a recurrent theme that needs to be taken in consideration when assessing autophagy contribution to a particular pathology.

4. Autophagy in other motor neuron diseases

ALS and SMA are notably the most studied MNDs for being the two most common types. As mentioned before, numerous studies have been published tackling different aspects of autophagy activity in ALS. Much fewer are available for SMA and almost none for other way less common MNDs, such as: Primary Lateral Sclerosis, Progressive Bulbar Palsy, Pseudobulbar Palsy or Progressive Muscular Atrophy.

Juvenile primary lateral sclerosis is a rare MND characterized by progressive weakness and spasticity of muscles in the arms, legs and face due to the loss of upper motor neurons. Most of the cases are caused by mutations in the *ALS2* gene, which codes for the ubiquitous protein ALSIN, highly abundant in MNs, which function is poorly understood. The mutations lead to truncated forms of the protein that are quickly degraded. As often the case, it is unclear why deficits in this protein damage particularly upper MNs. Possibly due to the lack of good animal models, which do not resemble the profound human motor deficits, and to the heterogeneity and complexity of the cerebral cortex, very little has been studied about autophagy involvement in this disease. We know from Özdinler´s group that the apical dendrites of corticospinal MNs of an *Alsin* KO mouse showed defects in mitochondria and Golgi apparatus, abnormal vacuolization and increased autophagy.[145] These data suggest that autophagy could be involved in the pathophysiology of the disease; however, much more research is needed to determine whether autophagy indeed plays a role or whether such alterations are just secondary phenomena to cellular processes unrelated to autophagy.

In 2017 Michael Sendtner´s group published a study that unraveled a mechanism through which autophagy regulates the degradation of synaptic vesicles in the axon terminals of MNs that could have an implication in multiple MNDs. They demonstrated that Plekhg5, a protein involved in the control of neuronal cell differentiation, acts as a guanine exchange factor (GEF) for Rab26.[146] It was previously shown that the small GTPase Rab26 is abundant in synaptic vesicles and that when in its GTP-bound form it promotes the recruitment of autophagy machinery to the synaptic vesicles by interacting with ATG16L.[147] Sendtner´ study demonstrated that Rab26 activation is triggered by Plekhg5, therefore unraveling Plekhg5´s role as regulator of synaptic vesicle degradation by autophagy.[146] Deletion of this gene in mice causes accumulation of undegraded synaptic vesicles in cultured MNs due to impaired autophagic degradation, axon terminal degeneration and MND phenotype. Interestingly, mutations in the *PLEKHG5* gene have been associated with several MNDs like distal spinal muscular atrophy type IV,[148] intermediate Charcot-Marie-Tooth disease[149,150] and ALS.[151] Mutations in *PLEKHG5* do not seem to impact global proteostasis in these MNs suggesting that its function is restricted to autophagic degradation of synaptic vesicles. It will be interesting to investigate whether other mechanisms of autophagic regulation of neurotransmission play also a major causative role in the neuromuscular degeneration that characterizes most MNDs.

5. Autophagy regulation of membraneless organelles in MNDs

In the last few years, liquid-liquid phase separation (LLPS) has emerged as a new field of research to understand subcellular organization and protein aggregation. LLPS is a phenomenon in which macromolecules (such as RNA and proteins) separate into a demixed liquid phase within a solvent. The proteins prone to undergo this process often contain low-complexity domains (LCD), which are regions that show a composition biased, as they contain fewer amino acid subtypes than an average protein composition while a subset of amino acids is overrepresented. These regions constitute domains called intrinsically disordered regions (IDRs) that lack secondary and tertiary structures. Upon various stimuli that are currently under intense investigation (signaling pathways, stress signals, disease mutations, etc.), these proteins can demix from the cytoplasm (solvent) and form membraneless organelles often called droplets, which are spherical structures whose constituents can diffuse within them and exchange with the cytoplasm.[152] These droplets or condensates assemble and disassemble under basal conditions and are believed to play key physiological cellular roles. Examples of these macromolecular complexes are P granules in *C. elegans* germline, nucleoli, the nucleopore complex, stress granules and Balbiani bodies. However, if their disassembly is defective, they can evolve into a less flexible state, or hydrogel, where proteins cross-link turning into aggregates, which are solid formations of proteins that fall out of solution. These aberrant condensates that turn into aggregates constitute hallmarks of most neurodegenerative diseases. This perception has nurtured the establishment of partnerships between biophysics and biochemistry-focused labs with neuroscience groups to understand the biology of such aggregates and their contribution to disease. Intracellular LLPS has therefore opened a new direction of research to understand the molecular etiology of pathological protein misfolding and misregulation in neurodegeneration.

Examples of disordered proteins phase separating into fibrils, plaques, tangles or inclusion bodies have been found in the majority of neurodegenerative diseases: Aβ and TAU in Alzheimer's disease, α-SYN in Parkinson's disease, Huntingtin in Huntington's disease, TDP-43, FUS, C9ORF72 or hnRNPA1 in ALS and AR in SBMA.[153] However, the identification of the events that lead to the formation of these protein aggregates and the mechanisms that can take over their clearance are very much unknowns that presently constitutes a blazing topic of investigation.

Stress granules (SGs) represent the first link between phase separated proteins and neurological disorders, particularly ALS. These structures are membraneless organelles composed of RNA-binding proteins (RBPs) and mRNA which form during cellular stress and are considered to exert a physiological function in the control of protein translation upon protein homeostasis loss. In 2013 Buchan *et al.*, by performing a genetic screen in yeast aimed at identifying the molecular players controlling the dynamics of these complexes, were the first to discover that SGs are targeted by autophagy in a process they termed granulophagy.[154] They showed that inhibiting autophagy in MEFS by *Atg7* or *Atg3* deletion, or upon wortmannin treatment, reduced SG clearance. Valosin-containing protein (VCP/p97)—which mutations have been linked to familial ALS and other neurological disorders- was firstly described as an adaptor that recognizes poly-ubiquitinated cargo and delivers it for proteasomal degradation[155,156]; however, growing evidence has demonstrated its additional role as mediator of selective autophagy.[154,157–160] In this study Buchan *et al.* further demonstrated that mutations in VCP/p97 prevent its normal function of promoting SG degradation by autophagy. This seminal work opened a new direction of research linking pathogenic ribonucleoprotein granules to disease, but also laid the foundation to study the potential key role that autophagy might play in the clearance of these protein complexes. In 2015 another breakthrough study showed that the ALS-linked protein FUS forms liquid compartments in the nucleus and the cytoplasm that turn into aggregates with time and upon disease mutations.[161] They were the first to link mutations in ALS-related proteins to aberrant phase transitions within membraneless organelles. Since the advent of these two major studies, there is increasing evidence supporting that aberrant RNP granules play a central role in the pathology of ALS. As such, TIA-1 is a key component of SGs and it has been found in pathological inclusions of post-mortem brain tissue of ALS patients carrying *TDP-43*, *FUS* or *hnRNPA1* mutations.[162] These may indicate that mutated or misfolded proteins accumulate in SGs impairing their cytoplasmic exchange of components and ultimately their dynamics. What are the mechanisms that control SG assembly and disassembly, what events alter these two processes—toxic stimuli, physiological aging and/or disease-causing mutations, and what are the molecular pathways that mediate their degradation once their LLPS properties are lost are intensively investigated questions nowadays.

Recent findings have started to shed some light into the molecular mechanisms of autophagic control of SG clearance in an ALS context. Several studies had previously reported a dysregulation on SG biology upon the presence of ALS-linked *C9ORF72* repeat expansions leading to the accumulation of their translated dipeptide repeat proteins (DRPs) and/or decreased levels of C9ORF72,[163–167] however the underlying mechanisms were still ignored. Chitiprolu *et al.* showed that C9orf72 complexes with p62 to mediate the degradation of SGs by selective autophagy.[168] Using a variety of cell types and mouse *in vivo* approaches, they showed that ubiquitination is not the signal that drives p62 recruitment to SG, but the presence of proteins within the SG that had been symmetrically methylated in arginines by the symmetric dimethylation enzyme PRMT5, being one of these proteins FUS.[169] Such interaction between p62 and SG-methylated proteins requires the binding of p62 with the Tudor domain of SMN, which directly binds PRMT5 and other protein arginine methyltransferases. In agreement with previous findings they showed that post-mortem brain regions of ALS patients carrying *C9ORF72* repeat expansions present accumulation of inclusion bodies positive for p62 and symmetric arginine dimethylated proteins, suggesting that the degradation of aberrant SGs is impaired due to the lack of C9ORF72. Multiple ALS associated proteins, such as HNRPA1 and FUS bear these arginine dimethylation sites which coat SGs; it is therefore possible that their mutations disrupt p62 ability to recognize them and mediate the degradation of these multiprotein complexes by autophagy. It would be also interesting to elucidate whether C9ORF72 could exert a similar function on SG autophagic control independent of p62, and whether p62 regulates the clearance of SG via a non-canonical autophagy. It is also still to determine whether autophagy engulfs entire SGs or just part of them.

A new pathway for mTORC1 regulation has been identified lately in relationship with LLPS. ATAXIN-2, a RNA binding protein involved in regulating mRNA translation and SG formation and which mutations result in neurodegenerative diseases such as ALS and spinocerebellar ataxia type, has been recently identified as a regulator of TORC1 signaling in yeast.[170] ATAXIN-2 had been previously linked to C9ORF72 deficiency and motor neuron dysfunction.[15] Yang *et al.'s* study has found that in conditions of respiratory growth and thank to its methionine-rich and low-complexity domains, ATAXIN-2 phase-separates into condensates required to inhibit mTORC1 and induce autophagy when cells rely on mitochondria for energy production.

Recent studies have validated the degradation of SGs and other cytoplasmic RNPs by autophagy,[171–175] contributing to restoring RNP homeostasis and partially restoring motor function in fly models of ALS.[176,177] However, most of the conclusions raised in these studies are based on the analysis of the number, size and composition of these granules upon pharmacological and/or genetic modulation of autophagy, but still lack molecular insights into how autophagy exerts such a role. There is also some level of controversy in this field. Ganassi *et al.* recently reported that only a small fraction of aberrant SGs is targeted by autophagy, and instead suggested that is the HSPB8-BAG3-HSP70 chaperone complex the one controlling SGs composition and mediating their disassembly. The authors coined granulostasis as a new term to describe the SG surveillance system mediated by this chaperone-based axis which main role is preventing SG conversion into the aberrant state.[178] Their autophagy studies were restricted to blocking lysosomal activity with NH4Cl or E64D plus leupeptin in HeLa cells treated with MG132—a general proteasomal inhibitor—to induce SG formation. It has been recently shown that 20% of SG diversity is stress or cell-type dependent and that neuronal SG are particularly complex in their composition.[174] It would be very interesting to explore whether this chaperone complex plays a similar predominant role in SG dynamics in comparison to genetic inhibition of autophagy machinery in human neurons, under basal conditions and when carrying ALS-causing mutations. Interestingly, this is not the only study suggesting that autophagy plays a minor role in the degradation of SGs. Other investigations have reported that chaperones of the HSP70, HSP110 and HSP100 families mediate the disaggregation of SGs and aggregates.[179–182]

Notably, several proteomic analyses have been lately performed to identify the interactome of SGs.[174,183–185] Through them, alternative molecular signaling pathways of SG disassembly regulated by autophagy have been recently proposed. ULK1/2 are two similar and highly redundant isoforms that form part of the ULK-1 complex together with FIP200, ATG13 and ATG101 which is involved in the initiation of the phagophore formation. ULK1 directly phosphorylates Beclin-1 and activates the PI3K type III VPS34 complex to promote autophagy induction.[186] Recent findings have identified a new function of ULK1/2 independent of autophagy. Kundu's group has discovered that ULK1/2 localize to SGs and phosphorylate the ATPase VCP increasing its activity at mediating SG disassembly by promoting the extraction of ubiquitinated proteins from the granule.[184] The authors further showed that mice lacking *Ulk1/2* develop a proteinopathy with muscle inclusions positive for ubiquitin and the SG markers TIA-1 and TDP-43. This new function of ULK1/2 in regulating SG dynamics through its binding and activation of VCP seems to be stress-dependent since under basal conditions the brains of *Ulk1/2* deficient mice do not present ubiquitin or TDP-43 intracellular deposits, and also cell-type specific since the authors found ULK1/2 localized to SGs in HeLa cells and MEFs but not in U2OS cells.[184] It was also surprising that given the well-studied role of ULK1/2 and VCP in autophagy and the growing evidences supporting clearance of SGs by autophagy, that pharmacological or genetic inhibition of this process by RNAi-mediated downregulation of key canonical genes (*Atg7*, *Atg13*, *Atg14*) did not alter TDP-43 and TIA-1 positive SGs, suggesting that in their system most of these structures are not degraded by autophagy. Instead, the authors proposed that the delayed or impaired clearance of SGs that they observed in *Atg7*-/- MEFs may be due to alterations in the composition or dynamic properties of the SGs caused by a prolonged autophagy defect. Indeed, the type of TDP-43 positive pathology observed in ALS patients and patients affected by related proteinopathy diseases (like frontotemporal dementia or Inclusion body myositis), as well as in *Ulk1/2* KO mice is not present in *Atg7* neuronal-conditional KO mice. Interestingly, the autophagy-related proteins that so far have been found mutated in ALS are related to ubiquitin signaling (like UBQLN2, SQSTM1 and OPTN), rather than being core autophagy proteins.[37,39,187,188] Also, they undergo phase transitions and the latest findings indicate that they can shuttle post-translationally modified proteins into SGs. These and other studies are therefore hinting that the role of autophagy in SG dynamics is complex and both canonical autophagy and autophagy-related proteins through autophagy-independent functions—likely in close partnership with chaperone-based quality control systems—regulate the biology of these macromolecular complexes. Hence, it could be postulated that the simultaneous dysregulation of both catalytic pathways would lead to the development of ALS and other proteinopathies. Fig. 3 summarizes the current knowledge on the interaction between autophagy-related proteins and SG biology. Understanding the cell intrinsic and extrinsic stimuli that impel the formation of these solid aggregates and untangling the intricate network of events that participate in their elimination are challenges that promise to provide new directions for treating MNDs.

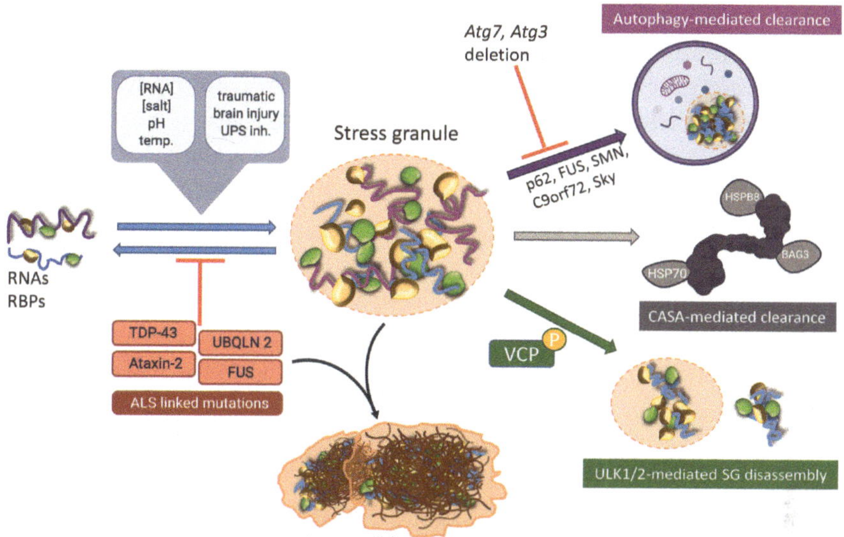

Fig. 3 Schematic representation of the interrelations among functional RNAs-RNPs associations, SG assembly, clearance and formation of permanent aggregates upon presence of ALS-causing mutations. A variety of stimuli drives the formation of SGs, which can dynamically exchange their components with the cytoplasm and disassemble. SGs may also be targeted for autophagic or CASA-mediated degradation or their ubiquitinated constituents extracted via the activity of ULK1/2-VCP in an autophagy independent process. (Figure created with BioRender).

6. Concluding remarks and future directions

Autophagy is a fundamental mechanism to ensure cellular homeostasis and is especially relevant to maintain neuronal health. The scarce number of APs that is normally observed in neurons was initially interpreted as a low dependence on this catabolic pathway by neuronal cells. However, the observation of large accumulations of APs upon lysosomal activity blockage reported soon after explained that basal autophagy is particularly efficient in neurons and hence the high AP turnover rate. Autophagy machinery is involved in regulating multiple key neuronal processes such as energy metabolism, metabolic switch upon differentiation, vesicle trafficking at the synapse, cytoskeleton remodeling and protein quality control. It is widely accepted that autophagy activity declines with age. Given the massive plethora of essential processes autophagy is involved in it is expected that a decrease in its performance would unleash a cascade of harmful events that would increase the risk of appearance of age-related neurodegenerative diseases, such as ALS. Interestingly, several familial ALS-linked genes code for proteins that in their wild-type form directly interact or modulate different aspects of the autophagy pathway or autophagy-related processes (p62, OPTN, TBK1, C9orf72, UBQLN2, TDP-43, SOD1 among others). It is not so evident though that autophagy could play a causative role in the pathology of early-onset motor neuron diseases, like SMA. It is intriguing that despite ALS, SMA, SBMA and the rest of rare MNDs are genetically heterogeneous diseases and etiologically very distinct from each other, the spinal motor neuron is the selectively affected neuronal type in all of them. The marked polarization and extensive axon length of MNs are special features that could make them particularly vulnerable to the malfunctioning of one of the autophagy-regulated processes. Still, like for the other neurodegenerative diseases, why MNs selectively degenerate in MNDs still represents a mystery that surely will engender exciting research in the future.

One of the blooming fields of neurodegeneration research is the study of the crosstalk among exosomes and autophagy. Exosomes are extracellular vesicles of endosomal origin and their biogenesis shares molecular machinery with autophagy. The signaling competences of their cargo as well as their biomarker potential make these extracellular vesicles especially interesting. How exosomes and autophagy communicate to maintain neuronal homeostasis upon basal or stress conditions by releasing (bulk/selective, beneficial/toxic) cargo or degrading cellular components constitutes an intriguing question with a broad impact in the understanding of neurodegenerative diseases biology. The discovery of the contribution of membraneless organelle homeostasis loss to the biology of neurodegenerative diseases has aroused a great degree of interest in the recent years. Particularly in the MND field, whether and how ALS-linked mutations in intrinsically disordered proteins drive the transformation of stress granules into insoluble structures and how this aggregation propagates from the primarily affected neurons into other cells types and brain regions are intensively investigated and fascinating questions. The identification of the protein states, cellular and environmental cues that exert a major causative role in hampering the dynamics of these membraneless organelles hold the potential to provide invaluable knowledge to understand the origin of MN degeneration and therefore finally develop efficacious treatments.

Acknowledgments

I apologize to those colleagues whose work could not be referenced due to space limitations. I would like to thank Tobias Grass and Patricia Vázquez for helpful comments on the manuscript. Work in the author's laboratory is funded by grants from the European

Research Council (ERC-802182-Neurovulnerability), the German Center for Neurodegenerative Diseases, the TU Dresden and the Max Planck Institute for Molecular Cell Biology and Genetics.

References

1. Li L, Zhang X, Le W. Altered macroautophagy in the spinal cord of SOD1 mutant mice. *Autophagy*. 2008;4:290–293. https://doi.org/10.4161/auto.5524.
2. Xie Y, Zhou B, Lin MY, et al. Endolysosomal deficits augment mitochondria pathology in spinal motor neurons of asymptomatic fALS mice. *Neuron*. 2015;87:355–370. https://doi.org/10.1016/j.neuron.2015.06.026.
3. DeJesus-Hernandez M, Mackenzie IR, Boeve BF, et al. Expanded GGGGCC hexanucleotide repeat in noncoding region of C9ORF72 causes chromosome 9p-linked FTD and ALS. *Neuron*. 2011;72:245–256. https://doi.org/10.1016/j.neuron.2011.09.011.
4. Renton AE, Majounie E, Waite A, et al. A hexanucleotide repeat expansion in C9ORF72 is the cause of chromosome 9p21-linked ALS-FTD. *Neuron*. 2011;72: 257–268. https://doi.org/10.1016/j.neuron.2011.09.010.
5. Balendra R, Isaacs AM. C9orf72-mediated ALS and FTD: multiple pathways to disease. *Nat Rev Neurol*. 2018;14:544–558. https://doi.org/10.1038/s41582-018-0047-2.
6. Almeida S, Gascon E, Tran H, et al. Modeling key pathological features of frontotemporal dementia with C9ORF72 repeat expansion in iPSC-derived human neurons. *Acta Neuropathol*. 2013;126:385–399. https://doi.org/10.1007/s00401-013-1149-y.
7. Farg MA, Sundaramoorthy V, Sultana JM, et al. C9ORF72, implicated in amytrophic lateral sclerosis and frontotemporal dementia, regulates endosomal trafficking. *Hum Mol Genet*. 2014;23:3579–3595. https://doi.org/10.1093/hmg/ddu068.
8. Levine TP, Daniels RD, Gatta AT, Wong LH, Hayes MJ. The product of C9orf72, a gene strongly implicated in neurodegeneration, is structurally related to DENN Rab-GEFs. *Bioinformatics*. 2013;29:499–503. https://doi.org/10.1093/bioinformatics/bts725.
9. Zhang D, Iyer LM, He F, Aravind L. Discovery of novel DENN proteins: implications for the evolution of eukaryotic intracellular membrane structures and human disease. *Front Genet*. 2012;3:283. https://doi.org/10.3389/fgene.2012.00283.
10. Xu J, Camfield R, Gorski SM. The interplay between exosomes and autophagy—partners in crime. *J Cell Sci*. 2018;131. https://doi.org/10.1242/jcs.215210.
11. Webster CP, Smith EF, Bauer CS, et al. The C9orf72 protein interacts with Rab1a and the ULK1 complex to regulate initiation of autophagy. *EMBO J*. 2016;35:1656–1676. https://doi.org/10.15252/embj.201694401.
12. Yang M, Liang C, Swaminathan K, et al. A C9ORF72/SMCR8-containing complex regulates ULK1 and plays a dual role in autophagy. *Sci Adv*. 2016;2: e1601167. https://doi.org/10.1126/sciadv.1601167.
13. Jung J, Nayak A, Schaeffer V, et al. Multiplex image-based autophagy RNAi screening identifies SMCR8 as ULK1 kinase activity and gene expression regulator. *Elife*. 2017;6. https://doi.org/10.7554/eLife.23063.
14. Sullivan PM, Zhou X, Robins AM, et al. The ALS/FTLD associated protein C9orf72 associates with SMCR8 and WDR41 to regulate the autophagy-lysosome pathway. *Acta Neuropathol Commun*. 2016;4:51. https://doi.org/10.1186/s40478-016-0324-5.
15. Sellier C, Campanari ML, Julie Corbier C, et al. Loss of C9ORF72 impairs autophagy and synergizes with polyQ Ataxin-2 to induce motor neuron dysfunction and cell death. *EMBO J*. 2016;35:1276–1297. https://doi.org/10.15252/embj.201593350.
16. Zhang Y, Burberry A, Wang JY, et al. The C9orf72-interacting protein Smcr8 is a negative regulator of autoimmunity and lysosomal exocytosis. *Genes Dev*. 2018;32: 929–943. https://doi.org/10.1101/gad.313932.118.

17. Behrends C, Sowa ME, Gygi SP, Harper JW. Network organization of the human autophagy system. *Nature*. 2010;466:68–76. https://doi.org/10.1038/nature09204.
18. Amick J, Roczniak-Ferguson A, Ferguson SM. C9orf72 binds SMCR8, localizes to lysosomes, and regulates mTORC1 signaling. *Mol Biol Cell*. 2016;27:3040–3051. https://doi.org/10.1091/mbc.E16-01-0003.
19. Lan Y, Sullivan PM, Hu F. SMCR8 negatively regulates AKT and MTORC1 signaling to modulate lysosome biogenesis and tissue homeostasis. *Autophagy*. 2019;15: 871–885. https://doi.org/10.1080/15548627.2019.1569914.
20. Settembre C, Di Malta C, Polito VA, et al. TFEB links autophagy to lysosomal biogenesis. *Science*. 2011;332:1429–1433. https://doi.org/10.1126/science.1204592.
21. Settembre C, Zoncu R, Medina DL, et al. A lysosome-to-nucleus signalling mechanism senses and regulates the lysosome via mTOR and TFEB. *EMBO J*. 2012;31: 1095–1108. https://doi.org/10.1038/emboj.2012.32.
22. Ugolino J, Ji YJ, Conchina K, et al. Loss of C9orf72 enhances autophagic activity via deregulated mTOR and TFEB signaling. *PLoS Genet*. 2016;12 e1006443. https://doi.org/10.1371/journal.pgen.1006443.
23. Liu Y, Wang T, Ji YJ, et al. A C9orf72-CARM1 axis regulates lipid metabolism under glucose starvation-induced nutrient stress. *Genes Dev*. 2018;32:1380–1397. https://doi.org/10.1101/gad.315564.118.
24. Corrionero A, Horvitz HR. A C9orf72 ALS/FTD ortholog acts in endolysosomal degradation and lysosomal homeostasis. *Curr Biol*. 2018;28:1522–1535. e1525. https://doi.org/10.1016/j.cub.2018.03.063.
25. Nassif M, Woehlbier U, Manque PA. The enigmatic role of C9ORF72 in autophagy. *Front Neurosci*. 2017;11:442. https://doi.org/10.3389/fnins.2017.00442.
26. Hergesheimer RC, Chami AA, de Assis DR, et al. The debated toxic role of aggregated TDP-43 in amyotrophic lateral sclerosis: a resolution in sight? *Brain*. 2019;142: 1176–1194. https://doi.org/10.1093/brain/awz078.
27. van Es MA, Hardiman O, Chio A, et al. Amyotrophic lateral sclerosis. *Lancet*. 2017;390:2084–2098. https://doi.org/10.1016/S0140-6736(17)31287-4.
28. Xia Q, Wang H, Hao Z, et al. TDP-43 loss of function increases TFEB activity and blocks autophagosome-lysosome fusion. *EMBO J*. 2016;35:121–142. https://doi.org/10.15252/embj.201591998.
29. Liu G, Coyne AN, Pei F, et al. Endocytosis regulates TDP-43 toxicity and turnover. *Nat Commun*. 2017;8:2092. https://doi.org/10.1038/s41467-017-02017-x.
30. Leibiger C, Deisel J, Aufschnaiter A, et al. TDP-43 controls lysosomal pathways thereby determining its own clearance and cytotoxicity. *Hum Mol Genet*. 2018;27: 1593–1607. https://doi.org/10.1093/hmg/ddy066.
31. Dormann D, Rodde R, Edbauer D, et al. ALS-associated fused in sarcoma (FUS) mutations disrupt Transportin-mediated nuclear import. *EMBO J*. 2010;29:2841–2857. https://doi.org/10.1038/emboj.2010.143.
32. Lattante S, Rouleau GA, Kabashi E. TARDBP and FUS mutations associated with amyotrophic lateral sclerosis: summary and update. *Hum Mutat*. 2013;34:812–826. https://doi.org/10.1002/humu.22319.
33. Sabatelli M, Moncada A, Conte A, et al. Mutations in the 3' untranslated region of FUS causing FUS overexpression are associated with amyotrophic lateral sclerosis. *Hum Mol Genet*. 2013;22:4748–4755. https://doi.org/10.1093/hmg/ddt328.
34. Ling SC, Dastidar SG, Tokunaga S, et al. Overriding FUS autoregulation in mice triggers gain-of-toxic dysfunctions in RNA metabolism and autophagy-lysosome axis. *Elife*. 2019;8https://doi.org/10.7554/eLife.40811.
35. Oakes JA, Davies MC, Collins MO. TBK1: a new player in ALS linking autophagy and neuroinflammation. *Mol Brain*. 2017;10:5. https://doi.org/10.1186/s13041-017-0287-x.

36. Duan W, Guo M, Yi L, et al. Deletion of Tbk1 disrupts autophagy and reproduces behavioral and locomotor symptoms of FTD-ALS in mice. *Aging (Albany NY)*. 2019;11:2457–2476. https://doi.org/10.18632/aging.101936.
37. Deng Z, Lim J, Wang Q, et al. ALS-FTLD-linked mutations of SQSTM1/p62 disrupt selective autophagy and NFE2L2/NRF2 anti-oxidative stress pathway. *Autophagy*. 2019;1–15. https://doi.org/10.1080/15548627.2019.1644076.
38. Filimonenko M, Stuffers S, Raiborg C, et al. Functional multivesicular bodies are required for autophagic clearance of protein aggregates associated with neurodegenerative disease. *J Cell Biol*. 2007;179:485–500. https://doi.org/10.1083/jcb.200702115.
39. Renaud L, Picher-Martel V, Codron P, Julien JP. Key role of UBQLN2 in pathogenesis of amyotrophic lateral sclerosis and frontotemporal dementia. *Acta Neuropathol Commun*. 2019;7:103. https://doi.org/10.1186/s40478-019-0758-7.
40. Senturk M, Lin G, Zuo Z, et al. Ubiquilins regulate autophagic flux through mTOR signalling and lysosomal acidification. *Nat Cell Biol*. 2019;21:384–396. https://doi.org/10.1038/s41556-019-0281-x.
41. Maruyama H, Morino H, Ito H, et al. Mutations of optineurin in amyotrophic lateral sclerosis. *Nature*. 2010;465:223–226. https://doi.org/10.1038/nature08971.
42. Wong YC, Holzbaur EL. Optineurin is an autophagy receptor for damaged mitochondria in parkin-mediated mitophagy that is disrupted by an ALS-linked mutation. *Proc Natl Acad Sci U S A*. 2014;111:E4439–E4448. https://doi.org/10.1073/pnas.1405752111.
43. Johnson JO, Mandrioli J, Benatar M, et al. Exome sequencing reveals VCP mutations as a cause of familial ALS. *Neuron*. 2010;68:857–864. https://doi.org/10.1016/j.neuron.2010.11.036.
44. Chen T, Huang B, Shi X, Gao L, Huang C. Mutant UBQLN2(P497H) in motor neurons leads to ALS-like phenotypes and defective autophagy in rats. *Acta Neuropathol Commun*. 2018;6:122. https://doi.org/10.1186/s40478-018-0627-9.
45. Morimoto N, Nagai M, Ohta Y, et al. Increased autophagy in transgenic mice with a G93A mutant SOD1 gene. *Brain Res*. 2007;1167:112–117. https://doi.org/10.1016/j.brainres.2007.06.045.
46. Crippa V, Sau D, Rusmini P, et al. The small heat shock protein B8 (HspB8) promotes autophagic removal of misfolded proteins involved in amyotrophic lateral sclerosis (ALS). *Hum Mol Genet*. 2010;19:3440–3456. https://doi.org/10.1093/hmg/ddq257.
47. Zhang X, Li L, Chen S, et al. Rapamycin treatment augments motor neuron degeneration in SOD1(G93A) mouse model of amyotrophic lateral sclerosis. *Autophagy*. 2011;7:412–425. https://doi.org/10.4161/auto.7.4.14541.
48. Zhang X, Chen S, Song L, et al. MTOR-independent, autophagic enhancer trehalose prolongs motor neuron survival and ameliorates the autophagic flux defect in a mouse model of amyotrophic lateral sclerosis. *Autophagy*. 2014;10:588–602. https://doi.org/10.4161/auto.27710.
49. Hetz C, Thielen P, Matus S, et al. XBP-1 deficiency in the nervous system protects against amyotrophic lateral sclerosis by increasing autophagy. *Genes Dev*. 2009;23:2294–2306. https://doi.org/10.1101/gad.1830709.
50. Castillo K, Nassif M, Valenzuela V, et al. Trehalose delays the progression of amyotrophic lateral sclerosis by enhancing autophagy in motoneurons. *Autophagy*. 2013;9:1308–1320. https://doi.org/10.4161/auto.25188.
51. Rusmini P, Cortese K, Crippa V, et al. Trehalose induces autophagy via lysosomal-mediated TFEB activation in models of motoneuron degeneration. *Autophagy*. 2019;15:631–651. https://doi.org/10.1080/15548627.2018.1535292.

52. Li Y, Guo Y, Wang X, et al. Trehalose decreases mutant SOD1 expression and alleviates motor deficiency in early but not end-stage amyotrophic lateral sclerosis in a SOD1-G93A mouse model. *Neuroscience*. 2015;298:12–25. https://doi.org/10.1016/j.neuroscience.2015.03.061.
53. Imamura K, Izumi Y, Watanabe A, et al. The Src/c-Abl pathway is a potential therapeutic target in amyotrophic lateral sclerosis. *Sci Transl Med*. 2017;9https://doi.org/10.1126/scitranslmed.aaf3962.
54. Wang Y, Liu FT, Wang YX, et al. Autophagic modulation by trehalose reduces accumulation of TDP-43 in a cell model of amyotrophic lateral sclerosis via TFEB activation. *Neurotox Res*. 2018;34:109–120. https://doi.org/10.1007/s12640-018-9865-7.
55. Donde A, Sun M, Jeong YH, et al. Upregulation of ATG7 attenuates motor neuron dysfunction associated with depletion of TARDBP/TDP-43. *Autophagy*. 2019;1–11. https://doi.org/10.1080/15548627.2019.1635379.
56. Wu C, Watts ME, Rubin LL. MAP4K4 activation mediates motor neuron degeneration in amyotrophic lateral sclerosis. *Cell Rep*. 2019;26:1143–1156. e1145. https://doi.org/10.1016/j.celrep.2019.01.019.
57. Nguyen DKH, Thombre R, Wang J. Autophagy as a common pathway in amyotrophic lateral sclerosis. *Neurosci Lett*. 2019;697:34–48. https://doi.org/10.1016/j.neulet.2018.04.006.
58. Rudnick ND, Griffey CJ, Guarnieri P, et al. Distinct roles for motor neuron autophagy early and late in the SOD1(G93A) mouse model of ALS. *Proc Natl Acad Sci U S A*. 2017;114:E8294–E8303. https://doi.org/10.1073/pnas.1704294114.
59. Freischmidt A, Wieland T, Richter B, et al. Haploinsufficiency of TBK1 causes familial ALS and fronto-temporal dementia. *Nat Neurosci*. 2015;18:631–636. https://doi.org/10.1038/nn.4000.
60. Cirulli ET, Lasseigne BN, Petrovski S, et al. Exome sequencing in amyotrophic lateral sclerosis identifies risk genes and pathways. *Science*. 2015;347:1436–1441. https://doi.org/10.1126/science.aaa3650.
61. Evans CS, Holzbaur ELF. Autophagy and mitophagy in ALS. *Neurobiol Dis*. 2019;122:35–40. https://doi.org/10.1016/j.nbd.2018.07.005.
62. Brenner D, Sieverding K, Bruno C, et al. Heterozygous Tbk1 loss has opposing effects in early and late stages of ALS in mice. *J Exp Med*. 2019;216:267–278. https://doi.org/10.1084/jem.20180729.
63. Catanese A, Olde Heuvel F, Mulaw M, et al. Retinoic acid worsens ATG10-dependent autophagy impairment in TBK1-mutant hiPSC-derived motoneurons through SQSTM1/p62 accumulation. *Autophagy*. 2019;15:1719–1737. https://doi.org/10.1080/15548627.2019.1589257.
64. Wang IF, Guo BS, Liu YC, et al. Autophagy activators rescue and alleviate pathogenesis of a mouse model with proteinopathies of the TAR DNA-binding protein 43. *Proc Natl Acad Sci U S A*. 2012;109:15024–15029. https://doi.org/10.1073/pnas.1206362109.
65. Barmada SJ, Serio A, Arjun A, et al. Autophagy induction enhances TDP43 turnover and survival in neuronal ALS models. *Nat Chem Biol*. 2014;10:677–685. https://doi.org/10.1038/nchembio.1563.
66. Scotter EL, Vance C, Nishimura AL, et al. Differential roles of the ubiquitin proteasome system and autophagy in the clearance of soluble and aggregated TDP-43 species. *J Cell Sci*. 2014;127:1263–1278. https://doi.org/10.1242/jcs.140087.
67. Tamaki Y, Shodai A, Morimura T, et al. Elimination of TDP-43 inclusions linked to amyotrophic lateral sclerosis by a misfolding-specific intrabody with dual proteolytic signals. *Sci Rep*. 2018;8:6030. https://doi.org/10.1038/s41598-018-24463-3.

68. Komatsu M, Waguri S, Koike M, et al. Homeostatic levels of p62 control cytoplasmic inclusion body formation in autophagy-deficient mice. *Cell.* 2007;131:1149–1163. https://doi.org/10.1016/j.cell.2007.10.035.
69. Fecto F, Yan J, Vemula SP, et al. SQSTM1 mutations in familial and sporadic amyotrophic lateral sclerosis. *Arch Neurol.* 2011;68:1440–1446. https://doi.org/10.1001/archneurol.2011.250.
70. Teyssou E, Takeda T, Lebon V, et al. Mutations in SQSTM1 encoding p62 in amyotrophic lateral sclerosis: genetics and neuropathology. *Acta Neuropathol.* 2013;125:511–522. https://doi.org/10.1007/s00401-013-1090-0.
71. Mitsui S, Otomo A, Nozaki M, et al. Systemic overexpression of SQSTM1/p62 accelerates disease onset in a SOD1(H46R)-expressing ALS mouse model. *Mol Brain.* 2018;11:30. https://doi.org/10.1186/s13041-018-0373-8.
72. Strohm L, Behrends C. Glia-specific autophagy dysfunction in ALS. *Semin Cell Dev BiolSemin Cell Dev Biol.* 2019;99:172–182. https://doi.org/10.1016/j.semcdb.2019.05.024.
73. Chaytow H, Huang YT, Gillingwater TH, Faller KME. The role of survival motor neuron protein (SMN) in protein homeostasis. *Cell Mol Life Sci.* 2018;75:3877–3894. https://doi.org/10.1007/s00018-018-2849-1.
74. Wen HL, Lin YT, Ting CH, et al. Stathmin, a microtubule-destabilizing protein, is dysregulated in spinal muscular atrophy. *Hum Mol Genet.* 2010;19:1766–1778. https://doi.org/10.1093/hmg/ddq058.
75. Villalon E, Kline RA, Smith CE, et al. AAV9-Stathmin1 gene delivery improves disease phenotype in an intermediate mouse model of Spinal Muscular Atrophy. *Hum Mol Genet.* 2019;28:3742–3754. https://doi.org/10.1093/hmg/ddz188.
76. Miller N, Feng Z, Edens BM, et al. Non-aggregating tau phosphorylation by cyclin-dependent kinase 5 contributes to motor neuron degeneration in spinal muscular atrophy. *J Neurosci.* 2015;35:6038–6050. https://doi.org/10.1523/JNEUROSCI.3716-14.2015.
77. Dimitriadi M, Derdowski A, Kalloo G, et al. Decreased function of survival motor neuron protein impairs endocytic pathways. *Proc Natl Acad Sci U S A.* 2016;113:E4377–E4386. https://doi.org/10.1073/pnas.1600015113.
78. Hosseinibarkooie S, Peters M, Torres-Benito L, et al. The power of human protective modifiers: PLS3 and CORO1C unravel impaired endocytosis in spinal muscular atrophy and rescue SMA phenotype. *Am J Hum Genet.* 2016;99:647–665. https://doi.org/10.1016/j.ajhg.2016.07.014.
79. Janzen E, Mendoza-Ferreira N, Hosseinibarkooie S, et al. CHP1 reduction ameliorates spinal muscular atrophy pathology by restoring calcineurin activity and endocytosis. *Brain.* 2018;141:2343–2361. https://doi.org/10.1093/brain/awy167.
80. Ripolone M, Ronchi D, Violano R, et al. Impaired muscle mitochondrial biogenesis and myogenesis in spinal muscular atrophy. *JAMA Neurol.* 2015;72:666–675. https://doi.org/10.1001/jamaneurol.2015.0178.
81. Miller N, Shi H, Zelikovich AS, Ma YC. Motor neuron mitochondrial dysfunction in spinal muscular atrophy. *Hum Mol GenetHum Mol Genet.* 2016;25:3395–3406. https://doi.org/10.1093/hmg/ddw262.
82. Boyd PJ, Tu WY, Shorrock HK, et al. Bioenergetic status modulates motor neuron vulnerability and pathogenesis in a zebrafish model of spinal muscular atrophy. *PLoS Genet.* 2017;13:e1006744. https://doi.org/10.1371/journal.pgen.1006744.
83. Bernabo P, Tebaldi T, Groen EJN, et al. In vivo translatome profiling in spinal muscular atrophy reveals a role for SMN protein in ribosome biology. *Cell Rep.* 2017;21:953–965. https://doi.org/10.1016/j.celrep.2017.10.010.

84. Custer SK, Androphy EJ. Autophagy dysregulation in cell culture and animals models of spinal muscular atrophy. *Mol Cell Neurosci.* 2014;61:133–140. https://doi.org/10.1016/j.mcn.2014.06.006.
85. Periyakaruppiah A, de la Fuente S, Arumugam S, et al. Autophagy modulators regulate survival motor neuron protein stability in motoneurons. *Exp Neurol.* 2016;283: 287–297. https://doi.org/10.1016/j.expneurol.2016.06.032.
86. Rodriguez-Muela N, Parkhitko A, Grass T, et al. Blocking p62-dependent SMN degradation ameliorates spinal muscular atrophy disease phenotypes. *J Clin Invest.* 2018; https://doi.org/10.1172/JCI95231.
87. Rochette CF, Gilbert N, Simard LR. SMN gene duplication and the emergence of the SMN2 gene occurred in distinct hominids: SMN2 is unique to Homo sapiens. *Hum Genet.* 2001;108:255–266. https://doi.org/10.1007/s004390100473.
88. Monani UR, Lorson CL, Parsons DW, et al. A single nucleotide difference that alters splicing patterns distinguishes the SMA gene SMN1 from the copy gene SMN2. *Hum Mol Genet.* 1999;8:1177–1183.
89. Lorson CL, Hahnen E, Androphy EJ, Wirth B. A single nucleotide in the SMN gene regulates splicing and is responsible for spinal muscular atrophy. *Proc Natl Acad Sci U S A.* 1999;96:6307–6311.
90. Singh RN, Singh NN. Mechanism of splicing regulation of spinal muscular atrophy genes. *Adv Neurobiol.* 2018;20:31–61. https://doi.org/10.1007/978-3-319-89689-2_2.
91. Garcera A, Mincheva S, Gou-Fabregas M, et al. A new model to study spinal muscular atrophy: neurite degeneration and cell death is counteracted by BCL-X(L) Overexpression in motoneurons. *Neurobiol Dis.* 2011;42:415–426. https://doi.org/10.1016/j.nbd.2011.02.003.
92. Piras A, Schiaffino L, Boido M, et al. Inhibition of autophagy delays motoneuron degeneration and extends lifespan in a mouse model of spinal muscular atrophy. *Cell Death Dis.* 2017;8:3223. https://doi.org/10.1038/s41419-017-0086-4.
93. Clarke PG. Developmental cell death: morphological diversity and multiple mechanisms. *Anat Embryol.* 1990;181:195–213.
94. Bialik S, Dasari SK, Kimchi A. Autophagy-dependent cell death—where, how and why a cell eats itself to death. *J Cell Sci.* 2018;131https://doi.org/10.1242/jcs.215152.
95. Berry DL, Baehrecke EH. Growth arrest and autophagy are required for salivary gland cell degradation in Drosophila. *Cell.* 2007;131:1137–1148.
96. Denton D, Shravage B, Simin R, et al. Autophagy, not apoptosis, is essential for midgut cell death in Drosophila. *Curr Biol.* 2009;19:1741–1746. https://doi.org/10.1016/j.cub.2009.08.042.
97. Arakawa S, Tsujioka M, Yoshida T, et al. Role of Atg5-dependent cell death in the embryonic development of Bax/Bak double-knockout mice. *Cell Death Differ.* 2017;24:1598–1608. https://doi.org/10.1038/cdd.2017.84.
98. Koike M, Shibata M, Tadakoshi M, et al. Inhibition of autophagy prevents hippocampal pyramidal neuron death after hypoxic-ischemic injury. *Am J Pathol.* 2008;172: 454–469.
99. Xing S, Zhang Y, Li J, et al. Beclin 1 knockdown inhibits autophagic activation and prevents the secondary neurodegenerative damage in the ipsilateral thalamus following focal cerebral infarction. *Autophagy.* 2012;8:63–76. https://doi.org/10.4161/auto.8.1.18217.
100. Tooze SA, Abada A, Elazar Z. Endocytosis and autophagy: exploitation or cooperation? *Cold Spring Harb Perspect Biol.* 2014;6:a018358. https://doi.org/10.1101/cshperspect.a018358.
101. Puri C, Renna M, Bento CF, Moreau K, Rubinsztein DC. Diverse autophagosome membrane sources coalesce in recycling endosomes. *Cell.* 2013;154:1285–1299. https://doi.org/10.1016/j.cell.2013.08.044.

102. Razi M, Chan EY, Tooze SA. Early endosomes and endosomal coatomer are required for autophagy. *J Cell Biol.* 2009;185:305–321. https://doi.org/10.1083/jcb.200810098.
103. Fraser J, Simpson J, Fontana R, et al. Targeting of early endosomes by autophagy facilitates EGFR recycling and signalling. *EMBO Rep.* 2019;20:e47734. https://doi.org/10.15252/embr.201947734.
104. Oprea GE, Krober S, McWhorter ML, et al. Plastin 3 is a protective modifier of autosomal recessive spinal muscular atrophy. *Science.* 2008;320:524–527. https://doi.org/10.1126/science.1155085.
105. Baumer D, Lee S, Nicholson G, et al. Alternative splicing events are a late feature of pathology in a mouse model of spinal muscular atrophy. *PLoS Genet.* 2009;5: e1000773. https://doi.org/10.1371/journal.pgen.1000773.
106. Nikoletopoulou V, Tavernarakis N. Regulation and roles of autophagy at synapses. *Trends Cell Biol.* 2018;28:646–661. https://doi.org/10.1016/j.tcb.2018.03.006.
107. Liang Y, Sigrist S. Autophagy and proteostasis in the control of synapse aging and disease. *Curr Opin Neurobiol.* 2018;48:113–121. https://doi.org/10.1016/j.conb.2017.12.006.
108. Birdsall V, Waites CL. Autophagy at the synapse. *Neurosci Lett.* 2019;697:24–28. https://doi.org/10.1016/j.neulet.2018.05.033.
109. Rossoll W, Jablonka S, Andreassi C, et al. Smn, the spinal muscular atrophy-determining gene product, modulates axon growth and localization of beta-actin mRNA in growth cones of motoneurons. *J Cell Biol.* 2003;163:801–812. https://doi.org/10.1083/jcb.200304128.
110. Fallini C, Zhang H, Su Y, et al. The survival of motor neuron (SMN) protein interacts with the mRNA-binding protein HuD and regulates localization of poly(A) mRNA in primary motor neuron axons. *J Neurosci.* 2011;31:3914–3925. https://doi.org/10.1523/JNEUROSCI.3631-10.2011.
111. Thompson LW, Morrison KD, Shirran SL, et al. Neurochondrin interacts with the SMN protein suggesting a novel mechanism for spinal muscular atrophy pathology. *J Cell Sci.* 2018;131. https://doi.org/10.1242/jcs.211482.
112. Torres-Benito L, Neher MF, Cano R, Ruiz R, Tabares L. SMN requirement for synaptic vesicle, active zone and microtubule postnatal organization in motor nerve terminals. *PLoS One.* 2011;6: e26164. https://doi.org/10.1371/journal.pone.0026164.
113. Ning K, Drepper C, Valori CF, et al. PTEN depletion rescues axonal growth defect and improves survival in SMN-deficient motor neurons. *Hum Mol Genet.* 2010;19: 3159–3168. https://doi.org/10.1093/hmg/ddq226.
114. Tseng YT, Chen CS, Jong YJ, Chang FR, Lo YC. Loganin possesses neuroprotective properties, restores SMN protein and activates protein synthesis positive regulator Akt/mTOR in experimental models of spinal muscular atrophy. *Pharmacol Res.* 2016;111:58–75. https://doi.org/10.1016/j.phrs.2016.05.023.
115. Kye MJ, Niederst ED, Wertz MH, et al. SMN regulates axonal local translation via miR-183/mTOR pathway. *Hum Mol Genet.* 2014;23:6318–6331. https://doi.org/10.1093/hmg/ddu350.
116. Jing Z, Han W, Sui X, Xie J, Pan H. Interaction of autophagy with microRNAs and their potential therapeutic implications in human cancers. *Cancer Lett.* 2015;356: 332–338. https://doi.org/10.1016/j.canlet.2014.09.039.
117. Gozuacik D, Akkoc Y, Ozturk DG, Kocak M. Autophagy-regulating microRNAs and cancer. *Front Oncol.* 2017;7:65. https://doi.org/10.3389/fonc.2017.00065.
118. Pircs K, Petri R, Jakobsson J. Crosstalk between MicroRNAs and autophagy in adult neurogenesis: implications for neurodegenerative disorders. *Brain Plast.* 2018;3: 195–203. https://doi.org/10.3233/BPL-180066.

119. Goncalves I, Brecht J, Thelen MP, et al. Neuronal activity regulates DROSHA via autophagy in spinal muscular atrophy. *Sci Rep*. 2018;8:7907. https://doi.org/10.1038/s41598-018-26347-y.
120. Millino C, Fanin M, Vettori A, et al. Different atrophy-hypertrophy transcription pathways in muscles affected by severe and mild spinal muscular atrophy. *BMC Med*. 2009;7:14. https://doi.org/10.1186/1741-7015-7-14.
121. Lieberman AP, Shakkottai VG, Albin RL. Polyglutamine repeats in neurodegenerative diseases. *Annu Rev Pathol*. 2019;14:1–27. https://doi.org/10.1146/annurev-pathmechdis-012418-012857.
122. Finsterer J. Bulbar and spinal muscular atrophy (Kennedy's disease): a review. *Eur J Neurol*. 2009;16:556–561.
123. Adachi H, Katsuno M, Minamiyama M, et al. Widespread nuclear and cytoplasmic accumulation of mutant androgen receptor in SBMA patients. *Brain*. 2005;128:659–670. https://doi.org/10.1093/brain/awh381.
124. Montie HL, Cho MS, Holder L, et al. Cytoplasmic retention of polyglutamine-expanded androgen receptor ameliorates disease via autophagy in a mouse model of spinal and bulbar muscular atrophy. *Hum Mol Genet*. 2009;18:1937–1950. https://doi.org/10.1093/hmg/ddp115.
125. Giorgetti E, Rusmini P, Crippa V, et al. Synergic prodegradative activity of Bicalutamide and trehalose on the mutant androgen receptor responsible for spinal and bulbar muscular atrophy. *Hum Mol Genet*. 2015;24:64–75. https://doi.org/10.1093/hmg/ddu419.
126. Doi H, Adachi H, Katsuno M, et al. p62/SQSTM1 differentially removes the toxic mutant androgen receptor via autophagy and inclusion formation in a spinal and bulbar muscular atrophy mouse model. *J Neurosci*. 2013;33:7710–7727. https://doi.org/10.1523/JNEUROSCI.3021-12.2013.
127. Takayama S, Xie Z, Reed JC. An evolutionarily conserved family of Hsp70/Hsc70 molecular chaperone regulators. *J Biol Chem*. 1999;274:781–786. https://doi.org/10.1074/jbc.274.2.781.
128. Takayama S, Reed JC. Molecular chaperone targeting and regulation by BAG family proteins. *Nat Cell Biol*. 2001;3:E237–E241. https://doi.org/10.1038/ncb1001-e237.
129. Lee JH, Takahashi T, Yasuhara N, et al. Bis, a Bcl-2-binding protein that synergizes with Bcl-2 in preventing cell death. *Oncogene*. 1999;18:6183–6190. https://doi.org/10.1038/sj.onc.1203043.
130. Gamerdinger M, Hajieva P, Kaya AM, et al. Protein quality control during aging involves recruitment of the macroautophagy pathway by BAG3. *EMBO J*. 2009;28:889–901. https://doi.org/10.1038/emboj.2009.29.
131. Sturner E, Behl C. The role of the multifunctional BAG3 protein in cellular protein quality control and in disease. *Front Mol Neurosci*. 2017;10:177. https://doi.org/10.3389/fnmol.2017.00177.
132. Rusmini P, Polanco MJ, Cristofani R, et al. Aberrant autophagic response in the muscle of a knock-in mouse model of spinal and bulbar muscular atrophy. *Sci Rep*. 2015;5:15174. https://doi.org/10.1038/srep15174.
133. Crippa V, Carra S, Rusmini P, et al. A role of small heat shock protein B8 (HspB8) in the autophagic removal of misfolded proteins responsible for neurodegenerative diseases. *Autophagy*. 2010;6:958–960. https://doi.org/10.4161/auto.6.7.13042.
134. Cicardi ME, Cristofani R, Crippa V, et al. Autophagic and proteasomal mediated removal of mutant androgen receptor in muscle models of spinal and bulbar muscular atrophy. *Front Endocrinol*. 2019;10:569. https://doi.org/10.3389/fendo.2019.00569.
135. Fontaine JM, Sun X, Hoppe AD, et al. Abnormal small heat shock protein interactions involving neuropathy-associated HSP22 (HSPB8) mutants. *FASEB J*. 2006;20:2168–2170. https://doi.org/10.1096/fj.06-5911fje.

136. Irobi J, Almeida-Souza L, Asselbergh B, et al. Mutant HSPB8 causes motor neuron-specific neurite degeneration. *Hum Mol Genet.* 2010;19:3254–3265. https://doi.org/10.1093/hmg/ddq234.
137. Ghaoui R, Palmio J, Brewer J, et al. Mutations in HSPB8 causing a new phenotype of distal myopathy and motor neuropathy. *Neurology.* 2016;86:391–398. https://doi.org/10.1212/WNL.0000000000002324.
138. Cristofani R, Crippa V, Rusmini P, et al. Inhibition of retrograde transport modulates misfolded protein accumulation and clearance in motoneuron diseases. *Autophagy.* 2017;13:1280–1303. https://doi.org/10.1080/15548627.2017.1308985.
139. Rusmini P, Cristofani R, Galbiati M, et al. The role of the heat shock protein B8 (HSPB8) in motoneuron diseases. *Front Mol Neurosci.* 2017;10:176. https://doi.org/10.3389/fnmol.2017.00176.
140. Arndt V, Dick N, Tawo R, et al. Chaperone-assisted selective autophagy is essential for muscle maintenance. *Curr Biol.* 2010;20:143–148. https://doi.org/10.1016/j.cub.2009.11.022.
141. Yu Z, Wang AM, Adachi H, et al. Macroautophagy is regulated by the UPR-mediator CHOP and accentuates the phenotype of SBMA mice. *PLoS Genet.* 2011;7: e1002321. https://doi.org/10.1371/journal.pgen.1002321.
142. Chua JP, Reddy SL, Merry DE, et al. Transcriptional activation of TFEB/ZKSCAN3 target genes underlies enhanced autophagy in spinobulbar muscular atrophy. *Hum Mol Genet.* 2014;23:1376–1386. https://doi.org/10.1093/hmg/ddt527.
143. Cortes CJ, Miranda HC, Frankowski H, et al. Polyglutamine-expanded androgen receptor interferes with TFEB to elicit autophagy defects in SBMA. *Nat Neurosci.* 2014;17:1180–1189. https://doi.org/10.1038/nn.3787.
144. Rocchi A, Milioto C, Parodi S, et al. Glycolytic-to-oxidative fiber-type switch and mTOR signaling activation are early-onset features of SBMA muscle modified by high-fat diet. *Acta Neuropathol.* 2016;132:127–144. https://doi.org/10.1007/s00401-016-1550-4.
145. Gautam M, Jara JH, Sekerkova G, et al. Absence of alsin function leads to corticospinal motor neuron vulnerability via novel disease mechanisms. *Hum Mol Genet.* 2016;25:1074–1087. https://doi.org/10.1093/hmg/ddv631.
146. Luningschror P, Binotti B, Dombert B, et al. Plekhg5-regulated autophagy of synaptic vesicles reveals a pathogenic mechanism in motoneuron disease. *Nat Commun.* 2017;8:678. https://doi.org/10.1038/s41467-017-00689-z.
147. Binotti B, Pavlos NJ, Riedel D, et al. The GTPase Rab26 links synaptic vesicles to the autophagy pathway. *Elife.* 2015;4 e05597. https://doi.org/10.7554/eLife.05597.
148. Maystadt I, Rezsohazy R, Barkats M, et al. The nuclear factor kappaB-activator gene PLEKHG5 is mutated in a form of autosomal recessive lower motor neuron disease with childhood onset. *Am J Hum Genet.* 2007;81:67–76. https://doi.org/10.1086/518900.
149. Kim HJ, Hong YB, Park JM, et al. Mutations in the PLEKHG5 gene is relevant with autosomal recessive intermediate Charcot-Marie-Tooth disease. *Orphanet J Rare Dis.* 2013;8:104. https://doi.org/10.1186/1750-1172-8-104.
150. Azzedine H, Zavadakova P, Plante-Bordeneuve V, et al. PLEKHG5 deficiency leads to an intermediate form of autosomal-recessive Charcot-Marie-Tooth disease. *Hum Mol Genet.* 2013;22:4224–4232. https://doi.org/10.1093/hmg/ddt274.
151. Ozoguz A, Uyan O, Birdal G, et al. The distinct genetic pattern of ALS in Turkey and novel mutations. *Neurobiol Aging.* 2015;36:1764.e9–1764.e18. https://doi.org/10.1016/j.neurobiolaging.2014.12.032.
152. Courchaine EM, Lu A, Neugebauer KM. Droplet organelles? *EMBO J.* 2016;35:1603–1612. https://doi.org/10.15252/embj.201593517.

153. Elbaum-Garfinkle S. Matter over mind: liquid phase separation and neurodegeneration. *J Biol Chem*. 2019;294:7160–7168. https://doi.org/10.1074/jbc.REV118.001188.
154. Buchan JR, Kolaitis RM, Taylor JP, Parker R. Eukaryotic stress granules are cleared by autophagy and Cdc48/VCP function. *Cell*. 2013;153:1461–1474. https://doi.org/10.1016/j.cell.2013.05.037.
155. Meyer H, Weihl CC. The VCP/p97 system at a glance: connecting cellular function to disease pathogenesis. *J Cell Sci*. 2014;127:3877–3883. https://doi.org/10.1242/jcs.093831.
156. Meyer H, Bug M, Bremer S. Emerging functions of the VCP/p97 AAA-ATPase in the ubiquitin system. *Nat Cell Biol*. 2012;14:117–123. https://doi.org/10.1038/ncb2407.
157. Tresse E, Salomons FA, Vesa J, et al. VCP/p97 is essential for maturation of ubiquitin-containing autophagosomes and this function is impaired by mutations that cause IBMPFD. *Autophagy*. 2010;6:217–227. https://doi.org/10.4161/auto.6.2.11014.
158. Guo X, Sun X, Hu D, et al. VCP recruitment to mitochondria causes mitophagy impairment and neurodegeneration in models of Huntington's disease. *Nat Commun*. 2016;7:12646. https://doi.org/10.1038/ncomms12646.
159. Papadopoulos C, Kirchner P, Bug M, et al. VCP/p97 cooperates with YOD1, UBXD1 and PLAA to drive clearance of ruptured lysosomes by autophagy. *EMBO J*. 2017;36:135–150. https://doi.org/10.15252/embj.201695148.
160. Bento AC, Bippes CC, Kohler C, et al. UBXD1 is a mitochondrial recruitment factor for p97/VCP and promotes mitophagy. *Sci Rep*. 2018;8:12415. https://doi.org/10.1038/s41598-018-30963-z.
161. Patel A, Lee HO, Jawerth L, et al. A liquid-to-solid phase transition of the als protein fus accelerated by disease mutation. *Cell*. 2015;162:1066–1077. https://doi.org/10.1016/j.cell.2015.07.047.
162. Liu-Yesucevitz L, Bilgutay A, Zhang YJ, et al. Tar DNA binding protein-43 (TDP-43) associates with stress granules: analysis of cultured cells and pathological brain tissue. *PLoS One*. 2010;5: e13250. https://doi.org/10.1371/journal.pone.0013250.
163. Wen X, Tan W, Westergard T, et al. Antisense proline-arginine RAN dipeptides linked to C9ORF72-ALS/FTD form toxic nuclear aggregates that initiate in vitro and in vivo neuronal death. *Neuron*. 2014;84:1213–1225. https://doi.org/10.1016/j.neuron.2014.12.010.
164. Maharjan N, Kunzli C, Buthey K, Saxena S. C9ORF72 regulates stress granule formation and its deficiency impairs stress granule assembly, hypersensitizing cells to stress. *Mol Neurobiol*. 2017;54:3062–3077. https://doi.org/10.1007/s12035-016-9850-1.
165. Lee KH, Zhang P, Kim HJ, et al. C9orf72 dipeptide repeats impair the assembly, dynamics, and function of membrane-less organelles. *Cell*. 2016;167:774–788. e717. https://doi.org/10.1016/j.cell.2016.10.002.
166. Boeynaems S, Bogaert E, Kovacs D, et al. Phase separation of C9orf72 dipeptide repeats perturbs stress granule dynamics. *Mol Cell*. 2017;65:1044–1055. e1045. https://doi.org/10.1016/j.molcel.2017.02.013.
167. Zhang YJ, Gendron TF, Ebbert MTW, et al. Poly(GR) impairs protein translation and stress granule dynamics in C9orf72-associated frontotemporal dementia and amyotrophic lateral sclerosis. *Nat Med*. 2018;24:1136–1142. https://doi.org/10.1038/s41591-018-0071-1.
168. Chitiprolu M, Jagow C, Tremblay V, et al. A complex of C9ORF72 and p62 uses arginine methylation to eliminate stress granules by autophagy. *Nat Commun*. 2018;9:2794. https://doi.org/10.1038/s41467-018-05273-7.
169. Boisvert FM, Cote J, Boulanger MC, Richard S. A proteomic analysis of arginine-methylated protein complexes. *Mol Cell Proteomics*. 2003;2:1319–1330. https://doi.org/10.1074/mcp.M300088-MCP200.

170. Yang YS, Kato M, Wu X, et al. Yeast ataxin-2 forms an intracellular condensate required for the inhibition of TORC1 signaling during respiratory growth. *Cell.* 2019;177:697–710. e617. https://doi.org/10.1016/j.cell.2019.02.043.
171. Krisenko MO, Higgins RL, Ghosh S, et al. Syk Is recruited to stress granules and promotes their clearance through autophagy. *J Biol Chem.* 2015;290:27803–27815. https://doi.org/10.1074/jbc.M115.642900.
172. Eshleman N, Liu G, McGrath K, Parker R, Buchan JR. Defects in THO/TREX-2 function cause accumulation of novel cytoplasmic mRNP granules that can be cleared by autophagy. *RNA.* 2016;22:1200–1214. https://doi.org/10.1261/rna.057224.116.
173. Dimasi P, Quintiero A, Shelkovnikova TA, Buchman VL. Modulation of p-eIF2alpha cellular levels and stress granule assembly/disassembly by trehalose. *Sci Rep.* 2017;7:44088. https://doi.org/10.1038/srep44088.
174. Markmiller S, Soltanieh S, Server KL, et al. Context-dependent and disease-specific diversity in protein interactions within stress granules. *Cell.* 2018;172:590–604. e513. https://doi.org/10.1016/j.cell.2017.12.032.
175. Anderson EN, Gochenaur L, Singh A, et al. Traumatic injury induces stress granule formation and enhances motor dysfunctions in ALS/FTD models. *Hum Mol Genet.* 2018;27:1366–1381. https://doi.org/10.1093/hmg/ddy047.
176. Marrone L, Poser I, Casci I, et al. Isogenic FUS-eGFP iPSC reporter lines enable quantification of FUS stress granule pathology that is rescued by drugs inducing autophagy. *Stem Cell Rep.* 2018;10:375–389. https://doi.org/10.1016/j.stemcr.2017.12.018.
177. Marrone L, Drexler HCA, Wang J, et al. FUS pathology in ALS is linked to alterations in multiple ALS-associated proteins and rescued by drugs stimulating autophagy. *Acta Neuropathol.* 2019;138:67–84. https://doi.org/10.1007/s00401-019-01998-x.
178. Ganassi M, Mateju D, Bigi I, et al. A surveillance function of the HSPB8-BAG3-HSP70 chaperone complex ensures stress granule integrity and dynamism. *Mol Cell.* 2016;63:796–810. https://doi.org/10.1016/j.molcel.2016.07.021.
179. Mateju D, Franzmann TM, Patel A, et al. An aberrant phase transition of stress granules triggered by misfolded protein and prevented by chaperone function. *EMBO J.* 2017;36:1669–1687. https://doi.org/10.15252/embj.201695957.
180. Mazroui R, Di Marco S, Kaufman RJ, Gallouzi IE. Inhibition of the ubiquitin-proteasome system induces stress granule formation. *Mol Biol Cell.* 2007;18:2603–2618. https://doi.org/10.1091/mbc.e06-12-1079.
181. Cherkasov V, Hofmann S, Druffel-Augustin S, et al. Coordination of translational control and protein homeostasis during severe heat stress. *Curr Biol.* 2013;23:2452–2462. https://doi.org/10.1016/j.cub.2013.09.058.
182. Kroschwald S, Maharana S, Mateju D, et al. Promiscuous interactions and protein disaggregases determine the material state of stress-inducible RNP granules. *Elife.* 2015;4:e06807. https://doi.org/10.7554/eLife.06807.
183. Jain S, Wheeler JR, Walters RW, et al. ATPase-modulated stress granules contain a diverse proteome and substructure. *Cell.* 2016;164:487–498. https://doi.org/10.1016/j.cell.2015.12.038.
184. Wang B, Maxwell BA, Joo JH, et al. ULK1 and ULK2 regulate stress granule disassembly through phosphorylation and activation of VCP/p97. *Mol Cell.* 2019;74:742–757. e748. https://doi.org/10.1016/j.molcel.2019.03.027.
185. Youn JY, Dunham WH, Hong SJ, et al. High-density proximity mapping reveals the subcellular organization of mRNA-associated granules and bodies. *Mol Cell.* 2018;69:517–532. e511. https://doi.org/10.1016/j.molcel.2017.12.020.
186. Levine B, Kroemer G. Biological functions of autophagy genes: a disease perspective. *Cell.* 2019;176:11–42. https://doi.org/10.1016/j.cell.2018.09.048.

187. Alexander EJ, Ghanbari Niaki A, Zhang T, et al. Ubiquilin 2 modulates ALS/FTD-linked FUS-RNA complex dynamics and stress granule formation. *Proc Natl Acad Sci U S A*. 2018;115:E11485–E11494. https://doi.org/10.1073/pnas.1811997115.
188. Toth RP, Atkin JD. Dysfunction of optineurin in amyotrophic lateral sclerosis and glaucoma. *Front Immunol*. 2018;9:1017. https://doi.org/10.3389/fimmu.2018.01017.

CHAPTER TEN

Strategies employed by viruses to manipulate autophagy

Nilima Dinesh Kumar[a,b], Jolanda M. Smit[b,*], Fulvio Reggiori[a,*]

[a]Department of Biomedical Sciences of Cells & Systems, University of Groningen, University Medical Center Groningen, Groningen, The Netherlands
[b]Department of Medical Microbiology and Infection Prevention, University of Groningen, University Medical Center Groningen, Groningen, The Netherlands
[*]Corresponding authors: e-mail address: jolanda.smit@umcg.nl; f.m.reggiori@umcg.nl

Contents

Abstract

Autophagy, originally described as a conserved bulk degradation pathway important to maintain cellular homeostasis during starvation, has also been implicated in playing a central role in multiple physiological processes. For example, autophagy is part of our

Progress in Molecular Biology and Translational Science, Volume 172
ISSN 1877-1173
https://doi.org/10.1016/bs.pmbts.2020.01.004

innate immunity by targeting intracellular pathogens to lysosomes for degradation in a process called xenophagy. Coevolution and adaptation between viruses and autophagy have armed viruses with a multitude of strategies to counteract the antiviral functions of the autophagy pathway. In addition, some viruses have acquired mechanisms to exploit specific functions of either autophagy or the key components of this process, the autophagy-related (ATG) proteins, to promote viral replication and pathogenesis. In this chapter, we describe several examples where the strategy employed by a virus to subvert autophagy has been described with molecular detail. Their stratagems positively or negatively target practically all the steps of autophagy, including the signaling pathways regulating this process. This highlights the intricate relationship between autophagy and viruses and how by commandeering autophagy, viruses have devised ways to fine-tune their replication.

1. The mechanisms of autophagy

1.1 Microautophagy, macroautophagy and chaperone-mediated autophagy

The term autophagy derives from two Greek words that combinatorially mean self-eating, and groups all those intracellular delivering pathways that lead to the lysosomal degradation and recycling of intracellular components. There are three major types of autophagy: microautophagy, macroautophagy and chaperone-mediated autophagy (CMA).[1] These three pathways deliver cytoplasmic cargoes into lysosomes through different mechanisms. Microautophagy involves the invagination of the lysosomal limiting membrane leading to a direct engulfment of cytoplasmic cargo by lysosomes.[2] In contrast, CMA requires the heat shock protein 70 (HSC70) chaperone protein to escort cytoplasmic proteins, mostly containing the conserved KFERQ pentapeptide, to a channel formed on the lysosomal surface by the lysosome-associated membrane protein (LAMP) type 2A for direct translocation into the interior of this compartment.[2] Macroautophagy, hereafter referred to as autophagy, is characterized by the formation of large double-membrane vesicles called autophagosomes that sequester the components destined for destruction in lysosomes.[3]

1.2 The molecular mechanism of autophagy and the autophagy-related (ATG) proteins

The entire process of autophagy can be subdivided in five sequential steps: initiation, elongation, maturation, fusion and degradation (Fig. 1). Upon initiation of autophagy, autophagosome biogenesis begins with the formation

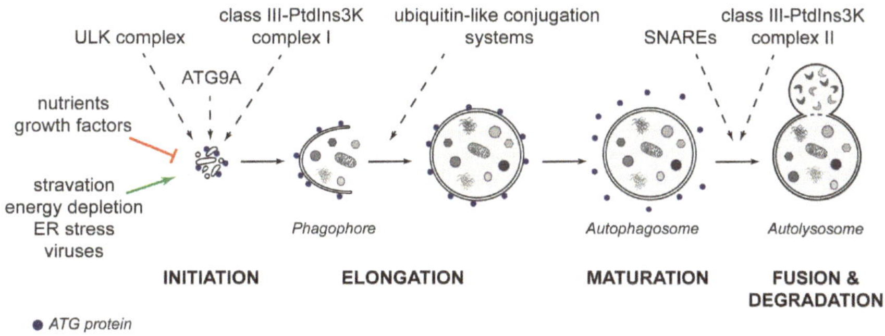

Fig. 1 Schematic model depicting the mechanism of autophagy. The key steps characterizing autophagy are initiation, elongation, maturation, fusion and degradation. Upon autophagy induction, the ULK complex, the class III PtdIns3K complex I and ATG9A catalyze the nucleation of the phagophore by probably mediating the heterotypic fusion of vesicles. Numerous signaling cascades regulating autophagy modulate the activity of these complexes. Thus, while nutrients and growth factors inhibit autophagy (red T-bar), nutrient starvation, energy depletion, ER stress and viral infections trigger this pathway (green arrow). The PtdIns3P generated on the surface of the phagophore is important for the recruitment of PtdIns3P-binding ATG proteins like WIPI2 and several components of the ATG12 and LC3 ubiquitination-like conjugation systems. Subsequently, these two ubiquitination-like conjugation systems drive the elongation and closure of phagophore to form the double-membrane autophagosome. The maturation step is marked by the release of ATG proteins and the turnover of PtdIns3P on the surface of the autophagosome. This is a pre-requisite for the engagement of the components of the machinery required for the fusion with endosomes and lysosomes, including SNARE proteins and other factor promoting this step such as the PLEKHM1 adaptor and the class III PtdIns3K complex II. After fusion, the autophagosomal cargo is degraded into metabolites by lysosomal hydrolases in autolysosomes.

of a membranous cistern known as the phagophore or isolation membrane. The phagophore subsequently elongates through the acquisition of extra lipids and the fusion of its extremities leads to the generation of the closed double-membrane autophagosome (Fig. 1). The maturation step entails the disengagement and dissociation of the autophagy-related (ATG) proteins (see below) from the surface of autophagosomes and the recruitment of machinery responsible for fusion with lysosomes (Fig. 1). The final steps comprise the fusion between the autophagosome and a lysosome and the subsequent exposure and degradation of the inner membrane of the autophagosome and the cargo by lysosomal hydrolases[4–7] (Fig. 1).

The principal players orchestrating autophagosome biogenesis are the so-called ATG proteins. Sixteen of them compose the highly conserved ATG core machinery and they have been subgrouped into five different functional clusters: (1) the ULK complex, (2) the autophagy-specific class III phosphatidylinositol 3-kinase (PtdIns3K) complex I, (3) ATG9A and its binding partners, and (4 and 5) two ubiquitin-like conjugation systems. The ULK complex is composed of ULK1 or ULK2 kinase, ATG13, RB1CC1/FIP200 and ATG101. This complex regulates the initiation of autophagosome formation by phosphorylating and interacting with several ATG components of the core machinery[4,5] (Fig. 1). The class III PtdIns3K complex I, which includes ATG14L, VPS34/PIK3C3, p150/PIK3R4 and BECLIN1 (BECN1), together with the ULK complex, is important for the phagophore nucleation and generation of phosphatidylinositol-3-phosphate (PtdIns3P) on autophagosomal membranes (Fig. 1). This lipid promotes the recruitment of PtdIns3P-binding proteins like WIPI1 and WIPI2 onto phagophores. The class III PtdIns3K complex I also directly interacts with proteins like AMBRA1 that are critical autophagy regulators.[4,8] ATG9A is the only transmembrane protein within the core ATG machinery, and it functions in both phagophore formation and elongation. ATG9A shuttles between the trans-Golgi network/endosomes and phagophores upon autophagy induction, and this trafficking is regulated by both the ULK and PtdIns3K complexes. The two ubiquitin-like conjugation systems play a role in expansion and closure of the phagophore (Fig. 1). The first ubiquitin-like conjugation system involves the covalent conjugation of ATG12 to ATG5, through the sequential action of the E1 activating enzyme ATG7 and the E2 conjugating enzyme ATG10. Subsequently, ATG16L1 binds non-covalently to the ATG12-ATG5 conjugate forming the ATG12-ATG5-ATG16L1 complex, which multimerizes through self-interaction. The second ubiquitin-like conjugation system begins with the cleavage of the members of the ATG8/LC3 protein family (i.e., LC3A, LC3B, LC3C, GABARAP, GABARAPL1 and GABARAPL2) by the cysteine proteases from the ATG4 protein family. This cleavage exposes a C-terminal glycine and this processed form of the LC3 proteins is referred to as LC3-I. Through the action of the E1 activating enzyme ATG7, the E2 conjugating enzyme ATG3 and guided by the ATG12-ATG5-ATG16L1 complex, LC3-I is covalently conjugated to the phosphotidylethanolamine (PE) molecules. LC3-I-PE is called LC3-II.[4,5]

Complete autophagosomes then tether, dock and finally fuse either first with endosomes and then lysosomes or directly with this latter organelle, to form autolysosomes (Fig. 1). The mechanism of fusion is not completely understood, but it requires the RAB7 GTPase, the homotypic fusion and protein sorting (HOPS) complex, and adaptor proteins such as pleckstrin homology domain containing protein family 1 (PLEKHM1), and is driven by interaction of SNAREs, in particular SYNTAXIN17 (STX17) or YKT6, which are located on autophagosomes, and SNAP29, VAMP7, VAMP8 or VAMP9, on lysosomes.[6,7] The PtdIns3K complex II, which differ from the autophagy-specific complex I in one subunit, i.e., UVRAG instead of ATG14L, has been shown to be important for fusion of autophagosome with lysosomes as well. UVRAG, through the binding to the HOPS complex, promotes RAB7 GTPase activity and stimulates the fusion between autophagosomes and lysosomes.[6,7] The activity of UVRAG is negatively regulated by RUBICON, leading to the inhibition of the fusion between autophagosomes and lysosomes.[6–8]

2. The signaling cascades regulating autophagy

Autophagy initiation is mainly controlled through the regulation of two ATG functional clusters, the ULK complex and the class III PtdIns3K complex I, orchestrated by metabolic and stress sensors in a rapamycin complex 1 (mTORC1)-dependent or -independent manner.[3] Now, we will present five main mechanisms regulating autophagy.

2.1 Nutrient-sensing pathways

Under nutrient-rich conditions, mTORC1 associates to and phosphorylates the ULK1 and ATG13 subunits of the ULK complex, leading to the inactivation of its kinase activity.[3,9] In parallel and through phosphorylation, mTORC1 also inhibits the class III PtdIns3K complex I and the nuclear transport of the transcription factor EB (TFEB), a master transcriptional regulator of autophagy and lysosome biogenesis.[3,9] Upon amino acid starvation, mTORC1 is inactivated and the factors modified by this kinase complex are dephosphorylated. This results in activation of both ULK and PtdIns3K complexes, triggering autophagosome biogenesis, and TFEB localization in the nucleus leads to the expression of genes involved in autophagy and lysosomal function.[3,4,9]

mTORC1 activity is also regulated by the signaling cascades triggered by insulin and growth factors, two types of molecules that indirectly indicate the presence of nutrients (Fig. 2A). Insulin and growth factors bind to their specific receptors, initiating a downstream signaling that leads to the activation of the class I PtdIns3K. The generated PtdIns3P enhances the activity of the protein kinase B (also known as AKT),[10] which inhibits TSC1/TSC2 complex through phosphorylation. mTORC1 activation requires GTP-bound Rheb. The TSC1/TSC2 complex acts as a GTPase-activating

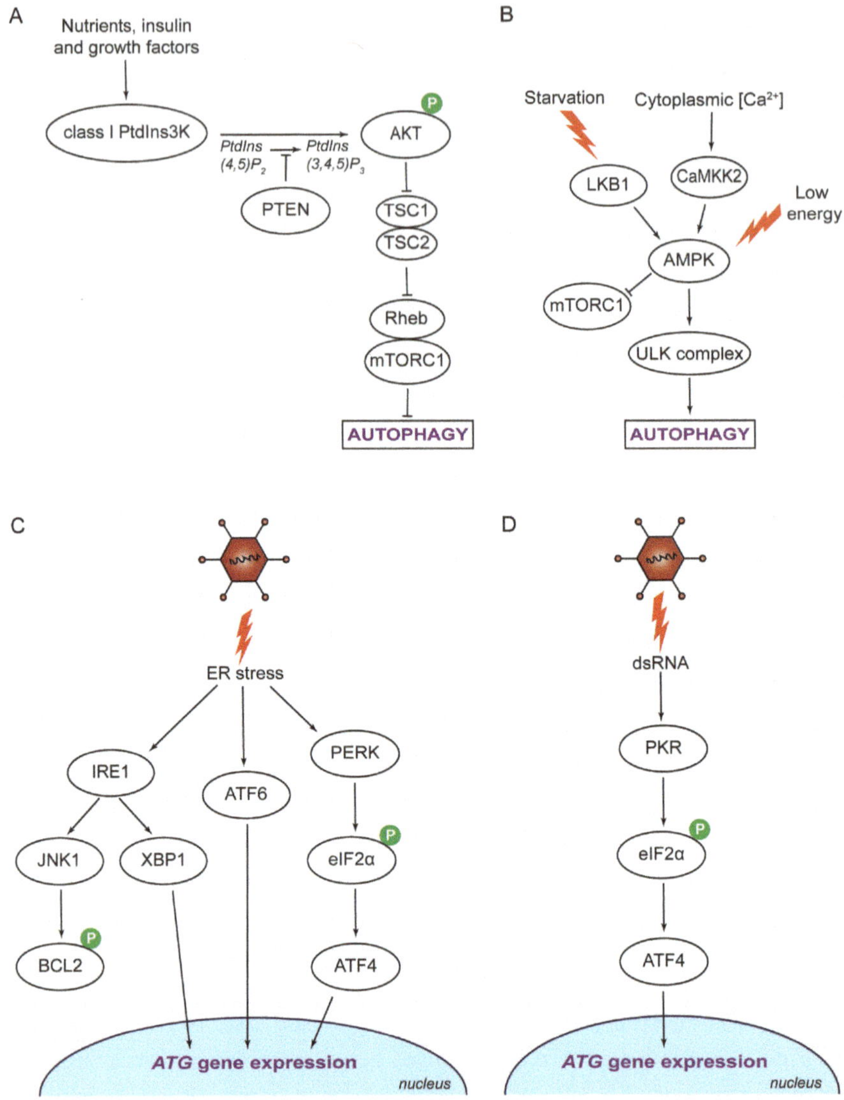

Fig. 2 The signaling cascades regulating autophagy. (A) Orchestration of autophagy by mTORC1. This kinase complex acts as a major regulator of autophagy and numerous signals converge at this kinase complex, either activating or inhibiting autophagy. For example, the class I PtdIns3K/AKT/mTORC1 signaling axis inhibits autophagy through the activation of mTORC1. Nutrients or growth factors induce the AKT kinase, which inhibits TSC1/TSC2 complex through phosphorylation. TSC1/TSC2 acts as GTPase-activating protein for the GTP-bound Rheb. GTP-bound Rheb interacts and activates mTORC1 inhibiting autophagy. (B) Regulation of autophagy by AMPK. AMPK, which gets activated by either a decrease in the intracellular AMP:ATP ratio, LKB1 kinase or calcium-activated CaMKK2, positively regulates autophagy by inhibiting mTORC1 through

protein (GAP) for Rheb and consequently its inhibition allows the stabilization of GTP-bound Rheb and activation of mTORC1, which inhibits autophagy[9] (Fig. 2A).

Another axis regulating autophagy in response to stress conditions is the one leading to the phosphorylation of eukaryotic translation initiation factor 2 alpha (eIF2α). Various stress conditions, including amino acid deprivation, viral infection, ER stress and heme deficiency, causes the activation of distinct kinases, including the general control non-derepressible 2 (GCN2), the double-stranded RNA-activated protein kinase (PKR), the PKR-like endoplasmic reticulum-resident kinase (PERK), and the heme-controlled inhibitor (HRI). Phosphorylation of eIF2α leads to decrease in the global protein synthesis by decreasing the recognition and translation of mRNA with the AUG initiation codon, and favoring translation of specific mRNAs containing short ORFs, like the ATF4 transcription factor (Fig. 2C). ATF4 subsequently drives the transcription and translation of many downstream proteins like SESTRIN2 and CHOP, which contributes to induction of autophagy by directly or indirectly inhibiting mTORC1.[3,8]

Autophagy can also be modulated through the class III PtdIns3K complex I activity. In presence of nutrients, binding of B-cell lymphoma 2 (BCL2) to BECN1 inhibits the activity of this complex. Upon amino acid starvation, BCL2 is phosphorylated by c-Jun N-terminal kinase (JNK1), triggering its dissociation from BECN1 and the concomitant class III PtdIns3K complex I activation.[3,8]

---

phosphorylation of TSC1/TSC2 and RAPTOR. AMPK also directly phosphorylates components of the ATG machinery, such as the ULK complex and the class III PtdIns3K complex I, and stimulates them. (C) ER stress-induced autophagy. ER stress caused for example by virus infections triggers autophagy through the induction of the UPR. Autophagy can be stimulated by all the three branches that characterize the UPR, which are centered around IRE1, ATF6 and PERK, and allow cells to maintain ER homeostasis under stress conditions. IRE1 oligomerization leads to the alternative splicing of the XBP1 mRNA and the resulting translated transcriptional factor induces the expression of *ATG* genes. IRE1 also activates JNK, which phosphorylates BCL2 releasing inhibition from the class III PtsIns3K complex I. Cleaved ATF6 translocates into nucleus where this transcription factor triggers the expression of *ATG* genes. PERK activates eIF2α through phosphorylation. Phosphorylated eIF2α inhibits general cellular translation while promoting that of ATF4, a transcription factor that induces the transcription of *ATG* genes. (D) Triggering of autophagy by viruses. Presence of dsRNA leads to the activation of the PKR kinase, which phosphorylates and activates eIF2α triggering downstream signaling (see panel C).

2.2 Energy-sensing pathways

The AMP-activated protein kinase (AMPK) is a kinase that senses energy levels by detecting augmentations in the AMP/ATP ratio (Fig. 2B). Binding of AMP to the γ-subunit of AMPK promotes its phosphorylation and activation by the upstream LKB1 kinase[11] (Fig. 2B). Activated AMPK, also inhibits mTORC1 through phosphorylation. Additionally, AMPK phosphorylates and activates the TSC1/TSC2 complex, which is a negative regulator of mTORC1. Further, AMPK upregulates autophagy by directly phosphorylating and activating BECN1, VPS34 and ULK1.[9,11]

2.3 ER stress

The endoplasmic reticulum (ER) is a central cell organelle that fulfills a multitude of functions, including protein and lipid synthesis, protein folding and sorting, and calcium homeostasis maintenance. Environmental adaptations and pathophysiological processes that increase demand for protein synthesis and folding, disrupt protein folding leading to an accumulation of misfolded proteins, which in turn causes ER stress and the consequent activation of the unfolded protein response (UPR).[12] UPR and autophagy are interconnected as autophagy alleviates ER stress by removing accumulated misfolded proteins (Fig. 2C). The UPR comprises three distinct protein sensors, which initiates a downstream signaling upon activation by accumulation of misfolded proteins: PERK, inositol-requiring enzyme 1 (IRE1), and activating transcription factor 6 (ATF6).[12] These three sensors positively regulate autophagy through different mechanisms (Fig. 2C). PERK acts by phosphorylating eIF2α and triggering the downstream autophagy activation mechanism. Signaling downstream of IRE1 causes phosphorylation and activation of JNK1, which in turn phosphorylates BCL2 to release BECN1 and upregulate autophagy. In parallel, IRE1, through its endoribonuclease activity, yields a spliced variant of XBP1 that acts as a transcriptional activator of genes important for ER homeostasis but also for autophagy. ATF6 is activated by a cleavage event induced by ER stress. Cleaved ATF6 translocates into the nucleus and activates the transcription of genes important for ER homeostasis and autophagy.[12,13]

2.4 Immune regulation of autophagy

Autophagy can be stimulated by many different immune signals, including interferon (IFN)-stimulated gene, PKR, Toll-like receptors (TLRs), IFN-γ, stimulator of IFN genes (STING), immunity-related GTPase family

M (IRGM), but also molecules such as tripartite motif (TRIM) proteins and the TANK-binding kinase 1 (TBK1) that have been linked to selective autophagy.[14–17] For example, the antiviral PKR kinase activates eIF2α-mediated autophagy upon sensing of dsRNA (Fig. 2D). Recognition of pathogen-associated molecular patterns (PAMPs) by cellular pattern recognition receptors (PRRs) either at the cell surface or in the lumen of endosomes, also initiate downstream signaling that is mediated by myeloid differentiation primary response protein 88 (MYD88) and TIR-domain-containing adapter-inducing interferon-β (TRIF). These two adaptor proteins regulate the tumor necrosis factor receptor 6 (TRAF6) and TRAF3 ligases, which ubiquitinate the Lys63 residue of BECN1 and also promote the release of BCL2 from the class III PtdIns3K complex I. Moreover, damage-associated molecular patterns (DAMPs) like DNA complexes, ATP and the high mobility group box 1 (HMGB1) protein can activate autophagy through AMPK-mediated mTORC1 inhibition or by releasing BCL2 from the class III PtdIns3K complex I[14,17]. Multiple other cytokines have been shown to modulate autophagy, including TNF-α, IL-1, IL-2, IL-6, and TGF-β, which are positive regulators, and IL-4, IL-10, and IL-13, which are inhibitors.[18] Furthermore, IRGM which has been characterized as a genetic risk factor for Crohn's disease[19] and tuberculosis,[20] has been shown to bind to ULK1 and BECN1, thereby promoting the orchestration of autophagy initiation complexes.[21] IRGM also interacts with several PRRs, including NOD1, RIG-I and TLR3, probably linking the innate immune response and autophagy.[21]

2.5 Autophagy and apoptosis

Autophagy has a dual role in cell survival and cell death. Autophagy-dependent cell death is a type of cell death requiring the ATG machinery for its execution.[22] During low levels of stress, cytoprotective function of autophagy tends to limit apoptosis by degrading damaged mitochondria or caspases, which are involved in intrinsic and extrinsic apoptosis and further supports cell growth by providing essential metabolites. Further, when stress exceeds beyond a certain threshold apoptosis is activated. Autophagy and apoptosis exhibit an extensive amount of cross talk.[23,24] Autophagy and apoptosis share several regulatory upstream pathways and regulatory proteins. ATG proteins like BECN1, ATG3 can be cleaved by caspases, which inactivate their autophagic function and activate their pro-apoptotic functions. In addition, autophagy can also be regulated by several apoptotic proteins like BCL2 and caspase 8/FADD-like IL-1β-converting enzyme (FLICE)-inhibitory proteins (FLIP) to inhibit autophagy and proteins like BCL2 interacting protein 3 (BNIP3) and BNIP3-like (NIX) plays a proautophagic role. In addition, several upstream pathways like JNK and AKT also regulate both autophagy and apoptosis.[24–26]

3. The physiological functions of autophagy

Although autophagy was initially thought to be a bulk degradation process principally dedicated to the generation of an internal pool of nutrients in response of starvation, a multitude of recent studies has revealed the physiological relevance of the selective targeting of specific cargo by autophagosomes, a process referred to as selective autophagy. Selective autophagy can occur in ubiquitin-dependent or -independent manner, and it requires the so-called autophagy receptors.[27] Autophagy receptors associate to the cargo and mediate both the initiation of autophagosome biogenesis locally by interacting and recruiting the ULK complex, and the exclusive sequestration of the targeted cargoes into autophagosomes by binding the LC3 pool present in the interior of the expanding phagophore.[27–30] Some of the known autophagy receptors include p62/SQSTM1, NDP52, NBR1, TAX1BP1, OPTN and NIX/BNIP3.[27] In addition, recent studies have highlighted that specific autophagy receptors such as OPTN, NDP52 and TAX1BP1 facilitate autophagosome maturation.[31,32]

Selective autophagy allows the specific removal of protein aggregates, dysfunctional or unnecessary complexes and damaged or superfluous organelles. Through these mechanisms, autophagy participates in numerous physiological processes such as adaptation to stress, programmed cell death, development and differentiation, and aging prevention.[4,33] Not surprisingly, dysregulation of autophagy is associated with numerous diseases, including cancer development and progression, many neurodegenerative disorders like Alzheimer's, Parkinson's and Huntington's disease, and other age-related disorders.[4,33,34]

Another key physiological function of autophagy is its role in eliminating cytoplasmic pathogens, including bacteria, virus and parasites, in a process termed xenophagy.[16] Here, innate immune receptors such as those belonging the TLRs family and autophagy receptors like p62, sense pathogens by recognizing various PAMPs and DAMPs, and host molecular tags such as ubiquitin and galectins, respectively, and guide autophagy in targeting microbes and/or microbe-related macromolecules to lysosomal degradation.[16,17]

In addition, autophagy also functions in controlling pro-inflammatory signaling.[14,17] Finally, this pathway participates in multiple important aspects of the adaptive immunity, including MHC class II presentation of intracellular antigens, T cell survival and polarization, as well as T cell selection and maturation.[14,16,17]

4. Autophagy and viruses

Coevolution between the host and pathogens has led microbes to develop different strategies to modulate autophagy to not only evade it, but also to subvert this pathway and the ATG machinery to successfully produce progeny particles. Here, we will summarize some of the strategies exploited by specific viruses to manipulate autophagy (Fig. 3 and Tables 1 and 2). The interaction between autophagy and viruses can be either indirect, when viruses modulate upstream pathways regulating autophagy, or direct, by targeting ATG proteins or complexes (Fig. 3).

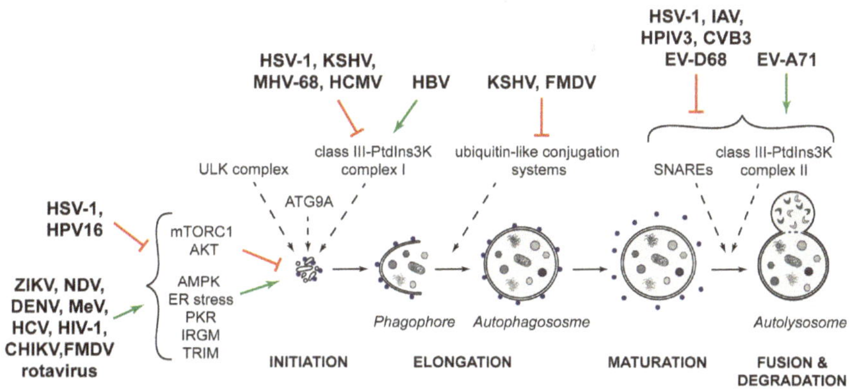

Fig. 3 Virus modulation of autophagy. Viruses have developed numerous unique strategies to activate (green arrows) or inhibit (red T-bars) autophagy, by either modulating upstream signaling cascades or interfering with autophagosome formation and fusion with lysosomes through interaction with key players of the autophagy machinery. This scheme provides an overview of the viruses discussed in the review and the autophagy step that they subvert to promote their intracellular replication or cell egression.

Table 1 Viral strategies to inhibit autophagic pathway.

Strategies to inhibit autophagic pathways

Virus family	Virus	Viral protein	Host target	Mechanism	References
Strategies to inhibit autophagy by hijacking signaling pathways					
Herpesviridae	HSV-1	Us11	PKR	Inhibition of PKR activity	Lussignol et al.[35]
	HSV-1	Us11	TRIM	Inhibition of TRIM- and TBK1-mediated autophagy	Liu et al.[36]
Papillomaviridae	HPV16	?	?	Inhibition of the PI3K-AKT–MTOR signaling cascade by inhibiting PTEN	Surviladze et al.[37]
Subversion of autophagy machinery					
Herpesviridae	HSV-1	ICP34.5	BECN1	Inhibition of the class III PtdIns3k complex I formation	Orvedahl et al.[38]
	KSHV	vBCL2	BECN1	Inhibition of the class III PtdIns3k complex I formation	Sarid et al.[39]
	MHV-68	vBCL2	BECN1	Inhibition of the class III PtdIns3k complex I formation	Virgin et al.[40] and Sinha et al.[41]
	HCMV	TRS1 and IRS1	BECN1	Inhibition of the class III PtdIns3k complex I formation and possibly also that of complex II	Chaumorcel et al.[42] and Mouna et al.[43]
	KSHV	vFLIP	ATG3	Inhibit the LC3 conjugation system	Lee et al.[44]
Picornaviridae	FMDV	3C protease	ATG12-ATG5 conjugate	Degradation of the ATG12-ATG5 conjugate	Fan et al.[45]

Disruption of selective types of autophagy					
Flaviviridae	DENV, ZIKV	NS3 protease and NS2B	FAM134B	Cleavage and inactivation of FAM134B	Lennemann and Coyne[46]
Picornaviridae	CVB3	2A protease	p62	Cleavage and inactivation of p62	Shi et al.[47,48] and Mohamud et al.[49]
	CVB3	2A and 3C protease	NBR1	Cleavage and inactivation of NBR1	Shi et al.[47]
	CVB3	3C protease	NDP52	Cleavage and inactivation of NDP52	Mohamud et al.[49]
The blockage of autophagosome-lysosome fusion					
Retroviridae	HIV-1	Nef	BECN1	Speculated to modulate of the class III PtdIns3K complex II	Kyei et al.[50]
Orthomyxoviridae	IAV	M2	BECN1	Speculated to modulate the class III PtdIns3K complex II	Gannagé et al.[51]
Paramyxoviridae	HPIV3	P	SNAP29	Interfering with SNAP29 function	Ding et al.[52]
Picornaviridae	CVB3	3C protease	SNAP29, PLEKHM1	Cleavage and inactivation of SNAP29 and PLEKHM1	Mohamud et al.[53]
Picornaviridae	EV-D68	3C protease	SNAP29	Cleavage and inactivation of SNAP29	Corona et al.[54]

Table 2 Viral strategies to induce autophagy.
Strategies to induce autophagy

Virus family	Virus	Viral protein	Host target	Mechanism	References
Strategies to trigger autophagy by hijacking signaling pathways					
Flaviviridae	ZIKV	NS4A and NS4B	?	Inhibits AKT phosphorylation	Liang et al.[55]
Flaviviridae	DENV2	?	?	IRE1α-JNK1 mediated phosphorylation of BECN1	Lee et al.[56]
Reoviridae	Rotavirus	NSP4	CaMKK2	CaMKK2 mediated activation of AMPK-dependent autophagy	Crawford et al.[57]
Picornaviridae	FMDV	VP1	HSPB1	Activation of ER stress-mediated autophagy via eIF2α-ATF4 cascade	Sun et al.[58]
Flaviviridae	DENV	?	?	Activation of AMPK-dependent lipophagy	Jordan and Randall[136]
Paramyxoviridae	Mev	C	IRGM	IRGM-mediated autophagy initiation	Grégoire et al.[59]
Flaviviridae	HCV	NS3	IRGM	IRGM-mediated autophagy initiation	Grégoire et al.[59] and Hansen et al.[60]
Retroviridae	HIV-1	Nef	IRGM	IRGM-mediated autophagy initiation	Grégoire et al.[59]
Subversion of the ATG machinery					
Hepadnaviridae	HBV	HBx	VPS34	Enhancement of class III PtdIns3k complex I activity	Sir et al.[61]
	HBV	HBx	DAPK	DAPK-mediated activation of BECN1	Zhang et al.[62]
	HBV	HBx	BECN1	Direct upregulation of BECN1	Tang et al.[63]
	HBV	HBx	miR-192-2p	Inhibition of miR-192-2p, which negatively regulates BECN1 via XIAP	Wang et al.[64]
Picornaviridae	PV	?	?	ULK complex-independent autophagy activation	Corona Velazquez et al.[65]
The enhancement of autophagosome-lysosome fusion					
Picornaviridae	EV-A71	2BC	STX17	Unknown	Lai et al.[66]
Paramyxoviridae	Mev	?	NDP52 and TAXBP1	NDP52 and TAX1BP1-mediated autophagosome maturation	Petkova et al.[67]

5. Strategies to inhibit autophagy

5.1 Hijacking signaling pathways

Autophagy induction as a consequence of viral infection was first reported in a study involving herpes simplex virus 1 (HSV-1), which belongs to the *Herpesviridae* family (*Alphaherpesvirinae* subfamily).[68] Specifically, the IFNα-inducible antiviral recognition pathway through PKR and eIF2α was found to upregulate autophagy in response to HSV-1 infection (Fig. 2D). In this study, Tallóczy and colleagues revealed that mouse embryonic fibroblasts (MEFs) that lack PKR or carry non-phosphorylable mutants of eIF2α, are unable to induce autophagy in response to HSV-1.[68] HSV-1 antagonizes PKR-mediated autophagy through its protein Us11, which directly interacts and inhibit PKR.[35] In particular, a carboxy-terminal region of Us11 was found to be crucial for the binding to PKR. Thus, HSV-1 inhibits eIF2α phosphorylation, which in turn leads to both an inhibition of autophagy upregulation and the translation of viral proteins.[35] In addition, HSV-1 Us11 was also recently reported to inhibit autophagy by disrupting TRIM23 and TBK1 mediated autophagy.[36] TRIM23 is a E3 ubiquitin ligase enzyme with a C-terminal ADP ribosylation factor (ARF) domain, which harbors a GTPase activity. Auto-ubiquitination of the ARF domain activates TRIM23 and TBK1-mediated selective autophagy.[69–71] The exact mechanism of TRIM23 subversion by Us11 remains to be identified but the binding of Us11 to the ARF domain of TRIM23 probably disrupts the activation of TBK1 by TRIM23 and ultimately autophagy induction.[36]

Human papillomavirus type 16 (HPV16), a DNA virus belonging to the *Papillomaviridae* family, inhibits autophagy during viral entry through activation of the cell surface epidermal growth factor receptor (EGFR).[37] EGFR activation inhibits autophagy by inducing the class I PtdIns3K-AKT-mTORC1 signaling cascade (Fig. 2A).[9,10] Binding of HPV16 virions to the cell surface causes a rapid AKT phosphorylation, which is mediated by EGFR as this does not occur when cells are treated with the specific EGFR inhibitor AG1478.[37] This negative regulation of autophagy does not require the expression of viral proteins as the study used pseudo-virions, which are just capable of entering in the cells and deliver a pseudogenome to the cells without expressing viral proteins. In parallel, HPV also inactivates PTEN through phosphorylation of its Ser380 residue,[72] which contributes to the inhibition of autophagy via class I PtdIns3K-AKT-mTORC1 signaling cascade (Fig. 2A). Thus, incubation of keratinocytes with HPV16 for just 1 h, caused a decrease in the LC3-II/I ratio, which may be indicative of autophagy suppression.[37] When autophagy is blocked by either treatment with 3-methyladenine (3-MA), class III PtdIns3K inhibitor, or knocking down of BECN1 or ATG7, infection was enhanced by 6- and 1.7–2-fold, respectively compared to the untreated control. Conversely, stimulation of autophagy by cell exposure to tamoxifen, which is commonly used as an anti-cancer drug and a known autophagy inducer in cells expressing the estrogen receptor,[73,74] abrogated HPV16 infection.[37] These results show that HPV16 interferes with autophagy induction by manipulating EGFR-mediated signaling early in infection, which is ultimately beneficial for the infection outcome.[37] A subsequent study revealed that 3-MA enhances infection by protecting the HPV16 capsid from being targeted and degraded by autophagy.[75]

5.2 Subversion of the ATG machinery

HSV-1 also inhibits autophagy initiation via the direct interaction of ICP34.5 with BECN1, which is a part of the class III PtdIns3k complex I.[38] Similar to BCL2,[76] binding of ICP34.5 to BECN1 appears to affect the formation of class III PtdIns3K complex I and prevents the generation of the PtdIns3P on autophagosomal membranes, which is necessary for the biogenesis of autophagosomes.[38] The interaction of ICP34.5 with BECN1 is mediated by its amino acids located at positions 68–87 and is independent of its GADD34 domain.[38] Studies in mice revealed that a mutant HSV-1 strain lacking 68–87 amino acids of ICP34.5 was severely neuroattenuated, because it leads to lower infectious titers and inflammation in the brain, resulting in a reduced animal mortality compared to control virus. This underlines the importance of the antiviral functions of autophagy, the crucial role of viral proteins in antagonizing autophagy and the impact that this has on disease pathophysiology.[38]

Herpesviruses belonging to the *Gammaherpesvirinae* subfamily such as the Kaposi's sarcoma-associated virus (KSHV) and the murine γ-herpesvirus-68 (MHV-68), carry genomic BCL2 homologs to inhibit autophagy by binding to the BH3 domain of BECN1.[39–41] Human cytomegalovirus (HCMV), which is part of the *Betaherpesvirinae* subfamily, encodes for two functional homologs of HSV-1 ICP34.5, called TRS1 and IRS1.[42,43] HCMV IRS1 binds to the coiled-coil region of BECN1, which is involved in binding to ATG14L1 and UVRAG.[43,77] Thus, one plausible scenario is that IRS1 blocks autophagy by interfering with the formation of the class III

PtdIns3K complexes I and II. Since several viruses have acquired BCL2 homologs and/or target BECN1, suggests there is a strong evolutionary pressure for viruses to be able to modulate autophagy and probably also apoptosis.[24,78]

In addition to BCL2, apoptosis is also regulated by several other proteins including FLIP, which negatively controls apoptosis by inhibiting caspase 8, but also autophagy through association to ATG3.[24,79,80] FLIP association to ATG3 outcompetes the binding of LC3 proteins to this E2-like enzyme, and therefore blocks their conjugation to PE and ultimately autophagy. KSHV encodes a viral FLIP (vFLIP) and like the cellular FLIP, binds and inhibits ATG3.[44] Cellular and viral FLIP proteins contain two N-terminal domains called the death effector domain 1 (DED1) and 2 (DED2).[44] Interestingly, 10 amino acids in the DED1 $\alpha 2$ helix and 12 in the DED2 $\alpha 4$ helix of vFLIP are sufficient to bind to ATG3. The $\alpha 2$ and $\alpha 4$ helices do not only bind ATG3 but also cellular FLIP with high affinity and when expressed in cells, they can induce autophagy and autophagy-mediated cell death by blocking the interaction between endogenous FLIP and ATG3.[44] The relevance of vFLIP-mediated inhibition of autophagy during KSHV infection, however, remains unknown.

Few studies have indicated that foot-and-mouth disease virus (FMDV), a member of the *Picornaviridiae* family, induces autophagy during early stages of infection, which was monitored by detection of an initial increase in both the number of GFP-LC3 puncta and the levels of LC3-II.[45,58,81] Interestingly, a recent study showed that UV-inactivated FMDV is also able to trigger autophagy.[58] The authors of this study demonstrated that capsid protein VP2 stimulates autophagy by a mechanism involving eIF2α-ATF4 and AKT-mTORC1 signaling pathways. Specifically, VP2 appears to block heat shock protein beta-1 (HSPB1) activity, which causes an ER stress that leads to the activation of the eIF2α-ATF4 cascade (Fig. 2C). Importantly, inhibition of autophagy by ATG5 depletion or 3-MA treatment significantly reduced FMDV genomic copies and titers.[58] However, a different investigation revealed that FMDV inhibits autophagy at the later time points of the infection by degrading ATG5-ATG12 complex using viral $3C^{pro}$.[45] In line with this finding, overexpression of ATG12-ATG5 suppressed viral mRNA and protein synthesis, and viral progeny, whereas depletion caused an overall enhancement of FMDV infection.[45] The ATG12-ATG5 conjugate was suggested to positively regulate NF-κB and IRF3 antiviral signaling thereby inhibiting FMDV replication.[45] However, this is probably not linked to autophagy since the block of this pathway significantly reduces FMDV production[58] and therefore it could be attributed to an unconventional, non-autophagy-related function of the ATG12-ATG5 conjugate.

5.3 Disruption of selective types of autophagy

The role of selective autophagy in targeting viral proteins was first reported by Beth Levine's group by demonstrating that the selective autophagy receptor p62 targets Sindbis virus capsid protein for autophagic degradation.[82] Later, a similar phenomenon was shown to take place during Chikungunya virus infection.[83] As autophagy receptors are also regulating immunity,[84] they represent an optimal target to simultaneously evade few host defenses.

The genome of flaviviruses, including Dengue (DENV) and Zika (ZIKV) virus, encode a NS3 protease and its cofactor NS2B.[85] The NS3-NS2B complex subverts reticulophagy, a selective type of autophagy leading to the turnover of ER, by cleaving FAM134B, one of the specific reticulophagy receptors.[46,86] Cleavage prevents FAM134B oligomerization,[46] which is critical for ER membrane curvature and selective autophagic targeting.[86,87] FAM134B cleavage occurs in the cytoplasmic loop at amino acid position 142. Mutation of this recognition site by changing the Arg into an Ala (i.e., FAM134B^{R142A}), impairs FAM134B processing by the NS3-NS2B complex. Importantly, knockdown of FAM134B increases DENV and ZIKV replication, and their titers. Further, co-expression of cleavage-resistant FAM134B^{R142A} variant together with the DENV NS2B3 complex resulted in an increase in the number of puncta positive for both FAM134B and NS3, suggesting that viral proteins are probably sequestered into reticulophagy-specific autophagosomes for degradation.[46]

The autophagy receptor p62/SQSTM1 targets ubiquitinated proteins, complexes and organelles for autophagic degradation.[88] In addition to its role in autophagy, the N-terminal PB1 and TB domains of p62 are also involved in activation of NF-κB, which is key in mediating antiviral innate immune response and cell survival.[89–91] The protease 2A^{pro} encoded by coxsackievirus B3 (CVB3), a picornavirus, cleaves p62 at Gly241. The resulting N- and C-terminal fragments are unlikely to sustain p62-mediated selective autophagy because contrary to the wild-type p62, these fragments very poorly colocalize with LC3. Furthermore, these two fragments fail to facilitate NF-κB signaling compared to wild-type p62.[48] On this line, a recent investigation has demonstrated that p62 knockdown increases CVB3 viral titers.[49] Interestingly, the 2A^{pro} and 3C^{pro} proteases of CVB3 cleave

NBR1, and 3C^{pro} also NDP52.[47,49] Immunoprecipitation of p62 and NDP52 revealed that these two autophagy receptors interact with CVB3 capsid protein VP1, which was also ubiquitinated providing the possible underlying mechanism of VP1 recognition by p62 and NDP52.[47] This study also showed that NDP52 has a CVB3 proviral role by suppressing type I IFN response through targeting the mitochondrial antiviral-signaling protein (MAVS) for autophagy-mediated degradation. Very interestingly, the authors demonstrated that although NDP52 is cleaved by viral 3C^{pro}, the resulting C-terminal fragment, which still contains the information to act as an autophagy receptor, retains this proviral role.[49] Further studies are required to determine whether this fragment has lost its capacity to target VP1 for virophagy. Of note, p62 is also processed by proteases encoded by other enteroviruses, including poliovirus, rhinovirus, and enterovirus D68 infection.[54] Thus, targeting selective autophagy receptors appear to be a common strategy adopted by enteroviruses to escape virophagy.

p62 has an antiviral role during DENV infection as well, as cell exposure to this virus leads to a time-dependent decrease in the levels of p62 while stable overexpression of this autophagy receptor results in a significant decrease in viral RNA and virus titers.[92] DENV, however, does not target p62 to lysosomal degradation but rather to proteasomal turnover.[92]

5.4 The blockage of autophagosome-lysosome fusion

In primary human macrophages infected with the human immunodeficiency virus 1 (HIV-1), the Nef protein of HIV-1 was found to control autophagy by preventing fusion of autophagosomes with lysosomes.[50] Thus, Nef protects HIV-1 particles from degradation into autolysosomes. Interestingly, Nef colocalizes and co-immunoprecipitates with BECN1 and knockdown of BECN1 reduced virus yields.[50] Mutational analysis of Nef revealed that the di-acidic Asp motif at positions 174 and 175 interact with BECN1.[50] Although Nef was initially also shown to interact with V-ATPase,[93] a subsequent study demonstrated that Nef does not control the acidification of endosomes/lysosomes.[94] With recent studies showing the role of UVRAG-containing class III PtdIns3K complex II in autophagosome maturation,[7] it is possible that Nef might prevent autophagosome-lysosome fusion by interfering with the formation of this complex. Nef may also interact with the RUBICON-positive PtdIns3K complex II, thereby increasing the inhibitory function of RUBICON on this complex.[77] In agreement with these data, another study showed that Nef expression in human astrocytes increases the accumulation of autophagy marker proteins such as LC3 and p62, confirming that Nef reduces the fusion of autophagosomes with lysosomes.[95]

Similar to HIV-1, multiple influenza A virus (IAV) strains were shown to inhibit fusion of autophagosomes with lysosomes as a cytoplasmic accumulation of autophagosomes was observed.[51] Ectopical expression of IAV matrix M2 protein was sufficient to cause an accumulation of autophagosomes and therefore the authors concluded that M2 is the protein responsible to interfere with autophagosome-lysosome fusion. Furthermore, silencing M2 or using a M2 knockout virus decreased autophagosome accumulation, thereby leading to an increased survival of IAV-infected cells. This suggests that a block in autophagosome-lysosome fusion promotes the IAV replication.[51] The M2 protein has a proton channel activity that is important for virion uncoating during virus cell entry.[96,97] The proton channel activity of M2 is, however, not responsible for the inhibition of fusion as autophagosomes also accumulated in cells in presence of amantadine hydrochloride, an inhibitor of the proton channel.[51,96] M2 co-immunoprecipitates with BECN1 and deletion of the first 60 N-terminal amino acids blocks this interaction.[51] Thus, similar to Nef, M2 might also target the PtdIns3K complex II, associated or not with RUBICON, to modulate the fusion of autophagosomes with lysosomes.

Human parainfluenza virus type 3 (HPIV3), a negative-strand RNA virus from the *Paramyxoviridae* family, also inhibits autophagosome-lysosome fusion but through a different mechanism.[52] Ectopic expression of the viral phosphoprotein (P) increased the cellular amounts of LC3-II and blocked fusion as tandem construct mCherry-GFP-LC3[98] accumulated in yellow puncta, which represent cytoplasmic autophagosomes.[52] To identify the mechanism by which P inhibits autophagosome-lysosome fusion, the authors performed a yeast-two hybrid screen and found that P specifically interacts with SNAP29.[52] This was additionally validated by co-immunoprecipitation of the overexpressed proteins. The mechanism by which viral P protein interferes with autophagosome-lysosome fusion is by competitively binding of P to SNAP29, which prevents the binding of STX17 to SNAP29 which is important for fusion.[52] Importantly, induced accumulation of autophagosomes in HPIV3-infected cells through knockdown of SNAP29 or treatment with bafilomycin A1, resulted in an increase in extracellular viral particles without enhancing viral protein synthesis or intracellular viral load.[52] The authors of this study speculated a possible mechanism how inhibition of this step of autophagy could promote virion egression. Their supposition is based on the observation that the M protein of HPIV3 has a crucial role in the process of viral particle release from host cells[99] and treatment of cells expressing M protein with bafilomycin A1 increased the ability of M to bind to membranes.[52] This suggests that the accumulation of autophagosomes might increase the association of this viral protein with the plasma membrane. As a result, a plausible scenario is that the hijacking of the fusion machinery by HPIV3 might help in increasing M protein binding to the plasma membrane, which in turn promotes virus budding.

Coxsackievirus CVB3 also inhibits autophagosome-lysosome fusion as yellow puncta accumulated in infected cells expressing the mRFP-GFP-LC3 authophagic flux reporter whereas co-localization between LC3 and LysoTracker, a dye staining lysosomes, was reduced.[53] The authors also showed an increase in the number of autophagosomes in vivo using a GFP-LC3 reporter mice,[100] but whether this in vivo result is due to enhanced autophagosome formation or inhibition of fusion with lysosomes remains to be established.[53] In vitro downregulation of the expression of SNAP29, STX17, VAMP8 and PLEKHM1, in combination with the cleavage of SNAP29 and PLEKHM1 by the viral 3C protease was shown to destabilizes the SNARE complex resulting in autophagosome-lysosome inhibition.[53] Furthermore, knockdown of SNAP29 and PLEKHM1 increases the expression of viral capsid protein (VP1) and intracellular CVB3 titers, supporting the notion that autophogosomes might provide membranes for viral replication and/or allow CVB3 escaping xenophagy. Accumulation of autophagosomes might also play a role in virus release as CVB3 triggers the extracellular release of LC3-II-positive microvesicles containing infectious viral particles.[101] In this context, several investigations have described the role of unconventional secretory autophagy in non-lytic release of picornaviruses.[101–104] Thus, CVB3 may hijack the fusion machinery to favor the virus particle release.[53] Similarly, enterovirus D68 (EV-D68), another picornavirus, also inhibits the autophagic flux by 3C protease-mediated cleavage of SNAP29.[54]

Interestingly, Nowag and coworkers have revealed that Epstein–Barr virus (EBV), which belongs to the *Gammaherpesvirinae* subfamily, hijacks autophagy machinery during its lytic cycle.[105] EBV has previously been shown to inhibit autophagic flux[106] and this study pointed out that EBV subverts the autophagy machinery to generate the envelope of viral particles.[105] Immunogold labeling of electron microscopy of viral particle-enriched preparations showed the presence of lipidated LC3 in virus particles.[105] A similar finding was also reported for HCMV, which initially triggers the formation of autophagosomes and then block their fusion with lysosomes.[42,107] In particular, several ATG proteins were detected in the extracellular viral particles, including some of the LC3 proteins.[107]

6. Strategies to trigger autophagy

6.1 The hijacking of signaling pathways

It has been shown that ZIKV induces autophagy in human fetal neural stem cells by interfering with AKT-mTORC1 pathway.[55] In particular, ZIKV replication leads to a reduction of AKT phosphorylation at Thr308 and Ser473, which are known to be modified by 3-phosphoinositide-dependent protein kinase 1 (PDK1) and mTORC2,[108–110] respectively. In turn, this causes a decrease in mTORC1 phosphorylation at Ser2448 that is followed by an induction of autophagy.[55] Co-expression of ZIKV nonstructural proteins NS4A and NS4B also suppresses the AKT-mTORC1 signaling axis and triggered autophagy similar to ZIKV infection, showing that these two viral proteins are responsible for the subversion of this signaling pathway. Interestingly, NS4A and NS4B failed to induce autophagy upon expression of a constitutively active form of AKT, which confirmed that these two viral proteins interfere with the activation of AKT-mTORC1 cascade.[55] The increase of autophagy during ZIKV infection is beneficial for this pathogen as the knockout of *ATG3* or *ATG5*, and the knockdown of ATG3 or ATG13, lead to a decrease in viral RNA and progeny virus production. The positive impact of autophagy on viral replication was also confirmed by treating cells exposed to ZIKV with several inducers and inhibitors of autophagy, including rapamycin, 3-MA and the lysosomal inhibitor chloroquine.[55]

Viruses are known to induce autophagy via ER stress elicited by the accumulation of viral proteins in the ER that trigger the UPR.[56,111–114] For example, initial studies on DENV infection have shown that autophagy induction is mediated by ER stress.[115,116] A recent study by Lee and colleagues demonstrated the impact of ER stress-mediated autophagy on DENV serotype 2 (DENV2) replication both in vitro and in vivo.[56] Induction of ER stress and autophagy upon DENV2 infection was shown by an increase in the levels of both glucose-regulated protein 78 kDa (GRP78), which is an ER chaperone and a marker for ER stress, and LC3-II. Alleviating ER stress using 4-phenyl butyric acid, which improves the ER folding capacity,[117] showed a dose-dependent reduction in the cellular levels of GRP78 and LC3-II, and reduced viral protein NS1 expression and the extracellular viral titers.[56] The UPR PERK-eIF2α signaling axis was transiently activated at an early stage upon cell exposure to DENV2 but had a minimal effect on autophagy. In contrast, IRE1 played a major role in the interplay between autophagy and DENV2 infection. Depletion of IRE1 using shRNA decreased the levels of LC3-II, viral NS1 and viral titers. IRE1 was shown to induce autophagy through JNK. DENV2 infection increases the levels of phosphorylated JNK in cells, which in turn phosphorylates BCL2, thereby releasing the BECN1-containing PtdIns3K complexes for autophagy activation (Fig. 2C). These observations were further confirmed using SP600125, a specific inhibitor of JNK, which reduced the levels of phosphorylated JNK, phosphorylated BCL2, LC3-II and virus titers. An in vivo *mice* study revealed an induction in ER stress and autophagy upon DENV2 infection by a slight increase in the levels of GRP78 and LC3-II, respectively, in brain tissues. Additionally, inhibition of JNK signaling using SP600125 after inoculation of DENV2 reduced the LC3-II levels in the animal brain tissue similar to those in the control mice.[56] Further studies are needed to confirm the direct link between ER stress and autophagy by utilizing alternative assays to monitor autophagy progression, in order to determine whether the in vivo impact on virus infection of the JNK inhibitor is due to the regulatory role of this kinase in autophagy and not in other pathways such as apoptosis and inflammation,[118] which might also contribute to the final positive outcome.

While DENV utilizes IRE1-JNK signaling axis during ER stress to induce autophagy,[56] other viruses appear to use other stratagems. For example, the Newcastle disease virus (NDV) from the *Paramyxoviridae* family, exploits the PERK and ATF6 branches of the UPR (Fig. 2C) with the same final beneficial effect for the virus replication cycle.[111] Chikungunya virus (CHIKV), which belongs to the *Togaviridae* family, uses the IRE1-XBP1 signaling cascade to induce autophagy and limit cell death.[119] Thus, although many viruses exploit the UPR to modulate autophagy, different viruses seem to subvert different branches of the UPR.

Rotavirus, which belongs to the *Reoviridae* family, stimulates autophagy but blocks its flux.[57] This was shown by measuring an increase in the amounts of LC3-II and the number of LC3 puncta upon cell exposure to this virus, which did not increase upon cell treatment with bafilomycin A1. Furthermore, yellow puncta were amassed in infected cells expressing the tandem mRFP-GFP-LC3 reporter construct revealing an accumulation of cytoplasmic autophagosomes.[57] The same authors previously showed that rotavirus nonstructural protein 4 (NSP4) acts as a viroporin that releases calcium into the cytoplasm from ER.[120] It is well known that an increase in cytoplasmic calcium leads to the activation of the resident calcium/calmodulin-dependent kinase kinase-β (CaMKK2), which in turn stimulates AMPK-dependent autophagy.[121] Interestingly, an NSP4 mutant protein with impaired viroporin activity was unable to induce autophagy, suggesting autophagy induction is mediated by higher levels of cytoplasmic calcium.[57] This notion was confirmed by using a calcium chelator, which also inhibited autophagy induction by NSP4. Importantly, this investigation also showed that calcium-activated, AMPK-mediated autophagy is important for the transport of NSP4 and the VP7 capsid protein from ER to the rotavirus replication/assembly sites as STO-609, an inhibitor of CaMKK2, blocked this event. The pharmacological and genetic impairment of autophagy or CaMKK2 inhibition decreased overall virus yield in host cells,[57] indicating that autophagy hijacking for the traffic of specific viral proteins is important for the virus life cycle.

Several RNA viruses such as the measles virus (MeV), the hepatitis C virus (HCV) and HIV-1 induce autophagy by interacting with IRGM.[59] Two studies reported that IRGM modulates autophagy by direct interaction with autophagy proteins such as BECN1, ATG5, ATG10, ATG14L1, ATG16L1, LC3C and ULK1.[21,59] Further, Chauhan and colleagues also reported that IRGM connects innate immune sensors like PRRs to autophagy.[21] Importantly, depletion of IRGM using siRNA and the consequent reduction of IRGM-mediated autophagy in response to the virus infection, caused a decrease in MeV, HCV and HIV-1 viral particle production to a similar extent as the ATG5 knockdown during infection.[59] The phenomenon is specific for these viruses as IAV was unaffected by IRGM knockdown. Furthermore, specific viral proteins, i.e., MeV C, HCV NS3 and HIV-1-Nef, directly interact with IRGM and are sufficient to trigger IRGM-mediated autophagy when ectopically expressed.[59] A recent study demonstrated that during HCV infection, IRGM positively regulates ULK1 activity by provoking its dephosphorylation at Ser757, which in turn stimulates ULK complex-dependent autophagosome biogenesis.[60] Altogether, these data show that RNA viruses belonging to different virus families hijack IRGM to induce autophagy and this leads to an increase in virus progeny.[59] It has been proposed that the increase in MeV virus production is associated with an autophagy-mediated protection against cell death.[122] In case of HCV, however, it has been proposed that autophagy plays a role in replication regulation,[123–125] virus release by the exosomal pathway[126,127] and inhibition of innate immune response.[128,129]

6.2 Subversion of the ATG machinery

So far there are not many examples showing subversion of the ATG machinery to promote viral infection. However, hepatitis B virus (HBV), which belongs so the *hepadnavirus* family, triggers the autophagic flux in multiple hepatoma cell lines but also in the liver tissues from HBV-infected patients.[61] As cell infection with this virus is not efficient, studies are often performed by transfecting HBV DNA into cells. Importantly, cell treatment with either 3-MA or siRNA targeting VPS34 or ATG7, inhibited HBV DNA replication. The multifunctional regulatory protein HBx encoded by HBV is responsible for autophagic flux induction as HBx mutant virus failed to stimulate this process. Interestingly, ectopically expressed HBx immunoprecipitated VPS34 indicating that HBx physically interacts with VPS34 and additionally these two factors colocalize shown by confocal microscopy. To determine whether HBx affects the VPS34 activity in vivo, the authors of this investigation took advantage of the p40phox-EGFP, which binds to PtdIns3P.[130] Co-expression of this reporter construct together with HBx showed an augmentation in GFP signal revealing that HBx might increase the intracellular PtdIns3P pool by enhancing the activity of one or more VPS34-containing class III PtdIns3K complexes.[61] A subsequent study, however, showed that ectopic expression of HBx inhibits autophagic degradation by impairing lysosomal functions.[131] It has additionally been shown that HBx induces autophagy by both direct upregulation of BECN1[63] and activation of the death-associated protein kinase (DAPK), which in turn phosphorylates and activates BECN1.[62] A recent study has highlighted that HBx also stimulates autophagy by inhibiting miR-192-3p.[64] This microRNA negatively regulates X-linked inhibitor of apoptosis (XIAP), a factor that upregulates BECN1 levels through the activation of the NF-κB signaling pathway.[64]

In addition to the HBV example, poliovirus (PV), which is part of the *Picornaviridae* family, could fall in the category of viruses that hijack the ATG machinery to promote viral infection. Previous studies have shown that PV infection induces autophagy and causes a rearrangement of the host membrane, which leads to the formation of double-membrane vesicles that are thought to being generated by the ATG machinery.[132,133] A recent study by the same team has uncovered that autophagy, also shown by the formation of GFP-WIPI2 in infected cells, is triggered by PV in an ULK complex-independent manner, as LC3-II accumulated in ULK1- or FIP200-depleted cells infected with PV to the same extent as the control.[65] Moreover, double-membrane vesicle formation was also not influenced by the absence of FIP200.[65] Additional studies, however, are needed to confirm that PV induces autophagy independently of the ULK complex by showing that under ULK and FIP200 knockdown conditions PV infection are still able to form GFP-WIPI2 puncta but not in the absence of other ATG genes. Although involving the viral proteins 2BC and 3A, the precise molecular details of the strategy employed by PV to trigger autophagy remains to be uncovered.[134,135]

6.3 Hijacking selective types of autophagy

DENV2 subverts the AMPK signaling to induce lipophagy, the selective degradation of lipid droplets by autophagy.[136] In particular, DENV2 infection causes activation of AMPK signaling by increasing AMPKα-1 phosphorylation, which in turn inhibits of mTORC1.[136] Silencing of AMPK or TSC2, reduces the total number of LC3-II positive puncta per cell and increases lipid droplets as a result of inhibition of DENV2 induced lipophagy. Importantly, DENV2-induced AMPK-mediated lipophagy is important for viral replication as silencing of AMPK or TSC2 also reduces viral RNA synthesis and titers. AMPK kinase activity is crucial for the viral replication cycle because a kinase-dead variant of AMPK did not complement the DENV2 replication defect of cells lacking wild-type AMPK. However, the mechanism by which DENV2 activates the AMPK signaling cascade remains unknown as well as the one used to selectively trigger lipophagy.[136] In addition, it remains to be determined whether the only currently known lipophagy receptor, the adipose triglyceride lipase (ATGL),[137] is critical for DENV2 infection.

A more recent study reported that bluetongue virus (BTV), which belongs to the *Reoviridae* family, evades IFN signaling by inducing the lysosomal degradation of signal transducer and activator of transcription 2 (STAT2) via autophagy, as inhibition of this pathway using 3-MA or by silencing ATG7 or BECN1 rescued STAT2 turnover.[138] Interesting, BTV-encoded NS3 binds and targets STAT2 for degradation.[138] As this event depends on NS3 ubiquitination and p62 colocalize with STAT2, a possible mechanistic scenario is that NS3 converts STAT2 into a substrate of selective autophagy.

As described above, some of the autophagy receptors like OPTN, NDP52 and TAX1BP1 facilitate autophagosome maturation.[31,32] This function of the autophagy receptors appears to be exploited by MeV as the individual or combined knockdown of NDP52 and TAX1BP1 impaired the production of MeV infectious particles, suggesting that NDP52 and TAX1BP1-mediated autophagosome maturation seems to be crucial for MeV replication.[67]

6.4 Enhancement of autophagosome-lysosome fusion

Opposite to CVB3 and EV-D68, enterovirus A71 (EV-A71), which also belongs to *Picornaviridae* family, enhances the formation of autolysosomes as cells exposed to this virus show an increase in both the autophagic flux and the number of mRFP-positive puncta when expressing the mRFP-GFP-LC3, compared to mock control.[66] The viral protein 2BC has been suggested to trigger autolysosome formation because its individual expression in cells leads to an accumulation of autolysosomes.[66] The mechanism by which 2BC increases fusion is unclear, but a yeast-two hybrid analysis uncovered an interaction between 2BC and STX17, which was validated by co-immunoprecipitation experiments in infected cells. This latter experimental approach also revealed that 2BC binds to SNAP29. Interestingly, immature viral capsid protein VP0 was also found to bind SNAP29 but the significance of this association remains to be explored. Enhanced autophagosome fusion with lysosomes might be beneficial for EV-A71 replication as inhibition of this step of autophagy by chloroquine, decreases viral protein and RNA synthesis, and viral titers.[66] This finding was confirmed by depleting different proteins involved in this fusion step, such as STX17, SNAP29 but also LC3B,[139,140] or the lysosomal protein LAMP1, which showed reduction in the ratio of mature/immature capsid proteins and virus titers.[66] Thus, autophagy, in particular autolysosomes formation, promotes EV-A71 replication and possibly maturation of the viral proteins and/or the generation of infectious virions. How an enhanced autophagic flux can be beneficial for EV-A71, however, remains to be investigated.

7. Conclusions and perspectives

Autophagy is a conserved catabolic pathway stimulated during cellular stress and is important to maintain cellular homeostasis by eliminating protein aggregates, damaged organelles and invading pathogens. Viruses have developed strategies to antagonize antiviral autophagy or ways to exploit this process for viral replication and progeny production. Antagonizing autophagy enables viruses to escape from degradation and in specific cases also limits cell death. In certain instances, it reduces cytokine production and immune signaling, which could be particularly crucial in vivo. Viruses that induce autophagy, on the other hand, appear to exploit autophagosome-derived membranes for replication or non-lytic virus particle release, but possibly also for transport of viral proteins between intracellular organelles.

Recent studies have started to highlight the intricate crosstalk between autophagy, ATG proteins, apoptosis, and inflammatory signaling. Frequently, viruses appear to subvert proteins that function as a regulatory hub with other cellular pathways such as BECN1, which limits cell death, or p62, which is also involved in autophagy-independent antiviral signaling. Thus, although each virus has evolved its unique hijacking mechanism, many of them subvert the same component of the ATG machinery like BECN1, which is targeted to interfere with autophagy initiation, or SNAP29, which is modulated to alter fusion.

In spite of the growing number of studies showing that autophagy is induced or inhibited by many viruses in different cell types, there are relatively few investigations thoroughly pinpointing the molecular determinants and the mechanism behind the hijacking stratagem. Importantly, because autophagy is both a stress and immune response, it remains unclear whether the observed change in autophagy progression is directly caused by the virus or whether it is a consequence of the infection. In this context, it will be crucial to extend the cell culture-based studies to animal models to determine the relevance of autophagy subversion for viral replication and pathogenesis in vivo. There is a clear need to better characterize the molecular level of the interaction between ATG proteins and viral factors during specific infections. This knowledge will also be crucial to design precise experiments to determine whether the studied interactions are relevant for viral replication in vivo and therefore, also of therapeutic relevance.

Acknowledgments

F.R. is supported by ZonMW VICI (016.130.606), ZonMW TOP (91217002), ALW Open Programme (ALWOP.310) and Marie Skłodowska-Curie Cofund (713660) and Marie Skłodowska-Curie ETN (765912) grants. J.S. is also supported by a Skłodowska-Curie Cofund (713660).

References

1. Parzych KR, Klionsky DJ. An overview of autophagy: morphology, mechanism, and regulation. *Antioxid Redox Signal*. 2014;20:460–473.
2. Tekirdag K, Cuervo AM. Chaperone-mediated autophagy and endosomal microautophagy: jointed by a chaperone. *J Biol Chem*. 2018;293:5414–5424.
3. Yin Z, Pascual C, Klionsky D. Autophagy: machinery and regulation. *Microb Cell*. 2016;3:588–596.

4. Dikic I, Elazar Z. Mechanism and medical implications of mammalian autophagy. *Nat Rev Mol Cell Biol*. 2018;19:349–364.
5. Lamb CA, Yoshimori T, Tooze SA. The autophagosome: origins unknown, biogenesis complex. *Nat Rev Mol Cell Biol*. 2013;14:759–774.
6. Reggiori F, Ungermann C. Autophagosome maturation and fusion. *J Mol Biol*. 2017;429:486–496.
7. Zhao YG, Zhang H. Autophagosome maturation: an epic journey from the ER to lysosomes. *J Cell Biol*. 2019;218:757–770.
8. Yang Z, Klionsky DJ. Mammalian autophagy: core molecular machinery and signaling regulation. *Curr Opin Cell Biol*. 2010;22:124–131.
9. Lahiri V, Hawkins WD, Klionsky DJ. Watch what you (self-) eat: autophagic mechanisms that modulate metabolism. *Cell Metab*. 2019;29:803–826.
10. Manning BD, Cantley LC. AKT/PKB signaling: navigating downstream. *Cell*. 2007;129:1261–1274.
11. Tamargo-Gómez I, Mariño G. AMPK: regulation of metabolic dynamics in the context of autophagy. *Int J Mol Sci*. 2018;19:3812.
12. Rashid H-O, Yadav RK, Kim H-R, Chae H-J. ER stress: autophagy induction, inhibition and selection. *Autophagy*. 2015;11:1956–1977.
13. Smith M, Wilkinson S. ER homeostasis and autophagy. *Essays Biochem*. 2017; 61:625–635.
14. Cadwell K. Crosstalk between autophagy and inflammatory signalling pathways: balancing defence and homeostasis. *Nat Rev Immunol*. 2016;16:661–675.
15. Sparrer KMJ, Gack MU. TRIM proteins: new players in virus-induced autophagy. *PLoS Pathog*. 2018;14. e1006787.
16. Virgin HW, Levine B. Autophagy genes in immunity. *Nat Immunol*. 2009;10:461–470.
17. Deretic V, Saitoh T, Akira S. Autophagy in infection, inflammation and immunity. *Nat Rev Immunol*. 2013;13:722–737.
18. Harris J. Autophagy and cytokines. *Cytokine*. 2011;56:140–144.
19. Burton PR, Clayton DG, Cardon LR, et al. Genome-wide association study of 14, 000 cases of seven common diseases and 3,000 shared controls. *Nature*. 2007;447:661–678.
20. Song JH, Kim SY, Chung KS, et al. Association between genetic variants in the IRGM gene and tuberculosis in a Korean population. *Infection*. 2014;42:655–660.
21. Chauhan S, Mandell MA, Deretic V. IRGM governs the core autophagy machinery to conduct antimicrobial defense. *Mol Cell*. 2015;58:507–521.
22. Galluzzi L, Vitale I, Aaronson SA, et al. Molecular mechanisms of cell death: recommendations of the Nomenclature Committee on Cell Death 2018. *Cell Death Differ*. 2018;25:486–541.
23. Levine B. Autophagy in cell death: an innocent convict? *J Clin Invest*. 2005; 115:2679–2688.
24. Mariño G, Niso-Santano M, Baehrecke EH, Kroemer G. Self-consumption: the interplay of autophagy and apoptosis. *Nat Rev Mol Cell Biol*. 2014;15:81–94.
25. Mukhopadhyay S, Panda PK, Sinha N, Das DN, Bhutia SK. Autophagy and apoptosis: where do they meet? *Apoptosis*. 2014;19:555–566.
26. Rubinstein AD, Kimchi A. Life in the balance—a mechanistic view of the crosstalk between autophagy and apoptosis. *J Cell Sci*. 2012;125:5259–5268.
27. Kirkin V. History of the Selective Autophagy research: how did it begin and where does it stand today? *J Mol Biol*. 2019;19:3–27.
28. Ravenhill BJ, Boyle KB, von Muhlinen N, et al. The cargo receptor NDP52 initiates selective autophagy by recruiting the ULK complex to cytosol-invading bacteria. *Mol Cell*. 2019;74:320–329.e6.
29. Turco E, Witt M, Abert C, et al. FIP200 claw domain binding to p62 promotes autophagosome formation at ubiquitin condensates. *Mol Cell*. 2019;74:330–346.e11.

30. Vargas JNS, Wang C, Bunker E, et al. Spatiotemporal control of ULK1 activation by NDP52 and TBK1 during selective autophagy. *Mol Cell.* 2019;74:347–362.e6.
31. Tumbarello DA, Waxse BJ, Arden SD, Bright NA, Kendrick-Jones J, Buss F. Autophagy receptors link myosin VI to autophagosomes to mediate Tom1-dependent autophagosome maturation and fusion with the lysosome. *Nat Cell Biol.* 2012; 14:1024–1035.
32. Verlhac P, Grégoire IP, Azocar O, et al. Autophagy receptor NDP52 regulates pathogen-containing autophagosome maturation. *Cell Host Microbe.* 2015;17:515–525.
33. Levine B, Kroemer G. Biological functions of autophagy genes: a disease perspective. *Cell.* 2019;176:11–42.
34. Van Beek N, Klionsky DJ, Reggiori F. Genetic aberrations in macroautophagy genes leading to diseases. *Biochim Biophys Acta, Mol Cell Res.* 2018;1865:803–816.
35. Lussignol M, Queval C, Bernet-Camard M-F, et al. The herpes simplex virus 1 Us11 protein inhibits autophagy through its interaction with the protein kinase PKR. *J Virol.* 2013;87:859–871.
36. Liu X, Matrenec R, Gack MU, He B. Disassembly of the TRIM23-TBK1 complex by the Us11 protein of herpes simplex virus 1 impairs autophagy. *J Virol.* 2019;93:e00497-19.
37. Surviladze Z, Sterk RT, DeHaro SA, Ozbun MA. Cellular entry of human papillomavirus type 16 involves activation of the phosphatidylinositol 3-kinase/Akt/mTOR pathway and inhibition of autophagy. *J Virol.* 2013;87:2508–2517.
38. Orvedahl A, Alexander D, Tallóczy Z, et al. HSV-1 ICP34.5 confers neurovirulence by targeting the Beclin 1 autophagy protein. *Cell Host Microbe.* 2007;1:23–35.
39. Sarid R, Sato T, Bohenzky RA, Russo JJ, Chang Y. Kaposi's sarcoma-associated herpesvirus encodes a functional Bcl-2 homologue. *Nat Med.* 1997;3:293–298.
40. Virgin HW, Latreille P, Wamsley P, et al. Complete sequence and genomic analysis of murine gammaherpesvirus 68. *J Virol.* 1997;71:5894–5904.
41. Sinha SC, Colbert CL, Becker N, Wei Y, Levine B. Molecular basis of the regulation of Beclin 1-dependent autophagy by the γ-herpesvirus 68 Bcl-2 homolog M11. *Autophagy.* 2008;4:989–997.
42. Chaumorcel M, Lussignol M, Mouna L, et al. The human cytomegalovirus protein TRS1 inhibits autophagy via its interaction with Beclin 1. *J Virol.* 2012;86:2571–2584.
43. Mouna L, Hernandez E, Bonte D, et al. Analysis of the role of autophagy inhibition by two complementary human cytomegalovirus BECN1/Beclin 1-binding proteins. *Autophagy.* 2016;12:327–342.
44. Lee J-S, Li Q, Lee J-Y, et al. FLIP-mediated autophagy regulation in cell death control. *Nat Cell Biol.* 2009;11:1355–1362.
45. Fan X, Han S, Yan D, et al. Foot-and-mouth disease virus infection suppresses autophagy and NF-κB antiviral responses via degradation of ATG5-ATG12 by 3Cpro. *Cell Death Dis.* 2018;8:e2561.
46. Lennemann NJ, Coyne CB. Dengue and Zika viruses subvert reticulophagy by NS2B3-mediated cleavage of FAM134B. *Autophagy.* 2017;13:322–332.
47. Shi J, Fung G, Piesik P, Zhang J, Luo H. Dominant-negative function of the C-terminal fragments of NBR1 and SQSTM1 generated during enteroviral infection. *Cell Death Differ.* 2014;21:1432–1441.
48. Shi J, Wong J, Piesik P, et al. Cleavage of sequestosome 1/p62 by an enteroviral protease results in disrupted selective autophagy and impaired NFKB signaling. *Autophagy.* 2013;9:1591–1603.
49. Mohamud Y, Qu J, Xue YC, Liu H, Deng H, Luo H. CALCOCO2/NDP52 and SQSTM1/p62 differentially regulate coxsackievirus B3 propagation. *Cell Death Differ.* 2019;26:1062–1076.
50. Kyei GB, Dinkins C, Davis AS, et al. Autophagy pathway intersects with HIV-1 biosynthesis and regulates viral yields in macrophages. *J Cell Biol.* 2009;186:255–268.

51. Gannagé M, Dormann D, Albrecht R, et al. Matrix protein 2 of influenza A virus blocks autophagosome fusion with lysosomes. *Cell Host Microbe*. 2009;6:367–380.
52. Ding B, Zhang G, Yang X, et al. Phosphoprotein of human parainfluenza virus type 3 Blocks autophagosome-lysosome fusion to increase virus production. *Cell Host Microbe*. 2014;15:564–577.
53. Mohamud Y, Shi J, Qu J, et al. Enteroviral infection inhibits autophagic flux via disruption of the SNARE complex to enhance viral replication. *Cell Rep*. 2018;22: 3292–3303.
54. Corona AK, Saulsbery HM, Corona Velazquez AF, Jackson WT. Enteroviruses remodel autophagic trafficking through regulation of host SNARE proteins to promote virus replication and cell exit. *Cell Rep*. 2018;22:3304–3314.
55. Liang Q, Luo Z, Zeng J, et al. Zika virus NS4A and NS4B proteins deregulate Akt-mTOR signaling in human fetal neural stem cells to inhibit neurogenesis and induce autophagy. *Cell Stem Cell*. 2016;19:663–671.
56. Lee Y-R, Kuo S-H, Lin C-Y, et al. Dengue virus-induced ER stress is required for autophagy activation, viral replication, and pathogenesis both in vitro and in vivo. *Sci Rep*. 2018;8:489.
57. Crawford SE, Hyser JM, Utama B, Estes MK. Autophagy hijacked through viroporin-activated calcium/calmodulin-dependent kinase kinase- signaling is required for rotavirus replication. *Proc Natl Acad Sci*. 2012;109:E3405–E3413.
58. Sun P, Zhang S, Qin X, et al. Foot-and-mouth disease virus capsid protein VP2 activates the cellular EIF2S1-ATF4 pathway and induces autophagy via HSPB1. *Autophagy*. 2018;14:336–346.
59. Grégoire IP, Richetta C, Meyniel-Schicklin L, et al. IRGM is a common target of RNA viruses that subvert the autophagy network. *PLoS Pathog*. 2011;7:e1002422.
60. Hansen MD, Johnsen IB, Stiberg KA, et al. Hepatitis C virus triggers Golgi fragmentation and autophagy through the immunity-related GTPase M. *Proc Natl Acad Sci*. 2017;114:E3462–E3471.
61. Sir D, Tian Y, Chen WL, Ann DK, Yen TSB, Ou JHJ. The early autophagic pathway is activated by hepatitis B virus and required for viral DNA replication. *Proc Natl Acad Sci*. 2010;107:4383–4388.
62. Zhang H-T, Chen GG, Hu B-G, et al. Hepatitis B virus x protein induces autophagy via activating death-associated protein kinase. *J Viral Hepat*. 2014;21:642–649.
63. Tang H, Da L, Mao Y, et al. Hepatitis B virus X protein sensitizes cells to starvation-induced autophagy via up-regulation of beclin 1 expression. *Hepatology*. 2009;49: 60–71.
64. Wang J, Chen J, Liu Y, et al. Hepatitis B virus induces autophagy to promote its replication by the axis of miR-192-3p-XIAP through NF kappa B signaling. *Hepatology*. 2019;69:974–992.
65. Corona Velazquez A, Corona AK, Klein KA, Jackson WT. Poliovirus induces autophagic signaling independent of the ULK1 complex. *Autophagy*. 2018;14: 1201–1213.
66. Lai J, Sam I-C, Verlhac P, et al. 2BC non-structural protein of enterovirus A71 interacts with SNARE proteins to trigger autolysosome formation. *Viruses*. 2017;9:169.
67. Petkova D, Verlhac P, Rozières A, et al. Distinct contributions of autophagy receptors in measles virus replication. *Viruses*. 2017;9:123.
68. Talloczy Z, Jiang W, Virgin HW, et al. Regulation of starvation- and virus-induced autophagy by the eIF2 kinase signaling pathway. *Proc Natl Acad Sci*. 2002;99: 190–195.
69. Liebel U, Kindler B, Pepperkok R. Bioinformatic "Harvester": a search engine for genome-wide human, mouse, and rat protein resources. *Methods Enzymol*. 2005;404: 19–26.

70. Matsumoto G, Shimogori T, Hattori N, Nukina N. TBK1 controls autophagosomal engulfment of polyubiquitinated mitochondria through p62/SQSTM1 phosphorylation. *Hum Mol Genet*. 2015;24:4429–4442.
71. Sparrer KMJ, Gableske S, Zurenski MA, et al. TRIM23 mediates virus-induced autophagy via activation of TBK1. *Nat Microbiol*. 2017;2:1543–1557.
72. Vazquez F, Ramaswamy S, Nakamura N, Sellers WR. Phosphorylation of the PTEN tail regulates protein stability and function. *Mol Cell Biol*. 2000;20:5010–5018.
73. Bilir A, Altinoz MA, Erkan M, Ozmen V, Aydiner A. Autophagy and nuclear changes in FM3A breast tumor cells after epirubicin, medroxyprogesterone and tamoxifen treatment in vitro. *Pathobiology*. 2001;69:120–126.
74. Bursch W, Hochegger K, Török L, Marian B, Ellinger A, Hermann RS. Autophagic and apoptotic types of programmed cell death exhibit different fates of cytoskeletal filaments. *J Cell Sci*. 2000;113:1189–1198.
75. Griffin LM, Cicchini L, Pyeon D. Human papillomavirus infection is inhibited by host autophagy in primary human keratinocytes. *Virology*. 2013;437:12–19.
76. Pattingre S, Tassa A, Qu X, et al. Bcl-2 antiapoptotic proteins inhibit Beclin 1-dependent autophagy. *Cell*. 2005;122:927–939.
77. Matsunaga K, Saitoh T, Tabata K, et al. Two Beclin 1-binding proteins, Atg14L and Rubicon, reciprocally regulate autophagy at different stages. *Nat Cell Biol*. 2009; 11:385–396.
78. Kvansakul M, Caria S, Hinds M. The Bcl-2 family in host-virus interactions. *Viruses*. 2017;9:290.
79. Adams JM, Cory S. Life-or-death decisions by the Bcl-2 protein family. *Trends Biochem Sci*. 2001;26:61–66.
80. Thome M, Tschopp J. Regulation of lymphocyte proliferation and death by FLIP. *Nat Rev Immunol*. 2001;1:50–58.
81. Berryman S, Brooks E, Burman A, et al. Foot-and-mouth disease virus induces autophagosomes during cell entry via a class III phosphatidylinositol 3-kinase-independent pathway. *J Virol*. 2012;86:12940–12953.
82. Orvedahl A, MacPherson S, Sumpter R, Tallóczy Z, Zou Z, Levine B. Autophagy protects against Sindbis virus infection of the central nervous system. *Cell Host Microbe*. 2010;7:115–127.
83. Judith D, Mostowy S, Bourai M, et al. Species-specific impact of the autophagy machinery on Chikungunya virus infection. *EMBO Rep*. 2013;14:534–544.
84. Wu Y, Cui J. Selective autophagy regulates innate immunity through cargo receptor network. *Adv Exp Med Biol*. 2019;1209:145–166.
85. Pierson TC, Diamond MS. Flaviviruses. In: Knipe DM, Howley PM, eds. *Fields Virology*. 6th ed. Philadelphia, PA, USA: Wolters Kluwer/Lippencott Williams & Wilkins; 2013:747–794.
86. Khaminets A, Heinrich T, Mari M, et al. Regulation of endoplasmic reticulum turnover by selective autophagy. *Nature*. 2015;522:354–358.
87. Shibata Y, Voss C, Rist JM, et al. The reticulon and DP1/Yop1p proteins form immobile oligomers in the tubular endoplasmic reticulum. *J Biol Chem*. 2008;283: 18892–18904.
88. Katsuragi Y, Ichimura Y, Komatsu M. p62/SQSTM1 functions as a signaling hub and an autophagy adaptor. *FEBS J*. 2015;282:4672–4678. d.
89. Moscat J, Diaz-Meco MT. p62 at the crossroads of autophagy, apoptosis, and cancer. *Cell*. 2009;137:1001–1004.
90. Le Negrate G. Viral interference with innate immunity by preventing NF-κB activity. *Cell Microbiol*. 2012;14:168–181.
91. Sanz L. The atypical PKC-interacting protein p62 channels NF-kappa B activation by the IL-1-TRAF6 pathway. *EMBO J*. 2000;19:1576–1586.

92. Metz P, Chiramel A, Chatel-Chaix L, et al. Dengue virus inhibition of autophagic flux and dependency of viral replication on proteasomal degradation of the autophagy receptor p62. *J Virol*. 2015;89:8026–8041.
93. Lu X, Yu H, Liu S-H, Brodsky FM, Peterlin BM. Interactions between HIV1 Nef and vacuolar ATPase facilitate the internalization of CD4. *Immunity*. 1998;8:647–656.
94. Jouve M, Sol-Foulon N, Watson S, Schwartz O, Benaroch P. HIV-1 buds and accumulates in "nonacidic" endosomes of macrophages. *Cell Host Microbe*. 2007;2:85–95.
95. Saribas AS, Khalili K, Sariyer IK. Dysregulation of autophagy by HIV-1 Nef in human astrocytes. *Cell Cycle*. 2015;14:2899–2904.
96. Hay AJ, Wolstenholme AJ, Skehel JJ, Smith MH. The molecular basis of the specific anti-influenza action of amantadine. *EMBO J*. 1985;4:3021–3024.
97. Pielak RM, Chou JJ. Influenza M2 proton channels. *Biochim Biophys Acta, Biomembr*. 2011;1808:522–529.
98. Kimura S, Noda T, Yoshimori T. Dissection of the autophagosome maturation process by a novel reporter protein, tandem fluorescent-tagged LC3. *Autophagy*. 2007;3:452–460.
99. Takimoto T, Portner A. Molecular mechanism of paramyxovirus budding. *Virus Res*. 2004;106:133–145.
100. Mizushima N. Chapter 2 methods for monitoring autophagy using GFP-LC3 transgenic mice. In: Klionsky D, ed. *Methods in Enzymology*. Academic Press; 2009:13–23.
101. Robinson SM, Tsueng G, Sin J, et al. Coxsackievirus B exits the host cell in shed microvesicles displaying autophagosomal markers. *PLoS Pathog*. 2014;10:e1004045.
102. Bird SW, Maynard ND, Covert MW, Kirkegaard K. Nonlytic viral spread enhanced by autophagy components. *Proc Natl Acad Sci*. 2014;111:13081–13086.
103. Chen Y-H, Du W, Hagemeijer MC, et al. Phosphatidylserine vesicles enable efficient en bloc transmission of enteroviruses. *Cell*. 2015;160:619–630.
104. Feng Z, Hensley L, McKnight KL, et al. A pathogenic picornavirus acquires an envelope by hijacking cellular membranes. *Nature*. 2013;496:367–371.
105. Nowag H, Guhl B, Thriene K, et al. Macroautophagy proteins assist epstein barr virus production and get incorporated into the virus particles. *EBioMedicine*. 2014;1:116–125.
106. Granato M, Santarelli R, Farina A, et al. Epstein-barr virus blocks the autophagic flux and appropriates the autophagic machinery to enhance viral replication. *J Virol*. 2014;88:12715–12726.
107. Taisne C, Lussignol M, Hernandez E, Moris A, Mouna L, Esclatine A. Human cytomegalovirus hijacks the autophagic machinery and LC3 homologs in order to optimize cytoplasmic envelopment of mature infectious particles. *Sci Rep*. 2019;9:4560.
108. Alessi DR, James SR, Downes CP, et al. Characterization of a 3-phosphoinositide-dependent protein kinase which phosphorylates and activates protein kinase Bα. *Curr Biol*. 1997;7:261–269.
109. Sarbassov DD. Phosphorylation and regulation of Akt/PKB by the rictor-mTOR complex. *Science*. 2005;307:1098–1101.
110. Stephens L, Anderson K, Stokoe D, et al. Protein kinase B kinases that mediate phosphatidylinositol 3,4,5-trisphosphate-dependent activation of protein kinase B. *Science*. 1998;279:710–714.
111. Cheng J-H, Sun Y-J, Zhang F-Q, et al. Newcastle disease virus NP and P proteins induce autophagy via the endoplasmic reticulum stress-related unfolded protein response. *Sci Rep*. 2016;6:24721.
112. Fung TS, Liu DX. The ER stress sensor IRE1 and MAP kinase ERK modulate autophagy induction in cells infected with coronavirus infectious bronchitis virus. *Virology*. 2019;533:34–44.
113. Joubert P-E, Werneke S, de la Calle C, et al. Chikungunya-induced cell death is limited by ER and oxidative stress-induced autophagy. *Autophagy*. 2012;8:1261–1263.

114. Wang J, Kang R, Huang H, et al. Hepatitis C virus core protein activates autophagy through EIF2AK3 and ATF6 UPR pathway-mediated MAP1LC3B and ATG12 expression. *Autophagy*. 2014;10:766–784.
115. Datan E, Roy SG, Germain G, et al. Dengue-induced autophagy, virus replication and protection from cell death require ER stress (PERK) pathway activation. *Cell Death Dis*. 2016;7:e2127.
116. Thepparit C, Khakpoor A, Khongwichit S, et al. Dengue 2 infection of HepG2 liver cells results in endoplasmic reticulum stress and induction of multiple pathways of cell death. *BMC Res Notes*. 2013;6:372.
117. Perlmutter DH. Chemical chaperones: a pharmacological strategy for disorders of protein folding and trafficking. *Pediatr Res*. 2002;52:832–836.
118. Johnson GL, Nakamura K. The c-jun kinase/stress-activated pathway: regulation, function and role in human disease. *Biochim Biophys Acta, Mol Cell Res*. 2007;1773: 1341–1348.
119. Joubert P-E, Werneke SW, de la Calle C, et al. Chikungunya virus-induced autophagy delays caspase-dependent cell death. *J Exp Med*. 2012;209:1029–1047.
120. Hyser JM, Collinson-Pautz MR, Utama B, Estes MK. Rotavirus disrupts calcium homeostasis by NSP4 viroporin activity. *MBio*. 2010;1:1–11.
121. Høyer-Hansen M, Bastholm L, Szyniarowski P, et al. Control of macroautophagy by calcium, calmodulin-dependent kinase kinase-beta, and Bcl-2. *Mol Cell*. 2007; 25:193–205.
122. Richetta C, Grégoire IP, Verlhac P, et al. Sustained autophagy contributes to measles virus infectivity. *PLoS Pathog*. 2013;9:e1003599.
123. Dreux M, Gastaminza P, Wieland SF, Chisari FV. The autophagy machinery is required to initiate hepatitis C virus replication. *Proc Natl Acad Sci*. 2009;106: 14046–14051.
124. Kim JY, Wang L, Lee J, Ou JJ. Hepatitis C virus induces the localization of lipid rafts to autophagosomes for its RNA replication. *J Virol*. 2017;91:e00541-17.
125. Sir D, Kuo C, Tian Y, et al. Replication of hepatitis c virus RNA on autophagosomal membranes. *J Biol Chem*. 2012;287:18036–18043.
126. Shrivastava S, Devhare P, Sujijantarat N, et al. Knockdown of autophagy inhibits infectious hepatitis V virus release by the exosomal pathway. *J Virol*. 2016;90:1387–1396.
127. Tanida I, Fukasawa M, Ueno T, Kominami E, Wakita T, Hanada K. Knockdown of autophagy-related gene decreases the production of infectious Hepatitis C virus particles. *Autophagy*. 2009;5:937–945.
128. Ke P-Y, Chen SSL. Activation of the unfolded protein response and autophagy after hepatitis C virus infection suppresses innate antiviral immunity in vitro. *J Clin Invest*. 2011;121:37–56.
129. Shrivastava S, Raychoudhuri A, Steele R, Ray R, Ray RB. Knockdown of autophagy enhances the innate immune response in hepatitis C virus-infected hepatocytes. *Hepatology*. 2011;53:406–414.
130. Kanai F, Liu H, Field SJ, et al. The PX domains of p47phox and p40phox bind to lipid products of PI(3)K. *Nat Cell Biol*. 2001;3:675–678.
131. Liu B, Fang M, Hu Y, et al. Hepatitis B virus X protein inhibits autophagic degradation by impairing lysosomal maturation. *Autophagy*. 2014;10:416–430.
132. Jackson WT, Giddings TH, Taylor MP, et al. Subversion of cellular autophagosomal machinery by RNA viruses. *PLoS Biol*. 2005;3:e156.
133. Richards AL, Jackson WT. Intracellular vesicle acidification promotes maturation of infectious poliovirus particles. *PLoS Pathog*. 2012;8:e1003046.
134. Suhy DA, Giddings TH, Kirkegaard K. Remodeling the endoplasmic reticulum by poliovirus infection and by individual viral proteins: an autophagy-like origin for virus-induced vesicles. *J Virol*. 2000;74:8953–8965.

135. Taylor MP, Kirkegaard K. Modification of cellular autophagy protein LC3 by poliovirus. *J Virol.* 2007;81:12543–12553.
136. Jordan TX, Randall G. Dengue virus activates the AMP kinase-mTOR axis to stimulate a proviral lipophagy. *J Virol.* 2017;91:e02020-16.
137. Sathyanarayan A, Mashek MT, Mashek DG. ATGL promotes autophagy/lipophagy via SIRT1 to control hepatic lipid droplet catabolism. *Cell Rep.* 2017;19:1–9.
138. Avia M, Rojas JM, Miorin L, et al. Virus-induced autophagic degradation of STAT2 as a mechanism for interferon signaling blockade. *EMBO Rep.* 2019;20:e48766.
139. McEwan DG, Popovic D, Gubas A, et al. PLEKHM1 regulates autophagosome-lysosome fusion through HOPS complex and LC3/GABARAP proteins. *Mol Cell.* 2015;57:39–54.
140. Nguyen TN, Padman BS, Usher J, Oorschot V, Ramm G, Lazarou M. Atg8 family LC3/GABARAP proteins are crucial for autophagosome–lysosome fusion but not autophagosome formation during PINK1/Parkin mitophagy and starvation. *J Cell Biol.* 2016;215:857–874.

CHAPTER ELEVEN

Crinophagy mechanisms and its potential role in human health and disease

Tamás Csizmadia[a], Gábor Juhász[a,b,*]
[a]Department of Anatomy, Cell and Developmental Biology, Eötvös Loránd University, Budapest, Hungary
[b]Institute of Genetics, Biological Research Centre, Szeged, Hungary
[*]Corresponding author: e-mail address: szmrt@elte.hu

Contents

Abstract

Autophagic-lysosomal degradation is essential for the maintenance of normal homeostasis in eukaryotic cells. Several types of such self-degradative and recycling pathways have been identified. From these, probably the least known autophagic process is crinophagy, during which unnecessary or obsolete secretory granules directly fuse with late endosomes/lysosomes as a means of rapid elimination of unused secretory material from the cytoplasm. This process was identified in 1966, but we are only beginning to understand the molecular mechanisms and regulation of crinophagy. In this review, we summarize the current examination methods and possible model systems, discuss the recently identified factors that are required for crinophagy, and give an overview of the potential medical relevance of this process.

Progress in Molecular Biology and Translational Science, Volume 172
ISSN 1877-1173
https://doi.org/10.1016/bs.pmbts.2020.02.002

1. The various forms of autophagy

The essential property of eukaryotic cells is their ability to renew themselves. Consequently, their own materials including all sorts of biomolecules and cellular organelles are continuously broken down and resynthesized. Distinct mechanisms evolved that ensure the removal of old, damaged, or obsolete cellular materials. These pathways not only maintain a steady-state homeostasis of cells but also enable their adaptation to the variable conditions. Thus, the breakdown of self-components is essential in the normal functioning of the eukaryotic cells, which involves various cellular lysosomal self-digestion (autophagy) mechanisms.[1,2] These degradative pathways are responsible for the breakdown and recycling of the largest amount of the cytoplasmic material in eukaryotic cells. Depending on how cytoplasmic components reach the lysosomal degradative milieu, we can distinguish four main forms of autophagic processes: macroautophagy, microautophagy, chaperone-mediated autophagy and crinophagy[1,2] (Fig. 1).

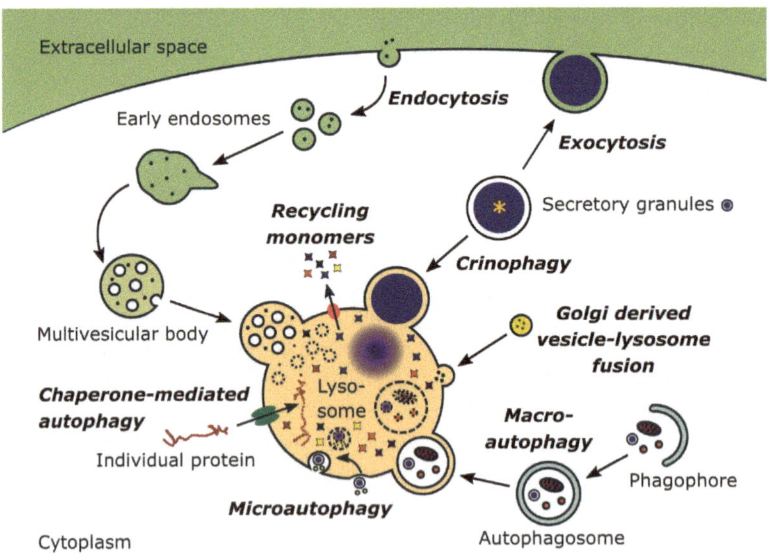

Fig. 1 Summary of lysosomal degradative and recycling pathways in eukaryotic cells. Secretory granules (also marked by yellow asterisk) can either be released or degraded via crinophagy (or macro/microautophagy depending on granule size). Blue vesicles represent secretory granules of various sizes.

Macroautophagy is the best known and most intensively researched among these degradative pathways. The process starts with the formation of a phagophore cistern, which sequesters various cellular components from the cytoplasm into double-membrane bound autophagosomes. These vesicles transport their cargo to the acidic lysosomes for the fast degradation of cytoplasmic material.[3] Autophagosome formation depends on the coordinated action of evolutionarily conserved Atg (Autophagy related) proteins, which form distinct protein complexes. The first Atg genes were discovered and characterized by *Yoshinori Ohsumi* and his research group in yeast,[4] the further study of which culminated in his 2016 Nobel Prize. The fusion of autophagosomes with endosomes and lysosomes is essential for successful degradation of cargo. The main factors mediating fusion are the tethering complex HOPS (homotypic fusion and vacuole protein sorting), the small GTPases Rab2 and Rab7, plus specific SNARE (Soluble NSF Attachment Protein Receptor) proteins including autophagosomal Syntaxin 17 and its binding partners SNAP29 and Vamp8 (Vamp7 in *Drosophila*).[5–13] As a result of autophagosome-lysosomal fusion, both the transported cargo and the inner membrane of the autophagosome are digested, while the outer membrane blends into the membrane of the resulting autolysosome. The membrane of lysosomes contains special efflux transporter proteins (permeases), which allow the monomers released from cargo degradation to be recycled into the cytoplasm where these fuel biosynthetic and energy production processes.

During microautophagy, part of the lysosomal membrane invaginates and engulfs a portion of the surrounding cytoplasm to create small intraluminal vesicles, which are then degraded.[14] Microautophagy may superficially resemble endocytosis, but it occurs in a topologically opposite direction (note that the lysosomal lumen can be considered as extracellular space). Although the capacity of this autophagic pathway lags behind that of macroautophagy, microautophagy can still mediate the lysosomal breakdown of mitochondria, peroxisomes and even parts of the nucleus.[14,15] Microautophagy has been mainly characterized in the lysosome equivalent vacuole of plant and yeast cells. A similar process is also observed in animal cells, where small amounts of cytoplasmic cargo are taken up into the lysosome or late endosome. Of note, endosomal microautophagy depends on ESCRT proteins that also mediate the sorting of internalized receptor-ligand complexes in multivesicular endosomes.[16]

The third lysosomal degradative pathway is chaperone-mediated autophagy (CMA), which only degrades individual soluble proteins.

During CMA, the cytoplasmic chaperone Hsc70 recognizes an exposed KFERQ-like amino acid sequence that is normally hidden within potential target proteins. Exposure of this sequence on the surface of old or damaged proteins is the signal for their delivery into the lysosomal lumen. These proteins, chaperoned by Hsc70, are transported into the lysosomal lumen through a channel formed by the lysosomal membrane protein Lamp2A for rapid breakdown by lysosomal hydrolases.[17]

Crinophagy is the least characterized autophagic process, during which secretory granules that are not released directly fuse with late endosomes or lysosomes, and then excess or obsolete secretory material is degraded and recycled.[2] Even though crinophagy was discovered more than 50 years ago, the molecular mechanism and regulation of this process remained largely unknown. Importantly, secretory vesicles can also degrade via macro- or microautophagy depending on their size. The insulin containing secretory granules are small enough (0.3–0.4 μm) to be degraded by macro- and microautophagy in addition to crinophagy.[18,19] However, the secretory glue granules in *Drosophila* larval salivary gland cells can only be degraded by crinophagy, because there are too large (3–3.5 μm) for macro- and macroautophagy.[20–22] In this paper, we summarize what is known about crinophagy as well as the available model systems and methods for its study. Finally, we also discuss the potential medical and pathological relevance of this autophagic process.

2. Pre-requisites for autophagic degradation

The last step of all autophagic processes is lysosomal degradation. Thus, functional lysosomes containing active acidic hydrolases (such as cathepsins) and a sufficiently low pH must be created and maintained. The acidity of lysosomes is essential for the proper activity of digestive enzymes. This is achieved by the action of lysosomal/vacuolar H^+ ATPase (v-ATPase).[23,24] Autophagic degradation is blocked if vacuolar/lysosomal degradation is impaired in yeast, *Drosophila* and mammalian cells.[24–26] Depending on the cell type, lysosomes have a specific enzyme composition that is transported to these organelles from the trans Golgi network. The Uvrag (UV-radiation Resistance Associated Gene) containing Vps34 (Vacuolar protein sorting 34) kinase complex promotes the trafficking of enzymes and membrane proteins to lysosomal compartments,[8,27–30] which is necessary for the final breakdown of cargo in all lysosomal degradation pathways. Finally, a critical function of autophagic pathways is to recycle the resulting monomers from the lysosomal lumen to the cytoplasm, which requires the active involvement of efflux transporters (permeases) in the lysosomal membrane.[31]

Nevertheless, little is known about the transcriptional regulation of genes whose products are necessary for lysosomal function. The best-known transcription factors that regulate the expression of lysosomal genes belong to the TFEB (Transcription Factor EB) family. These proteins regulate the transcription of several genes essential for lysosomal function, such as multiple subunits of the v-ATPase proton pump, Lamp (Lysosome associated membrane protein) and genes encoding permeases.[32,33]

In times of intensive degradation, there is increased need for lysosomal proteins and lysosomes. In parallel to boosting the production of such proteins, eukaryotic cells can also cope with this limitation via lysosomal reformation. During this process, the enzymes and membrane proteins of lysosomes are sorted into small tubules and vesicles that bud from the surface of the lysosome, followed by their recycling to help the quick production of new lysosomes.[34]

3. Crinophagy: A nearly forgotten cell biological pathway

One of the least known forms of autophagic processes is crinophagy (κρινοφαγία). The term is derived from the Greek word: krino (meaning secretion) and fagein (eating). The name of crinophagy was coined by Christian de Duve (the founder of the fields of autophagy, endocytosis and lysosomal biology), who named the process in reference to the selective degradation of secretory granules.[35] During crinophagy, unreleased secretory vesicles directly fuse with late endosomes/lysosomes, resulting in the rapid digestion and reuse of their contents.[2] Upon the fusion of secretory vesicles and lysosomes, a digestive secondary lysosome (called crinosome) is created.[36] In line with this, membrane lipids and proteins as well as the luminal components of crinosomes come from both compartments.[22,36] Crinophagy occurs in all cells that produce secreted material and it is particularly important in glandular (exocrine, endocrine and neuroendocrine) cells specialized in secretion.

Direct fusion of secretory granules with late endosomes or lysosomes was first suggested by Smith and Farquhar in 1966, based on their pioneering ultrastructural examination of the lactotroph hormone-producing cells in the anterior pituitary gland of female rats.[37] Authors of this work followed lysosomal activity in the lactotroph cells from lactating animals compared to female rats after weaning. The cytoplasm of lactotroph hormone-producing cells from lactating rats contains a highly developed secretory apparatus (extensive Golgi and RER cisterns), and many lactotroph hormone-containing secretory granules are seen in their cytoplasm. In contrast, the secretory apparatus in the lactotroph cells from non-lactating animals was greatly reduced: in fact, many secretory organelles (Golgi and RER cisterns as well as secretory vesicles) were observed inside autophagic vacuoles. Furthermore, the number of lactotroph hormone-containing secretory granules decreased significantly in cells from non-lactating animals.[37] Interestingly, these cells contained electron-dense secretory product in the lumen of heterogeneous lysosomes without having an extra membrane around it. This suggested that the membrane of secretory granules directly fused with the late endosomal/lysosomal membrane.[37] Smith and Farquhar thus discovered a previously unknown autophagic process that was later named crinophagy.

Other early studies also showed that lysosomes are necessary for the response of glandular and other cells to altered conditions. Crinophagy is involved in the regulation of the secretory cycle of salamander pituitary gland cells.[38] Secretory granule-lysosome fusions also occur in larval salivary gland cells during the metamorphosis of *Drosophila pseudoobscura* based on ultrastructural analysis.[21] The salivary glands of *Drosophila* larvae produce saliva to aid food ingestion and intestinal digestion. Then, at around 14 h before puparium formation, the gland changes its function and starts to produce glue that will help to attach the forming (pre)pupa to a solid surface. Glue production coincides with a behavioral switch, when larvae leave the food and start wandering in search of a dry and safe place where they can complete metamorphosis. After glue secretion and puparium formation, unreleased glue-containing secretory vesicles were found to be degraded via developmentally programmed crinophagy.[21]

Interestingly, prevention of secretion was found to strongly increase crinophagic activity in several types of glandular cells. Inhibition of exocytosis in the liver cells by vinblastine treatment induces crinophagic degradation of newly synthesized secretory proteins.[39] Along the same lines, genetic mutation of Rab3 (a small GTPase necessary for exocytosis), starvation and protein kinase D (PKD) deletion in β-cells of the endocrine pancreas in rats strongly increases intracellular degradation of insulin containing secretory granules.[18,40] Hypothetically, PKD may act at the Golgi to sort out lysosome-targeting factors from newly generated secretory vesicles to promote their secretion.[41] However, it remains to be seen whether these regulatory steps are specific to β-cells or they represent general modulators of crinophagy.

These observations suggest that crinophagy may continuously control the quantity and quality of secretory vesicles in the cytoplasm. As crinophagy ensures the breakdown of a subset of secretory granules, this way it shapes the extent of secretion by glandular cells.[2,18,21,37]

4. Molecular background of secretory granule-lysosome fusion in *Drosophila melanogaster*

Even though the phenomenon of crinophagy was discovered more than 50 years ago, its molecular mechanism and genetic regulation remained largely unknown in the absence of a suitable model system. We have recently established the larval salivary gland of *Drosophila melanogaster* as a powerful animal model for the molecular characterization of developmentally programmed crinophagy. Our work was motivated by early observations. First, the article describing mucopolysaccharide containing secretory vesicle degradation in the salivary gland of *Drosophila pseudoobscura* (mentioned earlier)[21] raised the possibility that the late larval salivary gland of *Drosophila melanogaster* may be suitable for the examination and molecular characterization of crinophagy. During the late larval and early prepupal period, salivary gland cells in fruit flies switch from secreting digestive enzymes to producing glue granules, which are found in very large (3–3.5 μm) secretory granules in salivary gland cells. This glue contains a set of special, hyperglycosylated adhesive proteins named Sgs (short for Salivary gland secretion). Components of glue granules including Sgs proteins stick the forming pupa to a solid surface, thus ensuring its normal pupal development in a dry and safe place.[42] Glue production begins in response to a small peak of the molting hormone ecdysone, which also induces the so-called wandering behavior in larvae. Another, much larger increase in ecdysone levels triggers the release of the majority of glue-containing secretory granules and at the same time induces puparium formation.[20,43] Subsequently, residual glue-containing secretory vesicles that remained in the cytoplasm are all degraded via crinophagy.[21,22] Second, the abnormal accumulation of secretory glue granules reported in Vps34-null mutant pupal salivary gland cells 4 h after puparium formation in *Drosophila*,[44] which was attributed to the potential role of macroautophagy in the (non-canonical) secretion of the cytoplasmic proteins.[45] However, glue granules in *Drosophila* larval salivary gland cells are produced in the canonical secretory pathway.[46] The large-scale accumulation of GFP labeled glue vesicles in the Vps34-null mutant animals raised the possibility that instead of or in addition to a secretory defect, secretory granule degradation may be impaired. Therefore, we started to develop transgenic fly reporter systems to follow secretory granule-lysosome fusion and crinophagic degradation of glue,[22] which are described in the current methods to analyze secretory granule-lysosome fusion paragraph. Using these newly developed assays in combination with clear-cut genetic analysis, we have identified the first set of genes involved in secretory granule-lysosome fusion 52 years after the initial discovery of this process.[22,37]

During crinophagy, a typical vesicle fusion process, the cell must be able to transport the participating vesicles (secretory granules and lysosomes) in specified directions, and then restrain the membranes of the two compartments at the optimal distance before their fusion (tethering). Once the optimal proximity of the opposing membranes is reached, fusion will be executed (docking and fusion).[47,48] On a molecular level, crinophagy depends on proper determination of vesicle identity and movement that is usually mediated by specific small GTPases, a tethering complex, and SNARE proteins.

Indeed, by testing the known factors of autophagosome-lysosome fusion that we have discovered in *Drosophila*,[7,8,11,13] we found a surprising similarity in these two fusion pathways. The small GTPases Rab2 and Rab7, the tethering complex HOPS (consisting of six subunits: Vps16a, Vps33a/carnation/car, Vps41/light/lt, Vps18/deep orange/dor, Vps11 and Vps39) and the SNARE proteins SNAP29 (also known as Ubisnap/Usnp) and Vamp7 are required for glue granule-lysosome fusion.[22] Importantly, the autophagosomal SNARE Syntaxin 17 turned out to be dispensable for crinophagy. In a small-scale screen for the missing SNARE protein, we have identified Syntaxin 13 as the likely glue granule-specific competence factor for lysosomal fusion. Of note, there is a potential human homolog of Syntaxin 13 called TSNARE1. Polymorphisms of this gene are risk variants for schizophrenia.[49] It will be interesting to see whether its function in crinophagy is conserved.

Following the initial identification of these fusion factors, our group characterized another two regulators of crinophagy in *Drosophila*. Overexpression of Vps8, a specific subunit of an early endosomal tether that shares its other subunits with HOPS,[50] was found to inhibit HOPS assembly and function, including the fusion of secretory granules with lysosomes.[51]

Furthermore, the small GTPase Arl8 (ADP ribosylation factor like-8) is not only involved in the periphery-directed motility of lysosomes but also in general lysosomal fusion events: endosome-lysosome fusion in garland nephrocytes, autophagosome-lysosome fusion in fat cells and crinophagy in larval salivary gland cells.[52]

As mentioned earlier, lysosomal hydrolases (such as cathepsins) and membrane proteins (e.g., Vacuolar H^+ ATPase complex subunits) are essential for the normal digestive function of the lysosomal compartment. Crinophagy also requires active lysosomes for rapid degradation and recycling of the secretory cargo. In line with this, silencing of a v-ATPase proton pump subunit prevented the normal degradation of glue.[22] Moreover, Uvrag RNAi or overexpression of a dominant negative, kinase dead form of Vps34 also showed glue degradation failure.[22] Importantly, loss of this class of genes did not prevent the fusion of glue granules and lysosomes. Thus, v-ATPase and the Uvrag-containing Vps34 kinase complex are not required for secretory granule-lysosome fusion and act at a later step: degradation of cargo within lysosomes by maintaining an acidic pH and the proper delivery of hydrolase enzymes, respectively.[22,44] Of note, core Atg genes acting in macroautophagy proved to be dispensable for crinophagy, which also supports that glue granules directly fuse with lysosomes without the need for autophagosome formation.[22]

5. Current methods to analyze secretory granule-lysosome fusion

Up until very recently, electron microscopy (EM) was a standard method to study crinophagy, including its first descriptions in the anterior pituitary gland of female rats and salamanders, and also in vinblastine treated liver cells and insulin producing pancreatic β-cells from rats.[18,36–38] EM is still important not only for ultrastructural analyses but also to follow lysosomal hydrolase activity: secretory vesicles acquire such enzymes upon fusion with lysosomes, and the black precipitate generated by acid phosphatase histochemistry can be detected in the newly formed crinosomes.[21,22,53] Another option is to carry out double immunogold labeling for secretory granule and lysosome specific proteins.[22,54]

In addition to these classical methods, we have also developed transgenic reporter systems to follow crinophagy in larval-prepupal salivary gland cells in *Drosophila*.[22] The first reporter system takes advantage of the known properties of green versus red fluorescent proteins: GFP and its relatives are sensitive to low pH and are quickly quenched in the acidic lumen of active lysosomes, while DsRed and its derivatives remain fluorescent and accumulate in lysosomes. This was the basis for the generation of so-called flux reporters for macroautophagy, when for example Atg8/LC3 is tagged with both fluorophores. Vesicles appearing positive for both in confocal microscopy are interpreted as autophagosomes, whereas structures positive for DsRed (or RFP) and lacking GFP fluorescence are identified as autolysosomes.[5,55,56] Using the same principles, we developed a reporter system that expresses two transgenes: the glue protein Sgs3 under its endogenous, late larval salivary gland-specific promoter tagged with either GFP or DsRed. This system is suitable to distinguish intact glue granules (positive both GFP and DsRed) and crinosomes (DsRed positive, GFP negative).[22] This GlueFlux reporter system allows the monitoring of glue granule degradation and acidification in larval-prepupal salivary gland cells, and time-course analyses clearly showed that glue granule degradation in the *Drosophila* late larval and early pupal salivary gland cells is developmentally programmed.[22] Moreover, in combination with a salivary gland-specific (fork head—fkh) Gal4 driver, the GlueFlux reporter system allows one to easily carry out loss and gain of function experiments by simply crossing this line to Gal4-induced UAS-RNAi or gene overexpression transgenics.[22] Importantly, this reporter system is excellent for the detection of a glue degradation defect, but it does not distinguish a fusion problem from a degradation/acidification defect (Fig. 2).

The second reporter system that we have developed is suitable for following glue granule-lysosome fusion more directly. In this system, secretory granules are labeled with Sgs3-DsRed, and the membrane of late endosomes/lysosomes is visualized by GFP-Lamp1 (note that this reporter uses a human lysosomal membrane protein fragment), so during crinophagy, green rings appear around red granules in confocal microscopy.[22] Of note, GFP is tagged to the N-terminal end of Lamp1 so it is located in the luminal side of the lysosomal membrane, thereby it will be continuously quenched in lysosomes.[57] Again, in combination with the fkh-Gal4 driver, it allows loss or gain of function experiments to follow the actual fusion of secretory granules with lysosomes (Fig. 2).

The third fluorescent reporter system that we developed to follow crinophagic activity uses Glue labeled with GFP and the lysosomal hydrolase Cathepsin B tagged with mCherry. When secretory granules fuse with late endosomes/lysosomes, the newly created crinosomes are initially positive for both GFP and mCherry. This way, we can make distinguish early

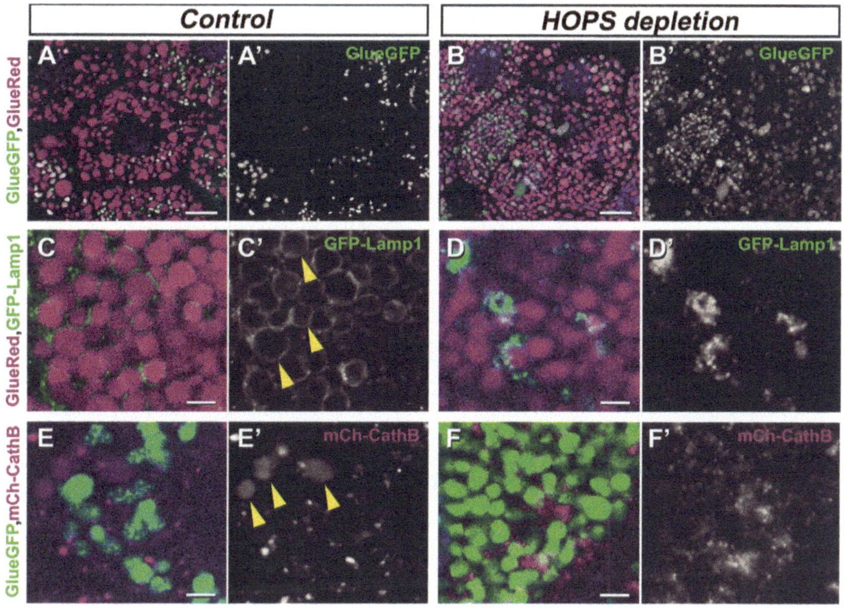

Fig. 2 Transgenic reporter systems for following crinophagy in *Drosophila* larval-prepupal salivary gland cells. Panels (A, C and E) show control cells, while panels (B, D and F) show HOPS loss-of-function (B–B′ and D–D′—Vps8 overexpression, F–F′—Vps16a RNAi) cells. (A–B′) The GlueGFP, GlueRed (aka. GlueFlux) system is suitable for examination of secretory granule degradation based on the quenching of GFP in acidic lysosomes/crinosomes. (C–D′) The GlueRed, GFP-Lamp1 system enables the monitoring of glue granule-lysosome fusions based on formation of a green ring around red granules. (E–F′) The GlueGFP, mCherry-Cathepsin-B (mCh-CathB) system allows one to distinguish normal vs blocked secretory granule-lysosome fusions. Yellow arrowheads point the forming crinosomes in (C′) and (E′). Scale bars: (A–B′): 20 μm, (C–F′): 5 μm.

(GFP and mCherry containing) from late (only mCherry positive) crinosomes,[22] and we can detect both degradation and fusion defects (Fig. 2). One disadvantage of this system, though, is the very low expression level of endogenous promoter-driven mCherry-Cathepsin B, which makes imaging a challenge.

Another fluorescent reporter system was used successfully in a mammalian insulinoma cell line under glucolipotoxic conditions, during which β-cells showed increased colocalization of young secretory granule specific (pro)insulin and LAMP2 (Lysosomal Associated Membrane Protein 2) compared to control cells. This clearly supports that glucolipotoxic conditions increase (pro)insulin granule degradation.[58]

Correlative light and electron microscopy method is also suitable for the detection of insulin granule—lysosome fusion. The GFP-tagged secretory granule marker phogrin and the lysosomal marker CD63 labeled with DsRed colocalized during the diabetes like glucolipotoxic conditions, detected both by fluorescent and electron microscopy.[58]

Lastly, genetic analysis can also be used to exclude the involvement of other lysosomal degradation pathways. For example, if degradation of secretory material is unaffected by the loss of various core Atg genes, this finding indicates that the process is independent of (macro)autophagy.[22]

6. Medical relevance and pathology

The rapid removal of unnecessary or damaged secretory products is essential for maintaining the balance of secretory activity in glandular cells. The human body contains several types of gland cells (exocrine, endocrine and neuroendocrine cells), which all rely on this autophagic process to shape their secretory activity, so the medical relevance of crinophagy is clear.[2,18,59] For example, hormones are already effective in very small amounts, so precise regulation of their levels is very important, including not only their synthesis but also degradation. Nevertheless, crinophagy has more functions than just quickly removing obsolete or damaged secretory products. In β-cells, the crinosome provides a transient energy source in response to short-term starvation. Moreover, it can recruit mTORC1 to suppress macroautophagy, an important adaptation for β-cells to avoid insulin secretion in the absence of nutrients.[40,58] Thus, the crinosome indeed acts as a signaling platform with direct impact on cellular metabolism.

Crinophagy has been extensively characterized in the adenohypophysis, where cells specialized to produce various hormones are found. Milk secretory hormone-producing cells (Mammotrophic hormone—MTH or prolactin) of the anterior pituitary gland are suitable for ultrastructural examination of crinophagy because they are easy to recognize among other cell types based on their size.[37] The ultrastructural identification of MTH cells is also possible by their secretory vesicles having the largest diameter (0.6–0.9 μm) compared to those in other cell types. Estrogen increases the number of secretory granules in prolactin-producing cells, and in the post-lactation period the excess secretory vesicles are degraded by crinophagy.[37] Interestingly, some pituitary gland tumors, such as spontaneous adenoma of MTH producing cells in rats contain a large number of crinosomes, indicating increased crinophagic activity. Along the same lines, adrenocorticotropic (ACTH-producing) cell adenomas also contain large numbers of crinosomes as a likely result of inhibited secretion of this hormone. Finally, increased crinophagy has also been reported in neuroblastomas.[60,61]

Crinophagy likely also contributes to the normal function of the endocrine pancreas. The endocrine gland cells are located in Langerhans islets. Insulin producing β-cells are in the middle of these islets, and they represent a validated model for mammalian crinophagy research. Secretory granule-lysosome fusion contributes to the fine-tuning of the intracellular amount of insulin containing secretory vesicles (β-granules) and controlling the quality and quantity of insulin granules.[18] Hyperactive crinophagy in β-cells was suggested to contribute to type II diabetes, which is characterized by insulin resistance, loss of β-cell function and low levels of insulin secretion.[2,58] However, it is important to highlight that insulin granules are quite small (with a diameter of about 0.3 μm), and thus these are also degraded by macroautophagy and microautophagy.[18] Of note, metabolic stress was recently reported to induce macroautophagy-independent degradation of insulin granules, leading to low insulin levels and acceleration of diabetes.[58] In this study, activation of protein kinase D (PKD) was found to counteract insulin degradation in this setting, raising the possibility that PKD is a negative regulator of crinophagy.

Crinophagy also occurs in the exocrine pancreas. In fact, several studies reported that lysosomal hydrolases (such as cathepsins) can activate the inactive form of trypsinogen (which is normally activated in the duodenal lumen) in the exocrine pancreas via trypsinogen granule-lysosome fusion.[62,63] This can lead to acute pancreatitis, when activated digestive enzymes are released from the secretory vesicles/crinosomes and enter the intracellular and then the extracellular space in this tissue. Uncontrolled digestive enzyme activity then causes the destruction of the pancreatic tissue. Importantly, the expression of Lamp1 and Lamp2 lysosomal membrane proteins (which are hyperglycosylated in the luminal faces) decrease during pancreatitis, which cause unstable lysosomal membranes.[64] Acute pancreatitis may be mild or chronic, and it can even lead to pancreatic cancer.

Activation of crinophagy is also seen in another inflammatory disorder: Crohn's disease, as revealed by ultrastructural analysis of secretory Paneth cells from the intestine of affected patients.[65]

7. Future perspectives

We strongly believe that the identification and characterization of the molecular mechanisms of secretory granule-lysosome fusion is necessary for the better understanding of human diseases. In many cases discussed earlier, uncontrolled crinophagy may contribute to the development of the given disorder, so patients may profit from a crinophagy-inhibiting therapy. However, it will be a long way from the recent identification of the first gene products involved in crinophagy to future therapeutical applications. Currently, the lack of in vivo model systems besides *Drosophila* salivary glands is a major hurdle: while many in vivo examples of crinophagy have been reported in various mammalian gland cells, none of these represent an easily accessible model for molecular genetic dissection. Although insulin-secreting cell cultures also use macroautophagy and microautophagy to turn over their relatively small secretory granules, the individual contribution of these pathways can be separated by using appropriate methods.[58] Ideally, a secretory cell line that mostly relies on crinophagy for the degradation of its products, or one with large enough secretory granules that exceeds the size limit of macro- and microautophagy (similar to the case of glue granules in *Drosophila* larval-prepupal salivary glands) and the development of in vivo mammalian models would boost crinophagy research in mammals.

8. Vocabulary

Crinophagy: direct fusion of unnecessary or obsolete (unreleased) secretory granules with late endosomes or lysosomes for the degradation and recycling of secretory material.

Crinosome: type of secondary lysosomes created by direct fusion of secretory granules and lysosomes. The crinosomal lumen contains secretory material together with lysosomal hydrolases that degrade cargo at a low pH.

Acknowledgments

This work was funded by the National Research, Development and Innovation Office of Hungary (GINOP-2.3.2-15-2016-00006 and -00032, K119842, KKP129797 to G.J.). The funders had no role in study design, decision to publish, or preparation of the manuscript.

References

1. Marzella L, Ahlberg J, Glaumann H. Autophagy, heterophagy, microautophagy and crinophagy as the means for intracellular degradation. *Virchows Arch B Cell Pathol Incl Mol Pathol*. 1981;36(2–3):219–234.
2. Weckman A, Di Ieva A, Rotondo F, et al. Autophagy in the endocrine glands. *J Mol Endocrinol*. 2014;52(2):R151–R163.
3. Feng Y, He D, Yao Z, Klionsky DJ. The machinery of macroautophagy. *Cell Res*. 2014;24(1):24–41.

4. Tsukada M, Ohsumi Y. Isolation and characterization of autophagy-defective mutants of *Saccharomyces cerevisiae*. *FEBS Lett.* 1993;333(1–2):169–174.
5. Kimura S, Noda T, Yoshimori T. Dissection of the autophagosome maturation process by a novel reporter protein, tandem fluorescent-tagged LC3. *Autophagy.* 2007;3(5): 452–460.
6. Itakura E, Kishi-Itakura C, Mizushima N. The hairpin-type tail-anchored SNARE syntaxin 17 targets to autophagosomes for fusion with endosomes/lysosomes. *Cell.* 2012;151(6):1256–1269.
7. Takáts S, Nagy P, Varga Á, et al. Autophagosomal Syntaxin17-dependent lysosomal degradation maintains neuronal function in Drosophila. *J Cell Biol.* 2013;201(4): 531–539.
8. Takáts S, Pircs K, Nagy P, et al. Interaction of the HOPS complex with Syntaxin 17 mediates autophagosome clearance in Drosophila. *Mol Biol Cell.* 2014;25(8): 1338–1354.
9. Jiang P, Nishimura T, Sakamaki Y, et al. The HOPS complex mediates autophagosome-lysosome fusion through interaction with syntaxin 17. *Mol Biol Cell.* 2014;25(8):1327–1337.
10. McEwan DG, Popovic D, Gubas A, et al. PLEKHM1 regulates autophagosome-lysosome fusion through HOPS complex and LC3/GABARAP proteins. *Mol Cell.* 2015;57(1):39–54.
11. Hegedűs K, Takáts S, Boda A, et al. The Ccz1-Mon1-Rab7 module and Rab5 control distinct steps of autophagy. *Mol Biol Cell.* 2016;27(20):3132–3142.
12. Fujita N, Huang W, Lin TH, et al. Genetic screen in Drosophila muscle identifies autophagy-mediated T-tubule remodeling and a Rab2 role in autophagy. *Elife.* 2017;6:e23367.
13. Lőrincz P, Tóth S, Benkő P, et al. Rab2 promotes autophagic and endocytic lysosomal degradation. *J Cell Biol.* 2017;216(7):1937–1947.
14. Mijaljica D, Prescott M, Devenish RJ. Microautophagy in mammalian cells: revisiting a 40-year-old conundrum. *Autophagy.* 2011;7(7):673–682.
15. Galluzzi L, Baehrecke EH, Ballabio A, et al. Molecular definitions of autophagy and related processes. *EMBO J.* 2017;36(13):1811–1836.
16. Mukherjee A, Patel B, Koga H, Cuervo AM, Jenny A. Selective endosomal microautophagy is starvation-inducible in Drosophila. *Autophagy.* 2016;12(11):1984–1999.
17. Cuervo AM, Wong E. Chaperone-mediated autophagy: roles in disease and aging. *Cell Res.* 2014;24(1):92–104.
18. Marsh BJ, Soden C, Alarcon C, et al. Regulated autophagy controls hormone content in secretory-deficient pancreatic endocrine beta-cells. *Mol Endocrinol.* 2007;21(9): 2255–2269.
19. Rorsman P, Renström E. Insulin granule dynamics in pancreatic beta cells. *Diabetologia.* 2003;46(8):1029–1045.
20. Burgess J, Jauregui M, Tan J, et al. AP-1 and clathrin are essential for secretory granule biogenesis in Drosophila. *Mol Biol Cell.* 2011;22(12):2094–2105.
21. Harrod MJ, Kastritsis CD. Developmental studies in Drosophila. II. Ultrastructural analysis of the salivary glands of *Drosophila pseudoobscura* during some stages of development. *J Ultrastruct Res.* 1972;38(5):482–499.
22. Csizmadia T, Lőrincz P, Hegedűs K, Széplaki S, Lőw P, Juhász G. Molecular mechanisms of developmentally programmed crinophagy in Drosophila. *J Cell Biol.* 2018;217(1):361–374.
23. Nelson N. The vacuolar proton-ATPase of eukaryotic cells. *Bioessays.* 1987;7(6):251–254.
24. Nakamura N, Matsuura A, Wada Y, Ohsumi Y. Acidification of vacuoles is required for autophagic degradation in the yeast, *Saccharomyces cerevisiae*. *J Biochem.* 1997;121(2): 338–344.

25. Mauvezin C, Nagy P, Juhász G, Neufeld TP. Autophagosome-lysosome fusion is independent of V-ATPase-mediated acidification. *Nat Commun.* 2015;6:7007.
26. Nakadera E, Yamashina S, Izumi K, et al. Inhibition of mTOR improves the impairment of acidification in autophagic vesicles caused by hepatic steatosis. *Biochem Biophys Res Commun.* 2016;469(4):1104–1110.
27. Brown WJ, DeWald DB, Emr SD, Plutner H, Balch WE. Role for phosphatidylinositol 3-kinase in the sorting and transport of newly synthesized lysosomal enzymes in mammalian cells. *J Cell Biol.* 1995;130(4):781–796.
28. Zeng X, Overmeyer JH, Maltese WA. Functional specificity of the mammalian Beclin-Vps34 PI 3-kinase complex in macroautophagy versus endocytosis and lysosomal enzyme trafficking. *J Cell Sci.* 2006;119(Pt 2):259–270.
29. Juhász G, Hill JH, Yan Y, et al. The class III PI(3)K Vps34 promotes autophagy and endocytosis but not TOR signaling in Drosophila. *J Cell Biol.* 2008;181(4):655–666.
30. Lőrincz P, Lakatos Z, Maruzs T, Szatmári Z, Kis V, Sass M. Atg6/UVRAG/Vps34-containing lipid kinase complex is required for receptor downregulation through endolysosomal degradation and epithelial polarity during Drosophila wing development. *Biomed Res Int.* 2014;2014:851349.
31. Rong Y, McPhee CK, McPhee C, et al. Spinster is required for autophagic lysosome reformation and mTOR reactivation following starvation. *Proc Natl Acad Sci USA.* 2011;108(19):7826–7831.
32. Zhang T, Zhou Q, Ogmundsdottir MH, et al. Mitf is a master regulator of the v-ATPase, forming a control module for cellular homeostasis with v-ATPase and TORC1. *J Cell Sci.* 2015;128(15):2938–2950.
33. Bouché V, Espinosa AP, Leone L, Sardiello M, Ballabio A, Botas J. Drosophila Mitf regulates the V-ATPase and the lysosomal-autophagic pathway. *Autophagy.* 2016;12(3):484–498.
34. Du W, Su QP, Chen Y, et al. Kinesin 1 drives autolysosome tubulation. *Dev Cell.* 2016;37(4):326–336.
35. Cd D. The lysosome in retrospect. In: Dingle JT, Fell HB, eds. *Lysosomes in Biology and Pathology*. Amsterdam and London: North Holland-Publishing Co; 1969: vol 3.
36. Ahlberg J, Beije B, Berkenstam A, Henell F, Glaumann H. Effects on in vivo and in vitro administration of vinblastine on the perfused rat liver—identification of crinosomes. *Exp Mol Pathol.* 1987;47(3):309–326.
37. Smith RE, Farquhar MG. Lysosome function in the regulation of the secretory process in cells of the anterior pituitary gland. *J Cell Biol.* 1966;31(2):319–347.
38. Masur SK, Holtzman E. Lysosomes and secretory granules in the normal and autotransplanted salamander pars distalis. *Gen Comp Endocrinol.* 1969;12(1):33–39.
39. Glaumann H, Sandberg PO, Marzella L. Degradation of secretory content in Golgi-enriched fractions from rat liver after vinblastine administration. *Exp Cell Res.* 1982;140(1):201–213.
40. Goginashvili A, Zhang Z, Erbs E, et al. Insulin granules. Insulin secretory granules control autophagy in pancreatic β cells. *Science.* 2015;347(6224):878–882.
41. Vivot K, Pasquier A, Goginashvili A, Ricci R. Breaking bad and breaking good: β-cell autophagy pathways in diabetes. *J Mol Biol.* 2019; in press.
42. Lehmann M. Drosophila Sgs genes: stage and tissue specificity of hormone responsiveness. *Bioessays.* 1996;18(1):47–54.
43. Biyasheva A, Do TV, Lu Y, Vaskova M, Andres AJ. Glue secretion in the Drosophila salivary gland: a model for steroid-regulated exocytosis. *Dev Biol.* 2001;231(1):234–251.
44. Shravage BV, Hill JH, Powers CM, Wu L, Baehrecke EH. Atg6 is required for multiple vesicle trafficking pathways and hematopoiesis in Drosophila. *Development.* 2013;140(6):1321–1329.

45. Subramani S, Malhotra V. Non-autophagic roles of autophagy-related proteins. *EMBO Rep*. 2013;14(2):143–151.
46. Jamieson JD, Palade GE. Intracellular transport of secretory proteins in the pancreatic exocrine cell. I. Role of the peripheral elements of the Golgi complex. *J Cell Biol*. 1967;34(2):577–596.
47. Hong W, Lev S. Tethering the assembly of SNARE complexes. *Trends Cell Biol*. 2014;24(1):35–43.
48. Baker RW, Hughson FM. Chaperoning SNARE assembly and disassembly. *Nat Rev Mol Cell Biol*. 2016;17(8):465–479.
49. Sleiman P, Wang D, Glessner J, et al. GWAS meta analysis identifies TSNARE1 as a novel Schizophrenia/Bipolar susceptibility locus. *Sci Rep*. 2013;3:3075.
50. Lőrincz P, Lakatos Z, Varga Á, et al. MiniCORVET is a Vps8-containing early endosomal tether in Drosophila. *Elife*. 2016;5:e14226.
51. Lőrincz P, Kenéz LA, Tóth S, et al. Vps8 overexpression inhibits HOPS-dependent trafficking routes by outcompeting Vps41/Lt. *Elife*. 2019;8:e45631.
52. Boda A, Lőrincz P, Takáts S, et al. Drosophila Arl8 is a general positive regulator of lysosomal fusion events. *Biochim Biophys Acta Mol Cell Res*. 2019;1866(4):533–544.
53. Glaumann H, Ahlberg J, Hultenby K, Jansson H, Mengarelli-Widholm S, Motakefi AM. Isolation and characterization of crinosomes—a subclass of secondary lysosomes. *Exp Mol Pathol*. 1989;50(2):167–182.
54. Kuriakose NR, Reifel CW, Bendayan M, Elce JS, Shin SH. Prolactin crinophagy is induced in the estrogen-stimulated male rat pituitary. *Histochemistry*. 1989;92(6):499–503.
55. Nezis IP, Shravage BV, Sagona AP, et al. Autophagic degradation of dBruce controls DNA fragmentation in nurse cells during late *Drosophila melanogaster* oogenesis. *J Cell Biol*. 2010;190(4):523–531.
56. Nagy P, Varga Á, Kovács AL, Takáts S, Juhász G. How and why to study autophagy in Drosophila: it's more than just a garbage chute. *Methods*. 2015;75:151–161.
57. Pulipparacharuvil S, Akbar MA, Ray S, et al. Drosophila Vps16A is required for trafficking to lysosomes and biogenesis of pigment granules. *J Cell Sci*. 2005;118 (pt 16):3663–3673.
58. Pasquier A, Vivot K, Erbs E, et al. Lysosomal degradation of newly formed insulin granules contributes to β cell failure in diabetes. *Nat Commun*. 2019;10(1):3312.
59. Koike H, Steer ML, Meldolesi J. Pancreatic effects of ethionine: blockade of exocytosis and appearance of crinophagy and autophagy precede cellular necrosis. *Am J Physiol*. 1982;242(4):G297–G307.
60. Mahe E, Nguyen C, Arredondo J. Crinophagy in neuroblastoma: a case report and review of the literature. *Ultrastruct Pathol*. 2014;38(3):237–241.
61. Weckman A, Rotondo F, Di Ieva A, et al. Autophagy in endocrine tumors. *Endocr Relat Cancer*. 2015;22(4):R205–R218.
62. Halangk W, Lerch MM, Brandt-Nedelev B, et al. Role of cathepsin B in intracellular trypsinogen activation and the onset of acute pancreatitis. *J Clin Invest*. 2000;106(6):773–781.
63. Van Acker GJ, Saluja AK, Bhagat L, Singh VP, Song AM, Steer ML. Cathepsin B inhibition prevents trypsinogen activation and reduces pancreatitis severity. *Am J Physiol Gastrointest Liver Physiol*. 2002;283(3):G794–G800.
64. Mareninova OA, Sendler M, Malla SR, et al. Lysosome associated membrane proteins maintain pancreatic acinar cell homeostasis: LAMP-2 deficient mice develop pancreatitis. *Cell Mol Gastroenterol Hepatol*. 2015;1(6):678–694.
65. Thachil E, Hugot JP, Arbeille B, et al. Abnormal activation of autophagy-induced crinophagy in Paneth cells from patients with Crohn's disease. *Gastroenterology*. 2012;142(5):1097–1099.e1094.

CHAPTER TWELVE

Autophagy role in environmental pollutants exposure

Gemma G. Martínez-García[a,b], Guillermo Mariño[a,b,*]
[a]Laboratorio "Autofagia y Metabolismo", Instituto de Investigación Sanitaria del Principado de Asturias, Oviedo, Spain
[b]Departamento de Biología Funcional, Universidad de Oviedo, Oviedo, Spain
*Corresponding author: e-mail address: marinoguillermo@uniovi.es

Contents

Abstract

During the last decades, the potential harmfulness derived from the exposure to environmental pollutants has been largely demonstrated, with associated damages ranging from geno- and cyto-toxicity to tissue malfunction and alterations in organism physiology. Autophagy is an evolutionarily-conserved cellular mechanism essential for cellular homeostasis, which contributes to protect cells from a wide variety of intracellular and extracellular stressors. Due to its pivotal importance, its correct functioning is directly linked to cell, tissue and organismal fitness. Environmental pollutants, particularly industrial compounds, are able to impact autophagic flux, either by increasing it as a protective response, by blocking it, or by switching its protective role toward a pro-cell death mechanism. Thus, the understanding of the effects of chemicals exposure on autophagy has become highly relevant, offering new potential approaches for risk assessment, protection and preventive measures to counteract the detrimental effects of environmental pollutants on human health.

Progress in Molecular Biology and Translational Science, Volume 172
ISSN 1877-1173
https://doi.org/10.1016/bs.pmbts.2020.02.003

1. Introduction

During evolution, eukaryotic cells and organisms have acquired strategies to cope with diverse types of stressors. Sometimes, these strategies are directly selected toward specific threats, such as the presence of pathogens, lack of nutrients or intracellular accumulation of misfolded proteins. Moreover, eukaryotic cells have developed an array of pro-survival mechanisms conferring them the plasticity required to counteract and resist adverse situations. One of these strategies involves autophagic degradation. Autophagy is a eukaryotic-specific catabolic pathway that keeps cellular and tissue homeostasis. This process is constitutively active at basal levels, allowing cells, especially non-proliferative ones, to get rid of noxious cellular products that would otherwise accumulate causing cellular damage. In addition to its basal activity, autophagic degradation can be enhanced as a pro-survival mechanism by a great variety of stimuli, including cellular stressors. In these circumstances, autophagy activation is able to reduce cell damage by removing potential sources of toxicity and increasing cellular fitness and adaptability.[1] Conversely, due to the essential nature of this housekeeping process, autophagy inhibition may greatly reduce cellular fitness, which in multicellular organisms usually impact tissue functions, leading to the development of severe pathologies.

According to the World Health Organization (WHO), environmental hazards are those physical, chemical or biological factors external to a person, excluding the natural environments that cannot be reasonably modified, and in 2012, they were related with more than 12 million deaths globally.[2] The last century of industrialization and economic development makes the exposure to environmental pollutants almost inevitable. Although some of them might not have a negative impact for living organisms, some others, are harmful, constituting novel types of cellular stressors, and may exert toxic effects. They can be grouped according to their nature, composition or properties, like heavy metals, persistent organic pollutants (POPs) or volatile organic compounds (VOCs), but also based on the effects they exert to the exposed organisms, like the endocrine disrupting compounds (EDCs). In the last decades, intense research and a growing number of epidemiological studies have improved our ability to detect and measure the presence of environmental pollutants in a broad range of biological matrices, confirming the idea that exposure to pollutants mainly occurs through three main routes: inhalation, dermal absorption and ingestion.[3]

Environmental pollutants effects not only depend on their effective concentrations but also on the exposure time, metabolization rates, derived metabolites and the number of the compounds acting together. However, and despite the large number of studies depicting the noxiousness of environmental pollutants, it is still not clear which are their effects, especially long-term ones, on human health. In the US, the number of contaminants of emerging concern was estimated to be more than 40,000 a decade ago, and that number has continuously increased since then.[4] In Europe, the EU regulation of chemical substances is dealing annually with more than 100,000 compounds (REACH regulation, EC 1907/2006). The standard methodology to assess toxicology of chemical substances includes cytotoxicity, carcinogenicity, mutagenicity or teratogenicity tests, among others, covering a broad spectrum of endpoints (Fig. 1). Still, the correct risk-characterization of such enormous number of chemicals is challenging, time consuming and requires appropriate resources. New approaches based on *in silico* behavioral simulations, as well as on information about genetic and epigenetic damages, have been proposed as alternative assessment tools. Among the potential cellular mechanisms whose disruption might provide consistent evidences of chemical toxicity is the autophagic-lysosomal system.[5] On one hand, autophagic activity can be increased upon exposure

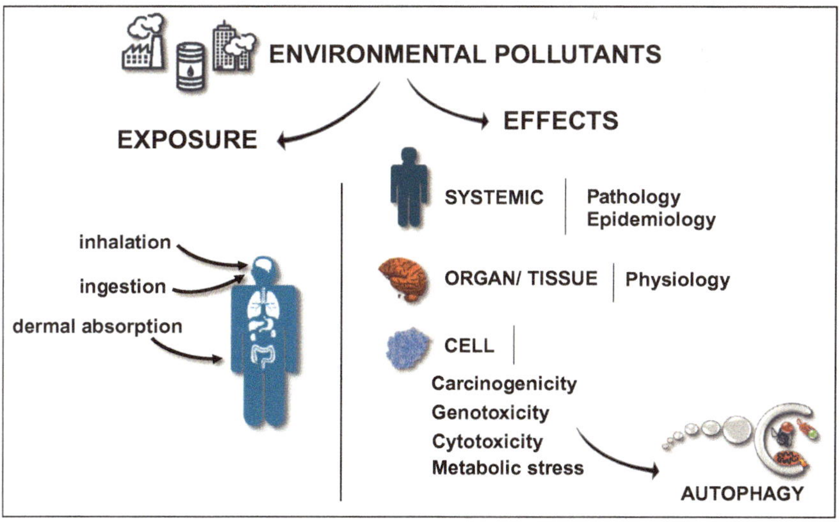

Fig. 1 Targets of environmental pollutants in human health. Scheme representing the main exposure routes and the scale at which pollutants can exert their noxious effects, which can impact body functions at the cell-, tissue- or organism-level.

to pollutants, limiting the detrimental effects of these toxic compounds. On the other hand, the correct autophagic-lysosomal degradation can be inhibited upon exposure to these substances. In fact, most of the effects linked to pollutants toxicity, such as mitochondrial or ER impairment, generation of oxidative stress/ROS or lysosomal dysfunction, have been shown to interfere with autophagy regulation or execution, altering its activity.[6] Studying the consequences of the exposure to environmental contaminants in the context of autophagy could therefore constitute a reliable tool to define their direct and/or indirect harmful effects at the cellular and molecular levels. Additionally, it could provide mechanistic clues in order to prevent the damaging effects of pollutants on human health.

In this review we discuss the potential role for autophagy as protective or cell death-inducing mechanism and gather information on how the exposure to different chemicals, representatives of the main industrial pollutant groups, is able to induce or hamper the autophagy process.

2. Autophagy molecular mechanism

Autophagic activity is a specific characteristic of nucleated cells and to date, autophagy is present in all known eukaryotic organisms.[7] Autophagic degradation starts with the formation of double-membrane vesicles, called autophagosomes. Cytoplasmic cargo is sequestered (both by specific and non-specific mechanisms) by autophagosomes, which move along cellular microtubule network until they eventually fuse with lysosomes.[8] Upon autophagosome-lysosome fusion, autophagosome content and autophagosomal inner membrane are degraded by lysosomal acidic hydrolases (Fig. 2). The resulting biomolecules of this degradation, such as amino acids, lipids or nucleotides are recycled back to the cytoplasm where they will be reused and incorporated into new biomolecules.[9] Autophagy-specific proteins are encoded by a group of evolutionarily-conserved genes called ATG (AuTophaGy-related).[10] ATG proteins were originally described in yeast and are required for the different steps of autophagic pathway (autophagosome formation, maturation, transport or degradation) in a hierarchically and temporally-coordinated fashion.[10,11]

From a molecular point of view, autophagy pathway starts with the formation of the pre-autophagosomal isolation membranes, also known as phagophores. Assembly of these structures until they become a fully-formed autophagosome implicates the coordinated temporal and spatial activation of several molecular components, namely, the Class III PI3K complex; the

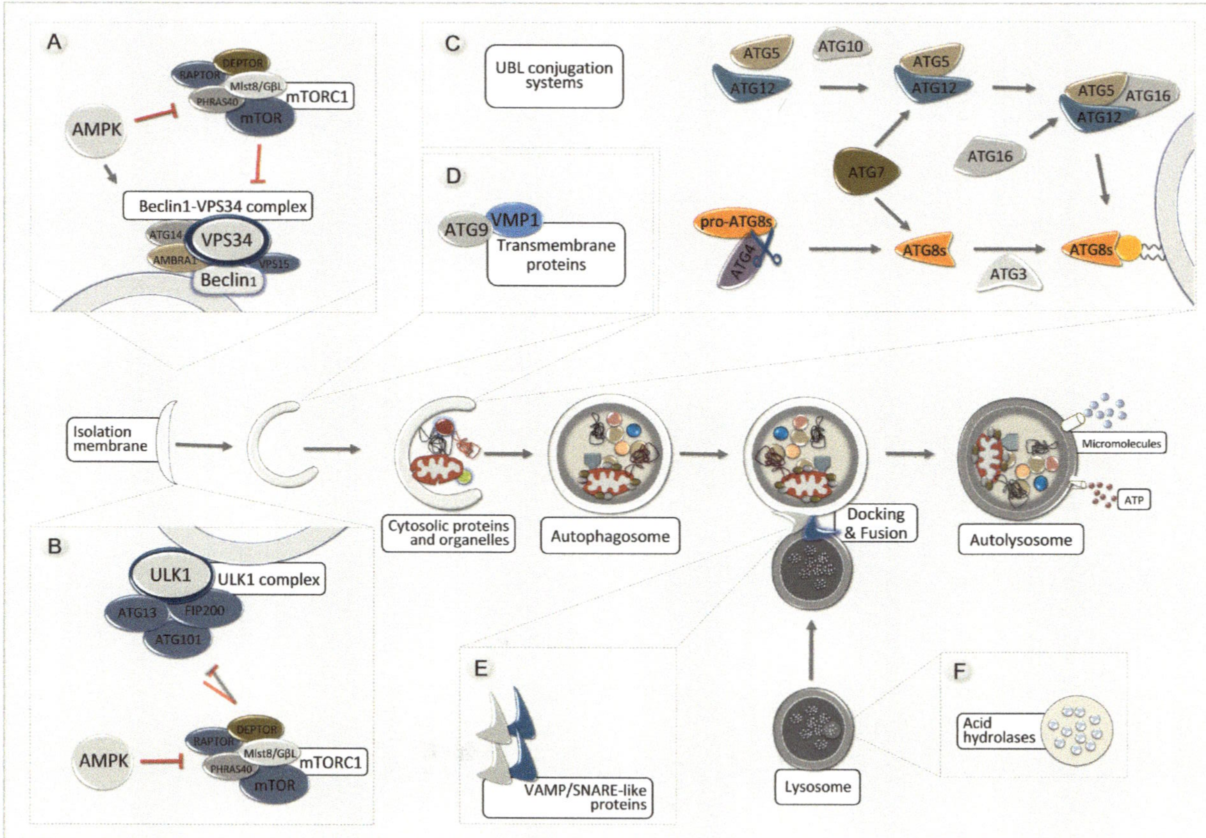

Fig. 2 Scheme of the main steps for autophagy and its regulation. The different protein complexes/groups involved in autophagy initiation (A and B), autophagosome biogenesis (C and D) and autophagosome/lysosome fusion and autolysosomal content degradation (E and F) are depicted. See main text for more details.

ULK1-FIP200-ATG13-ATG101 complex; the two ubiquitin-like (UBL) protein conjugation systems; autophagy-specific transmembrane proteins (VMP1 and ATG9) and the autophagy-specific SNARE/VAMP proteins. All these complex activities are required to fully execute autophagosome formation, maturation and degradation of autophagosomal cargo.[11,12]

The Class III PI3K complex formed by VPS34, Beclin-1, ATG14, AMBRA1 and other variable subunits creates a membrane domain enriched in PtsIns3P, which drives the nucleation of ATG proteins in the phagophore, either directly or indirectly. AMPK is able to increase pro-autophagic function of this complex and to enhance its formation, whereas mTORC1 activity negatively regulates its function (Fig. 2A). The ULK1-FIP200-ATG13-ATG101 complex is responsible for initiating the autophagic process. The activity of this protein complex is antagonistically regulated by mTORC1 (inhibitory phosphorylation) and by AMPK, which both activates ULK1 complex as well as inhibits the activity of mTORC1 complex (Fig. 2B). There are two ubiquitin-like (UBL) protein conjugation systems that participate in the autophagic process, namely ATG8- and ATG12-UBLs. The first involves the participation of ATG4 cysteine proteinases, (which activate ATG8s by cleaving their carboxyl terminus), the E1-like enzyme ATG7 (common to both UBLs), and the E2-like enzymes ATG3 (ATG8s system) and ATG10 (ATG12 system). In coordination, the activity of both systems is required to conjugate ATG8s (yeast ATG8 orthologues, namely GABARAP, GABARAPL1, GABARAPL2, LC3A, LC3B and LC3C in human cells) to a phosphatidyl-ethanolamine lipid at the nascent pre-autophagosomal membrane (Fig. 2C). Interestingly, LC3B, a member of the mammalian ATG8 family of proteins, is the most common marker for autophagy-related measurements. In mammalian cells, LC3B can be present as a soluble, non-autophagic-cytosolic form (LC3B-I), or bound to autophagosomal external and internal (LC3B-II). Given that LC3B-II is present in autophagic membranes from autophagosome formation to autophagosome-lysosome fusion, its cellular amount is indicative of the cellular amount of autophagosomes. Due to the complexity of autophagic degradation, the amount of LC3B-II, although proportional to cellular autophagosomes content, does not always correspond to the intensity of autophagic degradation, which has to be determined in autophagic flux experiments.[13,14] Two different transmembrane proteins, the vacuole membrane protein 1 (VMP1), involved in the first steps of autophagosome formation and ATG9, which participates in the recruitment of membranes to the phagophore are also essential for autophagosome biogenesis (Fig. 2D).

Upon completion, fully-formed autophagosomes move along microtubule network, eventually fusing with a lysosome, thus acquiring hydrolytic activity and becoming autolysosomes. Several SNARE-like proteins (*i.e.*, Syntaxin17 and VAMP8, among others) are required for efficient fusion between lysosomes and autophagosomes (Fig. 2E). Once content and inner membrane are degraded by acidic hydrolases, the resultant molecules (amino acids, nucleotides, lipids, etc.) are recycled back to the cytoplasm by membrane permeases (Fig. 2F).

In addition to the activity of these protein complexes and other ATG proteins, which are specifically involved in autophagy execution, multiple signaling cascades are able to regulate autophagic activity. The relative importance of the different regulatory inputs for autophagy execution is, in many cases, cell type-specific and tissue-dependent. However, there are some major autophagy regulatory circuits, which have remained conserved through evolution and are present in most tissues/cell types from most multicellular organisms, including mammals, such as AMPK or mTORC1.[8,15]

3. Autophagy at the intersection between cell protection and cell death

Although autophagy is constitutively active at basal levels, different types of stimuli (including nutrient availability, growth factors signaling or diverse types of cellular stressors) can increase autophagic activity to enhance the engulfment of cytoplasmic cargo, including entire organelles, which will be eventually degraded by lysosomal hydrolases.[16] Through a series of complex and intricate signaling pathways, cells are able to perceive multiple types of autophagy modulating extracellular and intracellular stimuli and adjust their autophagic activity accordingly.[1] Although an increase in autophagic degradation normally contributes to cell survival in adverse conditions, it has been evidenced that a high and sustained autophagy induction is also related to cell death.[17] However, the molecular mechanisms and signaling pathways underlying the switch from protective to detrimental autophagy remain unclear and still constitute a matter of controversial debate. The classical perspective of autophagy-dependent cell death, the so-called type II programmed cell death (PCD), involves cytoplasmic vacuolization, a massive accumulation of autophagosomal structures, usually preceding cell death. In fact, autophagy induction usually accompanies most of cell death types, in order to reduce pro-death signals, sometimes preventing cell death execution. Hence, a causative role for autophagy in cell death can only be confirmed when genetic blockage of essential *Atg* genes leads to increased cell survival.[18,19]

There is a prominent cross-talk between autophagy and apoptosis, also called type I PCD, often involving caspases, a cysteine protease family for the initiation and execution of apoptosis. Caspases modulate autophagic activity during PCD by cleaving different components of the autophagy molecular machinery, as several ATG proteins have caspase recognition domains. In most cases, caspase-mediated cleavage of ATG proteins leads to down-regulation or inhibition of autophagy, thus increasing cell death, as previously reported for the caspase-mediated cleavage of Beclin-1, ATG16L or AMBRA-1.[20–22] However, certain caspase-cleaved ATG products are able to enhance autophagic activity. Caspase-3, for example, can cleave the cysteine protease ATG4D, generating a processed form with higher activity toward its substrates, promoting autophagy under mild mitochondrial stress.[23] Similarly, the catalytic activity of caspase-9 is reduced upon interaction with the autophagic core protein ATG7. This ATG7/Caspase-9 complex shows a higher affinity for LC3B than ATG7 alone, which finally results in an increased autophagic activity.[24] Conversely, autophagy might counteract apoptosis signaling by sequestering Caspases for its degradation in specific contexts.[25]

Similarly to the caspase-mediated cleavage of other ATGs, ATG5 can be cleaved by calpains and act as an additional switch between autophagy and apoptosis. Calpains are Ca^{++} dependent cysteine proteases, capable of connecting intracellular Ca^{++} content with autophagy and apoptosis signaling. Inhibition of calpains activity in the context of pro-apoptotic Bax deficit have been shown to promote autophagy.[26] Likewise, either calpain inhibition or genetic knockdown reduced doxorubicin-induced apoptosis in an ATG5 dependent fashion.[27] Moreover, calpain-cleaved ATG5 form may relocate from cytosol to mitochondria, where it contributes to cytochrome *c* release, thus increasing apoptosis.[28,29] In addition to their specific effect on ATG5, calpains have also been shown to cleave other different ATG proteins, thus increasing the connections between autophagy and cell death execution.[30]

Mitophagy, the selective autophagic degradation of mitochondria, plays a protective role counteracting potential apoptosis by eliminating defective/damaged mitochondria that have lost their membrane potential. Interestingly, the anti-apoptotic BCL-2 and BCL-xL and the pro-apoptotic BAX/BAK factors also mediate the activation of both autophagy and apoptosis. Beclin-1, which contains a BCL-2 domain can be inactivated as autophagy promoter by interacting with BCL-2 and BCL-xL.[31] Conversely, phosphorylation of BCL-2 mediated by JNK1 prevents its association with Beclin-1 promoting autophagy. At the same time, this phosphorylated form is able of binding BAX and inhibit apoptosis.[32] Thus, autophagy plays a central part in fine-tuning and limiting unnecessary apoptosis, ensuring that it only takes place when cell death is inevitable.

A common target for pollutants exposure is the endoplasmic reticulum (ER). When ER stability is impaired, multiple signaling events trigger the "unfolded protein response" (UPR), which limits protein synthesis, and contributes to improve protein folding. Depending on the type of ER stressor, the signaling route for autophagy activation might vary, involving either IRE1/TRAF2/JNK or PERK/elF2α pathways, or even Ca^{2+}-dependent PKCθ.[33] Mild ER stress induces Beclin-1-dependent autophagosome formation as a protective measure,[34] but sustained high levels of ER stress can prompt apoptosis.[35] One regulatory step for switching between autophagy and apoptosis under conditions of prolonged ER injury involves the antagonistic activity of ATF4 and CHOP proteins within the PERP pathway. Some ER stressors induce CHOP expression as an apoptosis intermediate, and it might downregulates BCL-2 activity. CHOP siRNA treatment is not only able to suppress apoptosis, but to activate autophagy and increase the expression of ATF4, which can directly activate the transcription of ULK1.[36,37] Consistently, knockdown of CHOP is able to enhance tunicamycin-induced autophagy, diminish ER stress and reduce apoptosis.[38] Furthermore, under ER stress conditions, the chaperone BiP dissociates from the PERK/IRE1/ATF6 complex inducing apoptosis, while during non-stress conditions it has been shown to act as an autophagy inducer. Therefore, some specific ER stressors might regulate the balance between protective autophagy and apoptosis by dissociation of the BiP/Beclin-1/p62 complex.[39]

Environmental stressors might also impact the activity of the tumor suppressor p53, located in the cytoplasm under basal conditions, where it exerts an inhibitory reffect on FIP200 thus limiting autophagy initiation.[40] Mild cellular stress displaces p53 toward the nucleus, increasing the expression of the pro-autophagic factors AMPK, damage-regulated autophagy modulator 1 (DRAM1) and sestrins 1 and 2, in a pro-survival fashion. However, sustained high stress levels lead to p53 mitochondrial membrane localization triggering apoptosis.[41] A similar dual role is played by the nuclear factor-kappa B (NF-κB), with pro-autophagic and pro-apoptotic activities, which can be activated by certain neurotoxic chemicals.[42]

Thus, according to the enormous complexity of the interconnections between autophagy and cell death signals, it seems clear that cellular stress and cytotoxicity caused by many environmental pollutants would influence autophagy. The precise nature and implications, either pro-survival or pro-cell death, of this effect will likely depend on the intensity, dose and duration of the exposure.

4. Autophagy and environmental pollutants

Many of the environmental pollutants classified or considered toxic for animals and humans exert their actions through routes affecting either directly or indirectly autophagic and lysosomal systems. Autophagic activity, critical for cellular homeostasis, is ultimately linked to the efficiency of lysosomal degradation. Therefore, inhibition or blockage of this last stage derived from pollutants exposure might result in the accumulation of harmful content, otherwise expected to be degraded by autophagic-lysosomal mechanisms. Additionally, exposure to pollutants/xenobiotics might also affect the initial stages of autophagy and/or autophagy-related signaling routes, hindering the whole process and thus affecting cellular fitness. Conversely, exposure to pollutants and/or xenobiotics might lead to an autophagy enhancement as a protective cellular strategy in order to counteract potential damage. However, when damage is severe and persistent, the sustained induction of autophagy may, in certain cases, contribute to cell death execution. In order to better understand the complex and intricate relationships between pollutants/xenobiotics, autophagy and cell death, we summarize here the effects that some of the most important industrial pollutants exert on the autophagy pathway, modulating its activity from a protective to a pro-death role. The election of these groups of chemicals is based on the risk for human health due to their ubiquity, persistence, potential continuous exposure and high concentrations in almost all environmental compartments.[43] For reasons of space and complexity, we have not included information on engineered nano-materials (ENMs), but Wei and Duan reviewed extensively their relationship with autophagy in a recent publication.[44]

4.1 Flame retardants

Flame retardants (FRs), applied to delay combustion and limit flames spreading, are one of the main synthetic chemicals present in the environment. Brominate flame retardants (BFRs), intensively used in the past decades, are already withdrawn from the EU and the US markets and increasingly restricted almost elsewhere, but still can be measured at relevant concentrations in all environmental compartments.[45]

Polybrominated diphenyl ethers (PBDEs) are one of the major classes of BFRs. Ubiquitous and included in textiles, polyurethane foams and electronics, population is exposed to them mainly *via* dust (inhalation) and diet (ingestion). Additionally, there is also transfer through placenta, breastfeeding, hand to mouth behavior and dermal absorption, which explains why body concentrations are higher in children than in adults.[45,46] BFRs have been substituted by organophosphorus flameretardants (OPFRs). Mixed with plasticizers in furniture, automobiles or building materials, OPFRs are not covalently-bound to their matrix host and therefore easily leachable to the environment.[47]

FRs cytotoxicity has not been completely described, although it has been mainly related to developmental malformations, neuronal damage, anti-oestrogenic effects or metabolic disorders leading to weight gain.[48–50] PBDE-47 (2,2′, 4,4′tetrabromodiphenyl ether), one of the best studied FRs, has been shown to cause neurotoxicity through the disruption of thyroid hormones (THs) homeostasis during developmental stages. THs status is critical for fetal and postnatal neurodevelopment. In fact, maternal thyroid dysfunction causes cognitive and behavioral disorders.[51–53] The structural similarities between PBDE-47 and THs have been proposed as the potential reason for its toxicity.[54] Female Sprague-Dawley rats exposed to PBDE-47 from pre-pregnancy until offspring weaning showed ER stress, defective autophagy and apoptosis, all associated to thyroid structural and functional abnormalities. Serum concentrations of T3 and T4 hormones increased in a dose-response manner, as well as the apoptotic markers Caspase-3 and its cleaved substrate PARP. Prolonged exposure to PBDE-47 was shown to inhibit autophagy, as it significantly reduced LC3B-II levels. Indeed, a reduction of LC3B punctuated structures and an increase in p62 levels were detected in thyroid tissue. This was associated with an increase in apoptotic markers, such as TUNEL labeling, as well as with higher levels of the ER stress marker CHOP, potentially linking autophagy inhibition, ER stress increase and apoptotic cell death in the affected tissues.[55] In a similar study, rats perinatally exposed to PBDE-47 also showed developmental neurotoxicity, with higher levels of apoptosis and lower number of Nissl-stained neurons in CA3 hippocampal region, leading to memory impairment in adults. This was associated with autophagic flux blockade, as shown by increased LC3B positive structures and a high LC3B-II/LC3B-I ratio, together with high p62 levels in CA3 hippocampal neurons of treated animals. Tests in neuroendocrine pheochromocytoma (PC12) cells corroborated these results showing that PBDE-47 neurotoxicity is time dependent. In this context, autophagy induction seems to increase apoptosis, probably as a consequence of autophagic flux block. In fact, an increase in autophagosome formation, upon rapamycin treatment or by adenovirus-mediated ATG7 overexpression, significantly increased the levels of cleaved PARP and apoptosis. Accordingly, the use of the autophagy inhibitor wortmannin or siRNA-based ATG7 knockdown alleviated PBDE-47 effect, reducing apoptosis and increasing cell viability. Moreover, the use of the Caspase-3 inhibitor DEVD reduced LC3B puncta and prompted cell survival.[56] A similar scenario underlies female reproductive system damage after exposure to PBDE-47. In fact, prolonged exposure to this BFR reduced ovarian coefficient, oocyte number and leads to a thinner granular cell layer in postnatal exposed rats. ER stress, apoptotic and autophagy markers were altered, with increased levels of GRP78, IRE1, Caspase-12, Caspase-3, cleaved PARP, Beclin-1, LC3B-II and p62. Interestingly, the use of 4-BPA as ER stress inhibitor was able to attenuate all these effects, highlighting the close interrelation among ER stress, autophagy and apoptosis upon exposure to PBDE-47.[57]

Contrary to the case of PBDE-47, autophagy might play a protective role against the neurodevelopmental stress caused by the exposure to TDCIPP, another common OPFR. Zebrafish embryos exposed to TDCIPP between 2 and 120 h after fertilization showed a reduced hatching rate, an increased malformation rate and altered locomotive behavior. Lc3b-II levels were significantly increased in the exposed groups and *beclin-1, atg3, atg5* and *lc3b* transcription was upregulated. When TDCIPP was combined with rapamycin, an autophagy inducer, hatching and survival rates increased compared to those groups treated with a combination of TDCIPP and chloroquine, an autophagy inhibitor.[48] Rare minnows exposed to three different OPFRs (TBOEP, TDCIPP and TPHP) at environmentally relevant concentrations showed that both Caspase-3 activity and the percentage of apoptotic cells, were significantly increased in liver tissue.[58] Decabromodiphenyl ether (BDE-209), forbidden in the EU in 2019, is a still widely dispersed FR, whose use has been related to alterations in glycolipid metabolism. Mice exposed to environmentally relevant concentrations of BDE-209 presented higher levels of glucose and insulin, higher triglyceride and lower high-density lipoprotein cholesterol levels in serum, together with histological alterations in liver and adipose tissue.

Although specific studies on the effects of BDE-209 on autophagic activity have not been performed, prolonged exposure to this FR leads to an increase in the activity of mTOR, together with reduced PI3K and p-AKT levels, most likely as a consequence of upstream alterations of glucose and lipid metabolism, which could potentially impact autophagic activity.[59] Given the importance of autophagic degradation in liver physiology, autophagy modulation could be also involved in BDE-209 liver toxicity.

4.2 Phthalates

Phthalates (dialkyl or alkyl aryl esters of phthalatic acid) are plasticizers extensively used to enhance the flexibility, malleability and elasticity of rigid polymers such as polyvinyl chloride (PVC). They are non-covalently added to construction materials, clothing, food and beverage packaging, children goods, biomedical and pharmaceutical products, fixatives, solvents and cosmetics.[60] Di(2-ethylhexyl)phthalate (DEHP) is the most commonly used phthalate, appearing up to 40% in plastics by weight,[61] and the main PVC plastic softener. As in the case of FRs, phthalates are leachable from the polymer matrix and ubiquitously dispersed in the environment. Humans are exposed to phthalates *via* food, air or water, and several epidemiological studies have found concerning concentrations of these compounds in blood and urine.[62–64] Exposure to these chemicals has been primarily linked to disruption of thyroid regulation and changes in the physiological T4 and T3 concentrations.[65–67] During the past decades, several independent works have also demonstrated the genotoxic and cytotoxic effects of phthalates. DEHP, for example, was able to induce immunotoxicity[68] and to generate single chromatid aberrations, sister chromatid exchange in human lymphocytes and DNA damage in HepG2 cells[69,70] Similar effects were caused by its metabolite mono(2-ethylhexyl)phthalate (MEHP).[71] DEHP increased peroxisome proliferation and altered several antioxidant and hepatic enzymes activity.[72] Additionally, DEHP can interfere with pancreatic function, reducing pancreatic insulin content and causing the loss and abnormal substructure of ß-cells,[73,74] being also related to metabolic disorders.[75] DEHP and two closely related compounds, DEHA (Di(2-ethylhexyl)adipate) and DINP (Di-(isononyl)phthalate), are hepatotoxic[76] and cardiotoxic.[77] Exposure to DEHP leads to lipid metabolic disorders, weight gain and disruption of lipid serum concentrations, which might involve altered autophagic flux. In fact, Wistar rats exposed to growing concentrations of DEHP showed effects on lipid metabolism and autophagy, with a significantly rise in TYK2, STAT1 and fatty acids synthetase (FAS) levels. In parallel, an increased LC3B-II/LC3B-I ratio was measured in the adipose tissue of the exposed group.[78] At a cellular level, DEHP is able to induce dose-dependent oxidative stress and to impair mitochondrial function in pancreatic INS-1 cells, leading to an increase in apoptosis associated to upregulation of Caspase-3 and BAX levels and to BCL-2 downregulation. At the same time, exposure to DEHP seems to induce an increase in autophagosomal content, derived from autophagic flux block, as evidenced by higher levels of LC3B-II and p62 autophagy markers. In this case, autophagy modulation seems to be detrimental, as co-exposing INS cells to DEHP and rapamycin increased apoptosis, whereas co-treatment of INS cells with DEHP and the autophagy inhibitor 3-methyladenine (3-MA) reduced cell death.[79] DEHP and MEHP have also been reported to affect reproductive system.[80,81] Testis tissue from laboratory mice exposed to this compound showed significantly increased levels of LC3B-II, ATG5, Beclin-1, cleaved Caspase-8, cleaved Caspase-3 and BAX, along with decreased of BCL-2, indicating an activation of both autophagy and apoptosis. The increase in autophagic structures upon exposure to this phthalate was ultrastructurally confirmed by TEM analyses in mouse Leydig cell line (TM3 cells), presenting a high autophagosomal content after treatment with DEHP. Interestingly, cell viability was ameliorated upon addition of 3-MA (as previously mentioned for INS-1 cells), but also by co-exposure with N-acetyl-L-cysteine (NAC), which suggests a relevant participation of oxidative stress in DEHP-induced cell death.[82]

Di-n-butyl phthalate (DBP) also affects male reproductive system. Exposure to DBP seems to activate autophagic flux in urethral epithelial cells, increasing the levels of ATG5, Beclin-1 and LC3B-II.[55] Di-isononyl phthalate (DINP) has been proposed as a safer substitute for traditional phthalates but some concerns about its toxicity have recently emerged. DINP effects have been studied in the context of autoimmune thyroid disease (AITD) and autophagy, since it is known that AIDT may affect ROS levels in in the thyroid gland, altering AKT/mTOR pathways.[83] Cellular and histopathological markers used to assess AIDT were increased in treated groups of Wistar rats, as well as Caspase-3 levels. Decreased LC3B-II levels and lower number of autophagosomes but higher levels of p62 were measured in the same tissues, indicating a potential impairment of autophagy flux.[84]

4.3 Xenobiotic organic compounds: PAHs/dioxins/furans/PCBs

Polycyclic aromatic hydrocarbons (PAHs), dioxins (Ds), furans (Fs) and polychlorinated biphenyls (PCBs) are among the most common organic contaminants, with many compounds classified as PBT (persistent, bioaccumulative and toxic) substances and generating severely concerns due to their long-range transportable properties.[85] PAHs cause oxidative damage through the stimulation of antioxidant systems and ROS production.[86,87]

Benzo[*a*]pyrene (B[*a*]P) is one of the major PAHs derived from incomplete combustion and present in cigarette smoke, vehicle exhaust, industrial affluent, wood smoke or charcoal grilled foods. It is classified as group 1 carcinogenic by the International Agency for Research on Cancer and it has been related to inflammatory skin diseases and skin cancer.[88] B[*a*]P induces mitochondrial impairment and decreases mitochondrial membrane potential (MMP) leading to ATP depletion, drop of ATP/AMP ratio and a descent of oxygen consumption rate, finally resulting in the induction of apoptotic cell death. Human keratinocyte cells (HaCaT) exposed to B[*a*]P showed autophagy induction in a dose-dependent manner, with increased levels of ATG5, Beclin-1 and LC3B-II, together with a higher number of GFP-LC3 puncta. Likewise, mitophagy was enhanced and a higher number of acidic vesicles were also observed after exposure to this PAH. Additionally, cell viability upon B[*a*]P exposure was markedly decreased in autophagy-deficient HaCaT cells (in *Atg5*, *Beclin-1* and *Atg14* knocked-down cell lines and after 3-MA treatment), presenting a significant increase of Caspase-3/7 activity.[89] B[*a*]P-induced mitophagy involves the participation of both cytochrome P450 1B1 (CYP1B1) and aryl hydrocarbon receptor (AhR), in a mTOR/AMPK-dependent fashion. In this case, autophagy/mitophagy induction in HaCaT cells by B[*a*]P acts as a cellular protection mechanism against its toxicity, as Beclin-1 knockdown in this context led to reduced cell viability.[90]

Studies in the earthworm *C. elegans* show a significant impairment in lysosomal membrane stability in chloragogenous cells upon B[*a*]P exposure, together with an increase in neutral lipid accumulation, as well as an increase in the lysosomal/cytoplasmic volume ratio. Similar results, although with slightly lower toxicity, were obtained after treatment with 2,3,7,8-tetrachlorodibenzo-*para*-dioxin (TCDD).[91] In this sense, exposure to TCDD has been linked to carcinogenicity, the formation of neoplastic lesions in several tissues and to several physiological disorders, as hyperkeratosis, hepatotoxicity, altered lipid and glucose metabolism or neurotoxic effects.[92,93] TCDD neurotoxicity has been well described, but the mechanisms underlying it has not been fully addressed, although it is likely that ROS accumulation and ROS-dependent FOXO3a/$p27^{kip1}$ pathway activation play a pivotal role in this context.[94] This dioxin induces neural apoptosis through intensive ROS generation and may impair autophagic flux in neurons, as shown in human neuroblastoma SH-SY5Y cell line. TCDD exposure increased LC3B-II, Beclin-1 and p62 levels, as well as the number of autophagosomes, in a dose and time dependent manner. Pre-treatment with the antioxidant NAC inhibits the increase in LC3B-II, while 3MA-mediated autophagy inhibition increased TCDD cytotoxic effects, reducing cell viability in approximately 10%.[95] Paradoxically, exposure of Madin-Darby bovine kidney (MDBK) cells to TCDD caused cell proliferation and, despite of inducing morphological changes, enhanced cell viability. In this context, higher LC3B-II levels and an increase in LC3B punctuated structures were observed after TCDD exposure in a dose-dependent manner.[96]

Dibenzofuran (DBZ) is a common furan released into the air, mainly from coal gasification or carbonization processes but also from tobacco smoke.[97] Exposure to DBZ mostly occurs through inhalation and dermal contact, and it has been linked to obstructive pulmonary disease and lung cancer. A549 human epithelial lung cells exposed to DBZ showed inhibition of cell proliferation, reduced MMP, decreased mitochondrial activity and increased ROS generation. Additionally, DBZ exposure seems able to impact autophagy by increasing the number of LC3B punctuated structures, LC3B-II levels and also lysosomal content.[98]

Among PCBs and their metabolites, the PCB quinone metabolite, PCB29-pQ, is able to modulate autophagy and apoptosis in HepG2 cells in a dose-dependent manner. Exposure to high doses of PCB29-pQ also increased the expression and cytosolic localization of p53 and HMGB1, autophagy modulating proteins. Likewise, exposure to PCB29-pQ caused downregulation of the autophagy-related proteins DRAM and ULK1 and upregulation of HSPB1. Thus, PCB29-pQ might induce both autophagy and apoptosis, mechanistically-involving p53 and HMGB1, which indicates the possibility of an ambivalent apoptotic/autophagic signaling.[26]

4.4 Air pollution/particulate matter

Air pollution involves gaseous compounds and particulate matter (PM), being this last the most concerning. Long term exposure to air pollution has been linked to several respiratory, cardiovascular and cardiopulmonary diseases, lung cancer and atherosclerosis, as well as to increases in morbidity and mortality.[99,100] PM toxicity is determined mainly by its chemical composition, size, surface area/volume ratio and solubility. Fine particles with an average aerodynamic diameter equal or lower than 2.5 μm ($PM_{2.5}$) are able to induce inflammation and oxidative stress. In addition, as they usually interact and associate with organic and inorganic compounds such as PAHs, ions, elemental and organic carbon and metals, they may become more harmful.[101] In a survey performed in 29 EU countries, $PM_{2.5}$ concentrations in the range of 10 μg/m^3 were significantly associated to a 2% increase in respiratory diseases and to a reduction of the average life expectancy in 8.6 months.[102] Furthermore, it seems that $PM_{2.5}$ might cross the alveolar barrier affecting multiple organs.[103] Their cytotoxic effects on different cell lines pointed out their ability to generate a variety of alterations in cellular processes including autophagic flux.[104,105]

Exposure of human adenocarcinoma alveolar basal epithelial A549 cells to $PM_{2.5}$ induced significant and fast ROS production and reduced catalase and superoxide dismutase activities. These changes were associated to alterations in autophagic flux, with a higher number of LC3B puncta, higher levels of LC3B-II, ATG4A, ATG5, ATG7 and Beclin-1. Moreover, levels of SPAAR (small regulatory polypeptide of amino acid response, negative regulator of mTORC1) were significantly reduced after 6 h of exposure.[106] Exposure to $PM_{2.5}$ also induced a significant increase in cellular levels of inflammatory markers TNF-α, IL6 and IL8 in this cell line. Moreover, Toll like receptors (TLRs) TLR1, TLR4 and TLR6, which are involved in innate immune responses through the NFκB p65 pathway, were also upregulated, as well as NFκB p65. All these responses were associated to ER stress, with an increased expression of ATF6 and CHOP.[107] Human corneal epithelial cells (HCECs) exposed to PM $_{2.5}$ also showed similar results, with time and dose-dependent effects.[89] In human bronchial epithelial 16HBE140 cells, exposure to $PM_{2.5}$ also reduced cell viability, decreased MMP, increased ROS generation and apoptosis, as well as cell cycle arrest, evidenced by the downregulation of cyclin D1. Moreover, a significant increase in the levels of inflammatory markers IL-6 and TNF-α was also observed in this context, as well as an increase in p-AMPK levels and a reduction in mTOR signaling. These alterations were accompanied by an increase in the number of autophagosomes, higher levels of LC3B-II, ATG5, Beclin-1 and higher ULK1 activity. Interestingly, inhibition of both autophagy and AMPK activity restored cell viability and reversed most of $PM_{2.5}$ cytotoxic effects, suggesting that AMPK-mediated autophagy could be mechanistically involved in the noxious effects of these particles in lung cells.[108] A more recent study addressing the specific autophagy-related effects of PM exposure in both human bronchial epithelial cells and murine tracheal epithelial cells has shown that autophagosome accumulation upon exposure to PM is linked to autophagic flux blockade. In fact, LAMP2 levels were reduced after PM exposure, whereas cellular levels of p62 were significantly increased. Moreover, autophagic flux analysis by using the RFP-GFP-LC3 reporter[109] confirmed this idea.[110] Other particulate materials, such as carbon black (CB) or standardized urban dust (UD) have been reported to impact autophagic activity. When tested on naïve human alveolar epithelial cells, both compounds increased LC3B-II levels, indicative of high autophagosomal cellular content. While both increased LC3B-II levels, only UD led to a reduction in cell viability, with increased ROS generation and augmented heat shock protein 70 (HSP70) levels.[111] Although it appears clear that these pollutants affect autophagy dynamics, further studies are required to characterize whether the autophagosomal increase upon exposure to CB or UD corresponds to an increase in autophagic flux, or by contrast to an autophagic flux block, as reported for other types of PM.

Apart from lung tissue, several other organs can be affected by traffic related air pollution, especially diesel exhaust particles (DEPs). It is known that exposure to high levels of DEPs causes neuroinflammation and impaired neurogenesis in several brain regions.[112] Moreover, exposure to DEPs has been linked to nephrotoxicity.[113] These pathologies were associated to increased levels of inflammatory cytokines (such as IL-6 or TNF-α), ER stress, Caspase-3 activity, reduced cell viability and an increase in LC3B-II/LC3B-I ratio, both in cellular models and in Sprague-Dawley rats.[114] Despite the advances in the characterization of the molecular alterations elicited by these pollutants, it is still not clear if the observed increase of autophagosomes upon exposure is a consequence of an enhancement of autophagic flux or caused by a blockage in autophagic flux.

4.5 Phenolic compounds

Phenolic compounds have a huge range of industrial applications, such as wood preservatives, explosives intermediates or plastic hardeners. They are also used in the synthesis of polycarbonates and epoxy resins and are included in a myriad of products ranging from can linings, lunch boxes or electronics to building materials or water pipes.[115] Many phenols, along with their chlorinated derivatives and metabolites, have been proven to be subject of maternal transfer and have been linked to several health effects, due to their endocrine disruption properties.[116]

Bisphenol A (BPA), for example, has effects on steroidogenesis. BPA exposure during fetal development causes an increase of estrogen levels and cytochrome P450 (CYP) enzymes (CYP19A1 and CYP11A1).[117] BPA and BPS (bisphenol S, a compound proposed as BPA substitute) induce adipogenesis and alter the expression of genes specifically related to lipid metabolism.[118] Nonylphenol ethoxylates (NPnEO) and their metabolites, as the well-known nonylphenol (NP), are persistent and environmentally ubiquitous, and high concentrations these compounds have been found in human samples of urine, fetal cord blood or breast milk.[119] Phenolic compounds have been reported to affect reproductive, endocrine and immune systems, as well as central nervous system, where impairment of memory, learning and cognitive functions have emerged as consequences of exposure to these compounds.[120,121] BPA has been recently proposed, not only as an endocrine disrupting compound (EDC), but also as human carcinogen in breast and prostate.[122] In fact, different studies have linked its exposure with alterations in brain development, insulin signaling and diabetes, cardiovascular diseases, liver toxicity and hepatic lipid deposition.[123–125] Chlorinated derivatives of BPA (Cl_xBPA) have been found in different human samples and studies both *in vivo* and *in vitro* have shown their potential obesogenic and diabetogenic activities.[126]

In relation to autophagy, BPA exposure has been shown to induce autophagosome accumulation in HepG2 cells, accompanied by an increase in the levels of p-mTOR, p-p70S6 kinase and p-ULK1 and accumulation of p62. Interestingly, this autophagy activation was not linked to an increase of acidified compartments, which together with higher p62 levels, points to autophagic flux impairment. In liver hepatocytes, BPA exposure also led to autophagic flux impairment and hepatic lipid accumulation, with lipid content markers colocalized with LC3B.[127] In human fetal lung fibroblasts, BPA also increased DNA double-strand breaks and activated ATM signaling, as well as the expression of p53. In this context, LC3B-II levels and the number of cells with autophagic vacuoles were significantly higher after 24 h of exposure.[128]

Fluorene-9-bisphenol (BHPF) is a substitute for BPA largely used in the polyester polymers due to its high thermal stability.[129] BHPF has been proved to exert anti-oestrogenic activity, leading to endometrium atrophy and pregnancy damage in mice.[130] BHPF caused dose-dependent impairment in mice oocytes, with irregular spindle microtubule distribution, significant reduction in p-AMPK levels and defective chromosome alignment. In oocytes, exposure to this bisphenol induced ROS generation and reduced MMP in a dose-response manner, indicating mitochondrial damage that led to a significant increase in apoptosis after 14 h of exposure. Interestingly, the levels of autophagy markers LC3B-II, ATG12, ATG14 and Beclin-1 were also increased upon BHPF treatment, which suggests that BHPF exposure leads to a concomitant activation of early apoptosis and autophagy mechanisms.[131]

In rodents, 4-nonylphenol (NP) caused harmful effects on male reproductive function, ranging from testicular apoptosis, seminiferous tubule degeneration, reduction in the testicular germ cells and leads to a decrease in sperm quality.[132] In Sprague-Dawley rat Sertoli cells, exposure to environmentally and epidemiologically relevant NP concentrations reduced cell viability, increased necrosis and also apoptotic cell death, characterized by an activation of BAX, BAD, BIM, cytochrome *c*, APAF-1 and Caspase-3, as well as MMP disruption and G2/M phase arrest. Along with these cellular damages, NP enhanced autophagy, increasing the number of autophagosomes and autolysosomes, as well as LC3B-II/LC3B-I ratio and Beclin-1 levels. In line with these results, NP inhibits mTOR and AKT signaling, while increases AMPK and JNK activity. In this case, autophagy activation seems to counteract the detrimental effects of NP, as co-treatment with the autophagy inhibitor 3-MA strengthened NP-induced cell death.[133] Similarly, rat primary cortical neurons exposed to environmentally relevant concentrations of NP showed an increase in autophagic flux, but also ER stress, disrupted MMP, and activation of mitochondrial-dependent apoptosis, leading to reduced cell viability. Autophagy activation seems to act in this case as a protective mechanism, since autophagic flux inhibition by chloroquine increased apoptosis, whereas a further autophagy increased upon rapamycin co-treatment reduces cell death and increased neuronal viability.[134] Combined exposure to NP and BPA seems to cause similar but stronger effects than exposure to each of them alone, as shown in rat testis tissue.[84]

Pentachlorophenol (PCP) is a chlorinated phenolic compound largely used in pesticides, herbicides or wood preservatives, among other diverse industrial, agricultural and domestic applications. It has been shown to cause carcinogenic, genotoxic and teratogenic effects.[135,136] Exposure to this chemical during pregnancy led to fetal damages, with significantly higher number of abnormal fetus, resorption fetus per litter and fetus loss. Likewise, placenta pathological changes, as lower villous cells, trophoblast cells aggregates and villous interstitial atrophy, are associated to its exposure. When autophagy was measured in the labyrinthine region of placenta, the number of autophagosomes and the mRNA expression levels of *Lc3b, p62* and *Becn1* were higher in exposed tissues, while levels of *mTor* and *Bcl-2* did not shown relevant changes. Additionally, PCP exposure seems to alter miR-30a-5p levels, altering *Becn1* expression and thus modulating autophagy activation in this tissue.[137] Although it seems clear that exposure to PCP interferes with autophagic activity, further studies are required in order to understand the specific molecular effects of this phenolic compound on autophagic flux.

4.6 Perfluorinated compounds

Due to their huge chemical and thermal stability, that confers them valuable industrial properties, perfluorinated compounds (PFCs) have been largely applied in Teflon materials, water resistant textiles or firefighting foams synthesis, among other uses.[138] However, these same properties are the ones that make them highly concerning for health and environment, since they are very persistent. PFCs main target organs are digestive system, liver and lung tissue, since in addition to ingestion, they can be transported through air particulate matter.[139–141] Additionally, immunotoxicity, reproductive and developmental toxicity or endocrine disruption effects have been also linked to PFCs exposure.[142–144] Perfluorooctane sulfonate (PFOS) was able to induce neurotoxicity by altering the expression of neural developmental genes or the concentration of several neurotransmitters, such as dopamine, serotonin and γ-aminobutyric acid or in acetylcholinesterase (AChE) activity.[145] Perfluorooctanoic acid (PFOA), ubiquitously found at relevant concentrations in the environment and in biological samples,[146] might cause hepatomegaly and lipid metabolism disorders, characterized by peroxisome proliferator-activated receptor alpha (PPARα) activation, in rodents.[147] Moreover, PFOA has been shown to induce a significant increase in ROS levels, lactate dehydrogenase (LDH) leakage and Caspase-3/7 activity. Similarly, exposure to PFOA increases the expression of oxidative stress and apoptosis related genes (*Gsta2, Nq1, Ho-1, Caspase3, p53* and *p21*) in 3D spheroid models in a murine liver cell line.[80] The induction of mitophagy and ER-phagy in the liver seems to play a role in the development of hepatomegaly related to PFCs exposure.[148,149] Recently, additional concerns about PFOA harmful effects on immune system have emerged, especially since serum concentrations have been linked to lower antibody responses during childhood immunization campaigns and other immunodeficiencies.[150] Moreover, the effects of exposure to PFOA were tested in lymphocytes from the fish *Carassius auratus*, revealing its impact on growth rate and ROS generation. Glutathione (GSH) and superoxide dismutase (SOD) activities were significantly decreased in a dose-dependent manner, while that of malondialdehyde (MDA) was increased. The number of autophagosomes was significantly higher in exposed lymphocytes, as well as the expression of *atg5, atg7* and *becn1*. However, while *atg5* levels were significantly elevated at the highest concentrations, those of *atg7* and *becn1* were sensitive to even the lowest PFOA concentrations, decreasing at the highest ones. This differential dynamic expression of *Atg* genes upon exposure to PFOA could indicate a possible induction of autophagy process with an impairment in its final degradative steps.[151] In human lung epithelial cell lines, exposure to different PFCs, such as PFOA, perfluorodecanoic acid (PFDA) or perfluorononanoic acid (PFNA), increased the number of autophagosomes at cytotoxic concentrations. PFCs-induced autophagosomes mostly contained mitochondria and endoplasmic reticulum, as revealed by TEM and colocalization experiments, pointing out the specific induction of the selective types of autophagy (mitophagy and ER-phagy) presumably due to organelle-specific damage. Despite the observed high LC3B-II/LC3B-I ratio upon PFCs treatment, the increased levels of p62 and the lower content of lysosomal protein LAMP2 points to an impairment in autophagy last stages, although specific studies will be required to fully clarify the effects of these pollutants in autophagic flux. Interestingly, treatment with chloroquine or *Atg5* knockdown reduced cell viability, indicating a potential protective role of autophagy induction in this context. In fact, exposure to a cytotoxic concentration of these PFCs caused hyperpolarization of the mitochondrial membrane and ER stress. While treatment with NAC did not rescue cell viability, ER stress was apparently involved, since levels of p-IRE1, ATF4, p-p38, p-ERK1/2 and p-JNK1/2 (all of them associated to ER stress and the selective ER-phagy) were increased by the exposure to these compounds.[152]

5. Conclusions

During the last decades, the intensification of the study of effects on health caused by the exposure to environmental pollutants has shown that their cytotoxicity usually impacts mitochondria and/or ER stability, along with the generation and accumulation of ROS, all of them known to modulate autophagic activity.[138] Therefore, using autophagy as an endpoint for cytotoxicity upon exposure to environmental pollutants would be not only quite straight-forward, but highly informative. Autophagy, given its capacity for eliminating damaged cellular content, might play a protective role in pollutant exposure,[153] However, a persistent autophagy over-activation may contribute to accelerate cell injury, and could even lead to cell death. Although many contaminants seem to induce autophagy as protective reaction, some others specifically inhibit the latest steps of autophagic degradation, generally impairing lysosomal function, as we have reported herein. Environmental pollution might yet cause autophagy activation or lead to a blockage of autophagic flux (Fig. 3). Autophagic flux block is characterized by a massive autophagosome accumulation, which usually interferes with important cellular processes, rendering cells more sensitive to cellular stressors. In the case of environmental pollutants, most of the studies show

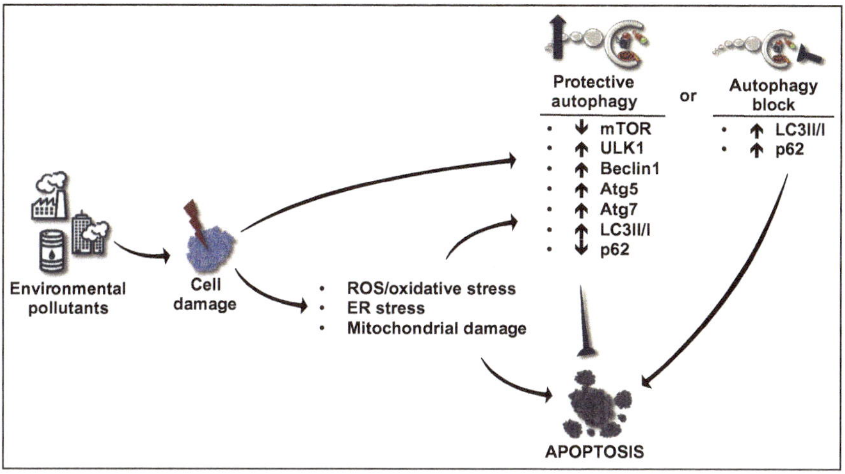

Fig. 3 Potential effects and signaling modulation on autophagy by environmental pollutants. Contaminants might act directly on autophagy, by increasing, impairing or blocking autophagic flux; indirectly, pollutants may elicit different types of cell damage, which may eventually lead to an increase in the autophagy process as a protective response. The severity of the insults will determine in many cases the shift between protective autophagy, apoptosis and occasionally, autophagic cell death.

co-activation of both autophagy and apoptosis, confirming the intricate relationship between these two processes. *In vivo* analyses of autophagy flux upon exposure to toxic pollutants are still indirect, complicating the generation of reliable data. In addition, most *in vitro* studies usually provide limited information, as they are either based upon LC3B puncta quantification or the measurement of the expression levels of certain *Atg* genes. In any case, it seems clear that many environmental pollutants affect autophagic flux, either by increasing or inhibiting it. This often impacts cellular signaling cascades, which ultimately leads to increased cellular damage, atrophy or even tissue failure at the target organs. Therefore, an assessment of potential autophagy impairments might provide quick and reliable information in order to address additional molecular alterations, constituting a good early biomarker. A better understanding of the mechanisms responsible for autophagy modulation, autophagy-dependent cell death and/or apoptosis upon exposure to environmental pollutants, will be required in order to fully understand how cells and organisms react to this type of molecular insults and to proper develop the guidance for using autophagy as biomarker. However, this could also lead to the design of new therapeutic strategies to counteract pollutants detrimental effects on human health and the development of biologically-relevant sensors for environmental pollutants toxicity (Tables 1 and 2).

Table 1 List of environmental pollutants causing autophagy activation and their associated effects on human health.

Role	Autophagy markers		Compound	Effects on health	Cell line/ animal model	References
Protective	↑LC3II/LC31 ratio ↑LC3B puncta	↓ p62	NP	Central nervous system damages	Rat primary cortical neurons	79
		↑ Beclin-1 ↑ Atg5	TDCIPP	Developmental neurotoxicity	Zebrafish larvae	48
			B[a]P	Inflammatory skin diseases and skin cancer	Human keratinocyte cells	89
			$PM_{2.5}$	Corneal cytotoxicity	Human corneal epithelial (HCE) cells	154
		↑ Beclin-1	NP	Male reproductive function impairment	Rat Sertoli cells	133
		↑ Atg5 ↑ Atg12	PCB29-PQ	Hepatotoxicity	Human HepG2 and MDA-MB-231 cells	26
		-	MEHP	Male infertility and spermatogonia damage	Porcine spermatogonia	81
Detrimental	↑ LC3II/LC3I ratio ↑ LC3B puncta	↓ p62	DEHP	Pancreatic alterations and insulin reduction	Pancreatic INS-1 cells	56
		↑ Beclin-1 ↑ Atg5	DEHP	Male germinal epithelium atrophy	Mouse epididymis and testis and TM3 cells	82
			DBP	Male congenital malformation (hypospadias)	Rat tubercle tissue; urethral epithelial cells	155
			$PM_{2.5}$	Respiratory system inflammation	Bronchial epithelial (16HBE140) cells	108
		↑ Beclin-1	$PM_{2.5}$	Respiratory system inflammation	Alveolar basal epithelial (A549) cells	107

Unknown	↑LC3II/LC3I ratio ↑ LC3B puncta	↓ p62	DBP	Male congenital malformation (hypospadias)	Rat genital tubercle tissue	134
		↑ Beclin-1 ↑Atg5	$PM_{2.5}$	Respiratory, cardio-vascular & cardiopulmonary diseases	Alveolar basal epithelial (A549) cells	106
			BPA + NP	Male reproductive system impairment	Rat testes tissue	156
			PFOA	Immunodeficiency	*Carassius auratus* lymphocytes	151
		↑ Beclin-1	DEPs	Nephrotoxicity	Kidney tubular epithelial (HK-2) cells	113
			BHPF	Endometrium atrophy and pregnancy damage	Mouse oocytes	131
			DEHP	Lipid metabolic disorder and weight gain	Rat adipose tissue	78
			TCDD	Neoplastic lesions	Madin-Darby Bovine Kidney (MDBK) cells	96
			DBZ	Obstructive pulmonal disease and lung cancer	Human epithelial lung (A549) cells	98
			PM	Brain inflammation and neurodegeneration	Bv2 cells and SD rats brain tissue	114
			BPA	Respiratory cytotoxicity	Foetal lung fibroblasts (HFLF) cells	128

The role of autophagy activation during the exposure can be protective, detrimental or has not been analysed (unknown). *: NP, 4-nonylphenol; TDCIPP: Tris(l,3-dichloro-2-propyl) phosphate; B[a]P: Benzo[a]pyrene; DBP: Di-n-buthyl phthalate; $PM_{2.5}$: Particulate matter $\leq$2.5 μm; PCB29-pQ: PCB quinone metabolite; DEHP: Di(2-ethylhexyl) phthalate; BPA: Bisphenol A; PFOA: Perfluorooctanoic acid; DEPs: Diesel exhaust particles; BHPF: Fluorene-9-bisphenol; MEHP: Mono(2-ethylhexyl) phthalate; TCDD: 2,3,7,8 tetrachlorodibenceno-para-dioxin; DBZ: Dibenzofuran; PM: Standardized particulate matter.

Table 2 Associated effects of environmental pollutants on autophagy and human health.

Compound*	Effects on autophagy	Autophagy markers	Effects on health	Cell line/animal model	References
PBDE-47	INHIBITION	↑ LC3II/LC3I ratio (↓ LC3B puncta) ↑ p62	Thyroid structural and functional abnormalities	Female SD rats (thyroid tissue)	48,55
PBDE-47	BLOCKAGE (Detrimental autophagy)	↓ LC3II/LC3I ratio (↓ LC3B puncta) ↓ Atg7 ↑ p62	Developmental neurotoxicity	Female SD rats (hippocampus) and PC12 cells	56
PBDE-47	BLOCKAGE	↑ LC3II/LC3I ratio ↑ Beclin-1 ↑ p62	Female reproductive system damages	Female SD rats (ovarian tissue)	57
DNIP	INHIBITION	↓ LC3II/LC3I ratio (↓ LC3B puncta) ↑ p62	Autoimmune thyroid disease (AITD)	Wistar rats (Thyroid tissue)	84
DBP	BLOCKAGE	↓ LC3B puncta ↑ p62	Male reproductive system damages	*R. macaque* testis tissue xenografted to mouse hosts	81
TCDD	BLOCKAGE (Protective autophagy)	↓ LC3II/LC3I ratio (↓ LC3B puncta) ↑ Beclin-1 ↑ p62	Neurotoxicity	Human neuroblastoma cells	95

BPA	BLOCKAGE	↑ LC3II/LC3I ratio ↑ p-mTOR ↑ p-p70S6 kinase ↑ p-ULK1 ↑ p62	Hepatic lipid accumulation	HepG2 cells and mice primary hepatocytes	127
PCP	BLOCKAGE	↑ LC3II/LC3I ratio ↓ Beclin-1 ↑ p62	Fetal damage and abnormalities	Mice (placental tissue)	137
PFOA + PFDA + PFNA	BLOCKAGE (Protective autophagy)	↑ LC3II/LC3I ratio ↑ Beclin-1 ↓ Lamp2 ↑ p62 ↓ p-Akt ↓ p-PI3K	Pulmonal cytotoxicity	Human lung epithelial (A549) cells	152
PM	BLOCKAGE (Detrimental autophagy)	↑ LC3II/LC3I ratio (↓ LC3B puncta) ↓ Lamp2 ↑ p62	Respiratory system inflammation	HBE cells and MTE cells	110

When data are available, the associated role of autophagy is indicated in parenthesis. *: 2,2′, 4,4′tetrabromodiphenyl ether (PBDE-47); DNIP: Di-isononyl phthalate; DBP: Di-n-buthyl phthalate; TCDD: 2,3,7,8 tetrachlorodibenceno-para-dioxin; BPA: Bisphenol A; PCP : Pentachlorophenol; PFOA: Perfluorooctanoic acid; PFDA: Perfluorodecanoic acid; PFNA: Perfluorononanoic acid; PM: Standardized particulate matter.

Acknowledgments

This work was supported by grants from Ministerio de Economía y Competitividad (Spain) (BFU2015-68539-R), Principality of Asturias Government (IDI/2018/000159) and by Instituto de Salud Carlos III (RTICC—Spain). G.G.M-G is supported by Ministerio de Economía y Competitividad (Spain) under the BES-2016-076977 fellowship. This work was supported also by FINBA (Fundación para la investigación y la innovación biosanitaria del Principado de Asturias).

References

1. Kroemer G, Marino G, Levine B. Autophagy and the integrated stress response. *Mol Cell*. 2010;40:280–293.
2. Fang SC, Rodrigues EG, Christiani DC. 26–Environmental health hazards in the tropics. In: Ryan ET, Hill DR, Solomon T, Aronson NE, Endy TP, eds. *Hunter's Tropical Medicine and Emerging Infectious Diseases*. 10th ed. London: Content Repository Only!; 2020:200–208.
3. Doucet J, Tague B, Arnold DL, Cooke GM, Hayward S, Goodyer CG. Persistent organic pollutant residues in human fetal liver and placenta from Greater Montreal, Quebec: a longitudinal study from 1998 through 2006. *Environ Health Perspect*. 2009;117:605–610.
4. Diamond JM, Latimer 2nd HA, Munkittrick KR, Thornton KW, Bartell SM, Kidd KA. Prioritizing contaminants of emerging concern for ecological screening assessments. *Environ Toxicol Chem*. 2011;30:2385–2394.
5. Moore MN, Allen JI, McVeigh A, Shaw J. Lysosomal and autophagic reactions as predictive indicators of environmental impact in aquatic animals. *Autophagy*. 2006;2:217–220.
6. Pesonen M, Vahakangas K. Autophagy in exposure to environmental chemicals. *Toxicol Lett*. 2019;305:1–9.
7. Klionsky DJ, Abdelmohsen K, Abe A, et al. Guidelines for the use and interpretation of assays for monitoring autophagy (3rd edition). *Autophagy*. 2016;12:1–222.
8. Yang Z, Klionsky DJ. Mammalian autophagy: core molecular machinery and signaling regulation. *Curr Opin Cell Biol*. 2010;22:124–131.
9. Singh R, Cuervo AM. Autophagy in the cellular energetic balance. *Cell Metab*. 2011;13:495–504.
10. Itakura E, Mizushima N. Characterization of autophagosome formation site by a hierarchical analysis of mammalian Atg proteins. *Autophagy*. 2010;6:764–776.
11. Koyama-Honda I, Itakura E, Fujiwara TK, Mizushima N. Temporal analysis of recruitment of mammalian ATG proteins to the autophagosome formation site. *Autophagy*. 2013;9:1491–1499.
12. Wesselborg S, Stork B. Autophagy signal transduction by ATG proteins: from hierarchies to networks. *Cell Mol Life Sci*. 2015;72:4721–4757.
13. Rubinsztein DC, Cuervo AM, Ravikumar B, et al. In search of an "autophagomometer" *Autophagy*. 2009;5:585–589.
14. Tanida I, Minematsu-Ikeguchi N, Ueno T, Kominami E. Lysosomal turnover, but not a cellular level, of endogenous LC3 is a marker for autophagy. *Autophagy*. 2005;1:84–91.
15. Russell RC, Yuan HX, Guan KL. Autophagy regulation by nutrient signaling. *Cell Res*. 2014;24:42–57.
16. Kenney DL, Benarroch EE. The autophagy-lysosomal pathway: general concepts and clinical implications. *Neurology*. 2015;85:634–645.

17. Liu Y, Shoji-Kawata S, Sumpter Jr RM, et al. Autosis is a Na^+,K^+-ATPase-regulated form of cell death triggered by autophagy-inducing peptides, starvation, and hypoxia-ischemia. *Proc Natl Acad Sci U S A*. 2013;110:20364–20371.
18. Liu Y, Levine B. Autosis and autophagic cell death: the dark side of autophagy. *Cell Death Differ*. 2015;22:367–376.
19. Marino G, Niso-Santano M, Baehrecke EH, Kroemer G. Self-consumption: the interplay of autophagy and apoptosis. *Nat Rev Mol Cell Biol*. 2014;15:81–94.
20. Luo S, Rubinsztein DC. Apoptosis blocks Beclin 1-dependent autophagosome synthesis: an effect rescued by Bcl-xL. *Cell Death Differ*. 2010;17:268–277.
21. Murthy A, Li Y, Peng I, et al. A Crohn's disease variant in Atg16l1 enhances its degradation by caspase 3. *Nature*. 2014;506:456–462.
22. Pagliarini V, Wirawan E, Romagnoli A, et al. Proteolysis of Ambra1 during apoptosis has a role in the inhibition of the autophagic pro-survival response. *Cell Death Differ*. 2012;19:1495–1504.
23. Betin VM, Lane JD. Caspase cleavage of Atg4D stimulates GABARAP-L1 processing and triggers mitochondrial targeting and apoptosis. *J Cell Sci*. 2009;122:2554–2566.
24. Jeong HS, Choi HY, Lee ER, et al. Involvement of caspase-9 in autophagy-mediated cell survival pathway. *Biochim Biophys Acta*. 2011;1813:80–90.
25. Zhu K, Dunner Jr K, McConkey DJ. Proteasome inhibitors activate autophagy as a cytoprotective response in human prostate cancer cells. *Oncogene*. 2010;29:451–462.
26. Shi Q, Wang Y, Dong W, Song E, Song Y. Polychlorinated biphenyl quinone-induced signaling transition from autophagy to apoptosis is regulated by HMGB1 and p53 in human hepatoma HepG2 cells. *Toxicol Lett*. 2019;306:25–34.
27. Xia H-G, Zhang L, Chen G, et al. Control of basal autophagy by calpain1 mediated cleavage of ATG5. *Autophagy*. 2010;6:61–66.
28. Codogno P, Meijer AJ. Atg5: more than an autophagy factor. *Nat Cell Biol*. 2006;8:1045–1047.
29. Lépine S, Allegood JC, Edmonds Y, Milstien S, Spiegel S. Autophagy induced by deficiency of sphingosine-1-phosphate phosphohydrolase 1 is switched to apoptosis by calpain-mediated autophagy-related gene 5 (Atg5) cleavage. *J Biol Chem*. 2011;286: 44380–44390.
30. Yousefi S, Perozzo R, Schmid I, et al. Calpain-mediated cleavage of Atg5 switches autophagy to apoptosis. *Nat Cell Biol*. 2006;8:1124–1132.
31. Sinha S, Levine B. The autophagy effector Beclin 1: a novel BH3-only protein. *Oncogene*. 2008;27(Suppl. 1):S137–S148.
32. Shimizu S, Kanaseki T, Mizushima N, et al. Role of Bcl-2 family proteins in a non-apoptotic programmed cell death dependent on autophagy genes. *Nat Cell Biol*. 2004;6:1221–1228.
33. Mori K. Tripartite management of unfolded proteins in the endoplasmic reticulum. *Cell*. 2000;101:451–454.
34. Orrenius S, Kaminskyy VO, Zhivotovsky B. Autophagy in toxicology: cause or consequence? *Annu Rev Pharmacol Toxicol*. 2013;53:275–297.
35. Schroder M, Kaufman RJ. ER stress and the unfolded protein response. *Mutat Res*. 2005;569:29–63.
36. Matsumoto H, Miyazaki S, Matsuyama S, et al. Selection of autophagy or apoptosis in cells exposed to ER-stress depends on ATF4 expression pattern with or without CHOP expression. *Biol Open*. 2013;2:1084–1090.
37. Pike LR, Singleton DC, Buffa F, et al. Transcriptional up-regulation of ULK1 by ATF4 contributes to cancer cell survival. *Biochem J*. 2013;449:389–400.
38. Lei Y, Wang S, Ren B, et al. CHOP favors endoplasmic reticulum stress-induced apoptosis in hepatocellular carcinoma cells via inhibition of autophagy. *PLoS One*. 2017;12: e0183680.

39. Song X, Lee DH, Dilly AK, et al. Crosstalk between apoptosis and autophagy is regulated by the arginylated BiP/Beclin-1/p62 complex. *Mol Cancer Res.* 2018;16:1077–1091.
40. Morselli E, Shen S, Ruckenstuhl C, et al. p53 Inhibits autophagy by interacting with the human ortholog of yeast Atg17, RB1CC1/FIP200. *Cell Cycle.* 2011;10: 2763–2769.
41. Kriel J, Loos B. The good, the bad and the autophagosome: exploring unanswered questions of autophagy-dependent cell death. *Cell Death Differ.* 2019;26:640–652.
42. Park A, Koh HC. NF-kappaB/mTOR-mediated autophagy can regulate diquat-induced apoptosis. *Arch Toxicol.* 2019;93:1239–1253.
43. Pose-Juan E, Fernández-Cruz T, Simal-Gándara J. State of the art on public risk assessment of combined human exposure to multiple chemical contaminants. *Trends Food Sci Technol.* 2016;55:11–28.
44. Wei F, Duan Y. Crosstalk between autophagy and nanomaterials: internalization, activation, termination. *Adv Biosys.* 2019;3:1800259.
45. Linares V, Belles M, Domingo JL. Human exposure to PBDE and critical evaluation of health hazards. *Arch Toxicol.* 2015;89:335–356.
46. Fromme H, Becher G, Hilger B, Volkel W. Brominated flame retardants—exposure and risk assessment for the general population. *Int J Hyg Environ Health.* 2016;219:1–23.
47. Zhang X, Suhring R, Serodio D, Bonnell M, Sundin N, Diamond ML. Novel flame retardants: estimating the physical-chemical properties and environmental fate of 94 halogenated and organophosphate PBDE replacements. *Chemosphere.* 2016;144:2401–2407.
48. Li R, Zhang L, Shi Q, Guo Y, Zhang W, Zhou B. A protective role of autophagy in TDCIPP-induced developmental neurotoxicity in zebrafish larvae. *Aquat Toxicol.* 2018;199:46–54.
49. Ma Z, Tang S, Su G, et al. Effects of tris (2-butoxyethyl) phosphate (TBOEP) on endocrine axes during development of early life stages of zebrafish (Danio rerio). *Chemosphere.* 2016;144:1920–1927.
50. Zhang Q, Lu M, Dong X, et al. Potential estrogenic effects of phosphorus-containing flame retardants. *Environ Sci Technol.* 2014;48:6995–7001.
51. Andersen SL, Laurberg P, Wu CS, Olsen J. Attention deficit hyperactivity disorder and autism spectrum disorder in children born to mothers with thyroid dysfunction: a Danish nationwide cohort study. *BJOG.* 2014;121:1365–1374.
52. Modesto T, Tiemeier H, Peeters RP, et al. Maternal mild thyroid hormone insufficiency in early pregnancy and attention-deficit/hyperactivity disorder symptoms in children. *JAMA Pediatr.* 2015;169:838–845.
53. Wang P, Gao J, Zhao S, Guo Y, Wang Z, Qi F. Maternal thyroxine levels during pregnancy and outcomes of cognitive development in children. *Mol Neurobiol.* 2016;53:2241–2248.
54. Costa LG, de Laat R, Tagliaferri S, Pellacani C. A mechanistic view of polybrominated diphenyl ether (PBDE) developmental neurotoxicity. *Toxicol Lett.* 2014;230: 282–294.
55. Li P, Liu L, Zhou G, et al. Perigestational exposure to low doses of PBDE-47 induces excessive ER stress, defective autophagy and the resultant apoptosis contributing to maternal thyroid toxicity. *Sci Total Environ.* 2018;645:363–371.
56. Li P, Ma R, Dong L, et al. Autophagy impairment contributes to PBDE-47-induced developmental neurotoxicity and its relationship with apoptosis. *Theranostics.* 2019;9:4375–4390.
57. Wang C, Zhang S, Ma R, et al. Roles of endoplasmic reticulum stress, apoptosis and autophagy in 2,2′,4,4′-tetrabromodiphenyl ether-induced rat ovarian injury. *Reprod Toxicol.* 2016;65:187–193.

58. Chen R, Hou R, Hong X, Yan S, Zha J. Organophosphate flame retardants (OPFRs) induce genotoxicity in vivo: a survey on apoptosis, DNA methylation, DNA oxidative damage, liver metabolites, and transcriptomics. *Environ Int.* 2019;130:104914.
59. Zhu Y, Li X, Liu J, et al. The effects of decabromodiphenyl ether on glycolipid metabolism and related signaling pathways in mice. *Chemosphere.* 2019;222:849–855.
60. Heudorf U, Mersch-Sundermann V, Angerer J. Phthalates: toxicology and exposure. *Int J Hyg Environ Health.* 2007;210:623–634.
61. Wittassek M, Heger W, Koch HM, Becker K, Angerer J, Kolossa-Gehring M. Daily intake of di(2-ethylhexyl)phthalate (DEHP) by German children—a comparison of two estimation models based on urinary DEHP metabolite levels. *Int J Hyg Environ Health.* 2007;210:35–42.
62. Durmaz E, Ozmert EN, Erkekoglu P, et al. Plasma phthalate levels in pubertal gynecomastia. *Pediatrics.* 2010;125:e122–e129.
63. Frederiksen H, Jensen TK, Jorgensen N, et al. Human urinary excretion of non-persistent environmental chemicals: an overview of Danish data collected between 2006 and 2012. *Reproduction.* 2014;147:555–565.
64. Serrano SE, Karr CJ, Seixas NS, et al. Dietary phthalate exposure in pregnant women and the impact of consumer practices. *Int J Environ Res Public Health.* 2014;11:6193–6215.
65. Meeker JD, Ferguson KK. Relationship between urinary phthalate and bisphenol A concentrations and serum thyroid measures in U.S. adults and adolescents from the National Health and Nutrition Examination Survey (NHANES) 2007–2008. *Environ Health Perspect.* 2011;119:1396–1402.
66. Morgenstern R, Whyatt RM, Insel BJ, et al. Phthalates and thyroid function in preschool age children: sex specific associations. *Environ Int.* 2017;106:11–18.
67. Weng TI, Chen MH, Lien GW, et al. Effects of gender on the association of urinary phthalate metabolites with thyroid hormones in children: a prospective cohort study in Taiwan. *Int J Environ Res Public Health.* 2017;14:123–145.
68. Huang Q, Chen Y, Chi Y, et al. Immunotoxic effects of perfluorooctane sulfonate and di(2-ethylhexyl) phthalate on the marine fish Oryzias melastigma. *Fish Shellfish Immunol.* 2015;44:302–306.
69. Muller-Tegethoff K, Kasper P, Muller L. Evaluation studies on the in vitro rat hepatocyte micronucleus assay. *Mutat Res.* 1995;335:293–307.
70. Turner JH, Petricciani JC, Crouch ML, Wenger S. An evaluation of the effects of diethylhexyl phthalate (DEHP) on mitotically capable cells in blood packs. *Transfusion.* 1974;14:560–566.
71. Choi S, Park SY, Jeong J, et al. Identification of toxicological biomarkers of di(2-ethylhexyl) phthalate in proteins secreted by HepG2 cells using proteomic analysis. *Proteomics.* 2010;10:1831–1846.
72. Erkekoglu P, Zeybek ND, Giray BK, et al. The effects of di(2-ethylhexyl)phthalate on rat liver in relation to selenium status. *Int J Exp Pathol.* 2014;95:64–77.
73. Lin LC, Wang SL, Chang YC, et al. Associations between maternal phthalate exposure and cord sex hormones in human infants. *Chemosphere.* 2011;83:1192–1199.
74. Lin Y, Wei J, Li Y, et al. Developmental exposure to di(2-ethylhexyl) phthalate impairs endocrine pancreas and leads to long-term adverse effects on glucose homeostasis in the rat. *Am J Physiol Endocrinol Metab.* 2011;301:E527–E538.
75. Desvergne B, Feige JN, Casals-Casas C. PPAR-mediated activity of phthalates: a link to the obesity epidemic? *Mol Cell Endocrinol.* 2009;304:43–48.
76. Zhang W, Shen XY, Zhang WW, Chen H, Xu WP, Wei W. The effects of di 2-ethyl hexyl phthalate (DEHP) on cellular lipid accumulation in HepG2 cells and its potential mechanisms in the molecular level. *Toxicol Mech Methods.* 2017;27:245–252.
77. Posnack NG. The adverse cardiac effects of Di(2-ethylhexyl)phthalate and bisphenol A. *Cardiovasc Toxicol.* 2014;14:339–357.

78. Zhang YZ, Zhang ZM, Zhou LT, et al. Di (2-ethylhexyl) phthalate disorders lipid metabolism via TYK2/STAT1 and autophagy in rats. *Biomed Environ Sci*. 2019;32: 406–418.
79. Li J, Zheng L, Wang X, et al. Taurine protects INS-1 cells from apoptosis induced by Di(2-ethylhexyl) phthalate via reducing oxidative stress and autophagy. *Toxicol Mech Methods*. 2019;29:445–456.
80. Sun S, Guo H, Wang J, Dai J. Hepatotoxicity of perfluorooctanoic acid and two emerging alternatives based on a 3D spheroid model. *Environ Pollut*. 2019;246:955–962.
81. Valenzuela-Leon P, Dobrinski I. Exposure to phthalate esters induces an autophagic response in male germ cells. *Environ Epigenet*. 2017;3.
82. Sun Y, Shen J, Zeng L, et al. Role of autophagy in di-2-ethylhexyl phthalate (DEHP)-induced apoptosis in mouse Leydig cells. *Environ Pollut*. 2018;243:563–572.
83. Fiorini C, Cordani M, Gotte G, Picone D, Donadelli M. Onconase induces autophagy sensitizing pancreatic cancer cells to gemcitabine and activates Akt/mTOR pathway in a ROS-dependent manner. *Biochim Biophys Acta*. 2015;1853:549–560.
84. Duan J, Deng T, Kang J, Chen M. DINP aggravates autoimmune thyroid disease through activation of the Akt/mTOR pathway and suppression of autophagy in Wistar rats. *Environ Pollut*. 2019;245:316–324.
85. Luchmann KH, Dafre AL, Trevisan R, et al. A light in the darkness: new biotransformation genes, antioxidant parameters and tissue-specific responses in oysters exposed to phenanthrene. *Aquat Toxicol*. 2014;152:324–334.
86. Bouraoui Z, Banni M, Ghedira J, Clerandeau C, Narbonne JF, Boussetta H. Evaluation of enzymatic biomarkers and lipoperoxidation level in Hediste diversicolor exposed to copper and benzo[a]pyrene. *Ecotoxicol Environ Saf*. 2009;72:1893–1898.
87. Xiu M, Pan L, Jin Q. Bioaccumulation and oxidative damage in juvenile scallop Chlamys farreri exposed to benzo[a]pyrene, benzo[b]fluoranthene and chrysene. *Ecotoxicol Environ Saf*. 2014;107:103–110.
88. Bansal V, Kim KH. Review of PAH contamination in food products and their health hazards. *Environ Int*. 2015;84:26–38.
89. Das DN, Naik PP, Mukhopadhyay S, et al. Elimination of dysfunctional mitochondria through mitophagy suppresses benzo[a]pyrene-induced apoptosis. *Free Radic Biol Med*. 2017;112:452–463.
90. Das DN, Naik PP, Nayak A, et al. Bacopa monnieri-induced protective autophagy inhibits benzo[a]pyrene-mediated apoptosis. *Phytother Res*. 2016;30:1794–1801.
91. Sforzini S, Moore MN, Boeri M, Bencivenga M, Viarengo A. Effects of PAHs and dioxins on the earthworm Eisenia andrei: a multivariate approach for biomarker interpretation. *Environ Pollut*. 2015;196:60–71.
92. Manikkam M, Tracey R, Guerrero-Bosagna C, Skinner MK. Dioxin (TCDD) induces epigenetic transgenerational inheritance of adult onset disease and sperm epimutations. *PLoS One*. 2012;7:e46249.
93. Ray S, Swanson HI. Activation of the aryl hydrocarbon receptor by TCDD inhibits senescence: a tumor promoting event? *Biochem Pharmacol*. 2009;77:681–688.
94. Xu G, Liu J, Yoshimoto K, et al. 2,3,7,8-Tetrachlorodibenzo-p-dioxin (TCDD) induces expression of p27(kip(1)) and FoxO3a in female rat cerebral cortex and PC12 cells. *Toxicol Lett*. 2014;226:294–302.
95. Zhao J, Tang C, Nie X, et al. Autophagy potentially protects against 2,3,7,8-tetrachlorodibenzo-p-dioxin induced apoptosis in SH-SY5Y cells. *Environ Toxicol*. 2016;31:1068–1079.
96. Fiorito F, Ciarcia R, Granato GE, et al. 2,3,7,8-Tetrachlorodibenzo-p-dioxin induced autophagy in a bovine kidney cell line. *Toxicology*. 2011;290:258–270.
97. Bhalla R, Tehrani R, Van Aken B. Toxicity of hydroxylated polychlorinated biphenyls (HO-PCBs) using the bioluminescent assay Microtox((R)). *Ecotoxicology*. 2016;25:1438–1444.

98. Duarte FV, Teodoro JS, Rolo AP, Palmeira CM. Exposure to dibenzofuran triggers autophagy in lung cells. *Toxicol Lett*. 2012;209:35–42.
99. Knaapen AM, Borm PJ, Albrecht C, Schins RP. Inhaled particles and lung cancer. Part A: mechanisms. *Int J Cancer*. 2004;109:799–809.
100. Xing YF, Xu YH, Shi MH, Lian YX. The impact of PM2.5 on the human respiratory system. *J Thorac Dis*. 2016;8:E69–E74.
101. Anderson JO, Thundiyil JG, Stolbach A. Clearing the air: a review of the effects of particulate matter air pollution on human health. *J Med Toxicol*. 2012;8:166–175.
102. Dai L, Zanobetti A, Koutrakis P, Schwartz JD. Associations of fine particulate matter species with mortality in the United States: a multicity time-series analysis. *Environ Health Perspect*. 2014;122:837–842.
103. Nemmar A, Al-Salam S, Zia S, Yasin J, Al Husseni I, Ali BH. Diesel exhaust particles in the lung aggravate experimental acute renal failure. *Toxicol Sci*. 2010;113:267–277.
104. Gao Y, Moten A, Lin HK. Akt: a new activation mechanism. *Cell Res*. 2014;24:785–786.
105. Yi S, Zhang F, Qu F, Ding W. Water-insoluble fraction of airborne particulate matter (PM10) induces oxidative stress in human lung epithelial A549 cells. *Environ Toxicol*. 2014;29:226–233.
106. Deng X, Zhang F, Rui W, et al. PM2.5-induced oxidative stress triggers autophagy in human lung epithelial A549 cells. *Toxicol In Vitro*. 2013;27:1762–1770.
107. Dai P, Shen D, Shen J, et al. The roles of Nrf2 and autophagy in modulating inflammation mediated by TLR4-NFκB in A549 cell exposed to layer house particulate matter 2.5 (PM2.5). *Chemosphere*. 2019;235:1134–1145.
108. Long F, Jiang H, Yi H, Su L, Sun J. Particulate matter 2.5 induced bronchial epithelial cell injury via activation of 5′-adenosine monophosphate-activated protein kinase-mediated autophagy. *J Cell Biochem*. 2019;120:3294–3305.
109. Kimura S, Noda T, Yoshimori T. Dissection of the autophagosome maturation process by a novel reporter protein, tandem fluorescent-tagged LC3. *Autophagy*. 2007;3:452–460.
110. Wu YF, Li ZY, Dong LL, et al. Inactivation of MTOR promotes autophagy-mediated epithelial injury in particulate matter-induced airway inflammation. *Autophagy*. 2019;16:1–16.
111. Lukaszewicz A, Cwiklinska M, Zarzecki M, Szoka P, Lachowicz J, Holownia A. Cytotoxicity, oxidative stress, and autophagy in human alveolar epithelial cell line (A549 cells) exposed to standardized urban dust. *Adv Exp Med Biol*. 2019;1176: 101–108.
112. Costa LG, Cole TB, Coburn J, Chang YC, Dao K, Roque PJ. Neurotoxicity of traffic-related air pollution. *Neurotoxicology*. 2017;59:133–139.
113. Hsu YH, Chuang HC, Lee YH, et al. Traffic-related particulate matter exposure induces nephrotoxicity in vitro and in vivo. *Free Radic Biol Med*. 2019;135:235–244.
114. Bai KJ, Chuang KJ, Chen CL, et al. Microglial activation and inflammation caused by traffic-related particulate matter. *Chem Biol Interact*. 2019;311:108762.
115. Lind PM, Lind L. Endocrine-disrupting chemicals and risk of diabetes: an evidence-based review. *Diabetologia*. 2018;61:1495–1502.
116. Chen J, Wu S, Wen S, et al. The mechanism of environmental endocrine disruptors (DEHP) induces epigenetic transgenerational inheritance of cryptorchidism. *PLoS One*. 2015;10:e0126403.
117. Bloom MS, Mok-Lin E, Fujimoto VY. Bisphenol A and ovarian steroidogenesis. *Fertil Steril*. 2016;106:857–863.
118. Boucher JG, Ahmed S, Atlas E. Bisphenol S induces adipogenesis in primary human preadipocytes from female donors. *Endocrinology*. 2016;157:1397–1407.
119. Peng F, Ji W, Zhu F, et al. A study on phthalate metabolites, bisphenol A and nonylphenol in the urine of Chinese women with unexplained recurrent spontaneous abortion. *Environ Res*. 2016;150:622–628.

120. Mao Z, Zheng YL, Zhang YQ. Behavioral impairment and oxidative damage induced by chronic application of nonylphenol. *Int J Mol Sci*. 2010;12:114–127.
121. Tabassum H, Ashafaq M, Parvez S, Raisuddin S. Role of melatonin in mitigating nonylphenol-induced toxicity in frontal cortex and hippocampus of rat brain. *Neurochem Int*. 2017;104:11–26.
122. Seachrist DD, Bonk KW, Ho SM, Prins GS, Soto AM, Keri RA. A review of the carcinogenic potential of bisphenol A. *Reprod Toxicol*. 2016;59:167–182.
123. Le Magueresse-Battistoni B, Multigner L, Beausoleil C, Rousselle C. Effects of bisphenol A on metabolism and evidences of a mode of action mediated through endocrine disruption. *Mol Cell Endocrinol*. 2018;475:74–91.
124. Lee MR, Park H, Bae S, et al. Urinary bisphenol A concentrations are associated with abnormal liver function in the elderly: a repeated panel study. *J Epidemiol Community Health*. 2014;68:312–317.
125. Ranciere F, Lyons JG, Loh VH, et al. Bisphenol A and the risk of cardiometabolic disorders: a systematic review with meta-analysis of the epidemiological evidence. *Environ Health*. 2015;14:46.
126. Andra SS, Charisiadis P, Arora M, van Vliet-Ostaptchouk JV, Makris KC. Biomonitoring of human exposures to chlorinated derivatives and structural analogs of bisphenol A. *Environ Int*. 2015;85:352–379.
127. Yang S, Zhang A, Li T, et al. Dysregulated autophagy in hepatocytes promotes bisphenol A-induced hepatic lipid accumulation in male mice. *Endocrinology*. 2017;158:2799–2812.
128. Mahemuti L, Chen Q, Coughlan MC, et al. Bisphenol A induces DSB-ATM-p53 signaling leading to cell cycle arrest, senescence, autophagy, stress response, and estrogen release in human fetal lung fibroblasts. *Arch Toxicol*. 2018;92:1453–1469.
129. Dai Z, Li Y, Yang S, Zhao N, Zhang X, Xu J. Kinetics and thermal properties of epoxy resins based on bisphenol fluorene structure. *Eur Polym J*. 2009;45:1941–1948.
130. Zhang Z, Hu Y, Guo J, et al. Fluorene-9-bisphenol is anti-oestrogenic and may cause adverse pregnancy outcomes in mice. *Nat Commun*. 2017;8:14585.
131. Jia Z, Wang H, Feng Z, et al. Fluorene-9-bisphenol exposure induces cytotoxicity in mouse oocytes and causes ovarian damage. *Ecotoxicol Environ Saf*. 2019;180:168–178.
132. Ponzo OJ, Silvia C. Evidence of reproductive disruption associated with neuroendocrine changes induced by UV–B filters, phtalates and nonylphenol during sexual maturation in rats of both gender. *Toxicology*. 2013;311:41–51.
133. Duan P, Hu C, Quan C, et al. 4-Nonylphenol induces apoptosis, autophagy and necrosis in Sertoli cells: involvement of ROS-mediated AMPK/AKT-mTOR and JNK pathways. *Toxicology*. 2016;341–343:28–40.
134. Li S, Jiang Z, Chai W, Xu Y, Wang Y. Autophagy activation alleviates nonylphenol-induced apoptosis in cultured cortical neurons. *Neurochem Int*. 2019;122:73–84.
135. Dorsey WC, Tchounwou PB, Sutton D. Mitogenic and cytotoxic effects of pentachlorophenol to AML 12 mouse hepatocytes. *Int J Environ Res Public Health*. 2004;1:100–105.
136. Zheng W, Yu H, Wang X, Qu W. Systematic review of pentachlorophenol occurrence in the environment and in humans in China: not a negligible health risk due to the re-emergence of schistosomiasis. *Environ Int*. 2012;42:105–116.
137. Huang X, Han X, Huang Z, et al. Maternal pentachlorophenol exposure induces developmental toxicity mediated by autophagy on pregnancy mice. *Ecotoxicol Environ Saf*. 2019;169:829–836.
138. Tsai MS, Lin CY, Lin CC, et al. Association between perfluoroalkyl substances and reproductive hormones in adolescents and young adults. *Int J Hyg Environ Health*. 2015;218:437–443.
139. Elcombe CR, Elcombe BM, Foster JR, Chang SC, Ehresman DJ, Butenhoff JL. Hepatocellular hypertrophy and cell proliferation in Sprague-Dawley rats from dietary exposure to potassium perfluorooctanesulfonate results from increased

expression of xenosensor nuclear receptors PPARalpha and CAR/PXR. *Toxicology*. 2012;293:16–29.
140. Liu B, Zhang H, Yao D, et al. Perfluorinated compounds (PFCs) in the atmosphere of Shenzhen, China: spatial distribution, sources and health risk assessment. *Chemosphere*. 2015;138:511–518.
141. Mao Z, Xia W, Wang J, et al. Perfluorooctane sulfonate induces apoptosis in lung cancer A549 cells through reactive oxygen species-mediated mitochondrion-dependent pathway. *J Appl Toxicol*. 2013;33:1268–1276.
142. Lau C, Butenhoff JL, Rogers JM. The developmental toxicity of perfluoroalkyl acids and their derivatives. *Toxicol Appl Pharmacol*. 2004;198:231–241.
143. Ren XM, Zhang YF, Guo LH, Qin ZF, Lv QY, Zhang LY. Structure-activity relations in binding of perfluoroalkyl compounds to human thyroid hormone T3 receptor. *Arch Toxicol*. 2015;89:233–242.
144. Yang Q, Abedi-Valugerdi M, Xie Y, et al. Potent suppression of the adaptive immune response in mice upon dietary exposure to the potent peroxisome proliferator, perfluorooctanoic acid. *Int Immunopharmacol*. 2002;2:389–397.
145. Yuan Z, Shao X, Miao Z, Zhao B, Zheng Z, Zhang J. Perfluorooctane sulfonate induced neurotoxicity responses associated with neural genes expression, neurotransmitter levels and acetylcholinesterase activity in planarians Dugesia japonica. *Chemosphere*. 2018;206:150–156.
146. Castiglioni S, Valsecchi S, Polesello S, et al. Sources and fate of perfluorinated compounds in the aqueous environment and in drinking water of a highly urbanized and industrialized area in Italy. *J Hazard Mater*. 2015;282:51–60.
147. Rosen MB, Das KP, Rooney J, Abbott B, Lau C, Corton JC. PPARalpha-independent transcriptional targets of perfluoroalkyl acids revealed by transcript profiling. *Toxicology*. 2017;387:95–107.
148. Yang JH. Perfluorooctanoic acid induces peroxisomal fatty acid oxidation and cytokine expression in the liver of male Japanese medaka (Oryzias latipes). *Chemosphere*. 2010;81:548–552.
149. Yao XF, Cao J, Xu LM, et al. Perfluorooctane sulfonate blocked autophagy flux and induced lysosome membrane permeabilization in HepG2 cells. *Food Chem Toxicol*. 2014;67:96–104.
150. Midgett K, Peden-Adams MM, Gilkeson GS, Kamen DL. In vitro evaluation of the effects of perfluorooctanesulfonic acid (PFOS) and perfluorooctanoic acid (PFOA) on IL-2 production in human T-cells. *J Appl Toxicol*. 2015;35:459–465.
151. Tang J, Lu X, Chen F, et al. Effects of perfluorooctanoic acid on the associated genes expression of autophagy signaling pathway of Carassius auratus lymphocytes in vitro. *Front Physiol*. 2018;9:1748.
152. Xin Y, Wan B, Yang Y, Cui XJ, Xie YC, Guo LH. Perfluoroalkyl acid exposure induces protective mitochondrial and endoplasmic reticulum autophagy in lung cells. *Arch Toxicol*. 2018;92:3131–3147.
153. Eskelinen EL. The dual role of autophagy in cancer. *Curr Opin Pharmacol*. 2011;11:294–300.
154. Fu Q, Lyu D, Zhang L, Qin Z, Tang Q, Yin H, Lou X, Chen Z, Yao K. Airborne particulate matter (PM2.5) triggers autophagy in human corneal epithelial cell line. *Environ Pollut*. 2017;227:314–322.
155. Zhao Z, Liu GC, Fu W. Role of PERK-eIF2alpha signaling pathway in fetal male rats with hypospadias induced by di-n-butyl phthalate. *Kaohsiung J Med Sci*. 2018;34(9): 487–493.
156. Su Y, Quan C, Li X, Shi Y, Duan P, Yang K. Mutual promotion of apoptosis and autophagy in prepubertal rat testes induced by joint exposure of bisphenol A and nonylphenol. *Environ Pollut*. 2018;243(Pt A):693–702.

CHAPTER THIRTEEN

Genetic defects of autophagy linked to disease

Milana Fraiberg*, Zvulun Elazar*
Department of Biomolecular Sciences, The Weizmann Institute of Science, Rehovot, Israel
*Corresponding authors: e-mail address: milana.fraiberg@weizmann.ac.il; zvulun.elazar@weizmann.ac.il

Contents

Progress in Molecular Biology and Translational Science, Volume 172
ISSN 1877-1173
https://doi.org/10.1016/bs.pmbts.2020.04.001

Abstract

Autophagy is a highly conserved lysosomal degradation pathway responsible for rapid elimination of unwanted cytoplasmic materials in response to stressful conditions. This cytoprotective function is essential for maintenance of cellular homeostasis and is mediated by conserved autophagy-related genes (*ATG*) and autophagic receptors. Impairment of autophagy frequently results in a wide variety of human pathologies. Recent studies have revealed direct links between diverse diseases and genetic defects of core autophagy genes, autophagy-associated genes, and genes encoding autophagic receptors. Here we provide a general description of autophagy-related genes and their mutations or polymorphisms that play a causative role in specific human disorders or may be risk factors for them.

1. Autophagy as a mechanism for intracellular homeostasis

Autophagy is a tightly regulated catabolic process that facilitates lysosomal degradation of unwanted cellular components including damage organelles, long-lived or aggregative proteins, and invasive microorganisms.[1,2] The autophagic pathway is essential for the maintenance of cellular homeostasis. The process is initiated by nucleation of an isolation membrane, which elongates into a cup-shaped phagophore that sequestrates cytoplasmic components and then closes on itself, sealing the contents inside a double-membrane mature vesicle, the autophagosome, which then fuses with the lysosome for cargo degradation.[3,4] Amino acids and fatty acids from the degraded cargoes are transported back to the cytosolic pool.[5] Autophagy is induced in response to different stressful conditions such as starvation, oxidative stress, ischemia, and growth factor deprivation.[6–9] In addition to the canonical autophagy pathway there is evidence for several types of noncanonical autophagy, which do not require the entire repertoire of AuTophaGy-related (ATG) proteins but also result in autophagosome formation.[10]

1.1 Core autophagy proteins

The core ATG proteins are key players in the initiation, nucleation, expansion and closure of phagophores to form autophagosomes, which then fuse with lysosomes.[11] The ATGs have been classified into five groups according to their functional and physical interactions[1,12]: (i) Autophagy initiation requires the ULK kinase complex (containing ULK1 or ULK2),

ATG13, FIP200 and ATG101. Upon autophagy activation ULK1 undergoes autophosphorylation, followed by phosphorylation of FIP200 and ATG13. (ii) Phagophore nucleation involves the class III phosphatidylinositol 3-kinase (PtdIns3K) complex comprising Beclin1, ATG14, VPS34 and p150 (Vps15). This complex is targeted to the phagophore by ATG14, possibly through phosphorylation by ULK1 and its interaction with ATG13. (iii) Autophagosome sealing and closure need two ubiquitin-like conjugation systems, one of which contains ATG12, ATG5, ATG7, ATG10 and ATG16L1. For efficient autophagosome sealing, ATG7 activates ATG12 while ATG10 promotes ATG12 conjugation to ATG5. ATG16L1 is then recruited to the phagophore by WIPI2 and undergoes conjugation to the ATG5-ATG12 complex, forming an ATG5-ATG12-ATG16L1 complex. The other conjugation system contains (iv) the ATG8 family conjugation system, consisting of microtubule-associated protein 1 light chain 3 (LC3A, LC3B and LC3C) and γ-aminobutyric acid receptor-associated proteins (GABARAP, GABARAP L1 and GABARAP L2), ATG3, ATG4 (ATG4A to ATG4D) cysteine proteases, ATG7, and WIPI2. Nascent pro-ATG8s are cleaved at the C-terminus by ATG4, exposing a glycine residue essential for conjugation to phosphatidylethanolamine (PE). Following this cleavage, ATG8s are activated by ATG7 and conjugated to PE by ATG3 activity. The lipidation of ATG8s promotes phagophore expansion. (v) The membrane supply for sealing is also regulated by the ULK1 complex and involves the ATG9 trafficking system, which includes ATG2A, ATG2B, WIPI4 and ATG9 (see Fig. 1).[1,15]

"Noncanonical autophagy" refers to processes that do not require the entire set of core ATG proteins in order to build up double-membrane autophagosomes.[10] Recently reported examples include ULK1-independent autophagy,[16] Beclin1-independent autophagy,[17,18] VPS34-independent autophagy,[19] p150/VPS15-independent autophagy,[20] and ubiquitin-like conjugation actors such as ATG5-, ATG7- and LC3-independent autophagy[21] (for details see Dupont et al.[10]). Another example of non-canonical autophagy is LC3-associated phagocytosis (LAP) in macrophages, where ATG8 proteins undergo lipidation to single-membrane phagosomes. This process too requires only a subset of the core ATG factors.[22–24]

1.2 Selective autophagy and cargo delivery receptors

Whereas bulk autophagy is responsible for the nonspecific engulfment and degradation of cytoplasmic material, selective autophagy contributes to intracellular homeostasis by selectively degrading distinct cellular structures

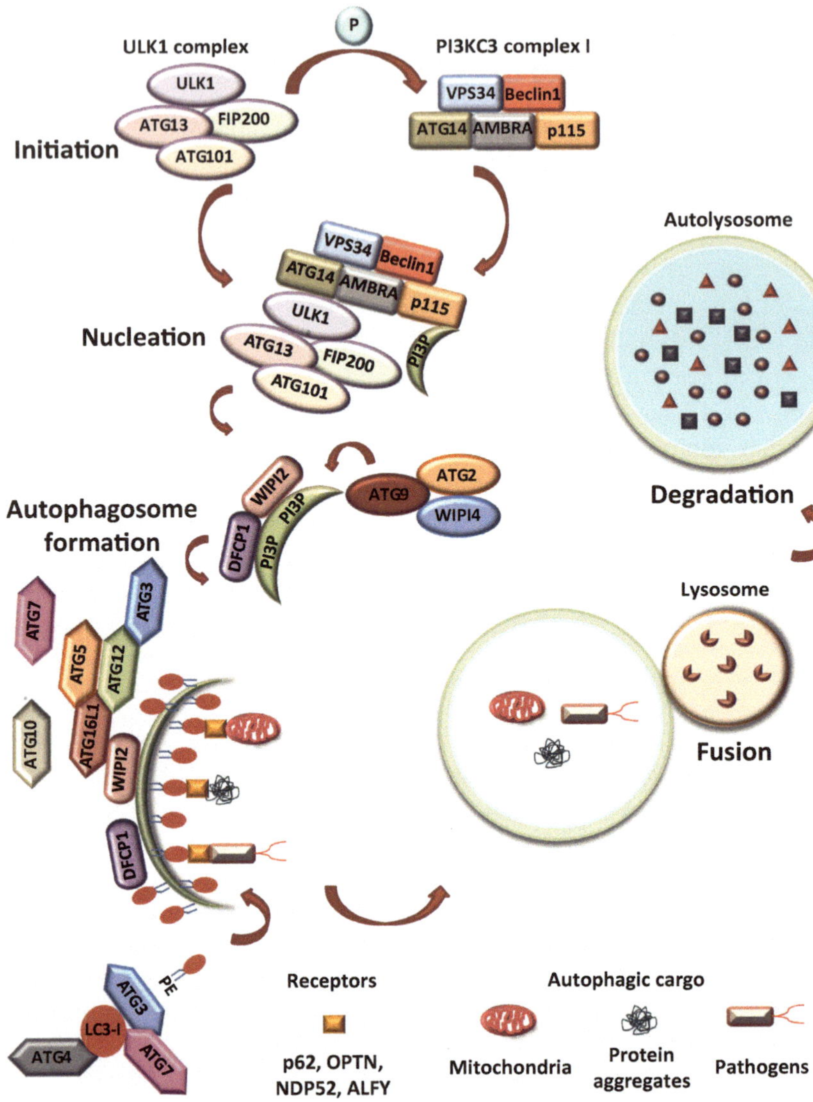

Fig. 1 Schematic overview of autophagy at different stages. Activation of autophagy is typically marked by formation of the ULK1 complex, comprising ULK1, ATG13, ATG101 and FIP200. This initiation complex, upon phosphorylation of components of the class III PI3K complex I (composed of class III PI3K, VPS34, Beclin1, ATG14, AMBRA and p115), triggers phagophore nucleation. In its active form this complex mediates the production of phosphatidylinositol-3-phosphate (PI3P) at a characteristic endoplasmic reticulum (ER) regions labeled by WIPI2 and DFCP1. PI3P-bound WIPI2 facilitates the recruitment of ATG16L1 to the ATG5-ATG12 conjugate to promote ATG7/ATG3 mediated conjugation of ATG8 proteins (represented here by LC3) to phosphatidylethanolamine (PE). Conjugated ATG8 proteins in turn regulate the expansion and possibly also its closure

such as damaged mitochondria (mitophagy) or endoplasmic reticulum (ERphagy), invading bacteria (xenophagy), aggregated proteins (aggrephagy) and others.[1,25] The molecular machinery for selective autophagy must efficiently recognize the cargo and sequester it within autophagosomes. Autophagic cargo receptors/adaptors, such as SQSTM1/p62, NDP52, NBR1, OPTN, and ALFY, specifically bind the cargo material to the autophagic membrane via their LC3-interacting domain (LIR) in an ubiquitin-dependent or -independent manner.[25–27]

2. Diseases caused by genetic mutations in core *ATG* genes

Wide-ranging studies over the past two decades have established links between aberrant autophagy and various diseases.[1] Dysfunctional autophagy due to genetic aberrations or even polymorphism (mostly identified by genome-wide association studies; GWAS) in conserved core autophagy-related genes (*ATG*) have been implicated in muscular, inflammatory, metabolic, neurodegenerative and immune system diseases, as well as in cancer.[28] Importantly, $ATG5^{-/-}$ and $ATG7^{-/-}$ (conditional knockout) mice exhibit progressive motor functional deficits accompanied by accumulation of protein aggregates in neurons.[29–31] Here we focus on the core autophagy genes, described according to the order of their action in the autophagic pathway (see Fig. 1), and on their genetic aberrations that result in human diseases. Note that in this chapter we do not elaborate on cancer-associated mutations with the exception of *Beclin1*.

2.1 Beclin1

Beclin1 is a key player in the induction of autophagy through its interaction with Vps34 to form the Beclin1-ATG14-Vps34-Vps15 complex, which is essential for localization of autophagic proteins to the autophagosome formation site.[32] Beclin1, through its role in autophagy initiation, is highly

to form the double-membrane autophagosome. Recent studies have suggested that the lipids for phagophore elongation are supplied by the ATG9-ATG2-WIPI4 complex.[13,14] In addition, ATG8 proteins play a central role in selective autophagy, where they mediate sequestration of labeled cargo into the autophagosome via the LIR domain on cargo receptors. ATG4, a cysteine protease specifically cleaves ATG8 proteins at their C-terminal glycine residues. The mature autophagosome finally fuses with the lysosome to form the autolysosome. Here the autophagic cargo undergoes degradation, and nutrients are returned from here to the cytoplasm.

involved in cancer, possessing both tumor-suppressive and tumor-supportive functions. Such functions point to Beclin1 as a target for the therapeutic drug development of Beclin1 inducers and inhibitors, according to its action.

Breast, ovarian, prostate and colorectal cancers. Relationships between autophagy and cancer are complex and they vary at different stages of the disease. Autophagy probably has a preventive effect against cancer initially, but once the tumor is progressing the cancer cells utilize autophagy for their own cytoprotection.[33] *BECN1* is among the most extensively studied of the autophagic genes implicated in many types of tumors. *BECN1* maps to a chromosome region 17q21, and its monoallelic deletion has been identified in human breast, ovarian and prostate tumors.[34–37] In addition, Beclin1 plays a role in the growth and metastasis of colorectal cancer. Under-expression of Beclin1 due to allelic loss or miRNA regulatory activity portends poor prognosis by promoting an anti-apoptotic pathway, while its overexpression is associated with tumor hypoxia and acidity, defining a subgroup of aggressive colorectal tumors.[38]

Celiac disease (CeD). *BECN1* (similarly to *ATG7* described below) was found to be implicated in CeD. Since its mRNA and protein levels are decreased, in line with its reduced regulatory miR-30a in intestinal epithelial cells of CeD patients, *BECN1* was suggested to serve as a biomarker for CeD.[39]

2.2 WDR45/WIPI4

The four members of the WIPI family specifically bind PI3P at the nascent autophagosome, acting as scaffold molecules that exhibit distinct interactions with different autophagic proteins. WIPI4 interacts with ATG2, mediating the regulation of autophagosome formation.[40,41] *WDR45/WIPI4* is located at Xp11.23 and includes 12 exons. It encodes a member of the WD40-repeat protein family with β-propeller structure, facilitating protein-protein interaction.[42] In general, *WDR45/WIPI4* is strongly associated with neurodegeneration with brain iron accumulation (NBIA), a heterogeneous group of neurodegenerative diseases characterized by iron accumulation in the basal ganglia and by extrapyramidal movement disorder.

β-propeller protein-associated neurodegeneration (BPAN), originally known as SENDA (static encephalopathy of childhood with neurodegeneration in adulthood), belongs to the group of NBIA disorders.[42] BPAN is characterized by developmental delay in childhood, which deteriorates in adulthood with the onset of muscle stiffness and dementia.[42] Dominant mutations in

WDR45/WIPI4 at Xp11.23 have been identified as a causative factor of BPAN. These mutations include a canonical splice-site (c.439 + 1G > T), synonymous mutation at the last base of exon 8 (c.516G > C), and c.1033_1034dupAA mutation (in addition to one nonsense and two frameshift mutations).[42,43] Another novel heterozygous missense mutation (c.626C > A p.Ala209Asp) in exon 9 of *WDR45/WIPI4* was identified in a female patient with psychomotor retardation following secondary decline.[44]

Rett syndrome is an NBIA-associated neurodevelopmental disorder, in which apparently normal development up to about 18 months of age is followed by degeneration, seizures, stereotypical hands movement, breathing disturbances, growth retardation, sleep disorder, spasticity, and loss of motor skills and language.[45] A splicing site mutation c.440-2A > G in *WDR45/WIPI4* was identified in an Argentinian girl, implicating WIPI4 in Rett syndrome.[45]

Developmental and epileptic encelopathy (DEE) is an NBIA-associated disorder characterized by refractory seizures and epileptic activity that results in developmental delay and early age regression. Seven mutations in *WDR45/WIPI4* (one truncation, five frameshift and one missense Gly205Asp) have been identified as causative factors in DEE.[46]

Early-onset epileptic encephalopathy (EOEE) is another NBIA-associated neurodegenerative disorder. Sequencing analysis revealed a de novo homozygous in-frame deletion c.752_754delCCT, p.(Ser251del) on chromosome Xp.11.23, affecting exon 11 of *WDR45/WIPI4*.[47] Notably, this mutation was found in males whereas most *WDR45/WIPI4* mutations affect women owing to the location of *WDR45/WIPI4* on the X chromosome.

2.3 ATG5

ATG5 is a key autophagy factor and an integral part of the ATG5-ATG12-ATG16L1 complex that catalyzes the ATG8 lipidation essential for autophagosome formation and expansion.[48,49] ATG5 is also indispensable for the fusion of autophagosomes with lysosomes in both canonical and non-canonical autophagy.[50] Mutations in the *ATG5* gene are strongly associated with a number of immune-related and neurodegenerative diseases as detailed below. However, the role of ATG5 in these diseases is based mainly on observation of different phenotypes. Evaluation of the precise functional mechanism of ATG5 in the disorders discussed below will require further study, and will be essential for the development of innovative therapies targeting ATG5.

Asthma is a chronic inflammatory disease of the airways characterized by episodes of coughing, wheezing, shortness of breath, and chest tightness.[3] A single nucleotide polymorphism (SNP) in *ATG5* (rs12201458), which localizes to the putative promoter region of *ATG5*, was linked to asthma patients.[51] Recently, ATG5 was also implicated in collagen deposits in the airway, leading to airway remodeling and asthma exacerbation.[52] Interestingly, the increased *ATG5* expression was found to be responsible for the described phenotypes.[51,52] It should be noted, however, that the exact link to autophagy in these cases has not yet been fully established.

Neuromyelitis optica (NMO) is a central nervous system (CNS) disease of the optic nerves and spinal cord, characterized by autoantibody production against the astrocyte water channel protein aquaporin-4 (AQP4), astrocyte inflammation, loss of astrocytes and neurons, and neuron demyelination.[53,54] In the southern Chinese Han population, several *ATG5* variants were found to be implicated in NMO. These SNPs (located in the *ATG5-PRDM1* intergenic region) in the C allele frequency of rs548234 and the G allele frequency of the rs6937876 SNPs were more abundant in NMO patients than in healthy controls.[54] Notably, the same SNPs have also been linked to other autoimmune diseases (see below).

Systemic lupus erythematosus (SLE) is a chronic autoimmune disease characterized by inflammation of multiple organs including skin, muscles, joints, kidney, heart, blood vessels and the CNS.[55,56] Aberrant autoimmunity in SLE patients includes defects in apoptotic cell clearance, abnormal antigen presentation, and production of autoantibodies usually against double-stranded DNA.[56,57] ATG5 was implicated in SLE through LC3-associated phagocytosis (LAP), which is important for the clearance of dying cells or pathogenic bacteria. LAP-deficient mice, which are characterized by *ATG5* deletion in myeloid cells, were found to exhibit an SLE-like phenotype with increased cytokine and autoantibody levels in the serum and decreased clearance of dead cells.[58] Several GWAS have identified SPNs in *ATG5*, including rs6937876, rs693912, rs9480642, and intergenic regions of *ATG5-PRDM1* (rs548234 and rs6568431), as genetic predisposing factors for SLE in European, Caucasian and Chinese populations.[19,59–62] In silico analyses revealed that these SNPs regulate *ATG5* expression, suggesting that deregulated *ATG5* expression may mediate SLE, a suggestion supported by increased ATG5 in SLE patients.[50] Likewise, since ATG5 accelerates the immune response and functions in antigen presentation, these pathways may be influenced by SLE-related *ATG5* SNPs.[50]

Childhood ataxia is a neurodegenerative disease caused by disruption of the cerebellum and Purkinje cells, leading to the absence of coordinated muscle movements.[63] Recently, whole-exome sequencing of cells from two young Turkish patients with congenital ataxia, mental retardation and developmental delay revealed the homozygous missense mutation E122D in *ATG5*. This mutation resulted in reduced conjugation of ATG5 to ATG12 and an overall decrease in autophagy.[63]

2.4 ATG12

ATG12 plays a crucial role during autophagosome formation. This protein undergoes conjugation to ATG5, mediated by ATG7 and ATG10 in an ATP-dependent manner. The ATG12-ATG5 complex then interacts with ATG16L1 to form an ATG12-ATG5-ATG16L1 conjugate (see Section 2.3).[48,49] ATG12 was found to be implicated in Parkinson's disease, but the mechanism of its activity in this disease requires further investigation.

Parkinson's disease (PD). As mentioned above, several genetic variants of *ATG* genes contribute to the development of sporadic PD. Genetic and functional analyses of the *ATG12* promoter in three PD patients revealed three DNA sequence variants: g.115842507G>T, g.115842394C>T and g.115841817_18del, all of which significantly affected the transcription activity of the *ATG12* gene promoter, possibly owing to aberrations in binding sites for transcriptional factors.[64]

2.5 ATG16L1

ATG16L1 is a core autophagy protein that interacts with ATG5 and ATG12 to form a large multimeric complex, which targets the ATG5-ATG12 conjugate to the phagophore where, together with ATG16L1, they facilitate the conjugation of Atg8s proteins to PE.[65]

Crohn's disease (CD) is an inflammatory bowel disease (IBD) characterized by chronic transmural intestinal inflammation that affects the gastrointestinal tract mucosa.[66] These patients suffer from abdominal pain, diarrhea, fatigue, strictures, abscesses and fistulas.[67] CD is associated with an inappropriate immune response to antigens in the intestinal lumen, defective bacterial recognition, lymphocyte activation and decreased epithelial protection, but it may also be triggered by environmental factors.[67,68] The GWAS identified an SNP at rs2241880 encoding an ATG16L1 (T300A) mutation that is strongly associated with CD.[69,70] This single amino acid mutation may lead to impaired intracellular trafficking of endosomes labeled by the type-I transmembrane protein TMEM59, which interacts with ATG16L1 and leads to an unconventional autophagic process that causes LC3 labeling of single-membrane endosomes.[71] In addition, the ATG16L1 (T300A) mutation increases susceptibility to caspase-3 processing of ATG16L1, leading to decreased levels of the full-length protein following dysfunctional autophagy.[71]

Chronic obstructive pulmonary disease (COPD) is a chronic inflammatory disease of mucosal tissues in which the respiratory system undergoes an accelerated decline in lung function.[72] This disease was found to be associated with the rs2241880 SNP in *ATG16L1* (the same location as that identified in CD). Interestingly, the allele that confers risk on COPD is A, which still encodes the original T300, whereas the risk-associated allele in CD is G, which encodes the mutation T300A.[73]

2.6 ATG7

ATG7 is an E1-like ligase that plays a central role in autophagosome biogenesis by conjugating ATG5 to ATG12. It also participates in lipidation of the ATG8 family.[74] The reduction in ATG7 protein level has been implicated as a causative factor in the autoimmune and neurodegenerative diseases described below, by leading to reduced autophagic function.

Celiac disease (CeD) is a chronic autoimmune disease of the small intestine, where intestinal damage results from gluten consumption.[39] Blood tests and intestinal biopsy are required for diagnosis. In a recent study it was suggested that ATG7 and its microRNA-17 (miR-17) activity may serve as biomarkers for CeD, owing to a decline in intestinal epithelial cell autophagy in correlation with a decrease in *ATG7* expression followed by reduction in miR-17.[39]

Parkinson's disease (PD). Several studies have pointed to deregulation of core autophagy genes in the development and progression of sporadic PD. Bidirectional sequencing analysis of the *ATG7* gene promoter in sporadic PD patients and healthy controls revealed four novel heterozygous variants, 11313449G>A, 11313811 T>C, 11313913G>A and 11314041G>A, which significantly decreased transcription from *ATG7* promoter. This finding implicates ATG7 level as a risk factor for sporadic PD.[75]

2.7 ATG4

The ATG4 family of cysteine proteases in mammals comprises four homologous proteins (ATG4A, ATG4B, ATG4C and ATG4D), all of which process ATG8s to expose a glycine residue at their C-terminus following ATG7-mediated lipidation of ATG8. In addition to its cleavage of ATG8s, ATG4 also plays an essential role in delipidation of ATG8-phosphatidylethanolamine (ATG8-PE) between the C-terminal carboxyl moiety and the amine group of PE. This cleavage is important for ATG8s recycling and promotes elongation of the phagophore.[76] Protease activity reportedly varies among ATG4 homologs, as does the susceptibility of ATG8 homologs to each ATG4 homolog.[77] Since ATG4 is the sole protease that regulates the processing and deconjugation of ATG8s, and because the activity of ATG4 is essential for autophagy, it is a target for the potential development of autophagy-specific inhibitors.[76]

Kashin-Beck disease (KB) is a chronic joint disorder (an osteoarthritis), characterized by functional limitations and disabilities aggravated by poor health conditions. The disease is endemic to eastern Siberia, northern Korea and central regions of China. KB causes degeneration of cartilage, chondrocytic necrosis, and apoptosis of articular cartilage.[78] Moreover, the affected chondrocytes exhibit dysfunctional autophagy. It was recently shown that *ATG4C* variants at rs11208030, rs4409690, rs12097658 and rs6587988 lead to lower mRNA and protein levels of ATG4C in KB chondrocytes, implicating ATG4C in KB.[78]

2.8 MAP1LC3B

MAP1LC3B is one of the six members of the mammalian ATG8 protein family, commonly used to monitor autophagy. LC3 is post-translationally modified into two forms: LC3-I, which is located in the cytosol, and its lipidated form LC3-II, which is conjugated to the autophagosomal membrane and remains there until the autophagosome is degraded following its fusion with the lysosome.[79] Several *MAP1LC3B* gene variants were found to be implicated in both the autoimmune and the neurodegenerative diseases described below, leading to *MAP1LC3B* overexpression in both cases. The exact mechanistic role of LC3 in these disorders, however, remains unclear.

Systemic lupus erythematosus (SLE). In addition to the strong association of *ATG5* with SLE described above, a rare Chinese variant rs933717 at the *FBXO31-MAP1LC3B* region was recently identified as a risk factor for SLE.[80] Similarly to the suggested implication of *ATG5* in SLE, analyses of LC3B expression suggest that it is increased in these patients and also in lupus-prone mouse models.[81,82]

Parkinson's disease (PD), a neurodegenerative disorder that primarily affects movement, is associated with motor deficits accompanied by tremors, cognitive impairment and mood disorders.[83] The disease can be familial or sporadic. PD pathology is characterized by the progressive loss of dopaminergic neurons in the *substantia nigra*, the presence of intraneuronal inclusions termed Lewy bodies, and Lewy neurites enriched with α-synuclein.[83,84] Aggregation of Lewy bodies leads to dysfunction of mitochondrial respiration, resulting in decreased ATP and increased levels of reactive oxygen species.[84] Mutations in PINK1 and in parkin are associated with autophagic proteins involved in mitophagy, and are risk factors for PD.[84] Recently, a novel 21-bp-insertion variant g.4993Ins, identified within the *MAP1LC3B* gene promoter, was found to enhance the promoter's activity (Table 1).[86]

Table 1 Core *ATG* genes and their mutation-related diseases.

Core *ATG* gene	Protein function in autophagy	Disease
Beclin1	Phagophore nucleation[1]	Breast, ovarian, prostate and colorectal cancers[33–38] Celiac disease (CeD)[39]
WDR45/ WIPI4	Lipid transfer[40,41,85]	β-propeller protein-associated neurodegeneration (BPAN)[42–44] Rett syndrome[45] Developmental and epileptic encephalopathy (DEE)[46] Early-onset epileptic encephalopathy (EOEE)[47]
ATG5	Complex of ATG5-ATG12-ATG16L1 is responsible for phagophore expansion[1]	Asthma[51,52] Neuromyelitis optica (NMO)[53,54] Systemic lupus erythematosus (SLE)[19,55–62] Childhood ataxia[63]
ATG12	Complex of ATG5-ATG12-ATG16L1 is responsible for phagophore expansion[1]	Parkinson's disease (PD)[64]
ATG16L1	Complex of ATG5-ATG12-ATG16L1 is responsible for phagophore expansion[1]	Crohn's disease (CD)[66–71] Chronic obstructive pulmonary disease (COPD)[43]

Table 1 Core *ATG* genes and their mutation-related diseases.—cont'd

Core *ATG* gene	Protein function in autophagy	Disease
ATG7	Conjugation of ATG12 to ATG5 and activation of ATG8[1]	Celiac disease (CeD)[39] Parkinson's disease (PD)[75]
ATG4	Cysteine proteases responsible for ATG8 processing and delipidation[76,77]	Kashin-Beck disease (KB)[78]
MAP1LC3B	Phagophore elongation and autophagosome closure[1,79]	Systemic lupus erythematosus (SLE)[80–82] Parkinson's disease (PD)[83,84]

3. Disease caused by mutations in genes associated with autophagy

Autophagic impairments leading to various human disorders may occur not only because of mutations in genes encoding the core autophagic machinery proteins, but also as a result of mutations or polymorphisms in genes encoding proteins associated with autophagic functions such as fusion between lysosomes and autophagosomes, lysosomal regeneration after cargo degradation, cargo targeting, or scaffolding for phagophore biogenesis. In this section we focus, in alphabetic order, on autophagy-associated genes and their genetic aberrations that may lead to diverse human diseases.

3.1 EPG5

EPG5 is a human homolog of the metazoan-specific autophagy gene *EPG5*, which encodes a large coiled-coil domain important for starvation-induced autophagy and the formation of degradative autolysosomes.[87]

Vici syndrome is a rare, recessively inherited multisystem disorder characterized by corpus callosum agenesis, cataracts, cardiomyopathy, immunodeficiency and hypopigmentation.[88] Exome and Sanger sequence analyses revealed 14 causative mutations in the *EPG5* gene located on chromosome 18q12.3, including five nonsense, two frameshift, three splicing and one missense mutation, one multi-exon deletion, and two initiation codon variants.[88,89] Vici syndrome thus serves as a paradigm of a human multisystem disorder linked to defective autophagy.[89]

3.2 HTT

HTT is a large (350-kDa) protein with a flexible structure that alters its activity, suggesting that it functions as a scaffold for different protein complexes.[90] It was found to be an important regulator of autophagy, to be active in autophagosomal transport in neurons, and to serve as a scaffold for selective autophagy-promoting cargo recognition and autophagy initiation by mediating the interaction between the autophagic receptor p62 and its initiating kinase, ULK1.[91,92]

Huntington's disease (HD) is an autosomal dominantly inherited neurodegenerative disease characterized mainly by motor and cognitive dysfunction and psychiatric disorders.[93] Neurodegeneration begins in the medium spiny neurons of the caudate putamen, and the affected regions broaden to the cortex, cerebellum, thalamus and white matter.[94] The trinucleotide repeat expansion mutation of cytosine, adenine and guanine (CAG) in *HD* leads to the production of polyglutamine (polyQ) repeats. Such polyQ expansion becomes pathogenic after it reaches a certain length: fewer than 35 polyQ repeats do not cause HD, 35–39 polyQ repeats may or may not result in the disease, 40–60 polyQ repeats lead to its adult onset, and more than 60 repeats cause juvenile forms of the disease.[95] This expansion of polyQ at the NH3-terminus of HTT leads to protein accumulation and aggregation that result in the intranuclear and intracytoplasmic neuronal inclusions considered to be a pathological hallmark of this disease.[96,97]

3.3 IRGM

IRGM, a member of the immunity-related GTPase family M proteins, plays a central role in organizing the core autophagy machinery by interacting with ULK1 and Beclin1 and leading to the formation of an autophagy initiation complex.[98] IRGM-mediated autophagy was also implicated in protection against *Escherichia coli and Salmonella typhimurium.*[99]

Non-alcoholic fatty liver disease (NAFLD) is a chronic liver disease that may lead to cirrhosis characterized by necroinflammation, fibrosis, alterations in liver vascular architecture, and reduction in liver functional mass.[100] Multiple logistic regression analysis points to the *IRGM* gene variant rs10065172 TT as a risk factor for NAFLD development.[101]

In *Crohn's disease* (CD), the *IRGM* variants rs13361189T/C (a 20-kb deletion polymorphism upstream of *IRGM*) and the exonic synonymous SNP rs10065172C/T on chromosome 5q33.1 have been identified as causal variants for CD.[102,103]

3.4 LAMP2

The *LAMP2* gene is located on the Xq24 chromosome. Lysosome-associated membrane protein-2 (LAMP2), together with LAMP1, comprises a major protein component of lysosomal membranes. LAMP2 is a type I transmembrane protein consisting of a transmembrane domain, a small C-terminal cytoplasmic tail with a lysosomal membrane-targeting signal, and a large intraluminal domain.[104]

Danon cardiomyopathy. A lysosomal glycogen storage disease with normal acid maltase was described in 1981.[105] Its pathology included cardiomyopathy, myopathy and mental retardation, and it was considered to be X-linked. These patients accumulate intracytoplasmic vacuoles with autophagic material and glycogen in the skeletal and cardiac muscle cells.[105,106] Eight mutations in *LAMP-2*, comprising three microdeletions/insertions, two nonsense point mutations, two intronic point mutations and one 10-bp deletion, were identified as causative factors for Danon cardiomyopathy.

3.5 RAB7A

RAB7A is a small ubiquitous protein of the GTPase family. It is responsible for regulating late endocytic transport and for controlling the final step of autophagosomal maturation into an autolysosome.[107]

Charcot-Marie-Tooth type 2B disease (CMT2B) is an inherited autosomal dominant axonal peripheral neuropathy, characterized by prominent sensory loss, atrophy of leg muscles, foot ulcers, and recurrent infections often requiring toe amputation.[108] Five mutations (L129F, K157N, N161T, V162M and N161I) in *RAB7A* were identified as causative factors for CMT2B disease.[109–112] These *RAB7A* mutants affect the ability of RAB7A to regulate autophagy.[108]

3.6 SPG11

The *SPG11* gene encodes spatacsin, a large protein consisting of 2443 amino acids and containing an N-terminal β-propeller-like domain. This protein was detected in autophagic lysosome reformation.[113]

SPG11 is a hereditary autosomal recessive spastic paraplegia characterized by thin corpus callosum and similar pathology to that described below for SPG15.[114–116] More than 100 mutations in the SPG11 locus located on chromosome 15q21.1, including deletions, insertions, nonsense and frameshift mutations, lead to loss of spatacsin function and have been implicated in SPG11 disease. SPG11-mutated fibroblasts, similarly to the SPG15 phenotype, accumulate autophagosomes and enlarged lysosomes, indicating defective autophagosome-to-lysosome fusion.[117,118]

3.7 TECPR2

Tectonin β-propeller containing protein 2 (TECPR2), a recently characterized large protein consisting of 1411 amino acids in humans, has been implicated in intracellular trafficking and autophagy. It is predicted to contain three WD repeats at the N-terminus, six TECPR repeats at the C-terminus, a nuclear localization signal in the middle, and an LIR motif at its C-terminus.[119] High-throughput proteomic analyses of autophagic pathways recognized TECPR2 as a component of the ATG8-interaction network involved in conjugation, vesicle elongation and autophagosome formation.[120]

SPG49 (spastic paraplegia) is a severe neurodegenerative disorder that appears in childhood. Several mutations in *TECPR2* from diverse Jewish origins have been identified as the basis of this unique neuronal disorder. These patients suffer from moderate to severe intellectual disability, evolving spasticity, infantile onset of chronic respiratory disease, instability of blood pressure, temperature and osmolarity, and decreased pain sensitivity.[119,121] All SPG49 patients share unique dysmorphic features including short chubby appearance, microcephaly, dental crowding, and short broad neck.[119] A frame-shift mutation (c.1319delT, c.3416delT) was found in both exon 8 and exon 16 of *TECPR2,* leading to a premature stop codon. Also found was a missense variant (c.C566T), which is expected to result in an amino-acid substitution (p.Thr189Ile) in a highly preserved position within the third WD domain of TECPR2.[119,121]

3.8 ZFYVE26

The *ZFYVE26* gene encodes spastizin, a large (285-kDa) protein containing a zinc-finger domain, an FYVE domain, and a leucine zipper domain.[122] Spastizin is involved in autophagosome maturation through its interaction with the Beclin 1-UVRAG-Rubicon complex.[123,124] Spastizin was also implicated in a process of lysosome biogenesis from autolysosomes after degradation of autophagic cargo.[113,125]

SPG15 is a form of hereditary autosomal recessive spastic paraplegia (HSP) characterized by thin corpus callosum, early onset of spastic paraplegia, pigmentary retinopathy-intellectual disability syndrome, cerebellar

Table 2 Autophagy-associated genes and their mutation-related diseases.

Autophagy-associated genes	**Protein function in autophagy**	**Disease**
EPG5	Formation of degradative autolysosomes[87]	Vici syndrome[88,89]
HTT	Autophagy initiation and cargo recognition[90–92]	Huntington's disease (HD)[96,97]
IRGM	*Formation of an autophagy initiation complex*[98]	Non-alcoholic fatty liver disease (NAFLD)[100,101] Crohn's disease (CD)[102,103]
LAMP2	Lysosomal protein[104]	Danon cardiomyopathy[105,106]
RAB7A	Formation of degradative autolysosomes[107]	Charcot-Marie-Tooth type 2B disease (CMT2B)[109–112]
SPG11	Autophagic lysosome reformation[113]	Spastic paraplegia (SPG11)[117,118]
TECPR2	Component of the ATG8-interaction network[119,120]	Spastic paraplegia (SPG49)[119–121]
ZFYVE26	Autophagosome maturation,[101,102] autophagic lysosome reformation[113,123–125]	Spastic paraplegia (SPG15)[126,127]

dysfunction, and distal amyotrophy.[114,115] More than 20 truncating and nonsense mutations of *ZFYVE26*, which is located on chromosome 14q and encodes spastizin protein, were identified as causative factors for SPG15.[126,127] These mutations interrupt the interaction of spastizin with Beclin1 and inhibit autophagosome maturation, leading to a phenotype of accumulated immature autophagosomes in fibroblasts and lymphoblasts from SPG15 patients (Table 2).[113,124]

4. Diseases caused by mutations in autophagic receptor genes

Selective autophagy is mediated by various autophagic receptors, including p62, OPTN, NDP52 and more, which lead to exclusive degradation of cargo following its specific recognition and sequestration into autophagosomes through binding to ATG8 family members in an ubiquitin-dependent and -independent manner.[26] Mutations in autophagy receptor genes are strongly linked to a wide range of human diseases, especially neurodegenerative and inflammatory disorders. In this section we focus, in alphabetic order, on autophagic receptor genes and their genetic aberrations that may lead to diverse human diseases.

4.1 CALCOCO2/NDP52

The *CALCOCO2* gene located on chromosome 17q21.32 encodes nuclear dot protein (NDP52), a selective autophagic receptor for xenophagy. It detects ubiquitinated cytosolic bacteria and directs the formation of autophagosomes around these pathogens.[128,129] In addition to its role in xenophagy, NDP52 is implicated in autophagosome fusion with endo-lysosomes by attachment of the autophagosomes to MYOSIN VI.[130]

Crohn's disease (CD), as described above, is an IBD characterized by intestinal inflammation.[66] Whole-genome sequencing analysis revealed that the missense SNP rs2303015 (V48A) of *CALCOCO2/NDP52* is implicated in CD. This mutation results in increased NF-κB signaling. Since NDP52 is a suppressor of NF-κB signaling through selective degradation of toll-like receptor signaling adaptors, its mutant is unable to negatively regulate this pathway. The NDP52(V248A) mutant was found to be more stable, leading to accumulative NF-κB activity.[131]

4.2 OPTN1

The optineurin gene (*OPTN*) is located on chromosome 10q13, and encodes a coiled-coil-containing protein with various cellular functions including autophagy, NF-κB regulation, membrane and vesicle trafficking, exocytosis, transcriptional activation, and reorganization of actin and microtubules.[132] OPTN is implicated in autophagosome formation by mediating recruitment of the ATG12-ATG5-ATG16L1 complex to WIPI2-positive phagophores.[133] OPTN also plays a central role in several forms of selective autophagy including xenophagy, mitophagy and aggrephagy.[134]

Amyotrophic lateral sclerosis (ALS, also described below) is a rapidly progressive severe neuropathology associated with loss of upper motor neurons in the cortex and brainstem and lower neurons in the ventral horn of the spinal cord.[134] More than 20 different mutations in *OPTN* were identified as causative for ALS. Since OPTN mediates parkin-dependent mitophagy of damaged mitochondria through recruitment of ZFYVE1/DFCP1 and LC3, its depletion or OPTN(E478G) mutation in its ubiquitin-binding UBAN domain disrupts autophagosome assembly around damaged mitochondria, leading to ALS.[135]

Paget disease of bone (PDP, also described below) is a metabolic bone disease in which osteoclasts contain osteolytic lesions that can lead to bone deformation and sclerosis.[136] Both the SNP rs1561570 on the *OPTN* gene and the SNP rs2234968 that leads to OPTN alternative splicing, resulting in a premature stop codon and loss of OPTN function, were identified as risk factors for PDB.[137,138] These data implicate the autophagic receptor OPTN in the development of PDB, although the etiology remains unclear.

Normal tension glaucoma (NTG) is a neurodegenerative disease that affects retinal ganglion cells and leads to bilateral blindness owing to degeneration of optic nerve fibers.[132] NTG is associated with several mutations in *OPTN*, but the variant OPTN(M98K) was identified as a risk factor for this disease.[132,139] This mutant mediates the increased uptake of transferrin receptor by phagophores, leading to an imbalance in iron metabolism that results in cell death.[132] Another glaucoma-associated mutant OPTN(E50K) inhibits autophagy, leading to apoptotic death of retinal cells.[140]

4.3 SQSTM1/p62

The *SQSTM1* gene, located in chromosome 5q35, includes eight exons and encodes the 440-amino-acid ubiquitin-binding protein p62. The latter is a multifunctional adaptor that is essential for selective autophagy and is responsive to oxidative stress, cell signaling, and DNA damage pathways.[27] p62 contains many different domains, enabling it to serve as a scaffold via various pathways, including selective autophagy. Post-translational modifications in p62 target this protein to its specific cargoes in response to stress. Dysfunction of p62-mediated protein degradation has been implicated in several forms of neurodegeneration, including ALS, PDB, frontotemporal dementia, and multiple system atrophy.[3] p62 was also detected in neuronal inclusions in Alzheimer's and Parkinson's diseases, but its actual role their pathogenesis remains largely unknown.[141]

Amyotrophic lateral sclerosis (ALS, also described above) is a severe neurodegenerative disease characterized by aggregation of cytotoxic protein in the motor neurons of the brain and spinal cord. As a consequence the motor neurons disconnect from their target muscles, leading to atrophy.[142] p62 has been detected in ubiquitinated protein aggregates in patients with ALS, as well as in patients with frontotemporal lobar degeneration (FTLD) (described below).[143] The main link between p62 and ALS-FTLD is the accumulation of p62 inclusions in patients' neurons and glia, suggesting that these disorders are protein aggregation diseases in which p62 aggregation represents a stress response involving selective autophagy.[27,144] More than 40 mutations have been associated with ALS-FTLD diseases, of which several examples are described here. Sequencing analyses of ALS patients revealed various mutations in *SQSTM1* encoding p62, including c.1175C > T, p.Pro392Leu in exon 8 and c.1165 + 1G > A in intron 7. Both mutations are located in the ubiquitin-associated (UBA) binding site and impair the binding of p62 to ubiquitin.[145,146] In addition, two missense mutations, c.259A > G p.Met87Val in exon 2 and c.304A > G p.Lys102Glu in exon 3, were found to lead to the accumulation of large round p62-positive inclusions in motor neurons and elevated p62 levels in spinal cord, pointing to these mutations as a susceptibility factor for ALS.[146] Another ALS-linked missense mutation, p.L341V, also identified in *SQSTM1*, was found to result in decreased affinity to LC3 following impaired autophagy.[147]

Frontotemporal dementia (FTD) and FTD-ALS. FTD is a frontal or temporal lobe atrophy that hampers cognitive abilities and the processing of sensory inputs, resulting in behavioral changes, loss of language, and dementia.[148] Neurons and glial cells of FTD patients were found to accumulate p62 and ubiquitin inclusions.[149] Approximately 5% of ALS cases also exhibit symptoms of FTD, generating a subtype known as FTD-ALS.[150] Thus, some of the mutations in p62 are similar for both diseases. Four heterozygous missense mutations in *SQSTM1* (c.1160C > T, p.P387L; c.1175C > T, p.Pro392L; c.1142C > T, p.A381V; and c.98C > T, p. A33V) were identified as risk factors for FTD. Another substitution in exon 7 K341V and K344E was found to interfere with p62 binding to LC3, leading to impaired autophagy in FTD patients.[151] Two mutants, P348L and G351L, in the KIR domain of p62, which lead to loss of Keap1 binding and decreased ability to activate Nrf2 signaling, were also found to be strongly associated with both FTD and ALS.[152]

Atypical apraxia of speech (AAS) has so far been described in members of only one family, who suffer from dysexecutive syndrome, visuoconstructional defects and apraxia of speech. Recently, the same mutation in p62 (p.Lys238del), found in ALS and FTD patients, was also identified in this family.[153] However, the AAS patients did not exhibit motor neuron symptoms and lacked most symptoms of FTD, suggesting that p62 mutation may result in heterogeneous phenotypes caused by its dysfunction.[153]

Paget disease of bone (PDB), as described above, is a metabolic bone disease. Three mutations in *SQSTM1* (c.1175C > T, p.Pro392Leu in exon 8;

c.1165 + 1G > A in intron 7; and a T insertion that introduced a stop codon at position 396 in exon 8) were found to be located in the UBA domain and to prevent ubiquitin binding, and each of them was identified as a cause of familial and sporadic PDB.[154,155]

Sporadic inclusion body myositis (SIBM), a common myopathy in people over the age of 45, is characterized by progressive muscle weakness and atrophy in proximal and distal muscles, especially in the wrist and finger flexors and knee extensors. This disorder combines inflammatory and degenerative features including rimmed vacuoles, sarcoplasmic inclusions, and depositions of degenerative proteins, one of which is p62.[156] Four missense mutations (A117V, G194R, K238E and P392L) in *SQSTM1*, were found in SIBM patients (part of them also associated with ALS, FTD and PDB). It was suggested that p62 aggregates in the affected muscles of these patients may result from increased p62 stability, leading in turn to dysfunction of proteasomal or lysosomal pathways.[156]

Distal myopathies with rimmed vacuoles (DMRV) belong to a heterogeneous group of muscle disorders characterized by weakness and atrophy of hands and feet muscles. Most of these myopathies exhibit vacuolation or myofibrillar disorganization on the affected muscles.[157] A recently detected mutation (c.1165 + 1 G > A transition) in *SQSTM1*, identified also in PDB (see above), leads to the generation of two abnormal p62 variants, one of them (p.G351_P388del) lacking the C-terminal PEST domain and the other (a truncated p.Glu389delinsAspLysTer) lacking the C-terminal UBA domain and hence unable to associate with ubiquitinated proteins. Both variants are transcribed and translated into abnormal p62, resulting in impaired autophagy.[157]

4.4 ZFYVE25/ALFY

The *ZFYVE25* gene located in chromosome 4q21.23 encodes the autophagic receptor ALFY, a large nuclear protein that consists of 3526 amino acids and is highly conserved throughout evolution. Upon formation of protein aggregates this autophagic receptor shuttles from the nucleus to the cytoplasm, where it serves as a scaffold that recruits autophagic machinery. It does this by interacting with ATG proteins and LC3-bound membranes, generating autophagosomes and facilitating, in a p62-dependent manner, removal of cytosolic ubiquitinated protein aggregates.[158,159]

Primary microcephaly is a dominantly hereditable neurodevelopmental disorder in which infants are born with reduced brain volume and small head circumference, coupled with development of mild to severe

Table 3 Autophagy receptor genes and their mutation-related diseases.

Autophagic receptor genes	Type of selective autophagy	Disease
CALCOCO2/ NDP52	Xenophagy,[128–130] mitophagy[161]	Crohn's disease (CD)[131]
OPTN	Xenophagy, mitophagy and aggrephagy[158,160–163]	Amyotrophic lateral sclerosis[91,134,135] Paget disease of bone (PDB)[136–138] Normal tension glaucoma (NTG)[132,139,140]
SQSTM1/p62	Mitophagy,[162] aggrephagy[3] xenophagy[163]	Amyotrophic lateral sclerosis (ALS)[143–147,150] Frontotemporal dementia (FTD) and FTD-ALS[151,152] Atypical apraxia of speech (AAS)[153] Paget disease of bone (PDB)[154,155] Sporadic inclusion body myositis (SIBM)[156] Distal myopathies with rimmed vacuoles (DMRV)[157]
ZFYVE25/ ALFY	Aggrephagy[158,159]	Primary microcephaly[160]

intellectual disabilities. The dominant missense mutation c.7909C > T g. Chr4:85636503G > A in *ALFY* was found to be responsible for autosomal dominant primary microcephaly.[160] It was further suggested that this primary microcephaly occurs as a result of ALFY mutation, and that the mutated ALFY is incapable of mediating the autophagic removal of DVL3 aggregates, owing to defective regulation of the Wnt signaling pathway, which determines human brain size and microcephaly (Table 3).[160]

5. Conclusions

This chapter records a diverse list of pathologies that are caused by mutations in human core autophagy genes, in genes associated with autophagy, and in genes encoding autophagic receptors. Since most of these genes function not only in autophagy but also in other biological pathways, a large spectrum of mutations has been implicated in the pathogenesis of human diseases. Notably, most of known genetic aberrations in autophagy result in neurological or inflammatory disorders. Since autophagy is essential for post-mitotic neurons, its imbalance may lead to a variety of neurodegenerative diseases including ALS, FTD, PD, HD, BPAN, HSP, Vici syndrome and others. Hallmarks of these disorders are cytoplasmic protein aggregates and dysfunctional organelles such as mitochondria that accumulate in neuronal cells, leading to severe disturbances and death.[1] In general, these neurodegenerative disorders are associated with a reduced autophagic function. As autophagy is also essential for immunity and immune functions such as xenophagy,[164] control of inflammation, antigen presentation,[165,166] cytokine secretion[167] and others, its impairment may cause inflammatory diseases such as SLE, CD, and COPD. In those disorders, however, as well as in some metabolic disorders and cancer, the autophagy impairment may be bilateral. Appropriate therapy may restore the function of the mutated protein, by neutralizing the pathogenic effect caused by the specific mutation and upregulating the autophagic pathway. Preclinical investigations in animal models with specific mutations in genes that impair autophagy will contribute to better understanding of the links between these genetic aberrations and the autophagy functions they affect, and may also contribute to the development of appropriate therapies for curing, ameliorating or delaying these illnesses.

Acknowledgments

Z.E. is the incumbent of the Harold L. Korda Professorial Chair of Biology and is supported by the Jeanne and Joseph Nissim Center for Life Sciences Research, and the Estate of Emile Mimran. We are grateful for funding from the Israel Science Foundation ISF (Grant #215/19), the Legacy Heritage Fund (Grant #1935/16), the Sagol Longevity Foundation, and the Yeda-Sela Center for Basic Research. We apologize to those authors whose works could not be cited here because of the journal's restriction of citations.

References

1. Dikic I, Elazar Z. Mechanism and medical implications of mammalian autophagy. *Nat Rev Mol Cell Biol*. 2018;19:349–364.
2. Xie W, Zhou J. Aberrant regulation of autophagy in mammalian diseases. *Biol Lett*. 2018;14:20170540.
3. van Beek N, Klionsky DJ, Reggiori F. Genetic aberrations in macroautophagy genes leading to diseases. *Biochim Biophys Acta Mol Cell Res*. 1865;2018:803–816.
4. Tooze SA, Abada A, Elazar Z. Endocytosis and autophagy: exploitation or cooperation? *Cold Spring Harb Perspect Biol*. 2014;6:a018358.
5. Ramesh J, Ronsard L, Gao A, Venugopal B. Autophagy intertwines with different diseases-recent strategies for therapeutic approaches. *Diseases*. 2019;7:15.

6. He C, Klionsky DJ. Regulation mechanisms and signaling pathways of autophagy. *Annu Rev Genet.* 2009;43:67–93.
7. Jiang P, Mizushima N. Autophagy and human diseases. *Cell Res.* 2014;24:69–79.
8. Mizushima N, Komatsu M. Autophagy: renovation of cells and tissues. *Cell.* 2011;147:728–741.
9. Mizushima N, Levine B. Autophagy in mammalian development and differentiation. *Nat Cell Biol.* 2010;12:823–830.
10. Dupont N, Nascimbeni AC, Morel E, Codogno P. Molecular mechanisms of noncanonical autophagy. *Int Rev Cell Mol Biol.* 2017;328:1–23.
11. Lamb CA, Yoshimori T, Tooze SA. The autophagosome: origins unknown, biogenesis complex. *Nat Rev Mol Cell Biol.* 2013;14:759–774.
12. Bestebroer J, V'Kovski P, Mauthe M, Reggiori F. Hidden behind autophagy: the unconventional roles of ATG proteins. *Traffic.* 2013;14:1029–1041.
13. Chowdhury S, Otomo C, Leitner A, et al. Insights into autophagosome biogenesis from structural and biochemical analyses of the ATG2A-WIPI4 complex. *Proc Natl Acad Sci USA.* 2018;115:E9792–E9801.
14. Gomez-Sanchez R, Rose J, Guimaraes R, et al. Atg9 establishes Atg2-dependent contact sites between the endoplasmic reticulum and phagophores. *J Cell Biol.* 2018;217:2743–2763.
15. Levine B, Kroemer G. Biological functions of autophagy genes: a disease perspective. *Cell.* 2019;176:11–42.
16. Cheong H, Lindsten T, Wu J, Lu C, Thompson CB. Ammonia-induced autophagy is independent of ULK1/ULK2 kinases. *Proc Natl Acad Sci USA.* 2011;108:11121–11126.
17. Zhu JH, Horbinski C, Guo F, Watkins S, Uchiyama Y, Chu CT. Regulation of autophagy by extracellular signal-regulated protein kinases during 1-methyl-4-phenylpyridinium-induced cell death. *Am J Pathol.* 2007;170:75–86.
18. Grishchuk Y, Ginet V, Truttmann AC, Clarke PG, Puyal J. Beclin 1-independent autophagy contributes to apoptosis in cortical neurons. *Autophagy.* 2011;7:1115–1131.
19. Zhou XJ, Lu XL, Lv JC, et al. Genetic association of PRDM1-ATG5 intergenic region and autophagy with systemic lupus erythematosus in a Chinese population. *Ann Rheum Dis.* 2011;70:1330–1337.
20. Nemazanyy I, Blaauw B, Paolini C, et al. Defects of Vps15 in skeletal muscles lead to autophagic vacuolar myopathy and lysosomal disease. *EMBO Mol Med.* 2013;5:870–890.
21. Nishida Y, Arakawa S, Fujitani K, et al. Discovery of Atg5/Atg7-independent alternative macroautophagy. *Nature.* 2009;461:654–658.
22. Sanjuan MA, Dillon CP, Tait SW, et al. Toll-like receptor signalling in macrophages links the autophagy pathway to phagocytosis. *Nature.* 2007;450:1253–1257.
23. Sanjuan MA, Milasta S, Green DR. Toll-like receptor signaling in the lysosomal pathways. *Immunol Rev.* 2009;227:203–220.
24. Jacquin E, Fletcher K, Florey O. Imaging noncanonical autophagy and LC3-associated phagocytosis in cultured cells. *Methods Mol Biol.* 1880;2019:295–303.
25. Zaffagnini G, Martens S. Mechanisms of selective autophagy. *J Mol Biol.* 2016;428:1714–1724.
26. Khaminets A, Behl C, Dikic I. Ubiquitin-dependent and independent signals in selective autophagy. *Trends Cell Biol.* 2016;26:6–16.
27. Deng Z, Purtell K, Lachance V, Wold MS, Chen S, Yue Z. Autophagy receptors and neurodegenerative diseases. *Trends Cell Biol.* 2017;27:491–504.
28. Shintani T, Klionsky DJ. Autophagy in health and disease: a double-edged sword. *Science.* 2004;306:990–995.
29. Hara T, Nakamura K, Matsui M, et al. Suppression of basal autophagy in neural cells causes neurodegenerative disease in mice. *Nature.* 2006;441:885–889.

30. Komatsu M, Waguri S, Ueno T, et al. Impairment of starvation-induced and constitutive autophagy in Atg7-deficient mice. *J Cell Biol.* 2005;169:425–434.
31. Komatsu M, Waguri S, Chiba T, et al. Loss of autophagy in the central nervous system causes neurodegeneration in mice. *Nature.* 2006;441:880–884.
32. Menon MB, Dhamija S. Beclin 1 phosphorylation—at the center of autophagy regulation. *Front Cell Dev Biol.* 2018;6:137.
33. White E. Deconvoluting the context-dependent role for autophagy in cancer. *Nat Rev Cancer.* 2012;12:401–410.
34. Liang XH, Jackson S, Seaman M, et al. Induction of autophagy and inhibition of tumorigenesis by beclin 1. *Nature.* 1999;402:672–676.
35. Saito H, Inazawa J, Saito S, et al. Detailed deletion mapping of chromosome 17q in ovarian and breast cancers: 2-cM region on 17q21.3 often and commonly deleted in tumors. *Cancer Res.* 1993;53:3382–3385.
36. Aita VM, Liang XH, Murty VV, et al. Cloning and genomic organization of beclin 1, a candidate tumor suppressor gene on chromosome 17q21. *Genomics.* 1999;59:59–65.
37. Gao X, Zacharek A, Salkowski A, et al. Loss of heterozygosity of the BRCA1 and other loci on chromosome 17q in human prostate cancer. *Cancer Res.* 1995;55: 1002–1005.
38. Koukourakis MI, Giatromanolaki A, Sivridis E, Pitiakoudis M, Gatter KC, Harris AL. Beclin 1 over- and underexpression in colorectal cancer: distinct patterns relate to prognosis and tumour hypoxia. *Br J Cancer.* 2010;103:1209–1214.
39. Comincini S, Manai F, Meazza C, et al. Identification of autophagy-related genes and their regulatory miRNAs associated with celiac disease in children. *Int J Mol Sci.* 2017;18:391.
40. Bakula D, Muller AJ, Zuleger T, et al. WIPI3 and WIPI4 beta-propellers are scaffolds for LKB1-AMPK-TSC signalling circuits in the control of autophagy. *Nat Commun.* 2017;8:15637.
41. Dooley HC, Razi M, Polson HE, Girardin SE, Wilson MI, Tooze SA. WIPI2 links LC3 conjugation with PI3P, autophagosome formation, and pathogen clearance by recruiting Atg12-5-16L1. *Mol Cell.* 2014;55:238–252.
42. Haack TB, Hogarth P, Kruer MC, et al. Exome sequencing reveals de novo WDR45 mutations causing a phenotypically distinct, X-linked dominant form of NBIA. *Am J Hum Genet.* 2012;91:1144–1149.
43. Saitsu H, Nishimura T, Muramatsu K, et al. De novo mutations in the autophagy gene WDR45 cause static encephalopathy of childhood with neurodegeneration in adulthood. *Nat Genet.* 2013;45:445–449. 9e1.
44. Tschentscher A, Dekomien G, Ross S, et al. Analysis of the C19orf12 and WDR45 genes in patients with neurodegeneration with brain iron accumulation. *J Neurol Sci.* 2015;349:105–109.
45. Hoffjan S, Ibisler A, Tschentscher A, Dekomien G, Bidinost C, Rosa AL. WDR45 mutations in Rett (-like) syndrome and developmental delay: case report and an appraisal of the literature. *Mol Cell Probes.* 2016;30:44–49.
46. Carvill GL, Liu A, Mandelstam S, et al. Severe infantile onset developmental and epileptic encephalopathy caused by mutations in autophagy gene WDR45. *Epilepsia.* 2018;59:e5–e13.
47. Redon S, Benech C, Schutz S, et al. Intragenic deletion of the WDR45 gene in a male with encephalopathy, severe psychomotor disability, and epilepsy. *Am J Med Genet A.* 2017;173:1444–1446.
48. Noda NN, Fujioka Y, Hanada T, Ohsumi Y, Inagaki F. Structure of the Atg12-Atg5 conjugate reveals a platform for stimulating Atg8-PE conjugation. *EMBO Rep.* 2013;14:206–211.
49. Noda NN, Inagaki F. Mechanisms of autophagy. *Annu Rev Biophys.* 2015;44:101–122.

50. Ye X, Zhou XJ, Zhang H. Exploring the role of autophagy-related gene 5 (ATG5) yields important insights into autophagy in autoimmune/autoinflammatory diseases. *Front Immunol*. 2018;9:2334.
51. Martin LJ, Gupta J, Jyothula SS, et al. Functional variant in the autophagy-related 5 gene promotor is associated with childhood asthma. *PLoS One*. 2012;7:e33454.
52. Poon AH, Choy DF, Chouiali F, et al. Increased autophagy-related 5 gene expression is associated with collagen expression in the airways of Refractory Asthmatics. *Front Immunol*. 2017;8:355.
53. Fujihara K. Neuromyelitis optica and astrocytic damage in its pathogenesis. *J Neurol Sci*. 2011;306:183–187.
54. Cai PP, Wang HX, Zhuang JC, et al. Variants of autophagy-related gene 5 are associated with neuromyelitis optica in the Southern Han Chinese population. *Autoimmunity*. 2014;47:563–566.
55. Mackillop LH, Germain SJ, Nelson-Piercy C. Systemic lupus erythematosus. *BMJ*. 2007;335:933–936.
56. Meas R, Burak MJ, Sweasy JB. DNA repair and systemic lupus erythematosus. *DNA Repair (Amst)*. 2017;56:174–182.
57. Gualtierotti R, Biggioggero M, Penatti AE, Meroni PL. Updating on the pathogenesis of systemic lupus erythematosus. *Autoimmun Rev*. 2010;10:3–7.
58. Martinez J, Cunha LD, Park S, et al. Corrigendum: noncanonical autophagy inhibits the autoinflammatory, lupus-like response to dying cells. *Nature*. 2016;539:124.
59. International Consortium for Systemic Lupus Erythematosus Genetics, Harley JB, Alarcon-Riquelme ME, et al. Genome-wide association scan in women with systemic lupus erythematosus identifies susceptibility variants in ITGAM, PXK, KIAA1542 and other loci. *Nat Genet*. 2008;40:204–210.
60. Gateva V, Sandling JK, Hom G, et al. A large-scale replication study identifies TNIP1, PRDM1, JAZF1, UHRF1BP1 and IL10 as risk loci for systemic lupus erythematosus. *Nat Genet*. 2009;41:1228–1233.
61. Han JW, Zheng HF, Cui Y, et al. Genome-wide association study in a Chinese Han population identifies nine new susceptibility loci for systemic lupus erythematosus. *Nat Genet*. 2009;41:1234–1237.
62. Zhang YM, Cheng FJ, Zhou XJ, Qi YY, Zhao MH, Zhang H. Rare variants of ATG5 are likely to be associated with Chinese patients with systemic lupus erythematosus. *Medicine (Baltimore)*. 2015;94:e939.
63. Kim M, Sandford E, Gatica D, et al. Mutation in ATG5 reduces autophagy and leads to ataxia with developmental delay. *Elife*. 2016;5:e12245.
64. Li Y, Huang J, Pang S, et al. Novel and functional ATG12 gene variants in sporadic Parkinson's disease. *Neurosci Lett*. 2017;643:22–26.
65. Fujita N, Itoh T, Omori H, Fukuda M, Noda T, Yoshimori T. The Atg16L complex specifies the site of LC3 lipidation for membrane biogenesis in autophagy. *Mol Biol Cell*. 2008;19:2092–2100.
66. Sartor RB. Microbial influences in inflammatory bowel diseases. *Gastroenterology*. 2008;134:577–594.
67. Liu TC, Stappenbeck TS. Genetics and pathogenesis of inflammatory bowel disease. *Annu Rev Pathol*. 2016;11:127–148.
68. Cho JH, Brant SR. Recent insights into the genetics of inflammatory bowel disease. *Gastroenterology*. 2011;140:1704–1712.
69. Hampe J, Franke A, Rosenstiel P, et al. A genome-wide association scan of nonsynonymous SNPs identifies a susceptibility variant for Crohn disease in ATG16L1. *Nat Genet*. 2007;39:207–211.
70. Weersma RK, Zhernakova A, Nolte IM, et al. ATG16L1 and IL23R are associated with inflammatory bowel diseases but not with celiac disease in the Netherlands. *Am J Gastroenterol*. 2008;103:621–627.

71. Boada-Romero E, Serramito-Gomez I, Sacristan MP, Boone DL, Xavier RJ, Pimentel-Muinos FX. The T300A Crohn's disease risk polymorphism impairs function of the WD40 domain of ATG16L1. *Nat Commun*. 2016;7:11821.
72. Donaldson GC, Seemungal TA, Bhowmik A, Wedzicha JA. Relationship between exacerbation frequency and lung function decline in chronic obstructive pulmonary disease. *Thorax*. 2002;57:847–852.
73. Chen CZ, Ou CY, Wang RH, et al. Association of Egr-1 and autophagy-related gene polymorphism in men with chronic obstructive pulmonary disease. *J Formos Med Assoc*. 2015;114:750–755.
74. Ohsumi Y, Mizushima N. Two ubiquitin-like conjugation systems essential for autophagy. *Semin Cell Dev Biol*. 2004;15:231–236.
75. Chen D, Pang S, Feng X, Huang W, Hawley RG, Yan B. Genetic analysis of the ATG7 gene promoter in sporadic Parkinson's disease. *Neurosci Lett*. 2013;534:193–198.
76. Maruyama T, Noda NN. Autophagy-regulating protease Atg4: structure, function, regulation and inhibition. *J Antibiot (Tokyo)*. 2017;71:72–78.
77. Li M, Hou Y, Wang J, Chen X, Shao ZM, Yin XM. Kinetics comparisons of mammalian Atg4 homologues indicate selective preferences toward diverse Atg8 substrates. *J Biol Chem*. 2011;286:7327–7338.
78. Wu C, Wen Y, Guo X, et al. Genetic association, mRNA and protein expression analysis identify ATG4C as a susceptibility gene for Kashin-Beck disease. *Osteoarthr Cartil*. 2017;25:281–286.
79. Slobodkin MR, Elazar Z. The Atg8 family: multifunctional ubiquitin-like key regulators of autophagy. *Essays Biochem*. 2013;55:51–64.
80. Qi YY, Zhou XJ, Nath SK, et al. A rare variant (rs933717) at FBXO31-MAP1LC3B in Chinese is associated with systemic lupus erythematosus. *Arthritis Rheumatol*. 2018;70:287–297.
81. Gros F, Arnold J, Page N, et al. Macroautophagy is deregulated in murine and human lupus T lymphocytes. *Autophagy*. 2012;8:1113–1123.
82. Clarke AJ, Ellinghaus U, Cortini A, et al. Autophagy is activated in systemic lupus erythematosus and required for plasmablast development. *Ann Rheum Dis*. 2015;74:912–920.
83. Spillantini MG, Schmidt ML, Lee VM, Trojanowski JQ, Jakes R, Goedert M. Alpha-synuclein in Lewy bodies. *Nature*. 1997;388:839–840.
84. Karabiyik C, Lee MJ, Rubinsztein DC. Autophagy impairment in Parkinson's disease. *Essays Biochem*. 2017;61:711–720.
85. Maeda S, Otomo C, Otomo T. The autophagic membrane tether ATG2A transfers lipids between membranes. *Elife*. 2019;8:e45777.
86. Xu J, Yang Y, Pang S, et al. Identification of a novel 21bp-insertion variant within the LC3B gene promoter in sporadic Parkinson's disease. *Transl Res*. 2013;161:441–443.
87. Tian Y, Li Z, Hu W, et al. *C. elegans* screen identifies autophagy genes specific to multicellular organisms. *Cell*. 2010;141:1042–1055.
88. Cullup T, Kho AL, Dionisi-Vici C, et al. Recessive mutations in EPG5 cause Vici syndrome, a multisystem disorder with defective autophagy. *Nat Genet*. 2013;45:83–87.
89. Hori I, Otomo T, Nakashima M, et al. Defects in autophagosome-lysosome fusion underlie Vici syndrome, a neurodevelopmental disorder with multisystem involvement. *Sci Rep*. 2017;7:3552.
90. MacDonald ME. Huntingtin: alive and well and working in middle management. *Sci STKE*. 2003;2003:pe48.
91. Wong YC, Holzbaur EL. The regulation of autophagosome dynamics by huntingtin and HAP1 is disrupted by expression of mutant huntingtin, leading to defective cargo degradation. *J Neurosci*. 2014;34:1293–1305.
92. Rui YN, Xu Z, Patel B, et al. Huntingtin functions as a scaffold for selective macroautophagy. *Nat Cell Biol*. 2015;17:262–275.

93. Sturrock A, Leavitt BR. The clinical and genetic features of Huntington disease. *J Geriatr Psychiatry Neurol.* 2010;23:243–259.
94. Vonsattel JP, Keller C, Cortes Ramirez EP. Huntington's disease—neuropathology. *Handb Clin Neurol.* 2011;100:83–100.
95. Bates GP, Dorsey R, Gusella JF, et al. Huntington disease. *Nat Rev Dis Primers.* 2015;1:15005.
96. Gutekunst CA, Li SH, Yi H, et al. Nuclear and neuropil aggregates in Huntington's disease: relationship to neuropathology. *J Neurosci.* 1999;19:2522–2534.
97. Croce KR, Yamamoto A. A role for autophagy in Huntington's disease. *Neurobiol Dis.* 2019;122:16–22.
98. Chauhan S, Mandell MA, Deretic V. IRGM governs the core autophagy machinery to conduct antimicrobial defense. *Mol Cell.* 2015;58:507–521.
99. Singh SB, Davis AS, Taylor GA, Deretic V. Human IRGM induces autophagy to eliminate intracellular mycobacteria. *Science.* 2006;313:1438–1441.
100. Li B, Zhang C, Zhan YT. Nonalcoholic fatty liver disease cirrhosis: a review of its epidemiology, risk factors, clinical presentation, diagnosis, management, and prognosis. *Can J Gastroenterol Hepatol.* 2018;2018:2784537.
101. Lin YC, Chang PF, Lin HF, Liu K, Chang MH, Ni YH. Variants in the autophagy-related gene IRGM confer susceptibility to non-alcoholic fatty liver disease by modulating lipophagy. *J Hepatol.* 2016;65:1209–1216.
102. Glas J, Seiderer J, Bues S, et al. IRGM variants and susceptibility to inflammatory bowel disease in the German population. *PLoS One.* 2013;8:e54338.
103. Prescott NJ, Dominy KM, Kubo M, et al. Independent and population-specific association of risk variants at the IRGM locus with Crohn's disease. *Hum Mol Genet.* 2010;19:1828–1839.
104. Eskelinen EL. Roles of LAMP-1 and LAMP-2 in lysosome biogenesis and autophagy. *Mol Aspects Med.* 2006;27:495–502.
105. Danon MJ, Oh SJ, DiMauro S, et al. Lysosomal glycogen storage disease with normal acid maltase. *Neurology.* 1981;31:51–57.
106. Nishino I, Fu J, Tanji K, et al. Primary LAMP-2 deficiency causes X-linked vacuolar cardiomyopathy and myopathy (Danon disease). *Nature.* 2000;406:906–910.
107. Hyttinen JM, Niittykoski M, Salminen A, Kaarniranta K. Maturation of autophagosomes and endosomes: a key role for Rab7. *Biochim Biophys Acta.* 1833;2013:503–510.
108. Colecchia D, Stasi M, Leonardi M, et al. Alterations of autophagy in the peripheral neuropathy Charcot-Marie-Tooth type 2B. *Autophagy.* 2018;14:930–941.
109. Meggouh F, Bienfait HM, Weterman MA, de Visser M, Baas F. Charcot-Marie-Tooth disease due to a de novo mutation of the RAB7 gene. *Neurology.* 2006;67:1476–1478.
110. Houlden H, King RH, Muddle JR, et al. A novel RAB7 mutation associated with ulcero-mutilating neuropathy. *Ann Neurol.* 2004;56:586–590.
111. Verhoeven K, De Jonghe P, Coen K, et al. Mutations in the small GTP-ase late endosomal protein RAB7 cause Charcot-Marie-Tooth type 2B neuropathy. *Am J Hum Genet.* 2003;72:722–727.
112. Wang X, Han C, Liu W, Wang P, Zhang X. A novel RAB7 mutation in a Chinese family with Charcot-Marie-Tooth type 2B disease. *Gene.* 2014;534:431–434.
113. Varga RE, Khundadze M, Damme M, et al. In vivo evidence for lysosome depletion and impaired autophagic clearance in hereditary spastic paraplegia type SPG11. *PLoS Genet.* 2015;11:e1005454.
114. Ebrahimi-Fakhari D, Saffari A, Wahlster L, et al. Congenital disorders of autophagy: an emerging novel class of inborn errors of neuro-metabolism. *Brain.* 2016;139: 317–337.
115. Kjellin K. Familial spastic paraplegia with amyotrophy, oligophrenia, and central retinal degeneration. *Arch Neurol.* 1959;1:133–140.

116. Stevanin G, Santorelli FM, Azzedine H, et al. Mutations in SPG11, encoding spatacsin, are a major cause of spastic paraplegia with thin corpus callosum. *Nat Genet.* 2007;39:366–372.
117. Renvoise B, Chang J, Singh R, et al. Lysosomal abnormalities in hereditary spastic paraplegia types SPG15 and SPG11. *Ann Clin Transl Neurol.* 2014;1:379–389.
118. Chang J, Lee S, Blackstone C. Spastic paraplegia proteins spastizin and spatacsin mediate autophagic lysosome reformation. *J Clin Invest.* 2014;124:5249–5262.
119. Oz-Levi D, Ben-Zeev B, Ruzzo EK, et al. Mutation in TECPR2 reveals a role for autophagy in hereditary spastic paraparesis. *Am J Hum Genet.* 2012;91:1065–1072.
120. Behrends C, Sowa ME, Gygi SP, Harper JW. Network organization of the human autophagy system. *Nature.* 2010;466:68–76.
121. Heimer G, Oz-Levi D, Eyal E, et al. TECPR2 mutations cause a new subtype of familial dysautonomia like hereditary sensory autonomic neuropathy with intellectual disability. *Eur J Paediatr Neurol.* 2016;20:69–79.
122. Hanein S, Martin E, Boukhris A, et al. Identification of the SPG15 gene, encoding spastizin, as a frequent cause of complicated autosomal-recessive spastic paraplegia, including Kjellin syndrome. *Am J Hum Genet.* 2008;82:992–1002.
123. Vantaggiato C, Crimella C, Airoldi G, et al. Defective autophagy in spastizin mutated patients with hereditary spastic paraparesis type 15. *Brain.* 2013;136:3119–3139.
124. Vantaggiato C, Clementi E, Bassi MT. ZFYVE26/SPASTIZIN: a close link between complicated hereditary spastic paraparesis and autophagy. *Autophagy.* 2014;10:374–375.
125. Yu L, McPhee CK, Zheng L, et al. Termination of autophagy and reformation of lysosomes regulated by mTOR. *Nature.* 2010;465:942–946.
126. Goizet C, Boukhris A, Maltete D, et al. SPG15 is the second most common cause of hereditary spastic paraplegia with thin corpus callosum. *Neurology.* 2009;73:1111–1119.
127. Pensato V, Castellotti B, Gellera C, et al. Overlapping phenotypes in complex spastic paraplegias SPG11, SPG15, SPG35 and SPG48. *Brain.* 2014;137:1907–1920.
128. Thurston TL, Ryzhakov G, Bloor S, von Muhlinen N, Randow F. The TBK1 adaptor and autophagy receptor NDP52 restricts the proliferation of ubiquitin-coated bacteria. *Nat Immunol.* 2009;10:1215–1221.
129. Verlhac P, Gregoire IP, Azocar O, et al. Autophagy receptor NDP52 regulates pathogen-containing autophagosome maturation. *Cell Host Microbe.* 2015;17:515–525.
130. Viret C, Rozieres A, Faure M. Novel insights into NDP52 autophagy receptor functioning. *Trends Cell Biol.* 2018;28:255–257.
131. Ellinghaus D, Zhang H, Zeissig S, et al. Association between variants of PRDM1 and NDP52 and Crohn's disease, based on exome sequencing and functional studies. *Gastroenterology.* 2013;145:339–347.
132. Minegishi Y, Nakayama M, Iejima D, Kawase K, Iwata T. Significance of optineurin mutations in glaucoma and other diseases. *Prog Retin Eye Res.* 2016;55:149–181.
133. Bansal M, Moharir SC, Sailasree SP, et al. Optineurin promotes autophagosome formation by recruiting the autophagy-related Atg12-5-16L1 complex to phagophores containing the Wipi2 protein. *J Biol Chem.* 2018;293:132–147.
134. Toth RP, Atkin JD. Dysfunction of optineurin in amyotrophic lateral sclerosis and glaucoma. *Front Immunol.* 2018;9:1017.
135. Wong YC, Holzbaur EL. Optineurin is an autophagy receptor for damaged mitochondria in parkin-mediated mitophagy that is disrupted by an ALS-linked mutation. *Proc Natl Acad Sci USA.* 2014;111:E4439–E4448.
136. Singer FR. Paget's disease of bone-genetic and environmental factors. *Nat Rev Endocrinol.* 2015;11:662–671.
137. Albagha OM, Visconti MR, Alonso N, et al. Genome-wide association study identifies variants at CSF1, OPTN and TNFRSF11A as genetic risk factors for Paget's disease of bone. *Nat Genet.* 2010;42:520–524.

138. Silva IAL, Conceicao N, Gagnon E, et al. Effect of genetic variants of OPTN in the pathophysiology of Paget's disease of bone. *Biochim Biophys Acta, Mol Basis Dis.* 1864;2018:143–151.
139. Rezaie T, Child A, Hitchings R, et al. Adult-onset primary open-angle glaucoma caused by mutations in optineurin. *Science.* 2002;295:1077–1079.
140. Chalasani ML, Kumari A, Radha V, Swarup G. E50K-OPTN-induced retinal cell death involves the Rab GTPase-activating protein, TBC1D17 mediated block in autophagy. *PLoS One.* 2014;9:e95758.
141. Bartlett BJ, Isakson P, Lewerenz J, et al. p62, Ref(2)P and ubiquitinated proteins are conserved markers of neuronal aging, aggregate formation and progressive autophagic defects. *Autophagy.* 2011;7:572–583.
142. Ajroud-Driss S, Siddique T. Sporadic and hereditary amyotrophic lateral sclerosis (ALS). *Biochim Biophys Acta.* 1852;2015:679–684.
143. Mizuno Y, Amari M, Takatama M, Aizawa H, Mihara B, Okamoto K. Immunoreactivities of p62, an ubiqutin-binding protein, in the spinal anterior horn cells of patients with amyotrophic lateral sclerosis. *J Neurol Sci.* 2006;249:13–18.
144. Lim J, Yue Z. Neuronal aggregates: formation, clearance, and spreading. *Dev Cell.* 2015;32:491–501.
145. Fecto F, Yan J, Vemula SP, et al. SQSTM1 mutations in familial and sporadic amyotrophic lateral sclerosis. *Arch Neurol.* 2011;68:1440–1446.
146. Teyssou E, Takeda T, Lebon V, et al. Mutations in SQSTM1 encoding p62 in amyotrophic lateral sclerosis: genetics and neuropathology. *Acta Neuropathol.* 2013;125:511–522.
147. Chen Y, Zheng ZZ, Chen X, et al. SQSTM1 mutations in Han Chinese populations with sporadic amyotrophic lateral sclerosis. *Neurobiol Aging.* 2014;35:726.e7–726.e9.
148. Hernandez I, Fernandez MV, Tarraga L, Boada M, Ruiz A. Frontotemporal lobar degeneration (FTLD): review and update for clinical neurologists. *Curr Alzheimer Res.* 2018;15:511–530.
149. Kuusisto E, Salminen A, Alafuzoff I. Ubiquitin-binding protein p62 is present in neuronal and glial inclusions in human tauopathies and synucleinopathies. *Neuroreport.* 2001;12:2085–2090.
150. Fecto F, Siddique T. UBQLN2/P62 cellular recycling pathways in amyotrophic lateral sclerosis and frontotemporal dementia. *Muscle Nerve.* 2012;45:157–162.
151. Rubino E, Rainero I, Chio A, et al. SQSTM1 mutations in frontotemporal lobar degeneration and amyotrophic lateral sclerosis. *Neurology.* 2012;79:1556–1562.
152. Goode A, Rea S, Sultana M, Shaw B, Searle MS, Layfield R. ALS-FTLD associated mutations of SQSTM1 impact on Keap1-Nrf2 signalling. *Mol Cell Neurosci.* 2016;76:52–58.
153. Boutoleau-Bretonniere C, Camuzat A, Le Ber I, et al. A phenotype of atypical apraxia of speech in a family carrying SQSTM1 mutation. *J Alzheimers Dis.* 2015;43:625–630.
154. Hocking LJ, Lucas GJ, Daroszewska A, et al. Domain-specific mutations in sequestosome 1 (SQSTM1) cause familial and sporadic Paget's disease. *Hum Mol Genet.* 2002;11:2735–2739.
155. Laurin N, Brown JP, Morissette J, Raymond V. Recurrent mutation of the gene encoding sequestosome 1 (SQSTM1/p62) in Paget disease of bone. *Am J Hum Genet.* 2002;70:1582–1588.
156. Gang Q, Bettencourt C, Machado PM, et al. Rare variants in SQSTM1 and VCP genes and risk of sporadic inclusion body myositis. *Neurobiol Aging.* 2016;47:218.e1–218.e9.
157. Bucelli RC, Arhzaouy K, Pestronk A, et al. SQSTM1 splice site mutation in distal myopathy with rimmed vacuoles. *Neurology.* 2015;85:665–674.
158. Deretic V. A master conductor for aggregate clearance by autophagy. *Dev Cell.* 2010;18:694–696.

159. Yamamoto A, Simonsen A. Alfy-dependent elimination of aggregated proteins by macroautophagy: can there be too much of a good thing? *Autophagy*. 2011;7:346–350.
160. Kadir R, Harel T, Markus B, et al. ALFY-controlled DVL3 autophagy regulates Wnt signaling, determining human brain size. *PLoS Genet*. 2016;12:e1005919.
161. Lazarou M, Sliter DA, Kane LA, et al. The ubiquitin kinase PINK1 recruits autophagy receptors to induce mitophagy. *Nature*. 2015;524:309–314.
162. Yamada T, Dawson TM, Yanagawa T, Iijima M, Sesaki H. SQSTM1/p62 promotes mitochondrial ubiquitination independently of PINK1 and PRKN/parkin in mitophagy. *Autophagy*. 2019;15:2012–2018.
163. Grasso D, Ropolo A, Lo Re A, et al. Zymophagy, a novel selective autophagy pathway mediated by VMP1-USP9x-p62, prevents pancreatic cell death. *J Biol Chem*. 2011;286:8308–8324.
164. Gomes LC, Dikic I. Autophagy in antimicrobial immunity. *Mol Cell*. 2014;54: 224–233.
165. Paludan C, Schmid D, Landthaler M, et al. Endogenous MHC class II processing of a viral nuclear antigen after autophagy. *Science*. 2005;307:593–596.
166. Loi M, Muller A, Steinbach K, et al. Macroautophagy proteins control MHC class I levels on dendritic cells and shape anti-viral CD8(+) T cell responses. *Cell Rep*. 2016;15:1076–1087.
167. Saitoh T, Fujita N, Jang MH, et al. Loss of the autophagy protein Atg16L1 enhances endotoxin-induced IL-1beta production. *Nature*. 2008;456:264–268.

CHAPTER FOURTEEN

C. elegans to model autophagy-related human disorders

Shi Quan Wong[†], Anita V. Kumar[†], Joslyn Mills[†], Louis R. Lapierre*
Department of Molecular Biology, Cell Biology, and Biochemistry, Brown University, Providence, RI, United States
*Corresponding author: e-mail address: louis_lapierre@brown.edu

Contents

Abstract

Autophagy is a highly conserved degradation process that clears damaged intracellular macromolecules and organelles in order to maintain cellular health. Dysfunctional autophagy is fundamentally linked to the development of various human disorders

[†] Equal contributions.

Progress in Molecular Biology and Translational Science, Volume 172
ISSN 1877-1173
https://doi.org/10.1016/bs.pmbts.2020.01.007

and pathologies. The use of the nematode *Caenorhabditis elegans* as a model system to study autophagy has improved our understanding of its regulation and function in organismal physiology. Here, we review the genetic, functional, and regulatory conservation of the autophagy pathway in *C. elegans* and we describe tools to quantify and study the autophagy process in this incredibly useful model organism. We further discuss how these nematodes have been modified to model autophagy-related human diseases and underscore the important insights obtained from such models. Altogether, we highlight the strengths of *C. elegans* as an exceptional tool to understand the genetic and molecular foundations underlying autophagy-related human diseases.

Abbreviations

AD Alzheimer's disease
AL autolysosomes
ALS amyotrophic lateral sclerosis
AP autophagosomes
APP amyloid precursor protein
Aβ amyloid beta
BiFC bimolecular fluorescence complementation
CLEM correlative light and electron microscopy
CR caloric restriction
DA dopaminergic
DNC dorsal nerve cord
dsRNA double-stranded RNA
EM electron microscopy
ER endoplasmic reticulum
FLCN folliculin
FUS fused in sarcoma
HD Huntington's disease
hnRNP heterogeneous nuclear ribonucleoprotein
Htt Huntingtin
LRO lysosome related organelles
miR microRNA
mitoUPR mitochondrial unfolded protein stress response
NC nucleocytoplasmic
NFT neurofibrillary tangles
p62 SQST-1/SQSTM1
PD Parkinson's disease
PolyQ polyglutamine
PPAR peroxisome proliferator-activated receptors
PTM post-transcriptional modifications
Q40 polyQ length of 40 glutamines
RNAi RNA interference
SCA spinocerebellar ataxia
TDP-43 transactive response DNA binding protein 43 kDa
TEM transmission electron microscopy

TFEB transcription factor EB
UPR unfolded protein response
VNC ventral nerve cord
α-syn α-synuclein

1. Introduction

Caenorhabditis elegans is an established multi-cellular model organism that has been used to study genetics and biology for the last five decades.[1] This soil nematode has been instrumental in defining evolutionarily conserved roles for autophagy in development, cell fate, pathogenesis, and aging. Over 80% of its proteome has human homologs,[2,3] allowing these nematodes to serve as a rapid and reproducible model system that offers mechanistic insights and pharmacological predictions for human diseases. As a self-fertilizing microscopic hermaphrodite, it possesses many practical advantages amenable to experimental biology and genetic screening.[1] Remarkably, it only takes 3 days for this tiny worm to grow from an egg to a ~1-mm-long adult nematode, and its embryonic and post-embryonic cellular lineage is completely identified.[4,5] Larval development leads to the differentiation into 959 cells of the somatic tissues, including intestine, hypodermis, muscle, and neuronal tissues as well as a reproductive system capable of generating about 300 progenies within the first week of adulthood. The mean lifespan of wild-type (N2) *C. elegans* is approximately 15–18 days at 20 °C, which makes it an outstanding model to study age-related diseases and the biology of aging. The transparent anatomy of this nematode allows for fluorescent microscopy and morphological analyses of cell proliferation, tissue development, and tissue integrity. Genetic manipulation from chemical mutagenesis[6] to CRISPR is routine, and mutant and transgenic strains are readily available and affordable (academic fee of $10) at repositories such as the Caenorhabditis Genetics Center in the United States of America (https://cgc.umn.edu/) or the National Bioresource Project in Japan (https://shigen.nig.ac.jp/c.elegans/). Genetic crosses can be performed using males created stochastically with low propensity (<0.1%) or by stress-driven chromosomal non-disjunction.[7] The genome of *C. elegans* is completely sequenced[8] and updated gene-specific information, homology, tools, and functional annotations are available on the Wormbase database (https://wormbase.org). Most recently, efforts to improve our current understanding of the nervous system have been put forth by mapping neuronal connectomes and gene expression at single neuron resolution in both *C. elegans* sexes.[9,10] Gene silencing is easily performed by feeding nematodes *E. coli* bacteria expressing double-stranded RNA (dsRNA) against a gene of interest,[11] and this can be done as large as a genome-wide screen by feeding them commercially-available bacteria expressing dsRNA against specific genes,[12,13] either seeded on a plate of agar or in liquid cultures. For instance, this unbiased approach has been extremely useful for discovering new and conserved lifespan modulators.[14,15] Overall, *C. elegans* provides a practical model to assess autophagy-related mechanisms as conserved autophagy genes mirrors functions observed in mammals.[16,17] Autophagy mediates complex nucleic, proteomic, and organellar remodeling during development, pathogenesis, and aging that alter the ability of animals to survive stress and thrive. Here, we compiled the original and latest *C. elegans* research on conserved autophagy-related biology, pathology, and lifespan modulation, and we present nematode tools available to the scientific community to contribute to the advancement of our understanding of autophagy in human biology and diseases.

2. Conserved functions and regulatory mechanisms

The original discovery of conserved autophagy genes in yeast paved the way for the use of model organisms to study autophagy.[18,19] Indeed, the rationale for using *C. elegans* to study autophagy was founded on the discovery of homologous autophagy genes between yeast, nematodes, and humans.[20–22] Since then, numerous studies have taken advantage of the simplicity of using *C. elegans* to uncover roles for autophagy in development, aging, and pathogenesis as well as conserved regulatory mechanisms for autophagy[23,24] (Table 1).

2.1 Development

Studies in *C. elegans* have resolved several developmental functions of autophagy from oocyte stage to adulthood, including synapse formation, miR-mediated silencing, and stem cell proliferation.[25] Prior to fertilization, germline cell proteostasis depends on proper lysosomal acidification,[26] and autophagy proteins, including BEC-1 (BECLIN1), modulate germline stem cell proliferation.[27] Following fertilization, allophagy, the specific degradation of paternally-provided organelles, takes place.[28–31]

Table 1 Evolutionarily conserved autophagy-related genes and functions.

C. elegans ATG genes	Mammalian ATG homologs	Function/step in autophagy
unc-51	*ULK1, ULK2*	Autophagy initiation (ATG1/ULK initiation Complex)
atg-2	*ATG2A, ATG2B*	Phagophore formation (PI3P Binding Complex)
atg-3	*ATG3*	Phagophore elongation (ATG8/LC3 Conjugation System)
atg-4.1, atg-4.2	*ATG4A, ATG4B, ATG4C, ATG4D*	Phagophore elongation (ATG8/LC3 Conjugation System)
atg-5	*ATG5*	Phagophore formation (ATG12 Conjugation System)
atg-7	*ATG7*	Phagophore formation (ATG12 Conjugation System)/Phagophore elongation (ATG8/LC3 Conjugation System)
lgg-1	*GABARAP, GABARAPL1, GABARAPL2*	Phagophore elongation (ATG8/LC3 Conjugation System)
lgg-2	*MAP1LC3*	Phagophore elongation (ATG8/LC3 Conjugation System)
atg-9	*ATG9A, ATG9B*	Phagophore formation (PI3P Binding Complex)
atg-10	*ATG10*	Phagophore formation (ATG12 Conjugation System)
lgg-3	*ATG12*	Phagophore formation (ATG12 Conjugation System)
epg-1	*ATG13*	Autophagy initiation (ATG1/ULK initiation Complex)
epg-8	*ATG14*	Membrane nucleation (PI3K III Nucleation Complex)
atg-16.1, atg-16.2	*ATG16L1, ATG16L2*	Phagophore formation (ATG12 Conjugation System)
epg-7	*RB1CC1*	Autophagy initiation (ATG1/ULK initiation Complex)
atg-18	*WIPI1, WIPI2, WIPI3*	Phagophore formation (PI3P Binding Complex)
epg-9	*ATG101*	Autophagy initiation (ATG1/ULK initiation Complex), Stabilizes ATG13/epg-1

Continued

Table 1 Evolutionarily conserved autophagy-related genes and functions.—cont'd

***C. elegans* ATG genes**	**Mammalian ATG homologs**	**Function/step in autophagy**
epg-6	*WIPI4 (WDR45)*	Phagophore formation (ATG12 Conjugation System)/Phagophore elongation (ATG8/LC3 Conjugation System)
bec-1	*BECN1*	Membrane nucleation (PI3K III Nucleation Complex)
vps-34	*PIK3C3*	Membrane nucleation (PI3K III Nucleation Complex)
epg-3	*VMP1*	Negative regulation of autophagosome assembly
epg-4	*EI24*	Negative regulation of autophagosome assembly
epg-5	*EPG5*	Negative regulation of autophagosome assembly
***C. elegans* autophagy-related genes**	**Mammalian autophagy-related homologs**	**Function in autophagy**
vps-4	*VPS4A, VPS4B/SKD1*	Autophagosome-LY/endosome fusion (Autophagolysosome maturation)
rab-11.1	*RAB11A*	Autophagosome-LY/endosome fusion (Autophagolysosome maturation)
rab-11.2	*RAB11B*	Autophagosome-LY/endosome fusion (Autophagolysosome maturation)
*che-3, dhc-1, dci-1, etc**	*dynein (DYN) family*	Endosome transport; Autophagosome-LY/endosome fusion (Autophagolysosome maturation)
rab-7	*RAB7A*	Endosome transport; Autophagosome-LY/endosome fusion (Autophagolysosome maturation)
let-363	*MTOR*	Regulates autophagy gene expression
hlh-30	*TFEB, TFE3*	Regulates autophagy gene expression
daf-16	*FOXO*	Regulates autophagy gene expression
xpo-1	*XPO-1/CRM1*	Modulates transcription factors localization that regulates autophagy gene expression
vps-11	*VSP11*	HOPS complex, tethering complex for vesicle fusion; Autophagosome-LY/endosome fusion (Autophagolysosome maturation)

Table 1 Evolutionarily conserved autophagy-related genes and functions.—cont'd

***C. elegans* autophagy-related genes**	**Mammalian autophagy-related homologs**	**Function in autophagy**
vps-33.1	*VPS33A*	HOPS complex, tethering complex for vesicle fusion; Autophagosome-LY/endosome fusion (Autophagolysosome maturation)
vps-16	*VSP16*	HOPS complex, tethering complex for vesicle fusion; Autophagosome-LY/endosome fusion (Autophagolysosome maturation)
vps-18	*VSP18*	HOPS complex, tethering complex for vesicle fusion; Autophagosome-LY/endosome fusion (Autophagolysosome maturation)
vps-41	*VSP41*	HOPS complex, tethering complex for vesicle fusion; Autophagosome-LY/endosome fusion (Autophagolysosome maturation)
sqst-1	*SQSTM1/p62*	Selective autophagy receptor for ubiquitinated cargo
pdr-1	*PRKN*	Selective autophagy receptor for mitophagy
pink-1	*PINK1*	Selective autophagy marker for mitophagy
fundc-1	*FUNC1A*	Selective autophagy receptor for allophagy
phb-2	*PHB2*	Selective autophagy receptor for allophagy
lea-1	*PLIN4*	Selective autophagy marker for lipophagy
atgl-1	*PNPLA1, PNPLA2, PNPLA3, PNPLA5*	Lipid droplet protein, Potential role in lipophagy
ipla-6	*PNPLA8*	Potential role in lipophagy

[a]The precise dynein genes necessary for autophagy have not been identified; however, functional dynein activity is required for effective autophagy.

HOPS: Homotypic fusion and protein sorting.

Sources: Data from Chun Y, Kim J. Autophagy: an essential degradation program for cellular homeostasis and life. *Cells*. 2018;7: 278; Eskelinen EL, Saftig P. Autophagy: a lysosomal degradation pathway with a central role in health and disease. Biochim Biophys Acta. 2009;1793: 664-673; Gelino S, Hansen M. Autophagy—an emerging anti-aging mechanism. J Clin Exp Pathol. 2012;Suppl 4; Kovacs AL, Zhang H. Role of autophagy in *Caenorhabditis elegans*. FEBS Lett. 2010;584: 1335-1341; Scrivo A, Bourdenx M, Pampliega O, Cuervo, AM. Selective autophagy as a potential therapeutic target for neurodegenerative disorders. Lancet Neurol. 2018;17: 802-815; Suzuki H, Kaizuka T, Mizushima N, Noda NN. Structure of the Atg101-Atg13 complex reveals essential roles of Atg101 in autophagy initiation. Nat Struct Mol Biol. 2015;22: 572-580; Zuchero JB, Barres BA. Glia in mammalian development and disease. Development. 2015;142, 3805-3809.

This process requires the polyubiquitination of the paternal membranous organelles,[32] the proteins LGG-1 (GABARAP) and LGG-2 (LC3),[33] and recognition by mitophagy receptors ALLO-1,[34] PHB-2 (PHB2),[35] and FNDC-1 (FUNDC1).[36] Autophagy also mediates the selective breakdown of other intracellular components such as ribosomal RNA during development,[37] and the proper degradation of maternally-derived germ granules via the protein SEPA-1, which facilitates the interaction between the selective autophagy receptor SQST-1 (SQSTM1/p62) and the autophagy protein LGG-1 (GABARAP).[38,39] Genome-wide RNA interference (RNAi) screening uncovered modulators of SQST-1 dynamics, highlighting new signaling pathways in development, such as the Sma/Mab TGF-β signaling cascade and the dauer TGF-β signaling pathway,[40] and other pathways such as those involved in *lin-35* (Rb) signaling, the ER stress response mediated by XBP-1, and the mitochondrial stress response mediated by ATFS-1.[41] Finally, autophagy proteins also mediate the clearance of necrotic and apoptotic cell corpses during development.[42–45]

C. elegans progress through four stages of larval development, defined L1–L4.[46] Mutations in autophagy genes generally results in larval developmental phenotypes ranging from developmental delay to lethality,[17] akin to effects of autophagy loss in higher eukaryotes.[47] The ability of newly hatched L1 larvae to survive starvation is markedly decreased in the absence of functional autophagy.[48–53] Mutagenesis screening in nematodes uncovered four new autophagy-related genes involved in starvation-induced autophagy, the metazoan-specific gene *epg-2*, and conserved genes *epg-3*, *-4*, and *-5*.[52,54] Under stressful conditions, L2 larval nematodes can enter a long-term survival mode called dauer. Dauer development is characterized by enhanced autophagy,[55] in part mediated by the activity of the methyltransferase PCM-1 (PCMT-1)[56,57] and by autophagy proteins LGG-1 (GABARAP) and LGG-2 (LC3).[58] In addition, appropriate lipid metabolism[59] and neuronal development[60,61] during later larval stages require functional autophagy. Evidently, development programs are associated with spatial-, timing-, and substrate-specific autophagic degradation that leads to proper cell function and tissue differentiation.

2.2 Adult lifespan

The role of autophagy in lifespan has been the subject of intense discussion and research in the past decade[62–65] and multiple nematode studies have found roles for different autophagy and lysosomal genes in longevity.[66]

Initial evidence of the role of autophagy in nematode lifespan came from a seminal study finding that loss of the protein BEC-1 (Beclin1) interferes with the long lifespan of *daf-2* (Insulin/IGF-1 receptor) mutant animals.[55,67] The loss-of-function *daf-2* mutant typically leads to more than a doubling of the mean lifespan of wild-type animals,[68] but loss of *bec-1* markedly reduced its lifespan back to wild-type longevity. Following this discovery, other groups demonstrated roles for other autophagy genes in the long lifespan of *daf-2* mutants.[69,70] In addition, loss of autophagy was found to markedly reduce the lifespan in other genetic models of longevity, including dietary-restricted pumping-defective *eat-2* animals,[70–72] germline-less *glp-1* (Notch1) mutants,[73] protein synthesis *rsks-1* (S6 Kinase) mutant,[59] *cnb-1* (Calcineurin) mutants,[74] mitochondrial respiration *clk-1* (COQ7) and *isp-1* (UQCRFS1) mutants[59,70], and in animals with reduced mTOR signaling.[59,70]

Another strategy shown to extend lifespan across phyla is caloric restriction (CR), which requires the upregulation of autophagy.[75,76] Certainly, it is clear that autophagy is required for adult animals to survive starvation,[48,49] which is accomplished by organ-specific upregulation of autophagy in the liver, pancreas, kidney, and muscle to support the brain with glucose and ketone bodies until the next feeding.[77,78] CR or intermittent fasting experiments from nematodes up to non-human primates have shown lifespan and health span improvements,[79,80] leading to the development of CR mimetics such as metformin to combat aging by increasing autophagy gene expression via the AMPK pathway.[81–85]

Studies about context-, tissue-, and time-dependent roles for autophagy in lifespan have recently emerged. Indeed, expression of Rubicon, a negative regulator of autophagy, is increased during aging, and neuronal-specific silencing of Rubicon extends lifespan in nematodes,[86] in line with a role for neuronal autophagy genes in systemic autophagy regulation and longevity.[87–89] Alternatively, a study suggested that late adulthood autophagy gene silencing in neurons can extend lifespan, but the efficiency of gene silencing in neurons of aged animals, particularly with negligible *E. coli* pumping, even in an RNAi sensitive strain, remains debatable.[90] Even so, autophagy may become detrimental for longevity in specific contexts, including when it is hyperactive,[48] induced by high glucose[91] or when mitochondrial membrane permeability is compromised.[92,93] Nonetheless, it appears that wild-type animals activate autophagy later in life[94] potentially as a last resort to prevent the proteostatic decline associated with aging.[95,96] For instance, loss of proteasome function in the deletion mutant *rpn-10(ok1865)* enhances autophagy as a compensatory mechanism to maintain proteostasis.[97] Overall, the vast majority of nematodes studies have found a requirement for autophagy in somatic maintenance and lifespan extension for multiple genetic models and across phyla.[66]

2.3 Conserved regulatory mechanisms

The activation of autophagy is tightly regulated by the localization, accessibility, and proper modification of ATGs and associated proteins. Genetic studies in nematodes, supported in mammalian models and cells, have shown intricate modulation of the dynamic process of autophagic degradation, commonly referred to as autophagic flux, by nutrient-sensing pathways such as insulin, AMPK, and mTOR signaling in part by coordinating the activity of several transcription factors including, but not limited to, HLH-30 (TFEB), DAF-16 (FOXO), and PHA-4 (FOXA)[98] (Fig. 1). Transcriptional modulation of autophagy and lysosome gene expression is primarily regulated by HLH-30 (TFEB).[59,99–102] This transcription factor is modulated by Mondo complex proteins[103] and by the transcriptional repressor MXL-3,[99] and the nuclear translocation and activity is controlled by mTOR (mTORC1) and insulin signaling.[59] As in mammals, nematode HLH-30 (TFEB) interacts with numerous nuclear proteins, including DAF-16 (FOXO) to regulate gene expression in a stress-specific manner.[104]

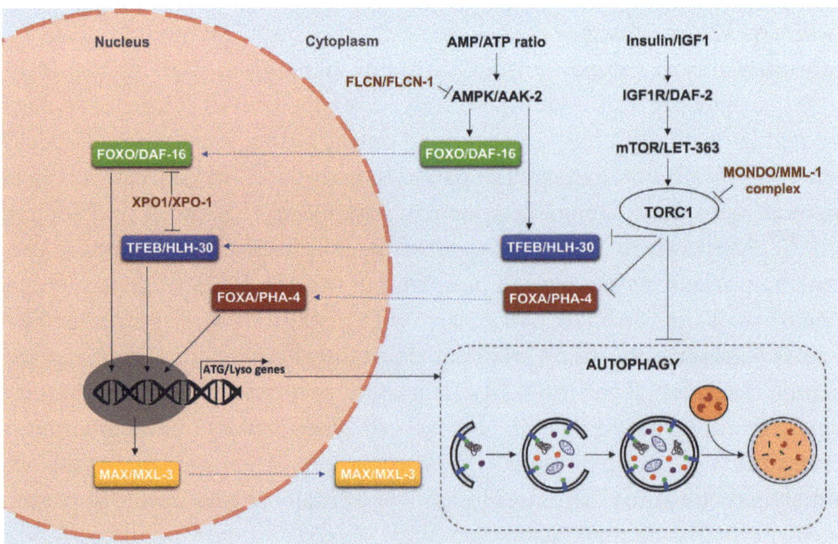

Fig. 1 Major transcriptional regulators of autophagy.

While cytosolic sequestration and nuclear import of HLH-30 (TFEB) is governed by 14-3-3 and importin 8, respectively,[105] export of HLH-30 (TFEB) out of the nucleus is regulated by the exportin protein XPO-1 (XPO1/CRM1).[106] A nuclear export sequence and adjacent phosphorylation sites dictate its recognition by XPO1 for export.[107,108] Autophagy gene expression can be stimulated via nuclear export inhibition by XPO1 inhibitor Leptomycin B or with selective inhibitors of nuclear export that target XPO1,[106] indicating that pharmacological stimulation of autophagy gene expression and autophagy flux itself is achievable in nematodes and mammalian cells and possibly implementable against autophagy-related and age-related diseases in humans.[109,110]

Autophagy proteins are also subjected to post-transcriptional modifications (PTMs) from protein modifying enzymes, which provides a regulatory entry point to modulate autophagic flux. In mammalian cells, initiation of autophagy is regulated by the inhibiting and activating phosphorylation of UNC-51 (ULK-1) protein by mTOR and AMPK, respectively.[111–113] Phosphorylation of LGG-1 (GABARAP) and LGG-2 (LC3) modulates autophagosome maturation.[114] Phosphorylation of lipids by class III PI3 Kinase VPS-34 (VPS34) is important in the recruitment of autophagy factors initiating the formation of the double-membrane autophagosome.[115] Changes in autophagy protein acetylation by compounds spermidine and resveratrol have been suggested to underlie the mechanism for their autophagy activating effects.[116,117] Altogether, the emergence of new studies on PTM of autophagy proteins has provided a deeper resolution of the molecular mechanisms modulating autophagic flux.

Novel conserved mechanisms of regulation of autophagy have also been discovered using the *C. elegans* model. For instance, clearance of protein aggregates by HLH-30-mediated autophagy is in part regulated by glutathione redox homeostasis, and this is conserved when tested in mammalian cells.[118] Additionally, changes in endoplasmic reticulum (ER) homeostasis affect autophagy, either by membrane phospholipid alterations that induce the IRE-1-XBP-1 axis of the unfolded protein response (UPR) or loss of N-linked glycan precursors via the hexosamine pathway.[119,120] Finally, dual roles of the heat-reactive kinase HPK-1 that promote cellular proteostasis were first identified in nematodes when its activation led to the transcriptional activation of both chaperone and autophagy gene expression.[121,122] Altogether, the successful use of nematodes to delineate autophagy-related mechanisms provides supportive evidence of the practicality of this model to understand conserved mechanisms in higher order organisms.

3. Tools for autophagy detection and flux analyses

In parallel with mammalian research, tools for visualization, functional analysis, and mechanistic studies of autophagy have been implemented in *C. elegans* (Fig. 2) (reviewed in Refs. 17,123–125).

3.1 Steady-state autophagosome visualization

Original research on autophagosome formation relied on visualizing a GFP protein fused to the autophagosome-related protein LGG-1 (GABARAP) under the control of the *lgg-1* promoter (GFP::LGG-1).[49,55] Under nutrient-rich conditions, the GFP signal generally appears diffuse through all tissues. This signal is enhanced and appears punctate under conditions of autophagy induction as LGG-1 gets incorporated into growing autophagosomes. The main tissues in which GFP puncta are generally quantified include the pharynx, seam cells, intestine, muscle, and hypodermis. Apart from GFP, LGG-1 has also been tagged with red fluorescence proteins such as DsRed and mCherry; however, their cytotoxicity and tendency to aggregate led to less extensive use of the strains expressing them (reviewed in Ref. 123). Because autophagy occurs in all tissues, it is useful to have tissue-specific reporters to avoid signal interference from other tissues during quantification. One such reporter strain expresses LGG-1 tagged to mCherry under the *nhx-2* promoter which is intestinally expressed.[126] Increase in LGG-1 levels could be indicative of autophagy induction, due to increased expression of LGG-1, or autophagy impairment, resulting in LGG-1 accumulation (reviewed by Ref. 17). Therefore, assessment of LGG-1 levels should be complemented with autophagic flux assays which measure autophagic capacity. In addition to LGG-1, its paralog, LGG-2 has also been tagged with GFP and used to monitor autophagy.[127] However, LGG-1 has been more of an acceptable marker for autophagosomes because LGG-1 is more functionally related to yeast ATG8[58] and GFP::LGG-1 vesicles correspond to autophagosomes observed by electron microscopy in long-lived *daf-2* mutants.[55] Overexpression models of LGG-1 have been used in a whole-body or tissue-specific context.[123] The GFP::LGG-1 and mCherry::LGG-1 strains have also been crossed into various disease model nematode mutant strains to quantify autophagy.[128,129] Several of these autophagy-related diseases are discussed in the following section on nematode models of autophagy-related human pathologies.

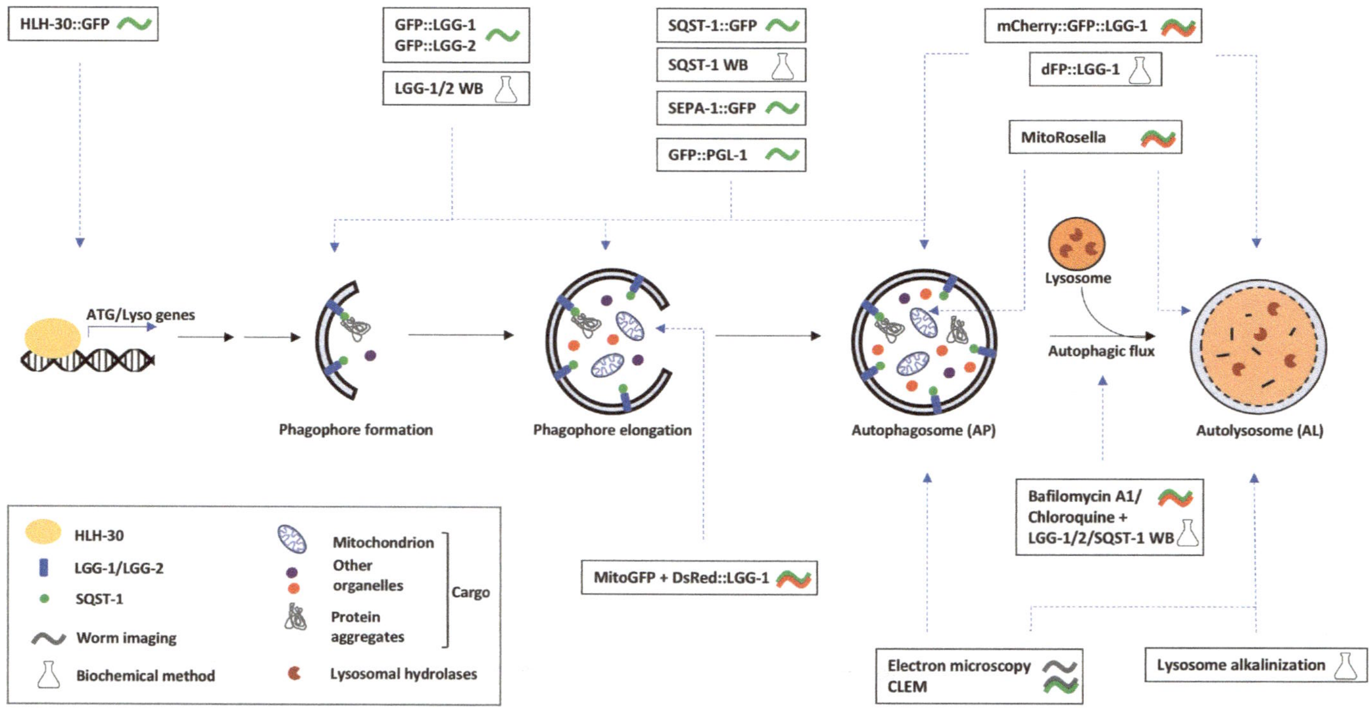

Fig. 2 Graphical representation of the toolkit for autophagy measurements in *C. elegans*. WB, Western Blot; CLEM, Correlated light and electron microscopy.

3.2 Selective autophagy visualization

Mechanistic interpretation of LGG-1 levels is substantially improved by complementary approaches. It is now regular practice to monitor levels of the selective autophagy receptor protein, p62/SQST-1 (reviewed in Ref. 17). The advantage of examining SQST-1 is that it is also an autophagic substrate which gets degraded in the process,[52] implying that it accumulates when autophagy is blocked. Traditionally, the whole-body expressing SQST-1::GFP strain has been used to visualize SQST-1 localization and accumulation,[52] with a few recent additions such as intestine-specific SQST-1::CemOrange2.[130]

3.3 Dynamic autophagic visualization and quantification

Autophagy is a multistep process requiring different sets of proteins, as cargo is selected, recruited, and passed on through autophagosomes and lysosomes until degradation. Enumeration of the number of autophagosomes (AP) alone does not inform about the rate at which AP are being formed or converted into autolysosomes (AL). Use of an autophagy tandem reporter can help discern the rates of AP formation and conversion. In this reporter, LGG-1 is tagged to mCherry and GFP in tandem such that AP appear yellow due to fluorescence signals from GFP and mCherry, while AL fluoresce red as GFP gets quenched in the acidic AL.[94] This reporter strain has been used in many recent studies and offers a more mechanistic readout of autophagy rates by quantification of AP and AL puncta.[106,131] In order to truly determine autophagic flux, this tandem reporter, or the GFP::LGG-1 strain is injected with bafilomycin, which prevents autophagic flux by blocking lysosomal acidification and AP-lysosome fusion. The readout is the extent of AP accumulation in test versus control conditions.[94] Chloroquine, a compound which blocks lysosomal acidification, can also be used in place of bafilomycin to assess autophagic flux.[132]

Other reporters of autophagosome formation have been developed, such as the dual fluorophore-tagged LGG-1 strain, the dFP::LGG-1, which provides a biochemical readout for autophagic flux. This reporter expresses LGG-1 tagged to two fluorescent proteins that are joined by a protease-sensitive linker that gets cleaved in AL to release monomeric fluorescent protein. When separated by immunoblotting, a higher ratio of monomeric fluorescent protein to full-length dFP::LGG-1 is indicative of increased autophagic flux.[133] This reporter has been used in a tissue-specific manner and also provides a readout of lysosomal protease activity.[133,134]

In *C. elegans* embryos, autophagy can also be detected using P granule degradation as a marker. P granules consist of ribonucleoproteins involved in germ cell specification and accumulation of their components, such as PGL proteins, is indicative of impaired autophagy.[135] Fluorescent reporter strains tagging PGL-1, and SEPA-1, the adaptor protein involved in P granule degradation, have been used to screen for novel autophagy genes.[39,52]

3.4 Nuclear translocation of HLH-30/TFEB

As mentioned earlier, genes involved in autophagy and lysosome biogenesis are under transcriptional regulation, and transcription factor EB (TFEB) is one of the main regulators.[100,102] Under conditions that induce autophagy, TFEB translocates to the nucleus and stimulates expression of several autophagy genes. The *C. elegans* orthologue for TFEB, HLH-30, tagged to GFP (HLH-30::GFP) has been used in screens to identify genes involved in transcriptional regulation of autophagy.[106] Nuclear enrichment of HLH-30::GFP or HLH-30::RFP has been used in several studies to report transcriptional activation of genes involved in autophagy, antimicrobial response, and cytoprotection.[59,103,106,134,136,137]

All of these autophagy reporter strains are very useful while performing genetic and compound screens to get a quick, yet robust readout of autophagic status. However, most of these reporters overexpress the tagged protein as they might have several copies of the transgene integrated into their genomes. Also, counting LGG-1 puncta could be erroneous because the number of GFP puncta are small even under autophagy-inducing conditions (reviewed in Ref. 17). With these caveats in mind, it is appropriate to complement the use of reporters with biochemical readouts for autophagy such as immunoblots and quantitative PCR.

3.5 Immunoblotting

In mammalian cells, P62 and LC3 immunoblotting are widely used to assess autophagic activity, with reduced levels of P62 and increased levels of lipidated to non-lipidated LC3 being indicative of increased autophagy (reviewed in Ref. 17). Furthermore, additional increase in lipidated LC3 upon bafilomycin treatment is indicative of autophagic flux. Traditionally, detection of LGG-1 or SQST-1/P62 on an immunoblot relied on using anti-GFP antibodies on lysates from GFP::LGG-1 or SQST-1/P62::GFP worms due to lack of specific antibodies (reviewed in Ref. 123).

A recently developed LGG-1 antibody reproducibly detected lipidated and non-lipidated forms of LGG-1 across all stages of worm development.[138] Studying levels of autophagy proteins using reporter stains and immunoblotting can be complemented by assessing autophagy and lysosomal gene transcript levels using quantitative RT-PCR. This is important to discern alterations in autophagy at the level of gene expression indicative of transcriptional regulation of autophagy.[59]

3.6 Electron microscopy

The subcellular structures and morphology associated with autophagy were first observed by electron microscopy (EM) (reviewed in Refs. 55,123). The advantages of using EM are that it can be independent of markers and provides better anatomical detail. Transmission electron microscopy (TEM) has been used to detect autophagy in WT and mutant stains, dauer animals that possess a detergent-resistant, thick cuticle,[55,73] and optimized and detailed protocols for the preparation of *C. elegans* samples for studying autophagy by EM have been described.[139,140] When combined with immunolabeling, EM has been used to confirm localization of LGG-1 and/or LGG-2 in AP membranes.[55,127] An extension of EM combined with light microscopy (correlative light and electron microscopy—CLEM), which uses fixation methods to preserve fluorescence, has been used to study localization of tagged LGG-1 and LGG-1 to vesicular structures in *C. elegans* embryos.[127]

3.7 Lysosome function

Beyond assessment of AP and AL, detection of lysosome number and activity also provides insights into the efficiency of the autophagic process as lysosomal fusion and degradation is the endpoint of autophagy. Lysosomal number has been used as a readout for the level of lysosome biogenesis, while lysosomal acidity informs about their activity and degradative capacity.[100,141] In the worm intestine, lysosomes and lysosome related organelles (LRO) correspond to lipofuscin granules, that are autofluorescent acidic compartments that accumulate with age.[142,143] Commonly used lysosome detection dyes such as lysotracker and lysosensor have also been extensively employed in worms to detect lysosomes and their activity, respectively.[26,141] Recently, Baxi and de Carvalho developed a protocol for detecting lysosomal alkalinization in the intestine of live worms using the pH sensitive dye, cDCFDA. The dye can detect small changes in pH and is useful for detecting loss of lysosomal acidity, a phenomenon observed as worms aged.[144] Similar to their use in AP detection, transgenic worm strains expressing fluorescent-tagged lysosomal membrane proteins such as the LMP-1::GFP, GFP::CUP-5, and intestine-specific LMP-1::CemOrange2 are commonly used as lysosomal markers.[130,131,145]

3.8 Mitophagy

Mitophagy mediates cellular homeostasis and contributes to lifespan extension in long-lived nematodes[146–148] and in nematodes treated with Urolithin A.[149] Although traditionally studied using EM, several quicker tools are now available to study mitophagy in nematodes. Colocalization between DsRed::LGG-1 and GFP targeted to mitochondria (usually the mitophagy receptor, DCT-1) in transgenic worms harboring both tags are indicative of mitophagy events.[147] Similar to the autophagy tandem reporter, delivery of mitochondria from AP to AL can be detected using MitoRosella. This consists of mitochondrial protein TOMM-20 tagged to pH-insensitive DsRed and pH-sensitive GFP in tandem in which yellow and red puncta indicate AP and AL, respectively.[150]

Altogether, there exist several robust tools to monitor bulk autophagy and some forms of selective autophagy in nematodes. The many fluorescent autophagy reporter strains, generated using tractable worm genetics methods, are attractive models to study perturbations in the autophagic process and screen genes and compounds for their ability to affect autophagy. Development of nematode-specific antibodies and optimization of biochemical methods has leveraged the utility of reporter strains for mechanistic research on autophagy. These tools can be used in combination with disease models in order to study autophagy-related human diseases as discussed in the next section.

4. Nematode models of autophagy-related human pathologies

Autophagy not only regulates development and aging, it is implicated in pathophysiological conditions such as neurodegenerative diseases, metabolic disorders, infectious diseases, and tumorigenesis. However, it should be emphasized that its causative role in these conditions is still unclear.[151] Hence, the development of these autophagy-related diseases in *C. elegans* provides insights into the role of autophagy in these conditions.[140] In particular, impairment in specific types of autophagies, including chaperone-mediated[152] and selective autophagy[153,154] have emerged as potentially relevant players in disease onset. Notably, a strength of *C. elegans* is its genetic tractability and easy adaptation for high throughput genetic[1,6] and chemical screens[155] to accelerate elucidation of disease mechanisms and drug discovery. The following sections describe a select number of autophagy-related conditions with emphasis on the most common neurodegenerative diseases, as well as ways they have been modeled in *C. elegans* and studies exemplifying their usage to understand the role of autophagy in these diseases. Fig. 3 provides an overview of autophagy-related disease models in nematodes.

4.1 Neurodegenerative diseases

Autophagic dysfunction and a consequential dysregulation of proteostasis are associated with various neurodegenerative diseases through genetics, pathology, and functional studies.[156,157] Indeed, aberrant proteostasis is reflected by the shared pathological hallmark of intra- and/or extracellular neuronal proteinaceous aggregates. Although the causative role of defective autophagy is unclear, studies have shown that autophagy can be directly impaired by various disease-associated proteins.[158–161] A short lifespan and completely mapped network of 302 neurons,[162] combined with a plethora of techniques to assess neuronal morphology and function, longevity, and age-related decline, makes *C. elegans* an appealing tool for modeling

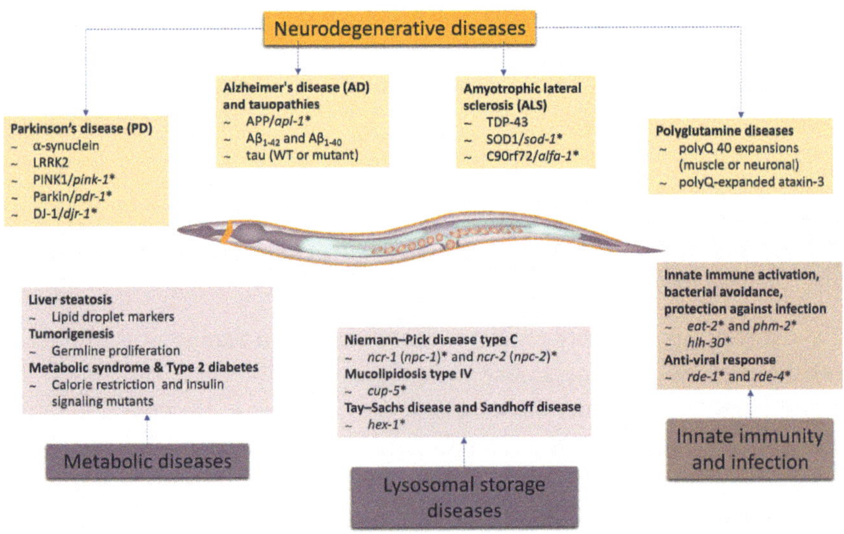

Fig. 3 Human disease models for neurodegenerative, metabolic, lysosomal, and immune diseases available in *C. elegans*.

age-related neurodegenerative disease.[163] For these reasons, numerous models have been developed mostly through genetic means, via transgenic expression of human disease-associated genes in body wall muscles, pan-neuronally, or in specific neuronal populations.[164,165] Less commonly, respective *C. elegans* orthologues were manipulated as an alternative to transgenic expression. Lastly, chemically-induced models are also available, such as in the case of Parkinson's disease. Taken together, these models recapitulate, albeit incompletely, various aspects of disease, such as neurodegeneration, behavioral deficits, and aggregations of disease-associated proteins.[166] Importantly, they have been employed by many studies for genetic and chemical screens to identify genetic and chemical disease modifiers dependently and/or independently of autophagy.[155,167]

4.1.1 Alzheimer's disease (AD)

AD is the most prevalent form of neurodegeneration and dementia, which is characterized pathologically by extracellular amyloid beta (Aβ) plaques and intracellular neurofibrillary tangles (NFTs) of hyperphosphorylated tau. These pathologies primarily afflict brain regions involved in memory and cognition such as the hippocampus. Preexisting efforts to recapitulate AD pathology in *C. elegans* was largely focused on overexpressing either the predominant aggregation-prone variants of Aβ ($A\beta_{1-42}$ and $A\beta_{1-40}$) or tau.[164,168,169] Aβ expression in preexisting Aβ models are either constitutive or conditionally induced by shifts to an expression-permissive temperature.[168] The latter tau models are based on mutations in the tau gene (*MAPT*) which are associated with primary tauopathies such as frontotemporal dementia, as no identified *MAPT* mutations were found to be causative for AD so far. It is important to note that AD is not considered a primary tauopathy as tau pathology is not considered the main initiating event. Furthermore, tau and Aβ pathologies coexist, and tau morphologies in primary tauopathies differ from AD NFTs.[170,171] As the product of an aberrant cleavage of the Aβ precursor, amyloid precursor protein (APP), by the enzyme β-secretase results in production of pathogenic Aβ in AD, APP-based models have additionally been established.[172] Models generated from overexpression of either human APP or the *C. elegans apl-1* orthologue presents various caveats, such as the lack of homology of *apl-1* to Aβ and the absence of β-secretase activity in *C. elegans*.[173] On the other hand, Aβ and tau models have been invaluable tools for understanding AD pathophysiology. These models exhibit disease-related features such as reduced lifespan, neurodegeneration, and Aβ or tau aggregation.[164,168,169] Locomotion defects such as paralysis, albeit not an overt clinical feature of AD, is a commonly observed phenotype in these models, and is used as a readout for Aβ or tau-induced toxicity. The utility of such models can be illustrated by an inducible muscle-specific Aβ system, which exhibits progressive paralysis, cytoplasmic Aβ aggregation, and enhanced protein oxidation.[174] Notably, autophagosome accumulation, as detected in dystrophic neurites of AD brains was observed.[175] Crosstalk between aging, autophagy, and AD were further illustrated whereby mutations in *daf-2* or *daf-7* mitigated paralysis, reduced autophagosome buildup, and enhanced autophagic breakdown of Aβ, while knockdown of autophagy-lysosomal function abolished these effects.[174]

Glutamate is the major excitatory neurotransmitter, and the roles of altered glutamatergic neurotransmission and excitotoxicity have been proposed in AD and other neurodegenerative diseases.[176–178] Glutamate-associated neurodegeneration was demonstrated with constitutive Aβ overexpression specifically in glutamatergic neurons in *C. elegans*, suggesting a direct toxic effect of Aβ on glutamate signaling in AD.[179] Expanding on this model, a separate study further co-expressed human Aβ and *APOE4*, the strongest risk allele for sporadic AD, in glutamatergic neurons in *C. elegans* to study the relationship between Aβ pathology, *APOE4*, and glutamatergic toxicity.[180] Findings suggest that *APOE4*-mediated glutamatergic toxicity occurs dependently of Aβ. In particular, glutamatergic neuronal-specific RNAi knockdown of the autophagy-related gene *atg-7* exacerbated degeneration of these neurons, suggesting that autophagy plays a protective role against *APOE4*-related AD. Taken together, these findings argue for the beneficial role of autophagy against AD.

Aside from Aβ models, findings from tau models have yielded valuable insights into tauopathies and may be informative for AD. For instance, forward genetic screens carried out on a pan-neuronal frontotemporal dementia mutant tau model identified novel *C. elegans* proteins SUT-1 and SUT-2 (*SUppressor of Tau pathology 1 and 2*) as enhancers and suppressers of tau pathology, respectively.[181–183] Notably, although SUT-2 is orthologous to mammalian SUT2 whereas SUT-1 has no known human orthologue and characterized function, interacting pathways regulated by both proteins appear to be conserved in humans, suggesting potential for discovering novel disease-related processes and therapeutic targets. In the same vein, a different tau model offers important insights into the selective neuronal vulnerability of AD and tauopathies, whereby pan-neuronal overexpression of the AD and tauopathy susceptibility A215T tau mutation selectively exacerbated glutamatergic neurodegeneration via necrosis, dependent on the lysosomal aspartic protease Cathepsin D.[184] Notably, the autophagy-lysosomal pathway has been implicated in various forms of cell death, including necrosis, highlighting the complex relationship between autophagy-lysosomal function and neurodegeneration.[185]

Of note, learning and memory deficits, which are clinical hallmarks of AD, can be assessed in *C. elegans* through various learning assays such as the chemotaxis assay.[163] Via such assays, a model of pan-neuronal Aβ expression was found to exhibit learning and memory deficits.[186] Given that the utility of *C. elegans* AD models as screening platforms have been demonstrated by various independent studies,[164,168] such models have the potential to be employed for early-stage drug screens to identify compounds with the potential to counteract clinical progression and development of dementia. Comparatively, early drug discovery phase performed in cellular systems does not inform about the behavioral effects of screened compounds. Furthermore, the use of AD *C. elegans* models for this purpose may improve the efficiency and cost-effectiveness of the AD drug discovery process, as behavioral readouts are currently only assessed at later-stage preclinical testing in rodent models. Combined with available tools to assess autophagy in *C. elegans* (refer to Section 3), these platforms may be adapted for high throughput early-stage screens to dissect the complex role of autophagy not only in disease pathology, but also clinically in the development of dementia.

4.1.2 Parkinson's disease (PD)

PD is the second most prevalent neurodegenerative disease which is characterized by dopaminergic (DA) neurodegeneration and movement deficits. Various PD models have been generated, either through chemical exposure to DA neurodegeneration-inducing chemicals, or by expression of disease-associated human genes such as *SCNA*, *LRRK2*, *PINK1*, *PRKN*, and *DJ-1*.[187,188] Of these genes, *SCNA* is notable as its protein product, α-synuclein (α-syn), is a main aggregate of the hallmark pathological Lewy Bodies found in dopamine neurons. Hence, α-syn PD models present great potential for understanding pathogenesis mechanisms common to both idiopathic and inherited PD.[189] One such model of body wall muscle α-syn::GFP expression allows the visualization of α-syn aggregation, and has enabled genetic suppressors of aggregation to be identified.[190] Notably, *C. elegans* ATG-7 (ATG7), required for autophagosome formation, was identified as a suppressor of DA neurodegeneration induced by α-syn overexpression in these neurons. Lending credence to the beneficial role of autophagy against neurodegeneration, separate findings suggest that the neuroprotective effects of low dose bafilomycin and spermidine against α-syn-induced DA neurodegeneration were likely attributed, at least in part, to autophagy induction.[191] The utility of α-syn PD models has additionally extended to the study of cell-to-cell spreading of the protein, a prion-like phenomenon reported in PD models.[192,193] With the completely mapped neuronal connectome of *C. elegans*, prion-like transmission can be studied through the expression of bimolecular fluorescence complementation (BiFC)-fused α-syn in anatomically-connected neuronal populations. In this model, fluorescence emission is observed as a readout of BiFC-fused α-syn spread and subsequent dimerization between synapses. Contrary to the reported neuroprotective role of autophagy, autophagy appears to play a role in propagating inter-neuron transmission of α-syn as knockdown of autophagy genes reduced BiFC fluorescence in this *C. elegans* model.[194]

Defects in mitochondrial function and quality control (mitophagy) are widely accepted PD pathogenic mechanisms.[147,195] Notably, mitophagy has a pro-longevity role,[196] in part by preventing protein aggregation.[197] The role of mitophagy in PD pathogenesis was first suggested by initial genetic studies which unraveled familial PD mutations in *PINK1* and *PRKN*.[198–210] Later studies revealed that defective mitochondria are recognized by the kinase PINK1, which localizes at and recruits the *PRKN*-encoded E3 ubiquitin ligase Parkin to ubiquitinate mitochondria for mitophagy removal.[211–215] Importantly, loss-of-function mutations induced in *C. elegans PINK1/pink-1*, *PRKN/pdr-1*, and *DJ-1/djr-1*,[216] have managed to mimic DA neurodegeneration and various aspects of dysregulated mitochondrial quality control such as altered mitochondrial morphology and function, and age-dependent accrual of defective mitochondria.[147,217–220] However, these mutants do not display discernible phenotypic defects, rendering them unsuitable for high content genetic and chemical screens.[218] Despite this caveat, these models are not nugatory as they may be used to assess if top hits from high throughput screens modify disease phenotypes through mitochondria homeostasis. Furthermore, their utility for understanding the mitochondria-PD relationship is invaluable, as exemplified by findings which suggest that induction of the mitochondrial unfolded protein stress response (mitoUPR) by *PINK-1/pink-1*- and *PRKN/pdr-1*-induced defective mitochondria accumulation was protective against DA neurodegeneration.[218] However, the neuroprotective role of the mitoUPR was called into question by contradictory reports of a toxic role of constitutive mitoUPR activation in *C. elegans* DA neurons.[197] Interestingly, α-syn-induced DA neurodegeneration and α-syn::GFP aggregation were exacerbated by mitoUPR activity. Taken together, these findings suggest that the interplay between mitochondria quality control and PD is context-dependent.

Another important PD gene is *LRRK2*, which encodes lysine-rich repeat kinase 2 (LRRK2), of which mutations constitute the most common cause of both familial and sporadic forms of PD. The role of LRRK2 appears to be pleiotropic, as numerous studies have suggested that it protects against oxidative stress, interacts with α-syn, and regulates other processes such as the phosphorylation of the microtubule-binding protein tau, inflammatory responses, mitochondrial homeostasis, and the autophagy-lysosomal function.[221] Although it remains unclear whether LRRK2 induces or inhibits autophagy, studying the role of LRRK2 in PD pathogenesis presents another valuable approach for bridging our understanding of autophagy in PD. In *C. elegans*, various models of *LRRK2*-associated PD were also developed mostly from overexpression of the human gene either pan-neuronally or specifically in DA neurons.[216] All of these models exhibited DA neurodegeneration, with some also shown to display deficits in DA-associated behavior and reduced DA levels. Complementarily, loss of function of the *C. elegans LRRK2/lrk-1* orthologue increased susceptibility to the known oxidative stress and DA neurodegeneration inducer rotenone,[222] and protected against human *LRRK2*-induced DA neurodegeneration,[223] supporting the mito- and neuroprotective role of LRRK2 in PD. LRRK2 *C. elegans* models have also shed light on autophagic dysfunction in PD, such as the model of LRRK2 and tau co-expression that probes the synergism of LRRK2 and tau in tau-associated LRRK2 PD. Co-expression of LRRK2 and tau was associated with enhanced oxidation levels of the lysosomal proton pump V-ATPase, implying defects in autophagy-lysosomal degradation. Further bolstering this is an amelioration of locomotion defects by the mTOR inhibitor and autophagy inducer ridaforolimus.[224,225] Taken together, existing *C. elegans* PD models has recapitulated aspects of the disease in humans. Complementing work performed in PD models generated in other organisms, they have provided insights into PD mechanisms, including the complex interplay between autophagy and PD.

4.1.3 Polyglutamine (PolyQ) diseases

PolyQ diseases are a group of neurodegenerative conditions caused by aberrant trinucleotide CAG repeat expansions in affected genes, resulting in abnormally-expanded polyQ tracts and aggregation of affected proteins.[226,227] These include Huntington's disease (HD) and different types of spinocerebellar ataxias (SCAs), with disease pathogenesis being dependent on the polyQ length. Despite the shared feature of polyQ aggregates, affected genes, proteins, and brain regions vary between conditions, suggesting that the CAG repeat expansion may be toxic per se. To study the pathogenic role of polyQ expansion by itself, pure polyQ models have been developed through expression of CAG repeat expansions.[227] For instance, overexpression of fluorescently-tagged polyQ-expanded tracts in body wall muscles or pan-neuronally resulted in polyQ length-dependent aggregation and corresponding toxicity, whereby a polyQ length of 40 glutamines (Q40) was found to be the toxicity threshold in both tissues.[228,229] Importantly, findings supported the protective role of autophagy against polyQ toxicity. RNAi inactivation of autophagy-related genes exacerbated muscle Q40 aggregation and associated locomotor defects.[230] In the same vein, separate findings suggested that autophagy induction by pharmacological activation of SIRT1/sir-2.1 alleviated similar polyQ-associated pathology in a neuronal Q67 model.[231] Notably, selective neuronal vulnerability appeared to be captured with pan-neuronal Q40 expression, whereby dorsal (DNC) and ventral nerve cord (VNC) neurons were found to be more susceptible to Q40 aggregates. Taken together with findings of autophagy interference by polyglutamine tracts in cellular and rodent systems,[158] this suggests that the Q40 pan-neuronal model could have potential utility for studying how the polyQ-autophagy interplay may contribute to selective neuronal vulnerability in polyQ diseases. Since these are pure polyQ models, extrapolation of findings for specific polyQ diseases may be an issue. It is hence important to have disease-specific polyQ models to understand how polyQ expansion impinges on specific proteins in disease pathogenesis.

Disease-specific polyQ models have been generated through overexpression of CAG repeat expansions in disease-affected genes. In HD, for instance, models have been developed by overexpressing polyQ-expanded truncated human huntingtin (Htt) protein.[232–234] Similarly, models for the most common SCA, SCA3, have been developed through the overexpression of polyQ-expanded truncated or full-length ataxin-3.[227] Findings from these disease-specific models highlight the protective role of autophagy against polyQ pathology. For instance, one such study on a HD polyQ model showed that lysine acetylation facilitated autophagic degradation and prevented polyQ proteotoxicity.[235] Additionally, RNAi screens identified autophagy-related genes as protective against polyQ toxicity in HD and SCA3 polyQ models.[230,236,237] Notably, this effect in the HD model corroborated observations in the aforementioned muscle-specific Q40 model. Taken together, these observations suggest that autophagy is an important common protective mechanism.[230]

Importantly, insights into disease mechanisms were gained from disease-specific polyQ systems. Pan-neuronal overexpression of polyQ-expanded ataxin-3 not only corroborated findings from the pan-neuronal Q40 overexpression model that DNC and VNC neurons were more susceptible to polyQ aggregation, it additionally enhanced the vulnerability of head sensory neurons.[238] Moreover, toxicity from polyQ-expanded ataxin was not only similarly polyQ length-dependent, but was further subjected to the ataxin-3 sequence affected by the expansion.[238] Collectively, these findings support that selectivity of neuronal vulnerabilities in polyQ diseases may be directly attributed to the affected genes and their proteins. The mechanisms for this selectivity remain to be established, but it will be important to elucidate the role of autophagy in this respect. Finally, it is noteworthy that Htt has been suggested by several studies in mammalian cells and *Drosophila melanogaster* to function as a scaffold for selective autophagy.[239,240]

4.1.4 Amyotrophic lateral sclerosis (ALS)

ALS is the most common motor neuron disorder.[241] The genetic etiology of ALS is broad, with mutations identified in genes such as *SOD1*, *FUS*, *TARDBP*, *C9orf72*, *HNRNPA1*, and those involved in autophagy such as *SQSTM1* and *OPTN*.[242] Of these, *C9orf72*-related ALS is the most prevalent. Despite such genetic diversity, the majority of ALS cases present with cytoplasmically-mislocalized inclusions of the *TARDBP*-encoded DNA/RNA-binding transactive response DNA binding protein 43 kDa (TDP-43) in affected motor neurons. Aberrant cytoplasmic mislocalization was similarly observed for other ALS-affected proteins such as fused in sarcoma (FUS) and the heterogeneous nuclear ribonucleoproteins (hnRNPs).[243]

Various models of ALS have been developed in *C. elegans*, mostly by overexpression of human disease-associated mutant genes.[164,165,244,245] For instance, pan-neuronal overexpression of mutant G85R SOD1 recapitulated locomotor defects and synaptic dysfunction.[246] To study toxicity in motor neurons specifically, a similar model was generated via overexpression of a different SOD1 mutant (G93A) in GABAergic motor neurons.[247] Notably, RNAi screens did not identify autophagy genes as protective against G85R SOD1 aggregation, whereas autophagy induction improved locomotor defects in the G93R SOD1 model. These opposing requirements for autophagy may be attributed to differences in biochemical and/or biophysical properties between SOD1 variants. In support of this, various single copy knock-in disease-associated mutations of endogenous *SOD1/sod-1* in *C. elegans* exerted differential susceptibilities of glutamatergic and cholinergic neurons to oxidative stress.[248]

ALS models have also been generated through human TDP-43 overexpression.[249–251] Given that TDP-43 pathology affects the majority of ALS cases, these models have potential utility for uncovering common pathophysiological mechanisms of disease. Although the use of these models to study the role of autophagy in ALS have not been reported as of now, studies in other model systems demonstrated cross regulation between TDP-43 pathology and autophagy.[252,253] It will be important to validate these findings in *C. elegans* ALS models for further utilization to elucidate the role of autophagy in ALS.

4.2 Interplay between nucleocytoplasmic (NC) transport and autophagy

Altered NC shuttling has been observed for not only ALS, but additionally for other neurodegenerative diseases such as AD and HD.[254] Studies in other model systems have revealed potential mechanisms of impaired NC transport. For instance, dipeptide repeats encoded by *C9orf72* repeat expansions in ALS have been shown to directly block nuclear pores through interactions with nuclear pore proteins.[255–257] In the context of AD and tauopathies, an FTD-associated mutant form of tau was shown to cause nuclear deformation in neuronal cell bodies.[258] Notably, intact NC transport mediate protection against toxicity in *C. elegans* neurodegenerative disease models via autophagy, as inhibiting NC export of transcription factor EB (TFEB)/HLH-30, the master transcriptional regulator of autophagy and lysosomal biogenesis, improved autophagic flux and ameliorated polyQ aggregation and Aβ-induced paralysis.[106] Whether and how NC transport regulates other aspects of autophagy and vice versa remains to be established but are important questions to be addressed in order to understand this interrelationship in normal aging and age-related neurodegeneration. In this respect, *C. elegans* is an excellent model to dissect these mechanisms as approaches to assess autophagy (see Section 3) and nuclear cell biology are readily available.[259]

Altogether, neurodegenerative models in *C. elegans* have provided rapidly testable tools to study proteostasis dysregulation and neuronal degeneration. In addition, studies using these proteotoxic models of neurodegenerative diseases have facilitated insights into the roles of autophagy in these conditions.[260,261] Collectively, such studies highlight the pleiotropy of autophagy as it appears to be important for neuronal survival, whereas hyperactivation may be detrimental through induction of cell death.[48,262–264] Glia-neuron interactions are important in the pathogenesis of neurodegenerative diseases,[265] and on-going efforts to characterize these interactions in *C. elegans* in physiology are essential to facilitate the development of tools to study them in disease.[266] Lastly, findings from different models suggest that autophagy-lysosomal function contributes to the selective neuronal vulnerability in these diseases, although the exact mechanisms remain to be elucidated. It is possible that different genetic and/or environmental factors may impinge on differences in autophagy-lysosomal and proteostasis capacity in distinct neuronal populations, affecting neuronal susceptibility to proteotoxic insults. Moreover, some disease-associated proteins such as tau have been shown to directly interfere with both bulk and selective forms of autophagy, an observation that is corroborated in *C. elegans*, whereby pan-neuronal overexpression of P301L mutant tau inhibited mitophagy in neurons.[160] Evidently, more studies are required to understand the cellular and genetic contexts affecting autophagy pleiotropy and the pathological impact of different proteins involved in neurodegeneration on autophagic function.

4.3 Metabolic disorders

Metabolic homeostasis is maintained in part by remodeling macromolecules for energy and protein metabolism via autophagy. Given its role in metabolic regulation, autophagic failure, in part due to age-related decreases in insulin signaling efficiency and mTOR hyperactivity, is an important contributor in the pathogenesis of metabolic disorders.[267] For instance, impaired autophagic degradation of lipids (lipophagy) plays a role in liver steatosis, a condition caused by elevated lipid droplets in hepatocytes, highlighting the importance of autophagy-mediated lipid remodeling.[268] Particularly, lipophagy has an important role in mediating responses to nutrient-related stressors such as starvation. For instance, survival to starvation is dependent on the ability of *C. elegans* HLH-30/TFEB to mobilize lipids via lipophagy and lysosomal acid lipases.[99,101] Furthermore, lipids bound for lipoprotein biogenesis[269,270] are rerouted to lipophagy for lysosomal-driven nuclear hormone receptor lipid signaling akin to PPARs in humans.[270–272] This stress response-associated lipid signaling pathway is generally enhanced in long-lived nematodes, and may underlie the ability of animals to sustain elevated lipid remodeling processes of lipogenesis and lipolysis.

Autophagy may further impinge on mitochondria-related processes to regulate metabolism. This may not necessarily be via the selective removal of damaged mitochondria via mitophagy. For instance, lipophagic processes can affect mitochondria metabolism, whereby lipophagy-dependent lipid turnover is integrated with *de novo* lipogenesis and lipid droplet formation, as well as breakdown, to provide the mitochondria fatty acids at an optimal rate for beta oxidation.[273] Furthermore, findings suggest a macroautophagy-dependent and mitophagy-independent degradation of mitochondria during starvation, putting forth the possibility of a cross regulation of bulk and selective forms of autophagy to modulate metabolism.[274]

Findings from metabolic studies in *C. elegans* further highlighted the importance of specific tissues for integrating autophagy and metabolism. For instance, specific neurons modulated the autophagy-mediated survival to starvation stress.[87] Conversely, survival was attenuated with dysregulated autophagy.[49] Another important tissue in *C. elegans* is the intestine. Intestinal autophagy was shown to be elevated in long-lived nematodes, and was important for mediating the extended lifespans of these animals.[133,275] Furthermore, the intestine is the major site of lipid storage and metabolism in *C. elegans*. Notably, the ability of animals to distribute lipids into intestinal lipid droplets depends on the level of yolk protein biogenesis, a process that is conserved in mice.[270] Taken together, this suggests that autophagy-dependent metabolic regulation in the intestine is important for maintaining youthfulness. Being the principal metabolic organ for lipid homeostasis, the intestine in *C. elegans* has been used as a proxy to model metabolic sites of dysregulation in humans such as the liver, intestine and adipose tissue. Furthermore, intestinal fat droplets are easily stainable and visualized. Studies looking at metabolism in *C. elegans* have therefore been extremely important in providing insights into how lipid homeostasis goes awry in lipid and metabolic disorders such as obesity and Type II diabetes.[276] Furthermore, *C. elegans* has been used as a system to study other contributing factors to metabolic disorders, such as oxidative stress[277] and the complex interactions between genes and diet.[278] Importantly, these platforms provide a means to elucidate the autophagy-metabolism relationship in disease pathogenesis.

4.4 Innate immunity and infectious diseases

Autophagy is not only important for mediating resistance to environmental stressors such as hypoxia,[279] UV irradiation,[280] heat stress,[281] but also pathogenic infection.[282–284] Numerous studies have demonstrated *C. elegans* as a model system to study immune responses against pathogen infection.[283,284] Notably, findings which suggest that *C. elegans* death in old age is propagated by intestinal pathogenic infiltration highlight its utility to model host-pathogen interactions.[285,286] Although *C. elegans* lacks an adaptive immune system devoid of phagocytic immune cells and the key adaptive immunity-regulating transcription factor NFKB, these "disadvantageous" characteristics may contrarily be beneficial in facilitating the discovery of novel immune factors and mechanisms, which may translate to new therapeutic targets for infectious diseases. Furthermore, it remains to be established if other cell types may adopt adaptive immune roles in nematodes. For instance, it is suggested that coelomocytes may serve as primordial macrophages in *C. elegans*.[287] Regardless, studies in *C. elegans* have been instrumental in demonstrating the requirement for autophagy for immune resistance to *Nematocida parisii*,[288] *Pseudomonas aeruginosa*,[289] *Staphylococcus aureus*[137] and *Salmonella typhimurium*.[290] Importantly, autophagy-dependent immunity was similarly conserved in cellular and murine systems, bolstering the strength of *C. elegans* as a tool to assess immune function.[282] Intriguingly, autophagy may mediate protection against pathogenic infection at the expense of neuronal and germline health and integrity, further evidencing the pleiotropic effects of autophagy in different contexts.[291,292]

Recent studies highlight the interrelationships between energy-sensing, metabolism, longevity-regulating pathways, and innate immunity. For instance, the link between longevity and innate immunity was exemplified by a study which attributed lifespan extension of long-lived *eat-2* and *phm-2* loss-of-function mutants to innate immune activation and bacterial avoidance induced by intestinal bacterial accumulation.[293] Furthermore, the autophagy-regulating transcription factor HLH-30/TFEB was shown to mediate lifespan extension and immunity induction of *phm-2* mutants, corroborating its longevity-promoting role in *eat-2* animals[59] and its requirement for the ability of nematodes to withstand pathogen infection.[137] The interplay between metabolic regulation and innate immune function was also suggested by separate findings that the lipid metabolite, fatty acid oleate, is important for innate immunity induction in *C. elegans*, and defects in oleate production augmented the vulnerability toward pathogenic infection.[294]

Lastly, crosstalk between nutrient-sensing and innate immune pathways are highlighted through a third study demonstrating that TFEB/HLH-30-mediated innate immune protection against pathogenic infection in *C. elegans* was induced by AMPK and inhibited by the tumor suppressor Folliculin (FLCN) independently of mTOR.[295] This crosstalk was not only conserved in murine macrophages, but knockdown of *FLCN* in these immune cells was further shown to augment cellular bioenergetics and phagocytic capacity.

Although the role of autophagy in innate immunity activation was not assessed per se in these studies, several lines of evidence suggest that it is likely to partake in the integration of these pathways with innate immune function. First, HLH-30/TFEB-dependent autophagy was shown to mediate resistance to pathogen infection.[137,296] Second, not only does autophagy regulate lipid metabolism, it is also known to be differentially regulated by fatty acids.[297] Of note, oleate was reported to stimulate autophagy in both *C. elegans* and cells.[298,299] Intriguingly, this autophagy-inducing effect was reported to be independent of AMPK and Beclin-1, implicating non-canonical autophagy in immune regulation.[299] Third, long-lived *eat-2* mutants require HLH-30/TFEB and autophagy for lifespan extension, and also exhibit increased autophagic activity.[70–72] These mutants have pharyngeal pumping defects and are widely considered to be a genetic model of dietary restriction, of which low nutrient availability is known to induce autophagy. Taking these considerations with findings from *phm-2* and *eat-2* mutants described above, it is possible that HLH-30/TFEB-induced autophagy may be induced by intestinal bacterial accumulation to activate the innate immune response.

The studies highlighted above present strong support for the suitability of *C. elegans* as a model platform to understand immune responses to pathogen infection, especially in the context of autophagy. Furthermore, such infection models have been adapted for high throughput screening to identify novel and FDA-approved compounds which mediate anti-infectious effects through immunostimulatory, anti-pathogenic, and anti-virulence means.[300] Conversely, newly-identified anti-infective compounds uncovered from such screens can be used to better understand autophagy-dependent and -independent mechanisms of immunity against pathogens.

4.5 Tumorigenesis and cancer

Various features such as an imbalance of cellular proliferation and death, genome instability, dysregulated metabolism, cellular immortality, and tumor invasiveness and metastasis, propagate tumorigenesis and cancer.[301,302] Additionally, non-coding microRNAs (miRs) and autophagy have been implicated in cancer due to their regulatory roles of cell proliferation and cell death.[302–307] Notably, while autophagy was shown to suppress tumor development, its role in metastasis appears to be paradoxical.[302,306–308]

The use of *C. elegans* as a tool to study tumorigenic mechanisms was met with reservations due to the post-mitotic nature of most cells aside from its germline.[309,310] However, its immortal germline proved to be a suitable model in this respect as it is the only tissue to undergo apoptotic cell death and exhibit tumor-like morphology, whereby the gonad expands with mitotic nuclei.[309] For instance, *C. elegans* mutants exhibiting loss-of-function mutations in the genes *glp-1* and *gld-1* exhibit germ cell hyperproliferation, and have been used as models to study tumorigenesis. Aside from the germ line, the vulva presents another tissue to study tumorigenesis as vulva developmental defects can be used as readouts for dysregulated signaling in cell proliferation pathways.[310] Genetic screens for modulators of vulval defects have uncovered regulators of cell division and proliferation such as the *let-60* Ras homolog, of which Ras proteins have been implicated in human tumorigenesis.[310] Moreover, miRs were found to regulate autophagy and vice versa.[311–313] Taken together with the plethora of tools to monitor autophagy (see Section 3), this highlights *C. elegans* as a suitable system to study tumorigenic mechanisms from various angles, such as how miR-autophagy crosstalk can promote or suppress tumor development.

Conserved genetic modulators of germline cell proliferation and miRs historically discovered in nematodes have been linked to tumorigenesis in humans, illustrating the applicability of *C. elegans* to study tumorigenesis.[314,315] Studies performed in *C. elegans* further demonstrated the regulation of autophagy by tumor promoters and suppressors. For instance, HLH-30/TFEB activity can be inhibited by both tumor-suppressing protein FLCN-1/FLCN[316] and oncogenic protein XPO1/CRM1[106,110] via AMPK inhibition and nuclear export, respectively. Conversely, autophagy can be activated through LGG-1/LC3 phosphorylation by the tumor-suppressing Hippo kinases STK3/4.[114] Collectively, these studies highlight the complexity of autophagy in tumorigenesis as tumor suppressors can regulate autophagy in opposing ways.

Autophagy was shown to stimulate germ cell apoptosis induced by DNA damage[317] and inhibit tumor proliferation in germline tumor-bearing animals[318] demonstrating a more direct role of autophagy on cell growth and death via the *C. elegans* germline. Inhibition of autophagy led to germline tumor invasion to the head and vulva, showing that autophagic regulation of tumor metastasis can be modeled and studied in nematodes.[318] Taken together, these studies highlight the emergence of *C. elegans* as a system to model and understand tumorigenic mechanisms that lead to cancer. Importantly, insights into the role of autophagy in cancer development and progression can be achieved.

In addition to the aforementioned diseases (Fig. 3), *C. elegans* has also been used to model and provide insights into multiple conditions of which autophagic alterations have been implicated. These include inflammatory diseases such as Crohn's disease,[319,320] mitochondrial disorders such as Friedreich's ataxia[195,321,322] and lysosomal storage diseases.[323–325] Despite the evolutionary divergence between *C. elegans* and humans, the presence of highly conserved molecular mechanisms such as autophagy combined with opportunities to model pathologies reiterates the practicality of nematodes to study autophagy-related diseases.

5. Future challenges and opportunities

Numerous tools and models are available to study autophagy in aging and disease in *C. elegans*. Furthermore, these platforms are adaptable for high throughput genetic and chemical screens to uncover autophagy-dependent disease modifiers. Autophagy inducers,[326,327] inhibitors,[328,329] and HLH-30/TFEB activators[106,330] have been uncovered through screens. These compounds may not only expedite drug discovery processes but can also be used in a reverse approach to understand the role of autophagy in various conditions.

Autophagy not only exhibits pleiotropy in development and aging (reviewed in Ref. 331), but also in different diseases. This highlights its complex role in physiology and pathophysiology, and emphasizes the need to parse out its timing, tissue- and context-dependent contribution. Autophagy may be regulated circadially,[332–334] and both cell and non-cell autonomously, the latter for instance by astrocytes.[335] Conversely, autophagy may also mediate cell and non-cell autonomous effects.[336,337] Conditional autophagy expression, knockdown, or knockout will be particularly useful in enabling tissue- and context-dependent functions of autophagy to be elucidated. For instance, tissue-restricted RNAi knockdown is achievable in *C. elegans*.[338–340] Notably, many humanized disease models, especially of neurodegenerative diseases, are overexpression systems, which may preclude their physiological relevance. For example,

SOD1 single-copy CRISPR knock-in models differ from overexpression ones in that they do not exhibit overt locomotor deficits.[248] It will be important for future studies to generate, characterize, and compare new single-copy knock-in neurodegenerative disease models with currently-available overexpression ones. Such models may provide more physiologically-relevant insights into the role of autophagy in these diseases.

Additional to conditional genetic manipulation of autophagy, new reporter tools to assess the dynamics of autophagy are equally important. For instance, LGG-1 and LGG-2 act synergistically and have partially overlapping roles during AP formation and maturation, making them both important markers while assessing autophagic perturbations.[58,127] Hence, a tandem reporter for LGG-2 would be a useful complement to the currently-available one for LGG-1 to dissect out dynamics during AP maturation. Similarly, tandem reporters to study the flux of various selective autophagy receptors will be informative about the dynamics of cargo recruitment and degradation. Further research on uncovering receptors involved in other forms of selective autophagy, such as ribophagy, lipophagy, nucleophagy, will aid in tool development for the monitoring of these processes.

As a final note, it is clear that genetic tractability, mechanistic conservation, and technical practicality of nematodes facilitate the investigation and elucidation of human disease mechanisms, and highlights the use of *C. elegans* as a remarkable model organism for the study of autophagy.

References

1. Brenner S. The genetics of Caenorhabditis elegans. *Genetics*. 1974;77:71–94.
2. Lai CH, Chou CY, Ch'ang LY, Liu CS, Lin W. Identification of novel human genes evolutionarily conserved in *Caenorhabditis elegans* by comparative proteomics. *Genome Res*. 2000;10:703–713.
3. Shaye DD, Greenwald I. OrthoList: a compendium of *C. elegans* genes with human orthologs. *PLoS One*. 2011;6:e20085.
4. Sulston JE. Neuronal cell lineages in the nematode *Caenorhabditis elegans*. *Cold Spring Harb Symp Quant Biol*. 1983;48(Pt 2):443–452.
5. Sulston JE, Horvitz HR. Post-embryonic cell lineages of the nematode, *Caenorhabditis elegans*. *Dev Biol*. 1977;56:110–156.
6. Jorgensen EM, Mango SE. The art and design of genetic screens: *Caenorhabditis elegans*. *Nat Rev Genet*. 2002;3:356–369.
7. Anderson JL, Morran LT, Phillips PC. Outcrossing and the maintenance of males within *C. elegans* populations. *J Hered*. 2010;101(Suppl 1):S62–S74.
8. C. elegans Sequencing Consortium. Genome sequence of the nematode *C. elegans*: a platform for investigating biology. *Science*. 1998;282:2012–2018.
9. Cook SJ, Jarrell TA, Brittin CA, et al. Whole-animal connectomes of both *Caenorhabditis elegans* sexes. *Nature*. 2019;571:63–71.

10. Hammarlund M, Hobert O, Miller 3rd DM, Sestan N. The CeNGEN project: the complete gene expression map of an entire nervous system. *Neuron*. 2018;99:430–433.
11. Palmisano NJ, Melendez A. RNAi-mediated inactivation of autophagy genes in *Caenorhabditis elegans*. *Cold Spring Harb Protoc*. 2016;2016(2):pdb.prot086520. https://doi.org/10.1101/pdb.prot086520.
12. Kamath RS, Fraser AG, Dong Y, et al. Systematic functional analysis of the *Caenorhabditis elegans* genome using RNAi. *Nature*. 2003;421:231–237.
13. Rual JF, Ceron J, Koreth J, et al. Toward improving *Caenorhabditis elegans* phenome mapping with an ORFeome-based RNAi library. *Genome Res*. 2004;14:2162–2168.
14. Hamilton B, Dong Y, Shindo M, et al. A systematic RNAi screen for longevity genes in *C. elegans*. *Genes Dev*. 2005;19:1544–1555.
15. Hansen M, Hsu AL, Dillin A, Kenyon C. New genes tied to endocrine, metabolic, and dietary regulation of lifespan from a *Caenorhabditis elegans* genomic RNAi screen. *PLoS Genet*. 2005;1:119–128.
16. Gelino S, Hansen M. Autophagy—an emerging anti-aging mechanism. *J Clin Exp Pathol*. 2012;(Suppl 4):pii: 006.
17. Zhang H, Chang JT, Guo B, et al. Guidelines for monitoring autophagy in *Caenorhabditis elegans*. *Autophagy*. 2015;11:9–27.
18. Harding TM, Morano KA, Scott SV, Klionsky DJ. Isolation and characterization of yeast mutants in the cytoplasm to vacuole protein targeting pathway. *J Cell Biol*. 1995;131:591–602.
19. Tsukada M, Ohsumi Y. Isolation and characterization of autophagy-defective mutants of Saccharomyces cerevisiae. *FEBS Lett*. 1993;333:169–174.
20. Kuroyanagi H, Yan J, Seki N, et al. Human ULK1, a novel serine/threonine kinase related to UNC-51 kinase of *Caenorhabditis elegans*: cDNA cloning, expression, and chromosomal assignment. *Genomics*. 1998;51:76–85.
21. Matsuura A, Tsukada M, Wada Y, Ohsumi Y. Apg1p, a novel protein kinase required for the autophagic process in Saccharomyces cerevisiae. *Gene*. 1997;192:245–250.
22. Yan J, Kuroyanagi H, Kuroiwa A, et al. Identification of mouse ULK1, a novel protein kinase structurally related to *C. elegans* UNC-51. *Biochem Biophys Res Commun*. 1998;246:222–227.
23. Kovacs AL, Zhang H. Role of autophagy in *Caenorhabditis elegans*. *FEBS Lett*. 2010;584:1335–1341.
24. Melendez A, Levine B. Autophagy in *C. elegans*. *WormBook*. 2009;1–26.
25. Palmisano NJ, Melendez A. Autophagy in *C. elegans* development. *Dev Biol*. 2019;447:103–125.
26. Bohnert KA, Kenyon C. A lysosomal switch triggers proteostasis renewal in the immortal *C. elegans* germ lineage. *Nature*. 2017;551:629–633.
27. Ames K, Da Cunha DS, Gonzalez B, et al. A non-cell-autonomous role of BEC-1/BECN1/Beclin1 in coordinating cell-cycle progression and stem cell proliferation during germline development. *Curr Biol*. 2017;27:905–913.
28. Al Rawi S, Louvet-Vallee S, Djeddi A, et al. Postfertilization autophagy of sperm organelles prevents paternal mitochondrial DNA transmission. *Science*. 2011;334:1144–1147.
29. Al Rawi S, Louvet-Vallee S, Djeddi A, et al. Allophagy: a macroautophagic process degrading spermatozoid-inherited organelles. *Autophagy*. 2012;8:421–423.
30. Sato M, Sato K. Degradation of paternal mitochondria by fertilization-triggered autophagy in *C. elegans* embryos. *Science*. 2011;334:1141–1144.
31. Zhou Q, Li H, Xue D. Elimination of paternal mitochondria through the lysosomal degradation pathway in *C. elegans*. *Cell Res*. 2011;21:1662–1669.
32. Molina P, Lim Y, Boyd L. Ubiquitination is required for the initial removal of paternal organelles in *C. elegans*. *Dev Biol*. 2019;453:168–179.

33. Djeddi A, Al Rawi S, Deuve JL, et al. Sperm-inherited organelle clearance in *C. elegans* relies on LC3-dependent autophagosome targeting to the pericentrosomal area. *Development*. 2015;142:1705–1716.
34. Sato M, Sato K, Tomura K, Kosako H, Sato K. The autophagy receptor ALLO-1 and the IKKE-1 kinase control clearance of paternal mitochondria in *Caenorhabditis elegans*. *Nat Cell Biol*. 2018;20:81–91.
35. Wei Y, Chiang WC, Sumpter Jr R, Mishra P, Levine B. Prohibitin 2 is an inner mitochondrial membrane mitophagy receptor. *Cell*. 2017;168. 224–238.e210.
36. Lim Y, Rubio-Pena K, Sobraske PJ, et al. Fndc-1 contributes to paternal mitochondria elimination in *C. elegans*. *Dev Biol*. 2019;454:15–20.
37. Liu Y, Zou W, Yang P, et al. Autophagy-dependent ribosomal RNA degradation is essential for maintaining nucleotide homeostasis during *C. elegans* development. *eLife*. 2018;7:pii: e36588. https://doi.org/10.7554/eLife.36588.
38. Zhang G, Lin L, Qi D, Zhang H. The composition of a protein aggregate modulates the specificity and efficiency of its autophagic degradation. *Autophagy*. 2017;13:1487–1495.
39. Zhang Y, Yan L, Zhou Z, et al. SEPA-1 mediates the specific recognition and degradation of P granule components by autophagy in *C. elegans*. *Cell*. 2009;136:308–321.
40. Zhang Y, Qi L, Zhang H. TGFbeta-like DAF-7 acts as a systemic signal for autophagy regulation in *C. elegans*. *J Cell Biol*. 2019;218:3998–4006.
41. Guo B, Huang X, Zhang P, et al. Genome-wide screen identifies signaling pathways that regulate autophagy during *Caenorhabditis elegans* development. *EMBO Rep*. 2014;15:705–713.
42. Borsos E, Erdelyi P, Vellai T. Autophagy and apoptosis are redundantly required for *C. elegans* embryogenesis. *Autophagy*. 2011;7:557–559.
43. Huang S, Jia K, Wang Y, Zhou Z, Levine B. Autophagy genes function in apoptotic cell corpse clearance during *C. elegans* embryonic development. *Autophagy*. 2013; 9:138–149.
44. Li W, Zou W, Yang Y, et al. Autophagy genes function sequentially to promote apoptotic cell corpse degradation in the engulfing cell. *J Cell Biol*. 2012;197:27–35.
45. Ruck A, Attonito J, Garces KT, et al. The Atg6/Vps30/Beclin 1 ortholog BEC-1 mediates endocytic retrograde transport in addition to autophagy in *C. elegans*. *Autophagy*. 2011;7:386–400.
46. Corsi AK, Wightman B, Chalfie M. A transparent window into biology: a primer on *Caenorhabditis elegans*. In: *WormBook. W. The* C. elegans *Research Community, ed*. 2015.
47. Cecconi F, Levine B. The role of autophagy in mammalian development: cell makeover rather than cell death. *Dev Cell*. 2008;15:344–357.
48. Kang C, Avery L. To be or not to be, the level of autophagy is the question: dual roles of autophagy in the survival response to starvation. *Autophagy*. 2008;4:82–84.
49. Kang C, You YJ, Avery L. Dual roles of autophagy in the survival of *Caenorhabditis elegans* during starvation. *Genes Dev*. 2007;21:2161–2171.
50. Lu Q, Yang P, Huang X, et al. The WD40 repeat PtdIns(3)P-binding protein EPG-6 regulates progression of omegasomes to autophagosomes. *Dev Cell*. 2011;21: 343–357.
51. Tian E, Wang F, Han J, Zhang H. epg-1 functions in autophagy-regulated processes and may encode a highly divergent Atg13 homolog in *C. elegans*. *Autophagy*. 2009;5:608–615.
52. Tian Y, Li Z, Hu W, et al. *C. elegans* screen identifies autophagy genes specific to multicellular organisms. *Cell*. 2010;141:1042–1055.
53. Yang P, Zhang H. The coiled-coil domain protein EPG-8 plays an essential role in the autophagy pathway in *C. elegans*. *Autophagy*. 2011;7:159–165.
54. McPhee CK, Baehrecke EH. Autophagy shows its animal side. *Cell*. 2010; 141:922–924.

55. Melendez A, Talloczy Z, Seaman M, Eskelinen EL, Hall DH, Levine B. Autophagy genes are essential for dauer development and life-span extension in *C. elegans*. *Science*. 2003;301:1387–1391.
56. Gomez TA, Banfield KL, Trogler DM, Clarke SG. The L-isoaspartyl-O-methyltransferase in *Caenorhabditis elegans* larval longevity and autophagy. *Dev Biol*. 2007;303:493–500.
57. Gomez TA, Clarke SG. Autophagy and insulin/TOR signaling in *Caenorhabditis elegans* pcm-1 protein repair mutants. *Autophagy*. 2007;3:357–359.
58. Alberti A, Michelet X, Djeddi A, Legouis R. The autophagosomal protein LGG-2 acts synergistically with LGG-1 in dauer formation and longevity in *C. elegans*. *Autophagy*. 2010;6:622–633.
59. Lapierre LR, De Magalhaes Filho CD, McQuary PR, et al. The TFEB orthologue HLH-30 regulates autophagy and modulates longevity in *Caenorhabditis elegans*. *Nat Commun*. 2013;4:2267.
60. Ogura K, Okada T, Mitani S, et al. Protein phosphatase 2A cooperates with the autophagy-related kinase UNC-51 to regulate axon guidance in *Caenorhabditis elegans*. *Development*. 2010;137:1657–1667.
61. Stavoe AK, Hill SE, Hall DH, Colon-Ramos DA. KIF1A/UNC-104 transports ATG-9 to regulate neurodevelopment and autophagy at synapses. *Dev Cell*. 2016;38:171–185.
62. Jia K, Levine B. Autophagy and longevity: lessons from *C. elegans*. *Adv Exp Med Biol*. 2010;694:47–60.
63. Lapierre LR, Hansen M. Lessons from *C. elegans*: signaling pathways for longevity. *Trends Endocrinol Metab*. 2012;23:637–644.
64. Madeo F, Tavernarakis N, Kroemer G. Can autophagy promote longevity? *Nat Cell Biol*. 2010;12:842–846.
65. Salminen A, Kaarniranta K. Regulation of the aging process by autophagy. *Trends Mol Med*. 2009;15:217–224.
66. Lapierre LR, Kumsta C, Sandri M, Ballabio A, Hansen M. Transcriptional and epigenetic regulation of autophagy in aging. *Autophagy*. 2015;11:867–880.
67. Riddle DL, Gorski SM. Shaping and stretching life by autophagy. *Dev Cell*. 2003;5:364–365.
68. Kenyon C, Chang J, Gensch E, Rudner A, Tabtiang R. A *C. elegans* mutant that lives twice as long as wild type. *Nature*. 1993;366:461–464.
69. Hars ES, Qi H, Ryazanov AG, et al. Autophagy regulates ageing in *C. elegans*. *Autophagy*. 2007;3:93–95.
70. Toth ML, Sigmond T, Borsos E, et al. Longevity pathways converge on autophagy genes to regulate life span in *Caenorhabditis elegans*. *Autophagy*. 2008;4:330–338.
71. Hansen M, Chandra A, Mitic LL, Onken B, Driscoll M, Kenyon C. A role for autophagy in the extension of lifespan by dietary restriction in *C. elegans*. *PLoS Genet*. 2008;4:e24.
72. Jia K, Levine B. Autophagy is required for dietary restriction-mediated life span extension in *C. elegans*. *Autophagy*. 2007;3:597–599.
73. Lapierre LR, Gelino S, Melendez A, Hansen M. Autophagy and lipid metabolism coordinately modulate life span in germline-less *C. elegans*. *Curr Biol*. 2011;21:1507–1514.
74. Dwivedi M, Song HO, Ahnn J. Autophagy genes mediate the effect of calcineurin on life span in *C. elegans*. *Autophagy*. 2009;5:604–607.
75. Madeo F, Carmona-Gutierrez D, Hofer SJ, Kroemer G. Caloric restriction Mimetics against age-associated disease: targets, mechanisms, and therapeutic potential. *Cell Metab*. 2019;29:592–610.
76. Madeo F, Zimmermann A, Maiuri MC, Kroemer G. Essential role for autophagy in life span extension. *J Clin Invest*. 2015;125:85–93.

77. Kuma A, Hatano M, Matsui M, et al. The role of autophagy during the early neonatal starvation period. *Nature*. 2004;432:1032–1036.
78. Rabinowitz JD, White E. Autophagy and metabolism. *Science*. 2010;330:1344–1348.
79. Martinez-Lopez N, Tarabra E, Toledo M, et al. System-wide benefits of intermeal fasting by autophagy. *Cell Metab*. 2017;26. 856–871.e855.
80. Mattison JA, Colman RJ, Beasley TM, et al. Caloric restriction improves health and survival of rhesus monkeys. *Nat Commun*. 2017;8:14063.
81. Cabreiro F, Au C, Leung KY, et al. Metformin retards aging in *C. elegans* by altering microbial folate and methionine metabolism. *Cell*. 2013;153:228–239.
82. Castillo-Quan JI, Blackwell TK. Metformin: restraining nucleocytoplasmic shuttling to fight cancer and aging. *Cell*. 2016;167:1670–1671.
83. Chen J, Ou Y, Li Y, Hu S, Shao LW, Liu Y. Metformin extends *C. elegans* lifespan through lysosomal pathway. *eLife*. 2017;6:pii: e31268. https://doi.org/10.7554/eLife.31268.
84. De Haes W, Frooninckx L, Van Assche R, et al. Metformin promotes lifespan through mitohormesis via the peroxiredoxin PRDX-2. *Proc Natl Acad Sci U S A*. 2014;111:E2501–E2509.
85. Onken B, Driscoll M. Metformin induces a dietary restriction-like state and the oxidative stress response to extend *C. elegans* Healthspan via AMPK, LKB1, and SKN-1. *PLoS One*. 2010;5:e8758.
86. Nakamura S, Oba M, Suzuki M, et al. Suppression of autophagic activity by Rubicon is a signature of aging. *Nat Commun*. 2019;10:847.
87. Kang C, Avery L. Systemic regulation of autophagy in *Caenorhabditis elegans*. *Autophagy*. 2009;5:565–566.
88. Kang C, Avery L. Systemic regulation of starvation response in *Caenorhabditis elegans*. *Genes Dev*. 2009;23:12–17.
89. Minnerly J, Zhang J, Parker T, Kaul T, Jia K. The cell non-autonomous function of ATG-18 is essential for neuroendocrine regulation of *Caenorhabditis elegans* lifespan. *PLoS Genet*. 2017;13:e1006764.
90. Wilhelm T, Byrne J, Medina R, et al. Neuronal inhibition of the autophagy nucleation complex extends life span in post-reproductive *C. elegans*. *Genes Dev*. 2017;31:1561–1572.
91. Franco-Juarez B, Mejia-Martinez F, Moreno-Arriola E, et al. A high glucose diet induces autophagy in a HLH-30/TFEB-dependent manner and impairs the normal lifespan of *C. elegans*. *Aging (Albany NY)*. 2018;10:2657–2667.
92. Savini M, Wang MC. Does autophagy promote longevity? It depends. *Cell*. 2019;177:221–222.
93. Zhou B, Kreuzer J, Kumsta C, et al. Mitochondrial permeability uncouples elevated autophagy and lifespan extension. *Cell*. 2019;177:299–314.e216.
94. Chang JT, Kumsta C, Hellman AB, Adams LM, Hansen M. Spatiotemporal regulation of autophagy during *Caenorhabditis elegans* aging. *Elife*. 2017;6:pii: e18459. https://doi.org/10.7554/eLife.18459.
95. Ben-Zvi A, Miller EA, Morimoto RI. Collapse of proteostasis represents an early molecular event in *Caenorhabditis elegans* aging. *Proc Natl Acad Sci U S A*. 2009;106:14914–14919.
96. Kim DK, Kim TH, Lee SJ. Mechanisms of aging-related proteinopathies in *Caenorhabditis elegans*. *Exp Mol Med*. 2016;48:e263.
97. Keith SA, Maddux SK, Zhong Y, et al. Graded proteasome dysfunction in *Caenorhabditis elegans* activates an adaptive response involving the conserved SKN-1 and ELT-2 transcription factors and the autophagy-lysosome pathway. *PLoS Genet*. 2016;12:e1005823.

98. Denzel MS, Lapierre LR, Mack HID. Emerging topics in *C. elegans* aging research: transcriptional regulation, stress response and epigenetics. *Mech Ageing Dev.* 2019;177:4–21.
99. O'Rourke EJ, Ruvkun G. MXL-3 and HLH-30 transcriptionally link lipolysis and autophagy to nutrient availability. *Nat Cell Biol.* 2013;15:668–676.
100. Sardiello M, Palmieri M, di Ronza A, et al. A gene network regulating lysosomal biogenesis and function. *Science.* 2009;325:473–477.
101. Settembre C, De Cegli R, Mansueto G, et al. TFEB controls cellular lipid metabolism through a starvation-induced autoregulatory loop. *Nat Cell Biol.* 2013;15:647–658.
102. Settembre C, Di Malta C, Polito VA, et al. TFEB links autophagy to lysosomal biogenesis. *Science.* 2011;332:1429–1433.
103. Nakamura S, Karalay O, Jager PS, et al. Mondo complexes regulate TFEB via TOR inhibition to promote longevity in response to gonadal signals. *Nat Commun.* 2016;7:10944.
104. Lin XX, Sen I, Janssens GE, et al. DAF-16/FOXO and HLH-30/TFEB function as combinatorial transcription factors to promote stress resistance and longevity. *Nat Commun.* 2018;9:4400.
105. Li X, Yu W, Qian X, et al. Nucleus-Translocated ACSS2 promotes gene transcription for Lysosomal biogenesis and autophagy. *Mol Cell.* 2017;66:684–697.e689.
106. Silvestrini MJ, Johnson JR, Kumar AV, et al. Nuclear export inhibition enhances HLH-30/TFEB activity, autophagy, and lifespan. *Cell Rep.* 2018;23:1915–1921.
107. Li J, Zhu R, Chen K, et al. Potent and specific Atg8-targeting autophagy inhibitory peptides from giant ankyrins. *Nat Chem Biol.* 2018;14:778–787.
108. Napolitano G, Esposito A, Choi H, et al. mTOR-dependent phosphorylation controls TFEB nuclear export. *Nat Commun.* 2018;9:3312.
109. Gandhi UH, Senapedis W, Baloglu E, et al. Clinical implications of targeting XPO1-mediated nuclear export in multiple myeloma. *Clin Lymphoma Myeloma Leuk.* 2018;18:335–345.
110. Kumar AV, Thakurta TG, Silvestrini MJ, Johnson JR, Reenan RA, Lapierre LR. Give me a SINE: how selective inhibitors of nuclear export modulate autophagy and aging. *Mol Cell Oncol.* 2018;5(5):e1502511. https://doi.org/10.1080/23723556.2018.1502511.
111. Egan D, Kim J, Shaw RJ, Guan KL. The autophagy initiating kinase ULK1 is regulated via opposing phosphorylation by AMPK and mTOR. *Autophagy.* 2011;7:643–644.
112. Egan DF, Shackelford DB, Mihaylova MM, et al. Phosphorylation of ULK1 (hATG1) by AMP-activated protein kinase connects energy sensing to mitophagy. *Science.* 2011;331:456–461.
113. Russell RC, Tian Y, Yuan H, et al. ULK1 induces autophagy by phosphorylating Beclin-1 and activating VPS34 lipid kinase. *Nat Cell Biol.* 2013;15:741–750.
114. Wilkinson DS, Jariwala JS, Anderson E, et al. Phosphorylation of LC3 by the hippo kinases STK3/STK4 is essential for autophagy. *Mol Cell.* 2015;57:55–68.
115. Roggo L, Bernard V, Kovacs AL, et al. Membrane transport in *Caenorhabditis elegans*: an essential role for VPS34 at the nuclear membrane. *EMBO J.* 2002;21:1673–1683.
116. Marino G, Morselli E, Bennetzen MV, et al. Longevity-relevant regulation of autophagy at the level of the acetylproteome. *Autophagy.* 2011;7:647–649.
117. Morselli E, Marino G, Bennetzen MV, et al. Spermidine and resveratrol induce autophagy by distinct pathways converging on the acetylproteome. *J Cell Biol.* 2011;192:615–629.
118. Guerrero-Gomez D, Mora-Lorca JA, Saenz-Narciso B, et al. Loss of glutathione redox homeostasis impairs proteostasis by inhibiting autophagy-dependent protein degradation. *Cell Death Differ.* 2019;26:1545–1565.

119. Denzel MS, Storm NJ, Gutschmidt A, et al. Hexosamine pathway metabolites enhance protein quality control and prolong life. *Cell*. 2014;156:1167–1178.
120. Koh JH, Wang L, Beaudoin-Chabot C, Thibault G. Lipid bilayer stress-activated IRE-1 modulates autophagy during endoplasmic reticulum stress. *J Cell Sci*. 2018;131(22):pii: jcs217992. https://doi.org/10.1242/jcs.217992.
121. Berber S, Wood M, Llamosas E, et al. Homeodomain-interacting protein kinase (HPK-1) regulates stress responses and ageing in *C. elegans*. *Sci Rep*. 2016;6:19582.
122. Das R, Melo JA, Thondamal M, et al. The homeodomain-interacting protein kinase HPK-1 preserves protein homeostasis and longevity through master regulatory control of the HSF-1 chaperone network and TORC1-restricted autophagy in *Caenorhabditis elegans*. *PLoS Genet*. 2017;13:e1007038.
123. Chen Y, Scarcelli V, Legouis R. Approaches for studying autophagy in Caenorhabditis elegans. *Cells*. 2017;6:27.
124. Jenzer C, Simionato E, Legouis R. Tools and methods to analyze autophagy in *C. elegans*. *Methods*. 2015;75:162–171.
125. Palmisano NJ, Melendez A. Detection of autophagy in *Caenorhabditis elegans*. *Cold Spring Harb Protoc*. 2016;2016(2):pdb.top070466. https://doi.org/10.1101/pdb.top070466.
126. Miedel MT, Graf NJ, Stephen KE, et al. A pro-cathepsin L mutant is a luminal substrate for endoplasmic-reticulum-associated degradation in *C. elegans*. *PLoS One*. 2012;7:e40145.
127. Manil-Segalen M, Lefebvre C, Jenzer C, et al. The *C. elegans* LC3 acts downstream of GABARAP to degrade autophagosomes by interacting with the HOPS subunit VPS39. *Dev Cell*. 2014;28:43–55.
128. Li J, Huang KX, Le WD. Establishing a novel *C. elegans* model to investigate the role of autophagy in amyotrophic lateral sclerosis. *Acta Pharmacol Sin*. 2013;34:644–650.
129. Schmeisser K, Parker JA. Nicotinamide-N-methyltransferase controls behavior, neurodegeneration and lifespan by regulating neuronal autophagy. *PLoS Genet*. 2018;14:e1007561.
130. Thomas BJ, Wight IE, Chou WYY, et al. CemOrange2 fusions facilitate multifluorophore subcellular imaging in *C. elegans*. *PLoS One*. 2019;14:e0214257.
131. Imanikia S, Ozbey NP, Krueger C, Casanueva MO, Taylor RC. Neuronal XBP-1 activates intestinal lysosomes to improve Proteostasis in *C. elegans*. *Curr Biol*. 2019;29:2322–2338.e2327.
132. Palmisano NJ, Rosario N, Wysocki M, Hong M, Grant B, Melendez A. The recycling endosome protein RAB-10 promotes autophagic flux and localization of the transmembrane protein ATG-9. *Autophagy*. 2017;13:1742–1753.
133. Chapin HC, Okada M, Merz AJ, Miller DL. Tissue-specific autophagy responses to aging and stress in *C. elegans*. *Aging (Albany NY)*. 2015;7:419–434.
134. Bennett CF, Kwon JJ, Chen C, et al. Transaldolase inhibition impairs mitochondrial respiration and induces a starvation-like longevity response in *Caenorhabditis elegans*. *PLoS Genet*. 2017;13:e1006695.
135. Zhao Y, Tian E, Zhang H. Selective autophagic degradation of maternally-loaded germline P granule components in somatic cells during *C. elegans* embryogenesis. *Autophagy*. 2009;5:717–719.
136. Murphy JT, Liu H, Ma X, et al. Simple nutrients bypass the requirement for HLH-30 in coupling lysosomal nutrient sensing to survival. *PLoS Biol*. 2019;17:e3000245.
137. Visvikis O, Ihuegbu N, Labed SA, et al. Innate host defense requires TFEB-mediated transcription of Cytoprotective and antimicrobial genes. *Immunity*. 2014;40:896–909.
138. Springhorn A, Hoppe T. Western blot analysis of the autophagosomal membrane protein LGG-1/LC3 in *Caenorhabditis elegans*. *Methods Enzymol*. 2019;619:319–336.

139. Kovacs AL. The application of traditional transmission electron microscopy for autophagy research in *Caenorhabditis elegans*. *Biophys Rep*. 2015;1:99–105.
140. Melendez A, Neufeld TP. The cell biology of autophagy in metazoans: a developing story. *Development*. 2008;135:2347–2360.
141. Baxi K, Ghavidel A, Waddell B, Harkness TA, de Carvalho CE. Regulation of Lysosomal function by the DAF-16 Forkhead transcription factor couples reproduction to aging in *Caenorhabditis elegans*. *Genetics*. 2017;207:83–101.
142. O'Rourke EJ, Soukas AA, Carr CE, Ruvkun G. *C. elegans* major fats are stored in vesicles distinct from lysosome-related organelles. *Cell Metab*. 2009;10:430–435.
143. Soukas AA, Carr CE, Ruvkun G. Genetic regulation of *Caenorhabditis elegans* lysosome related organelle function. *PLoS Genet*. 2013;9:e1003908.
144. Baxi K, de Carvalho CE. Assessing lysosomal alkalinization in the intestine of live *Caenorhabditis elegans*. *J Vis Exp*. 2018;(134). https://doi.org/10.3791/57414.
145. Gee K, Zamora D, Horm T, et al. Regulators of lysosome function and dynamics in *Caenorhabditis elegans*. *G3*. 2017;7:991–1000.
146. Fang EF, Waltz TB, Kassahun H, et al. Tomatidine enhances lifespan and healthspan in *C. elegans* through mitophagy induction via the SKN-1/Nrf2 pathway. *Sci Rep*. 2017;7:46208.
147. Palikaras K, Lionaki E, Tavernarakis N. Coordination of mitophagy and mitochondrial biogenesis during ageing in *C. elegans*. *Nature*. 2015;521:525–528.
148. Palikaras K, Lionaki E, Tavernarakis N. Mitophagy: in sickness and in health. *Mol Cell Oncol*. 2016;3:e1056332.
149. Ryu D, Mouchiroud L, Andreux PA, et al. Urolithin A induces mitophagy and prolongs lifespan in *C. elegans* and increases muscle function in rodents. *Nat Med*. 2016;22:879–888.
150. Fang EF, Palikaras K, Sun N, et al. In vitro and in vivo detection of mitophagy in human cells, *C. elegans*, and mice. *J Vis Exp*. 2017;(129). https://doi.org/10.3791/56301.
151. Mizushima N, Levine B, Cuervo AM, Klionsky DJ. Autophagy fights disease through cellular self-digestion. *Nature*. 2008;451:1069–1075.
152. Kaushik S, Cuervo AM. The coming of age of chaperone-mediated autophagy. *Nat Rev Mol Cell Biol*. 2018;19:365–381.
153. Johansen T, Lamark T. Selective autophagy mediated by autophagic adapter proteins. *Autophagy*. 2011;7:279–296.
154. Scrivo A, Bourdenx M, Pampliega O, Cuervo AM. Selective autophagy as a potential therapeutic target for neurodegenerative disorders. *Lancet Neurol*. 2018;17:802–815.
155. Rangaraju S, Solis GM, Petrascheck M. High-throughput small-molecule screening in *Caenorhabditis elegans*. *Methods Mol Biol*. 2015;1263:139–155.
156. Menzies FM, Fleming A, Caricasole A, et al. Autophagy and Neurodegeneration: pathogenic mechanisms and therapeutic opportunities. *Neuron*. 2017;93:1015–1034.
157. Wong E, Cuervo AM. Autophagy gone awry in neurodegenerative diseases. *Nat Neurosci*. 2010;13:805–811.
158. Ashkenazi A, Bento CF, Ricketts T, et al. Polyglutamine tracts regulate beclin 1-dependent autophagy. *Nature*. 2017;545:108–111.
159. Caballero B, Wang Y, Diaz A, et al. Interplay of pathogenic forms of human tau with different autophagic pathways. *Aging Cell*. 2018;(1). https://doi.org/10.1111/acel.12692.
160. Cummins N, Tweedie A, Zuryn S, Bertran-Gonzalez J, Gotz J. Disease-associated tau impairs mitophagy by inhibiting Parkin translocation to mitochondria. *EMBO J*. 2019;38(3):pii: e99360. https://doi.org/10.15252/embj.201899360.

161. Tanik SA, Schultheiss CE, Volpicelli-Daley LA, Brunden KR, Lee VM. Lewy body-like alpha-synuclein aggregates resist degradation and impair macroautophagy. *J Biol Chem.* 2013;288:15194–15210.
162. Hobert O. Neurogenesis in the nematode *Caenorhabditis elegans. WormBook.* 2010;1–24. https://doi.org/10.1895/wormbook.1.12.2.
163. Hart AC. *Behavior.* The *C. elegans* Research Community; 2006.
164. Chen X, Barclay JW, Burgoyne RD, Morgan A. Using *C. elegans* to discover therapeutic compounds for ageing-associated neurodegenerative diseases. *Chem Cent J.* 2015;9:65.
165. Ma L, Zhao YD, Chen YC, Cheng B, Peng AL, Huang K. *Caenorhabditis elegans* as a model system for target identification and drug screening against neurodegenerative diseases. *Eur J Pharmacol.* 2018;819:169–180.
166. Li J, Le W. Modeling neurodegenerative diseases in *Caenorhabditis elegans. Exp Neurol.* 2013;250:94–103.
167. Sin O, Michels H, Nollen EA. Genetic screens in *Caenorhabditis elegans* models for neurodegenerative diseases. *Biochim Biophys Acta.* 2014;1842:1951–1959.
168. Griffin EF, Caldwell KA, Caldwell GA. Genetic and pharmacological discovery for Alzheimer's disease using *Caenorhabditis elegans. ACS Chem Nerosci.* 2017;8:2596–2606.
169. Pir GJ, Choudhary B, Mandelkow E. *Caenorhabditis elegans* models of tauopathy. *FASEB J.* 2017;31:5137–5148.
170. Josephs KA. Current understanding of neurodegenerative diseases associated with the protein Tau. *Mayo Clin Proc.* 2017;92:1291–1303.
171. Selkoe DJ, Hardy J. The amyloid hypothesis of Alzheimer's disease at 25 years. *EMBO Mol Med.* 2016;8:595–608.
172. Chow VW, Mattson MP, Wong PC, Gleichmann M. An overview of APP processing enzymes and products. *Neuromolecular Med.* 2010;12:1–12.
173. Link CD. *C. elegans* models of age-associated neurodegenerative diseases: lessons from transgenic worm models of Alzheimer's disease. *Exp Gerontol.* 2006;41:1007–1013.
174. Florez-McClure ML, Hohsfield LA, Fonte G, Bealor MT, Link CD. Decreased insulin-receptor signaling promotes the autophagic degradation of beta-amyloid peptide in *C. elegans. Autophagy.* 2007;3:569–580.
175. Nixon RA, Wegiel J, Kumar A, et al. Extensive involvement of autophagy in Alzheimer disease: an immuno-electron microscopy study. *J Neuropathol Exp Neurol.* 2005;64:113–122.
176. Hynd MR, Scott HL, Dodd PR. Glutamate-mediated excitotoxicity and neurodegeneration in Alzheimer's disease. *Neurochem Int.* 2004;45:583–595.
177. Lewerenz J, Maher P. Chronic glutamate toxicity in neurodegenerative diseases-what is the evidence? *Front Neurosci.* 2015;9:469.
178. Wang R, Reddy PH. Role of glutamate and NMDA receptors in Alzheimer's disease. *J Alzheimers Dis.* 2017;57:1041–1048.
179. Treusch S, Hamamichi S, Goodman JL, et al. Functional links between Abeta toxicity, endocytic trafficking, and Alzheimer's disease risk factors in yeast. *Science.* 2011;334: 1241–1245.
180. Griffin EF, Scopel SE, Stephen CA, et al. ApoE-associated modulation of neuroprotection from Abeta-mediated neurodegeneration in transgenic *Caenorhabditis elegans. Dis Model Mech.* 2019;12(2):pii: dmm037218. https://doi.org/10.1242/dmm.037218.
181. Guthrie CR, Schellenberg GD, Kraemer BC. SUT-2 potentiates tau-induced neurotoxicity in *Caenorhabditis elegans. Hum Mol Genet.* 2009;18:1825–1838.
182. Kraemer BC, Schellenberg GD. SUT-1 enables tau-induced neurotoxicity in C-elegans. *Hum Mol Genet.* 2007;16:1959–1971.

183. Kraemer BC, Zhang B, Leverenz JB, Thomas JH, Trojanowski JQ, Schellenberg GD. Neurodegeneration and defective neurotransmission in a *Caenorhabditis elegans* model of tauopathy. *Proc Natl Acad Sci U S A*. 2003;100:9980–9985.
184. Choudhary B, Mandelkow E, Mandelkow EM, Pir GJ. Glutamatergic nervous system degeneration in a *C. elegans* Tau(A152T) tauopathy model involves pathways of excitotoxicity and Ca(2+) dysregulation. *Neurobiol Dis*. 2018;117:189–202.
185. Ghavami S, Shojaei S, Yeganeh B, et al. Autophagy and apoptosis dysfunction in neurodegenerative disorders. *Prog Neurobiol*. 2014;112:24–49.
186. Dosanjh LE, Brown MK, Rao G, Link CD, Luo Y. Behavioral phenotyping of a transgenic *Caenorhabditis elegans* expressing neuronal amyloid-beta. *J Alzheimers Dis*. 2010;19:681–690.
187. Gaeta AL, Caldwell KA, Caldwell GA. Found in translation: the utility of *C. elegans* Alpha-Synuclein models of Parkinson's disease. *Brain Sci*. 2019;9(4):pii: E73. https://doi.org/10.3390/brainsci9040073.
188. Koopman M, Seinstra RI, Nollen EAA. *C. elegans* as a model for synucleinopathies and other neurodegenerative diseases: tools and techniques. *Methods Mol Biol*. 2019; 1948:93–112.
189. Hamamichi S, Rivas RN, Knight AL, Cao S, Caldwell KA, Caldwell GA. Hypothesis-based RNAi screening identifies neuroprotective genes in a Parkinson's disease model. *Proc Natl Acad Sci U S A*. 2008;105:728–733.
190. Tyson T, Senchuk M, Cooper JF, George S, Van Raamsdonk JM, Brundin P. Novel animal model defines genetic contributions for neuron-to-neuron transfer of alpha-synuclein. *Sci Rep*. 2017;7:7506.
191. Buttner S, Broeskamp F, Sommer C, et al. Spermidine protects against alpha-synuclein neurotoxicity. *Cell Cycle*. 2014;13:3903–3908.
192. Steiner JA, Quansah E, Brundin P. The concept of alpha-synuclein as a prion-like protein: ten years after. *Cell Tissue Res*. 2018;373:161–173.
193. Vargas JY, Grudina C, Zurzolo C. The prion-like spreading of alpha-synuclein: from in vitro to in vivo models of Parkinson's disease. *Ageing Res Rev*. 2019;50:89–101.
194. Tyson T, Senchuk M, Cooper JF, George S, Van Raamsdonk JM, Brundin P. Novel animal model defines genetic contributions for neuron-to-neuron transfer of α-synuclein. *Sci Rep*. 2017;7:1–10.
195. Schiavi A, Maglioni S, Palikaras K, et al. Iron-starvation-induced mitophagy mediates lifespan extension upon mitochondrial stress in *C. elegans*. *Curr Biol*. 2015;25:1810–1822.
196. Civelek M, Mehrkens JF, Carstens NM, Fitzenberger E, Wenzel U. Inhibition of mitophagy decreases survival of *Caenorhabditis elegans* by increasing protein aggregation. *Mol Cell Biochem*. 2019;452:123–131.
197. Martinez BA, Petersen DA, Gaeta AL, Stanley SP, Caldwell GA, Caldwell KA. Dysregulation of the mitochondrial unfolded protein response induces non-apoptotic dopaminergic neurodegeneration in *C. elegans* models of Parkinson's disease. *J Neurosci*. 2017;37:11085–11100.
198. Bentivoglio AR, Cortelli P, Valente EM, et al. Phenotypic characterisation of autosomal recessive PARK6-linked parkinsonism in three unrelated Italian families. *Mov Disord*. 2001;16:999–1006.
199. Hattori N, Kitada T, Matsumine H, et al. Molecular genetic analysis of a novel Parkin gene in Japanese families with autosomal recessive juvenile parkinsonism: evidence for variable homozygous deletions in the Parkin gene in affected individuals. *Ann Neurol*. 1998;44:935–941.
200. Hattori N, Matsumine H, Asakawa S, et al. Point mutations (Thr240Arg and Gln311Stop) [correction of Thr240Arg and Ala311Stop] in the Parkin gene. *Biochem Biophys Res Commun*. 1998;249:754–758.

201. Hristova VA, Beasley SA, Rylett RJ, Shaw GS. Identification of a novel Zn2+–binding domain in the autosomal recessive juvenile Parkinson-related E3 ligase parkin. *J Biol Chem*. 2009;284:14978–14986.
202. Kitada T, Asakawa S, Hattori N, et al. Mutations in the parkin gene cause autosomal recessive juvenile parkinsonism. *Nature*. 1998;392:605–608.
203. Leroy E, Anastasopoulos D, Konitsiotis S, Lavedan C, Polymeropoulos MH. Deletions in the Parkin gene and genetic heterogeneity in a Greek family with early onset Parkinson's disease. *Hum Genet*. 1998;103:424–427.
204. Lucking CB, Abbas N, Durr A, et al. Homozygous deletions in parkin gene in European and North African families with autosomal recessive juvenile parkinsonism. The European Consortium on Genetic Susceptibility in Parkinson's Disease and the French Parkinson's Disease Genetics Study Group. *Lancet (England)*. 1998;352:1355–1356.
205. Matsumine H, Saito M, Shimoda-Matsubayashi S, et al. Localization of a gene for an autosomal recessive form of juvenile Parkinsonism to chromosome 6q25.2-27. *Am J Hum Genet*. 1997;60:588–596.
206. Shimura H, Hattori N, Kubo S, et al. Familial Parkinson disease gene product, parkin, is a ubiquitin-protein ligase. *Nat Genet*. 2000;25:302–305.
207. Valente EM, Abou-Sleiman PM, Caputo V, et al. Hereditary early-onset Parkinson's disease caused by mutations in PINK1. *Science*. 2004;304:1158–1160.
208. Valente EM, Bentivoglio AR, Dixon PH, et al. Localization of a novel locus for autosomal recessive early-onset parkinsonism, PARK6, on human chromosome 1p35-p36. *Am J Hum Genet*. 2001;68:895–900.
209. Valente EM, Brancati F, Caputo V, et al. PARK6 is a common cause of familial parkinsonism. *Neurol Sci*. 2002;23(Suppl 2):S117–S118.
210. Valente EM, Brancati F, Ferraris A, et al. PARK6-linked parkinsonism occurs in several European families. *Ann Neurol*. 2002;51:14–18.
211. Matsuda N, Sato S, Shiba K, et al. PINK1 stabilized by mitochondrial depolarization recruits Parkin to damaged mitochondria and activates latent Parkin for mitophagy. *J Cell Biol*. 2010;189:211–221.
212. Narendra D, Tanaka A, Suen DF, Youle RJ. Parkin is recruited selectively to impaired mitochondria and promotes their autophagy. *J Cell Biol*. 2008;183: 795–803.
213. Narendra DP, Jin SM, Tanaka A, et al. PINK1 is selectively stabilized on impaired mitochondria to activate Parkin. *PLoS Biol*. 2010;8:e1000298.
214. Pickrell AM, Youle RJ. The roles of PINK1, Parkin and mitochondrial Fidelity in Parkinson's disease. *Neuron*. 2015;85:257–273.
215. Vives-Bauza C, Zhou C, Huang Y, et al. PINK1-dependent recruitment of Parkin to mitochondria in mitophagy. *Proc Natl Acad Sci U S A*. 2010;107:378–383.
216. Cooper JF, Van Raamsdonk JM. Modeling Parkinson's disease in *C. elegans*. *J Parkinsons Dis*. 2018;8:17–32.
217. Bornhorst J, Chakraborty S, Meyer S, et al. The effects of pdr1, djr1.1 and pink1 loss in manganese-induced toxicity and the role of alpha-synuclein in *C. elegans*. *Metallomics*. 2014;6:476–490.
218. Cooper JF, Machiela E, Dues DJ, Spielbauer KK, Senchuk MM, Van Raamsdonk JM. Activation of the mitochondrial unfolded protein response promotes longevity and dopamine neuron survival in Parkinson's disease models. *Sci Rep*. 2017;7:16441.
219. Luz AL, Rooney JP, Kubik LL, Gonzalez CP, Song DH, Meyer JN. Mitochondrial morphology and fundamental parameters of the mitochondrial respiratory chain are altered in *Caenorhabditis elegans* strains deficient in mitochondrial dynamics and homeostasis processes. *PLoS one*. 2015;10:e0130940.

220. Samann J, Hegermann J, von Gromoff E, Eimer S, Baumeister R, Schmidt E. Caenorhabditits elegans LRK-1 and PINK-1 act antagonistically in stress response and neurite outgrowth. *J Biol Chem*. 2009;284:16482–16491.
221. Li JQ, Tan L, Yu JT. The role of the LRRK2 gene in Parkinsonism. *Mol Neurodegener*. 2014;9:47.
222. Saha S, Guillily MD, Ferree A, et al. LRRK2 modulates vulnerability to mitochondrial dysfunction in *Caenorhabditis elegans*. *J Neurosci*. 2009;29:9210–9218.
223. Yao C, El Khoury R, Wang W, et al. LRRK2-mediated neurodegeneration and dysfunction of dopaminergic neurons in a *Caenorhabditis elegans* model of Parkinson's disease. *Neurobiol Dis*. 2010;40:73–81.
224. Di Domenico F, Sultana R, Ferree A, et al. Redox proteomics analyses of the influence of co-expression of wild-type or mutated LRRK2 and Tau on *C. elegans* protein expression and oxidative modification: relevance to Parkinson disease. *Antioxid Redox Signal*. 2012;17:1490–1506.
225. Ferree A, Guillily M, Li H, et al. Regulation of physiologic actions of LRRK2: focus on autophagy. *Neurodegener Dis*. 2012;10:238–241.
226. Fan HC, Ho LI, Chi CS, et al. Polyglutamine (PolyQ) diseases: genetics to treatments. *Cell Transplant*. 2014;23:441–458.
227. Rudich P, Lamitina T. Models and mechanisms of repeat expansion disorders: a worm's eye view. *J Genet*. 2018;97:665–677.
228. Brignull HR, Moore FE, Tang SJ, Morimoto RI. Polyglutamine proteins at the pathogenic threshold display neuron-specific aggregation in a pan-neuronal *Caenorhabditis elegans* model. *J Neurosci*. 2006;26:7597–7606.
229. Morley JF, Brignull HR, Weyers JJ, Morimoto RI. The threshold for polyglutamine-expansion protein aggregation and cellular toxicity is dynamic and influenced by aging in *Caenorhabditis elegans*. *Proc Natl Acad Sci U S A*. 2002;99:10417–10422.
230. Jia K, Hart AC, Levine B. Autophagy genes protect against disease caused by polyglutamine expansion proteins in *Caenorhabditis elegans*. *Autophagy*. 2007;3:21–25.
231. Shin BH, Lim Y, Oh HJ, et al. Pharmacological activation of Sirt1 ameliorates polyglutamine-induced toxicity through the regulation of autophagy. *PLoS one*. 2013;8:e64953.
232. Faber PW, Alter JR, MacDonald ME, Hart AC. Polyglutamine-mediated dysfunction and apoptotic death of a *Caenorhabditis elegans* sensory neuron. *Proc Natl Acad Sci U S A*. 1999;96:179–184.
233. Lee AL, Ung HM, Sands LP, Kikis EA. A new *Caenorhabditis elegans* model of human huntingtin 513 aggregation and toxicity in body wall muscles. *PLoS One*. 2017;12: e0173644.
234. Parker JA, Connolly JB, Wellington C, Hayden M, Dausset J, Neri C. Expanded polyglutamines in *Caenorhabditis elegans* cause axonal abnormalities and severe dysfunction of PLM mechanosensory neurons without cell death. *Proc Natl Acad Sci U S A*. 2001;98:13318–13323.
235. Jeong H, Then F, Melia Jr TJ, et al. Acetylation targets mutant huntingtin to autophagosomes for degradation. *Cell*. 2009;137:60–72.
236. Khan LA, Yamanaka T, Nukina N. Genetic impairment of autophagy intensifies expanded polyglutamine toxicity in *Caenorhabditis elegans*. *Biochem Biophys Res Commun*. 2008;368:729–735.
237. Lejeune F-X, Mesrob L, Parmentier F, et al. Large-scale functional RNAi screen in *C. elegans* identifies genes that regulate the dysfunction of mutant polyglutamine neurons. *BMC Genomics*. 2012;13:1–14.
238. Teixeira-Castro A, Ailion M, Jalles A, et al. Neuron-specific proteotoxicity of mutant ataxin-3 in *C. elegans*: rescue by the DAF-16 and HSF-1 pathways. *Hum Mol Genet*. 2011;20:2996–3009.

239. Ochaba J, Lukacsovich T, Csikos G, et al. Potential function for the Huntingtin protein as a scaffold for selective autophagy. *Proc Natl Acad Sci U S A*. 2014;111: 16889–16894.
240. Rui YN, Xu Z, Patel B, et al. Huntingtin functions as a scaffold for selective macroautophagy. *Nat Cell Biol*. 2015;17:262–275.
241. Brown RH, Al-Chalabi A. Amyotrophic lateral sclerosis. *N Engl J Med*. 2017;377: 162–172.
242. Kenna KP, Landers JE, Ticozzi N, O'Connor D. Chapter 3—genetics of amyotrophic lateral sclerosis A2—Boulis, Nicholas. In: Donsante A, ed. *Molecular and Cellular Therapies for Motor Neuron Diseases*. Academic Press; 2017:43–59.
243. Kim HJ, Taylor JP. Lost in transportation: nucleocytoplasmic transport defects in ALS and other neurodegenerative diseases. *Neuron*. 2017;96:285–297.
244. Patten SA, Parker JA, Wen XY, Drapeau P. Simple animal models for amyotrophic lateral sclerosis drug discovery. *Expert Opin Drug Discov*. 2016;11:797–804.
245. Therrien M, Parker JA. Worming forward: amyotrophic lateral sclerosis toxicity mechanisms and genetic interactions in *Caenorhabditis elegans*. *Front Genet*. 2014;5:85.
246. Wang J, Farr GW, Hall DH, et al. An ALS-linked mutant SOD1 produces a locomotor defect associated with aggregation and synaptic dysfunction when expressed in neurons of *Caenorhabditis elegans*. *PLoS Genet*. 2009;5:e1000350.
247. Li X, Xu HL, Liu YX, An N, Zhao S, Bao JK. Autophagy modulation as a target for anticancer drug discovery. *Acta Pharmacol Sin*. 2013;34:612–624.
248. Baskoylu SN, Yersak J, O'Hern P, et al. Single copy/knock-in models of ALS SOD1 in *C. elegans* suggest loss and gain of function have different contributions to cholinergic and glutamatergic neurodegeneration. *PLoS Genet*. 2018;14:e1007682.
249. Ash PE, Zhang YJ, Roberts CM, et al. Neurotoxic effects of TDP-43 overexpression in *C. elegans*. *Hum Mol Genet*. 2010;19:3206–3218.
250. Liachko NF, Guthrie CR, Kraemer BC. Phosphorylation promotes neurotoxicity in a *C. elegans* model of TDP-43 proteinopathy. *J Neurosci*. 2010;30:16208–16219.
251. Veriepe J, Fossouo L, Parker JA. Neurodegeneration in *C. elegans* models of ALS requires TIR-1/Sarm1 immune pathway activation in neurons. *Nat Commun*. 2015;6:7319.
252. Barmada SJ, Serio A, Arjun A, et al. Autophagy induction enhances TDP43 turnover and survival in neuronal ALS models. *Nat Chem Biol*. 2014;10:677–685.
253. Xia Q, Wang H, Hao Z, et al. TDP-43 loss of function increases TFEB activity and blocks autophagosome-lysosome fusion. *EMBO J*. 2016;35:121–142.
254. Hutten S, Dormann D. Nucleocytoplasmic transport defects in neurodegeneration—cause or consequence? *Semin Cell Dev Biol*. 2019; pii: S1084-9521(18)30190-3. https://doi.org/10.1016/j.semcdb.2019.05.020.
255. Lee KH, Zhang P, Kim HJ, et al. C9orf72 dipeptide repeats impair the assembly, dynamics, and function of membrane-less organelles. *Cell*. 2016;167:774–788.e717.
256. Lin Y, Mori E, Kato M, et al. Toxic PR poly-dipeptides encoded by the C9orf72 repeat expansion target LC domain polymers. *Cell*. 2016;167:789–802.e712.
257. Shi KY, Mori E, Nizami ZF, et al. Toxic PRn poly-dipeptides encoded by the C9orf72 repeat expansion block nuclear import and export. *Proc Natl Acad Sci U S A*. 2017;114:E1111–e1117.
258. Paonessa F, Evans LD, Solanki R, et al. Microtubules deform the nuclear membrane and disrupt nucleocytoplasmic transport in Tau-mediated frontotemporal dementia. *Cell Rep*. 2019;26:582–593.e585.
259. Cohen-Fix O, Askjaer P. Cell biology of the *Caenorhabditis elegans* nucleus. *Genetics*. 2017;205:25–59.
260. Wang P, Hanover JA. Nutrient-driven O-GlcNAc cycling influences autophagic flux and neurodegenerative proteotoxicity. *Autophagy*. 2013;9:604–606.

261. Wang P, Lazarus BD, Forsythe ME, Love DC, Krause MW, Hanover JA. O-GlcNAc cycling mutants modulate proteotoxicity in *Caenorhabditis elegans* models of human neurodegenerative diseases. *Proc Natl Acad Sci U S A*. 2012;109:17669–17674.
262. Takacs-Vellai K, Bayci A, Vellai T. Autophagy in neuronal cell loss: a road to death. *Bioessays*. 2006;28:1126–1131.
263. Toth ML, Simon P, Kovacs AL, Vellai T. Influence of autophagy genes on ion-channel-dependent neuronal degeneration in Caenorhabditis elegans. *J Cell Sci*. 2007;120:1134–1141.
264. Vellai T, Toth ML, Kovacs AL. Janus-faced autophagy: a dual role of cellular self-eating in neurodegeneration? *Autophagy*. 2007;3:461–463.
265. Zuchero JB, Barres BA. Glia in mammalian development and disease. *Development*. 2015;142:3805–3809.
266. Singhvi A, Shaham S. Glia-neuron interactions in *Caenorhabditis elegans*. *Annu Rev Neurosci*. 2019;42:149–168.
267. Kaushik S, Singh R, Cuervo AM. Autophagic pathways and metabolic stress. *Diabetes Obes Metab*. 2010;12(Suppl 2):4–14.
268. Singh R, Kaushik S, Wang Y, et al. Autophagy regulates lipid metabolism. *Nature*. 2009;458:1131–1135.
269. Harvald EB, Sprenger RR, Dall KB, et al. Multi-omics analyses of starvation responses reveal a central role for lipoprotein metabolism in acute starvation survival in *C. elegans*. *Cell Syst*. 2017;5:38–52.e34.
270. Seah NE, de Magalhaes Filho CD, Petrashen AP, et al. Autophagy-mediated longevity is modulated by lipoprotein biogenesis. *Autophagy*. 2016;12:261–272.
271. Folick A, Oakley HD, Yu Y, et al. Aging. Lysosomal signaling molecules regulate longevity in *Caenorhabditis elegans*. *Science*. 2015;347:83–86.
272. Ramachandran PV, Savini M, Folick AK, et al. Lysosomal signaling promotes longevity by adjusting mitochondrial activity. *Dev Cell*. 2019;48:685–696.e685.
273. Nguyen TB, Louie SM, Daniele JR, et al. DGAT1-dependent lipid droplet biogenesis protects mitochondrial function during starvation-induced autophagy. *Dev Cell*. 2017;42:9–21.e25.
274. Hibshman JD, Leuthner TC, Shoben C, et al. Nonselective autophagy reduces mitochondrial content during starvation in *Caenorhabditis elegans*. *Am J Physiol Cell Physiol*. 2018;315:C781–C792.
275. Gelino S, Chang JT, Kumsta C, et al. Intestinal autophagy improves healthspan and longevity in *C. elegans* during dietary restriction. *PLoS Genet*. 2016;12:e1006135.
276. Bouyanfif A, Jayarathne S, Koboziev I, Moustaid-Moussa N. The nematode *Caenorhabditis elegans* as a model organism to study metabolic effects of omega-3 polyunsaturated fatty acids in obesity. *Adv Nutr*. 2019;10:165–178.
277. Moreno-Arriola E, Cardenas-Rodriguez N, Coballase-Urrutia E, Pedraza-Chaverri J, Carmona-Aparicio L, Ortega-Cuellar D. *Caenorhabditis elegans*: a useful model for studying metabolic disorders in which oxidative stress is a contributing factor. *Oxid Med Cell Longev*. 2014;2014:705253.
278. Yen CA, Curran SP. Gene-diet interactions and aging in *C. elegans*. *Exp Gerontol*. 2016;86:106–112.
279. Samokhvalov V, Scott BA, Crowder CM. Autophagy protects against hypoxic injury in *C. elegans*. *Autophagy*. 2008;4:1034–1041.
280. Edifizi D, Nolte H, Babu V, et al. Multilayered reprogramming in response to persistent DNA damage in *C. elegans*. *Cell Rep*. 2017;20:2026–2043.
281. Kumsta C, Chang JT, Schmalz J, Hansen M. Hormetic heat stress and HSF-1 induce autophagy to improve survival and proteostasis in *C. elegans*. *Nat Commun*. 2017;8:14337.

282. Germic N, Frangez Z, Yousefi S, Simon HU. Regulation of the innate immune system by autophagy: monocytes, macrophages, dendritic cells and antigen presentation. *Cell Death Differ.* 2019;26:715–727.
283. Kuo CJ, Hansen M, Troemel E. Autophagy and innate immunity: insights from invertebrate model organisms. *Autophagy.* 2018;14:233–242.
284. Powell JR, Ausubel FM. Models of *Caenorhabditis elegans* infection by bacterial and fungal pathogens. *Methods Mol Biol.* 2008;415:403–427.
285. Podshivalova K, Kerr RA, Kenyon C. How a mutation that slows aging can also disproportionately extend end-of-life decrepitude. *Cell Rep.* 2017;19:441–450.
286. Sanchez-Blanco A, Kim SK. Variable pathogenicity determines individual lifespan in *Caenorhabditis elegans. PLoS Genet.* 2011;7:e1002047.
287. Fares H, Greenwald I. Genetic analysis of endocytosis in *Caenorhabditis elegans*: Coelomocyte uptake defective mutants. *Genetics.* 2001;159:133–145.
288. Troemel ER, Felix MA, Whiteman NK, Barriere A, Ausubel FM. Microsporidia are natural intracellular parasites of the nematode *Caenorhabditis elegans. PLoS Biol.* 2008;6:2736–2752.
289. Zou CG, Ma YC, Dai LL, Zhang KQ. Autophagy protects *C. elegans* against necrosis during Pseudomonas aeruginosa infection. *Proc Natl Acad Sci U S A.* 2014;111:12480–12485.
290. Jia K, Thomas C, Akbar M, et al. Autophagy genes protect against Salmonella typhimurium infection and mediate insulin signaling-regulated pathogen resistance. *Proc Natl Acad Sci U S A.* 2009;106:14564–14569.
291. Aballay A, Ausubel FM. Programmed cell death mediated by ced-3 and ced-4 protects *Caenorhabditis elegans* from Salmonella typhimurium-mediated killing. *Proc Natl Acad Sci U S A.* 2001;98:2735–2739.
292. Lezi E, Zhou T, Koh S, et al. An antimicrobial peptide and its neuronal receptor regulate dendrite degeneration in aging and infection. *Neuron.* 2018;97:125–138.
293. Kumar S, Egan BM, Kocsisova Z, et al. Lifespan extension in *C. elegans* caused by bacterial colonization of the intestine and subsequent activation of an innate immune response. *Dev Cell.* 2019;49:100–117.e106.
294. Anderson SM, Cheesman HK, Peterson ND, Salisbury JE, Soukas AA, Pukkila-Worley R. The fatty acid oleate is required for innate immune activation and pathogen defense in *Caenorhabditis elegans. PLoS Pathog.* 2019;15:17.
295. El-Houjeiri L, Possik E, Vijayaraghavan T, et al. The transcription factors TFEB and TFE3 link the FLCN-AMPK signaling axis to innate immune response and pathogen resistance. *Cell Rep.* 2019;26:3613–3628.
296. Chen HD, Kao CY, Liu BY, et al. HLH-30/TFEB-mediated autophagy functions in a cell-autonomous manner for epithelium intrinsic cellular defense against bacterial pore-forming toxin in *C. elegans. Autophagy.* 2017;13:371–385.
297. Mei S, Ni HM, Manley S, et al. Differential roles of unsaturated and saturated fatty acids on autophagy and apoptosis in hepatocytes. *J Pharmacol Exp Ther.* 2011;339:487–498.
298. Chu KY, O'Reilly L, Mellet N, Meikle PJ, Bartley C, Biden TJ. Oleate disrupts cAMP signaling, contributing to potent stimulation of pancreatic beta-cell autophagy. *J Biol Chem.* 2019;294:1218–1229.
299. Niso-Santano M, Malik SA, Pietrocola F, et al. Unsaturated fatty acids induce non-canonical autophagy. *EMBO J.* 2015;34:1025–1041.
300. Peterson ND, Pukkila-Worley R. *Caenorhabditis elegans* in high-throughput screens for anti-infective compounds. *Curr Opin Immunol.* 2018;54:59–65.
301. Hanahan D, Weinberg RA. The hallmarks of cancer. *Cell.* 2000;100:57–70.
302. Hanahan D, Weinberg RA. Hallmarks of cancer: the next generation. *Cell.* 2011;144:646–674.

303. Hayes J, Peruzzi PP, Lawler S. MicroRNAs in cancer: biomarkers, functions and therapy. *Trends Mol Med.* 2014;20:460–469.
304. Lee YS, Dutta A. MicroRNAs in cancer. *Annu Rev Pathol.* 2009;4:199–227.
305. Peng Y, Croce CM. The role of MicroRNAs in human cancer. *Signal Transduct Target Ther.* 2016;1:15004.
306. Sun K, Deng W, Zhang S, et al. Paradoxical roles of autophagy in different stages of tumorigenesis: protector for normal or cancer cells. *Cell Biosci.* 2013;3:35.
307. Yun CW, Lee SH. The roles of autophagy in cancer. *Int J Mol Sci.* 2018;19:3446.
308. Dower CM, Wills CA, Frisch SM, Wang HG. Mechanisms and context underlying the role of autophagy in cancer metastasis. *Autophagy.* 2018;14:1110–1128.
309. Gartner A, Boag PR, Blackwell KT. Germline survival and apoptosis. In: Strome JKAS, ed. *WormBook.* The *C. elegans* Research Community; 2008.
310. Kyriakakis E, Markaki M, Tavernarakis N. *Caenorhabditis elegans* as a model for cancer research. *Mol Cell Oncol.* 2015;2:e975027.
311. Gozuacik D, Akkoc Y, Ozturk DG, Kocak M. Autophagy-regulating microRNAs and cancer. *Front Oncol.* 2017;7:65.
312. Jing Z, Han W, Sui X, Xie J, Pan H. Interaction of autophagy with microRNAs and their potential therapeutic implications in human cancers. *Cancer Lett.* 2015;356:332–338.
313. Zhang P, Zhang H. Autophagy modulates miRNA-mediated gene silencing and selectively degrades AIN-1/GW182 in *C. elegans. EMBO Rep.* 2013;14:568–576.
314. Chandra S, Vimal D, Sharma D, Rai V, Gupta SC, Chowdhuri DK. Role of miRNAs in development and disease: lessons learnt from small organisms. *Life Sci.* 2017; 185:8–14.
315. Pinkston-Gosse J, Kenyon C. DAF-16/FOXO targets genes that regulate tumor growth in *Caenorhabditis elegans. Nat Genet.* 2007;39:1403–1409.
316. Possik E, Jalali Z, Nouet Y, et al. Folliculin regulates ampk-dependent autophagy and metabolic stress survival. *PLoS Genet.* 2014;10:e1004273.
317. Min H, Lee YU, Shim YH, Kawasaki I. Autophagy of germ-granule components, PGL-1 and PGL-3, contributes to DNA damage-induced germ cell apoptosis in *C. elegans. PLoS Genet.* 2019;15:e1008150.
318. Gomes LC, Odedra D, Dikic I, Pohl C. Autophagy and modular restructuring of metabolism control germline tumor differentiation and proliferation in *C. elegans. Autophagy.* 2016;12:529–546.
319. Grussendorf KA, Trezza CJ, Salem AT, et al. Facilitation of endosomal recycling by an IRG protein homolog maintains apical tubule structure in *Caenorhabditis elegans. Genetics.* 2016;203:1789–1806.
320. Jiminez JA, Uwiera TC, Douglas Inglis G, Uwiera RR. Animal models to study acute and chronic intestinal inflammation in mammals. *Gut Pathog.* 2015;7:29.
321. Maglioni S, Ventura N. *C. elegans* as a model organism for human mitochondrial associated disorders. *Mitochondrion.* 2016;30:117–125.
322. Ventura N, Rea SL, Handerson ST, Condo I, Testi R, Johnson TE. *C. elegans* as a model for Friedreich ataxia. *Faseb J.* 2006;20:1029–1030.
323. de Voer G, Peters D, Taschner PE. *Caenorhabditis elegans* as a model for lysosomal storage disorders. *Biochim Biophys Acta.* 2008;1782:433–446.
324. Lieberman AP, Puertollano R, Raben N, Slaugenhaupt S, Walkley SU, Ballabio A. Autophagy in lysosomal storage disorders. *Autophagy.* 2012;8:719–730.
325. Seranova E, Connolly KJ, Zatyka M, et al. Dysregulation of autophagy as a common mechanism in lysosomal storage diseases. *Essays Biochem.* 2017;61:733–749.
326. Pietrocola F, Castoldi F, Markaki M, et al. Aspirin recapitulates features of caloric restriction. *Cell Rep.* 2018;22:2395–2407.

327. Shamalnasab M, Gravel SP, St-Pierre J, Breton L, Jager S, Aguilaniu H. A salicylic acid derivative extends the lifespan of *Caenorhabditis elegans* by activating autophagy and the mitochondrial unfolded protein response. *Aging Cell*. 2018;17:e12830.
328. Li L, Friedrichsen HJ, Andrews S, et al. A TFEB nuclear export signal integrates amino acid supply and glucose availability. *Nat Commun*. 2018;9:2685.
329. O'Reilly LP, Long OS, Cobanoglu MC, et al. A genome-wide RNAi screen identifies potential drug targets in a *C. elegans* model of alpha1-antitrypsin deficiency. *Hum Mol Genet*. 2014;23:5123–5132.
330. Wang C, Niederstrasser H, Douglas PM, et al. Small-molecule TFEB pathway agonists that ameliorate metabolic syndrome in mice and extend *C. elegans* lifespan. *Nat Commun*. 2017;8:2270.
331. Schmeisser K, Parker AJ. Pleiotropic effects of mTOR and autophagy during development and aging. *Front Cell Dev Biol*. 2019;7:192.
332. Ma D, Li S, Molusky MM, Lin JD. Circadian autophagy rhythm: a link between clock and metabolism? *Trends Endocrinol Metab*. 2012;23:319–325.
333. Maiese K. Moving to the rhythm with clock (circadian) genes, autophagy, mTOR, and SIRT1 in degenerative disease and cancer. *Curr Neurovasc Res*. 2017;14:299–304.
334. Ryzhikov M, Ehlers A, Steinberg D, et al. Diurnal rhythms spatially and temporally organize autophagy. *Cell Rep*. 2019;26:1880–1892.e1886.
335. Alirezaei M, Kiosses WB, Flynn CT, Brady NR, Fox HS. Disruption of neuronal autophagy by infected microglia results in neurodegeneration. *PLoS One*. 2008;3:e2906.
336. Fenouille N, Nascimbeni AC, Botti-Millet J, Dupont N, Morel E, Codogno P. To be or not to be cell autonomous? Autophagy says both. *Essays Biochem*. 2017;61:649–661.
337. Hansen M, Rubinsztein DC, Walker DW. Autophagy as a promoter of longevity: insights from model organisms. *Nat Rev Mol Cell Biol*. 2018;19:579–593.
338. Qadota H, Inoue M, Hikita T, et al. Establishment of a tissue-specific RNAi system in *C. elegans*. *Gene*. 2007;400:166–173.
339. Zhuang JJ, Hunter CP. Tissue specificity of *Caenorhabditis elegans* enhanced RNA interference mutants. *Genetics*. 2011;235–237.
340. Zou L, Wu D, Zang X, Wang Z, Wu Z, Chen D. Construction of a germline-specific RNAi tool in C elegans. *Sci Rep*. 2019;9:2354.

CHAPTER FIFTEEN

Drosophila as a model to understand autophagy deregulation in human disorders

Donna Denton[a],*, Louise O'Keefe[b,c], Sharad Kumar[a],*
[a]Centre for Cancer Biology, University of South Australia and SA Pathology, Adelaide, SA, Australia
[b]Department of Molecular and Biomedical Science, School of Biological Sciences, The University of Adelaide, Adelaide, SA, Australia
[c]Hopwood Centre for Neurobiology, South Australian Health and Medical Research Institute, Adelaide, SA, Australia
*Corresponding authors: e-mail address: donna.denton@unisa.edu.au; sharad.kumar@unisa.edu.au

Contents

Progress in Molecular Biology and Translational Science, Volume 172
ISSN 1877-1173
https://doi.org/10.1016/bs.pmbts.2020.01.005

Abstract

Autophagy has important functions in normal physiology to maintain homeostasis and protect against cellular stresses by the removal of harmful cargos such as dysfunctional organelles, protein aggregates and invading pathogens. The deregulation of autophagy is a hallmark of many diseases and therapeutic targeting of autophagy is highly topical. With the complex role of autophagy in disease it is essential to understand the genetic and molecular basis of the contribution of autophagy to pathogenesis. The model organism, *Drosophila*, provides a genetically amenable system to dissect out the contribution of autophagy to human disease models. Here we review the roles of autophagy in human disease and how autophagy studies in *Drosophila* have contributed to the understanding of pathophysiology.

1. Introduction

The pathology of human diseases is remarkably complex and in many cases even though genetic mutations have been identified, there are limited effective treatment options available. In the last decade, there has been substantial advancement in understanding the molecular mechanisms of autophagy, and this has uncovered the contribution of autophagy to the pathophysiology of many diseases.[1–3] Understanding the contribution of autophagy not only to disease but also for the development of treatments to these diseases will require the use of animal models such as *Drosophila*. In this review we will describe the contribution of autophagy to several pathological conditions. We discuss *Drosophila* models of human genetic diseases including, cancer, neurodegenerative diseases and other complex diseases; where these models will aid in understanding the complex roles of autophagy in pathogenesis and the challenges involved in effective therapeutic targeting of autophagy (Fig. 1).

1.1 Autophagy machinery

The evolutionarily conserved catabolic process of autophagy plays an essential role in cellular, tissue, and organismal homeostasis. Macroautophagy (here after referred to as autophagy) mediates the lysosomal degradation and recycling of cytoplasmic material that is engulfed by a double membrane vesicle, the autophagosome, for eventual delivery to the lysosome. Other distinct types of autophagy also exist, where substrates are directly translocated into the lysosome without vesicle transport including microautophagy and chaperone-mediated autophagy (CMA). Microautophagy

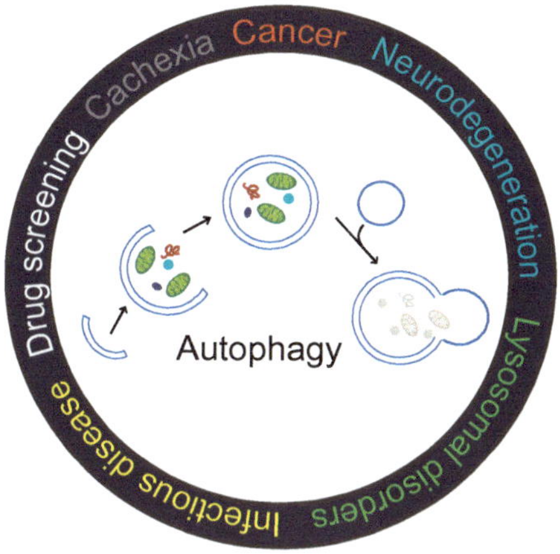

Fig. 1 Studies of autophagy in human disease. Examples of human disorders where the contribution of autophagy can be studied using *Drosophila* models.

involves the transfer of cytosolic cargo components directly into the lysosome through internalization in small vesicles at the surface of the lysosome.[4] CMA involves the direct delivery of selective cytosolic proteins targeted for degradation to the lysosome.[5] The substrates of this type of autophagy are soluble proteins containing a KFERQ-like motif bound to HSPA8 which are translocated into the lysosomal lumen.[5] For this review, we will be focusing on macroautophagy.

Autophagy is a tightly coordinated multi-step process regulated by distinct *Autophagy-related (Atg)* gene products that assemble into specific complexes. Genetic studies in yeast identified a number of *ATG* genes that are evolutionarily conserved in higher organisms including *Drosophila* and humans.[3] Distinct Atg proteins function in each step of the process that can be divided into: (i) induction, requiring the Atg1/ULK1 kinase complex; (ii) autophagosome nucleation, involving the class III phosphatidylinositol 3-kinase (PI3K) complex (including Atg6/Beclin1); (iii) autophagosome expansion and completion, which requires the Atg8 family and conjugation systems; and (iv) lysosomal fusion and degradation/recycling.[4]

Autophagy has essential functions in development, metabolism, cell growth, cell survival, infection and immunity.[2] Under normal physiological conditions basal levels of autophagy function to maintain cellular homeostasis and protect cells by eliminating toxic protein aggregates, unwanted cellular contents, as well as removing old and dysfunctional or damaged organelles. High levels of autophagy are induced in response to numerous types of extracellular and intracellular stresses, such as nutrient starvation, intracellular metabolic stress, genotoxic stress, hypoxia, organelle dysfunction, protein aggregation and pathogen infection. Normally, these functions of autophagy act to protect cells, however, in specific contexts autophagy activation can lead to cell death.

1.2 Autophagy in disease

Aberrant regulation of autophagy, including by mutation of autophagy genes, has significant adverse consequences on normal cellular functions and contributes to numerous human pathologies, including cancer, neurodegenerative diseases and inflammatory disorders.[1,2] For example, *Beclin 1* mutations are associated with breast and ovarian cancer[6]; *SQSTM1* mutation is linked to amyotrophic lateral sclerosis[7]; defective mitophagy due to the *PINK1* mutation is associated with Parkinson's disease[8]; and Crohn's disease is related to a *ATG16L1* mutation.[9] Under pathological conditions excessive autophagy can lead to death of neurons or cardiomyocytes, highlighting the need for the identification of cell type specific modulators of autophagy.[10,11] Moreover, there is substantial evidence for a contribution of dysregulated autophagy to Alzheimer's disease, Parkinson's disease and other neurological disorders yet whether this is protective or pathogenic remains to be established. Autophagy impacts many aspects of cancer; it can both suppress and promote tumor progression and metastasis at several stages. In addition, cancer therapies can induce autophagy, which can contribute to the development of resistance. The interest in manipulating autophagy for cancer treatment and other therapies clearly illustrates the importance of understanding the fundamental mechanisms that control this process in order to determine effective therapeutic strategies and for improving individual treatments of diseases.[12]

1.3 *Drosophila* as a model

Research using *Drosophila* has made seminal contributions to the discovery and characterization of several biological pathways, thus contributing to the understanding of many aspects of human diseases. The importance of findings from *Drosophila* models for medical science has been recognized by the award of several Nobel Prizes in Physiology or Medicine. *Drosophila* has been used as a genetically tractable model organism for research in diverse fields such as tumorigenesis and neuroscience. They provide a versatile model with well-established resources, experimental techniques, approaches and systems.

With a short life cycle and the capacity to generate large numbers of progeny, *Drosophila* is a time- and cost-effective model system to study complex biological pathways. Sophisticated genetic manipulation strategies are highly effective with the required research tools readily available to generate mutant or transgenic *Drosophila* lines.[13] Spatiotemporal specific targeted gene expression can be achieved using the UAS/Gal4 system[14,15] together with the temperature sensitive repressor of Gal4, Gal80^{ts},[16] or the *Gene-Switch* inducible expression system. In addition there are complementary binary expression systems such as LexA and Q system.[17,18] Another powerful tool available in *Drosophila* is mosaic clonal analysis that utilizes FLP/FRT mediated recombination in somatic cells thus allowing for the comparison of genetically modified cells (mutant, knockdown or overexpression) to the surrounding wild type cells in the same tissue under equivalent conditions. Mosaic clonal analysis enables functional studies of mutations that would cause lethality if present in the whole animal as well as allowing for the assessment of cell-autonomous or non cell-autonomous gene functions. Further genetic techniques, such as phiC31 integrase-mediated targeted insertion and CRISPR/Cas9 mutagenesis together provide a wide repertoire to study loss- or gain-of-function of multiple genes in a specific tissue or cell type.[19] The ability to undertake genetic modifier screens in *Drosophila* has been invaluable to determine the impact of a mutation on phenotypes quickly and enables the identification of novel regulators associated with different diseases. In addition to their genetic amenability, *Drosophila* also provide the ability for high resolution *in vivo* cell biological analyses.

More than 75% of known human disease-associated genes are represented in the *Drosophila* genome and there is lower genetic redundancy compared to humans.[20,21] Importantly, core molecular pathways are highly conserved and with many organ systems in mammals having equivalent systems in *Drosophila*, this enables the findings from research in *Drosophila* to be translated to humans. In fact studies in *Drosophila* have already made significant contributions to the recently established rare-diseases network, as clinicians using genomic analyses of undiagnosed human diseases collaborate with biomedical researchers who have assay systems (such as *Drosophila*) for functional verification of candidate disease-causing genes.[22] In addition, *Drosophila* has also been successfully used as a platform for personalized medicine to screen for effective drug therapy in colorectal cancer.[23] The regulation of autophagy and the *Atg* genes are highly conserved in *Drosophila*, and the role of autophagy has been characterized during different stages of *Drosophila* development as well as following stress.[24,25] This high level of conservation together with the availability of powerful genetic tools makes *Drosophila* an ideal model to study the contribution of autophagy in models relevant to human diseases.

2. Autophagy in tumorigenesis

The contribution of autophagy to tumorigenesis is remarkably complex, and the exact role of autophagy in cancer cells is strongly context dependent. During cancer progression, autophagy is known to play a critical role, however, this can either be protective, by acting to suppress tumor growth, or it can be tumor-promoting required for cancer cell survival. High levels of autophagic activity contribute to suppression of cancer initiation yet in primary tumors, autophagy is often upregulated to overcome stresses resulting from increased growth as well as low nutrient availability.[1] While the predominant function of autophagy is to promote cellular survival, autophagy can also promote the destruction of cells, during both normal development and in response to stress.[26] The processes for removal of superfluous cells of the interdigital webs during murine development[27] and removal of obsolete tissues during *Drosophila* development both require autophagy.[28] In response to replicative crisis, telomeric DNA damage leads to activation of the cGAS–STING (cyclic GMP-AMP synthase–stimulator of interferon genes) pathway and subsequent induction of autophagy responsible for the removal of potentially harmful cells to prevent oncogenic transformation.[29] In the clinical setting, there are numerous trials examining the modulation of autophagy in combination with standard chemotherapeutics.[30] Given that autophagy can mediate cell death in specific contexts and with the paradoxical roles of autophagy in tumorigenesis it is essential to fully understand the regulation and function of autophagy using *in vivo* models for effective disease treatment.

2.1 Modeling cancer in *Drosophila*

Many of the fundamental processes that are deregulated during tumorigenesis are conserved in *Drosophila* models of cancer, including increased cell proliferation, evasion of cell death, increased invasion, altered metabolism and dysregulation of autophagy.[31] The ability to model cancers in *Drosophila* has also led to the use of these models in drug screening for cancer therapeutics. The findings from research using *Drosophila* models of tumorigenesis have made significant contributions to our understanding of fundamental processes that promote initiation and progression of cancer.

The development of the genetic mosaic technique[32] to establish the mosaic analysis with a repressible cell marker (MARCM) system,[33] allows for the coexistence of cells with different genotypes in a single organism. This powerful model generates populations of mutant clonal cells that are surrounded by wild type cells within the same tissue and resembles epithelial tumorigenesis (as cancer arises from mutations that occur in single cells surrounded by wild type cells). In addition, as mutations in tumor suppressors and oncogenes are often lethal or generate severe morphological defects this system permits their analysis at later developmental stages. The mosaic clonal approach led to the identification of tumor suppressor pathways (including cell polarity pathway genes and the Hippo pathway) as well as mechanisms promoting tumorigenesis (including regulation of cell polarity, cell-cell cooperation, and cell competition). This sophisticated genetic approach pioneered the use of *Drosophila* models that recapitulate fundamental aspects of human cancers.

Drosophila models of epithelial tumors are well-established, and tumor-specific models for lung, colon, thyroid, and brain cancers as well as leukemias have been generated by tissue specific genetic manipulations. These models have provided important new insights into molecular pathways, including autophagy, that promote initiation and progression of cancer. In addition, novel therapeutic strategies have been pioneered using *Drosophila* cancer models. The use of whole animal models has enabled the analysis of not only the tumor but its microenvironment and tumor-host interactions. Here we will focus on epithelial tumors, particularly the eye imaginal disc and adult gut, cancer stem cells in the adult intestine and the cachexia model of tumor-host interaction.

2.2 Epithelial tumors: Eye disc hyperplasia and nonautonomous contributions to tumor growth

The *Drosophila* larval imaginal discs are developing epithelial tissues that, during metamorphosis, differentiate into adult structures. The larval epithelial tissue of the eye-antennal disc and wing disc have been used to model cooperative interactions between cells, and between mutations in oncogenes or tumor suppressor genes. For an extensive review on *Drosophila* models of cooperative tumorigenesis see Ref. 34.

Mutations that result in the activation of Ras signaling pathway genes are found in approximately 30% of human cancers.[35] Oncogenic Ras is a potent inducer of tissue growth, yet high activity can lead to senescence or differentiation, thereby limiting tumorigenesis with additional mutations acquired for Ras-driven transformation.[36] In *Drosophila*, activated oncogenes (such as *Ras* or *Notch*) or loss of function of the cell polarity regulators (*scribble (scrib), discs large (dlg), and lethal giant larvae (lgl)*) promote hyperplasia in epithelial tissues such as larval eye imaginal discs.[37,38] *scrib, dlg* and *lgl* encode key components of the basolateral polarity complex and act together in cell polarity functions during morphogenesis.[39]

The pioneering and now well-established model of *Drosophila* cooperative tumorigenesis makes use of clonal expression of oncogenic *Ras* (Ras^{V12}) or *Notch* (N^{ACT}) combined with loss of cell polarity mutants (*scrib* or *dlg* mutants) in larval eye-antennal imaginal discs.[31,38] The Ras^{V12} or N^{ACT} MARCM epithelial cell clones that are mutant for *scrib* have reduced cell death and increased proliferation producing overgrown, undifferentiated and invasive tumors.[38,40–43] This is a well-established *Drosophila* model of cooperative tumorigenesis, where second site mutations causing polarity impairment (*scrib* mutation) cooperate with activated oncogenes (Ras signaling driven growth) to cause cellular transformation. Since then, additional cooperative tumor models have revealed interactions between Ras and other pathways including actin cytoskeletal and JNK signaling, as well as other novel cooperative interactions, causing hyperplastic or neoplastic overgrowth.[34]

It has been proposed that the tumor-suppressive role of autophagy in cancer occurs during the early stages of tumor development, while later stages require autophagy to sustain their growth.[1,44,45] An advantage of the *Drosophila* system is the ability to model early and later stages of tumorigenesis, where the contribution of autophagy to various stages can be examined. In particular, the expression of oncogenic Ras^{V12} alone and the cooperative model of Ras^{V12} combined with loss of cell polarity mutants have been used to examine the contribution of autophagy to tumor progression. A key feature of these tumor models is that they are metabolically stressed with increased levels of reactive oxygen species (ROS) and inflammatory cytokines.

In a screen to identify tumor suppressors that cooperate with oncogenic Ras^{V12}, autophagy genes were identified.[46] Disruption to autophagy at any stage in Ras^{V12} expressing eye imaginal epithelium results in overgrowth.[46] Similarly, in eye epithelial tissue, ablation of genes required for autophagosome-lysosome function (*Vps33, Vps16, Vps18/dor*) cooperate with Ras^{V12} to cause neoplastic overgrowth.[47]

In *Ras*^{V12} eye epithelial clones (MARCM) inhibition of autophagy also increased clone size, however, increased proliferation was not only detected in the tumor but also in the surrounding wild type tissue that had increased cell death. Restricted expression of *Ras*^{V12} in wing disc epithelium increased autophagy in both the *Ras*^{V12} cells and surrounding cells, and despite inhibiting autophagy in the Ras expressing cells, there was still increased autophagy in the wild-type surrounding cells. This suggests that oncogenic Ras signaling induces both cell autonomous and non-cell autonomous autophagy in both the Ras expressing cells and the surrounding wild-type cells.[46]

Autophagy has a tumor-promoting effect in more aggressive models of tumorigenesis. In a *Ras*^{V12} *scrib* mutant clonal model of cooperative tumorigenesis, high levels of autophagy were induced in the wild-type epithelial cells surrounding the tumor cells.[48] Furthermore, autophagy was also induced in distant tissues including the muscle, fat body and midgut.[48] In a series of elegant allograft experiments, this model also revealed that tumor growth is dependent on autophagy in the tumor microenvironment. Blocking autophagy specifically in the *Ras scrib* mutant cells reduced tumor overgrowth and this was further reduced when both the tumor and surrounding cells were autophagy-deficient.[48] When the autophagy-deficient *Ras scrib* mutant tumors were transplanted into an autophagy-competent host, the cells resumed growth and formed large tumors, whereas transplantation into an autophagy-deficient host resulted in small tumors. In this model, autophagy acts as a tumor suppressor within the tumor, while non cell-autonomous induction of autophagy in the tumor microenvironment, and in distant peripheral tissues, promotes tumor growth.

The induction of autophagy in the tumor microenvironment (local and peripheral non-tumor tissues) has been proposed to "fuel" tumor growth by mobilizing nutrients from the non-tumor cells and tissues. In a *Ras*^{V12} *scrib* mutant clonal model, knockdown of the amino acid transporter *Slimfast* in the tumor reduced its growth suggesting that autophagy in the tumor microenvironment may utilize amino acids (and nutrients, glucose) to support growth of the tumor.[48] Under starvation conditions, autophagy induced by Desat1-dependent Myc activity may act non cell-autonomously to promote tumor growth.[49] Interestingly, in a model of cell competition, "loser" cells that are eliminated from epithelial tissue have reduced protein synthesis and induce autophagy that promotes cell death.[50] This suggests that autophagy-mediated cell competition may contribute to tumorigenesis. A greater understanding of the contribution of environmental factors, including diet, on tumor progression will be of future importance and *Drosophila* can provide *in vivo* models for this research.[51]

The tumor models discussed here have revealed distinct requirements for autophagy and highlight the importance of genetic background. Genotype differences between neighboring epithelial cells in a tumor affect the growth, and stage of the tumor. In a premalignant stage, such as Ras activation, inhibiting autophagy promotes tumor growth, whereas in the tumor inhibiting autophagy reduces its growth. Context specific differences in the consequence of autophagy on tissue growth and survival, both autonomously and non-autonomously occur in the tissue as well as systemically in the whole animal. Therefore, it will be important to understand the influence of autophagy on appropriate models for the desired therapeutic strategy. The fundamental discoveries in regulation of epithelial biology from *Drosophila* have made seminal contributions to the current understanding of tumor progression. With the sophisticated genetic techniques and advanced *in vivo* cell biological analysis, *Drosophila* will continue to play an important role in understanding the contribution of autophagy in tumorigenesis.

2.3 Organ specific cancer models: The gut system

The *Drosophila* adult gut is composed of three main regions, the foregut, the midgut (similar to the mammalian intestine) and the hindgut (similar to the colon). The adult gut is an established model system to study stem cells, intestinal homeostasis and tumorigenesis.[52–56] The single layer midgut epithelium has a high rate of cellular turnover and is maintained by newly differentiated cells produced from intestinal stem cells (ISCs).[54] The *Drosophila* adult ISCs divide to self-renew and to produce enteroblast cells (EBs) that differentiate into enterocytes (EC, large absorptive cells) and enteroendocrine cells that form the epithelial monolayer of the adult gut.

Several studies have focused on directed modeling in *Drosophila* of specific human cancers, such as colorectal cancer. The tumor suppressor gene, *Adenomatous polyposis coli (APC)* is commonly found to be mutated in colon cancers, as is the oncogenic activation of K-Ras.[57] Germ-line mutations in *APC* are responsible for the dominantly inherited disease, familial adenomatous polyposis (FAP), that is characterized by intestinal polyps that if untreated develop into carcinoma. These pathways and others that have been implicated in colon cancer are highly conserved in *Drosophila*. Importantly, mutations that are involved in cancer, including *Apc*, have also been found to lead to overgrowth and tumor formation in the *Drosophila* adult intestine, recapitulating key events in human cancers.[58,59] In a *Drosophila* model of cooperative tumorigenesis, knockdown of *Apc* and overexpression of Ras^{V12} in the adult midgut results in a loss of cell polarity and hyperplasia, highlighting the conserved functional cooperation between Ras and APC in controlling cell growth in the gut.[60]

More recently, genetically complex models of colon cancer that are based on frequently observed mutations in human colorectal cancers have been established in *Drosophila*. Using data from The Cancer Genome Atlas, specific combinations of multiple gene knockdowns and overexpression were generated specifically in the *Drosophila* adult hindgut epithelium, the functional equivalent of the mammalian colon. These include models that combine Ras^{V12} expression with *p53, Apc*, and *pten* knockdown (and *dSmad4, pten, Apc* knockdown), and result in invasive tumors.[61] Importantly, these models replicate several of the properties of the equivalent human cancer and were utilized for drug screening to identify novel therapeutic combinations that were shown to be effective at limiting tumor growth in the *Drosophila* model as well as in human cancer.[23,61]

Recent studies have indicated that autophagy is required for the maintenance of ISC and the progression of CRC in the mouse intestine. However, the cell autonomous contribution of autophagy in the ISCs or potential functions of autophagy in other intestinal cell types cannot be differentiated. Thus, how autophagy acts to restrict cancer stem cell transformation and cancer progression remains inconclusive. The *Drosophila* adult midgut epithelium has also been established to model stem cell behavior in tumorigenesis, as ISCs have molecular, physiological, and cellular features similar to their mammalian equivalents.[54] The *Drosophila* genetic platform makes it possible to study ISC functions of autophagy that are cell autonomous. Several recent studies have shown that autophagy acts as a tumor suppressor in ISCs by regulating their division and maintaining cell numbers. Inhibition of autophagy by knockdown of multiple *Atg* genes demonstrates its involvement in various steps of the autophagy pathway.[62] Autophagy is required to maintain ISC function, important for the continued renewal of the intestinal epithelium to sustain the normal gut anatomy and lifespan.[62]

Disruption of the endocytosis-autophagy network by knockdown of genes that function at multiple stages drastically deregulates *Drosophila* ISC proliferation and promotes hyperplasia.[63,64] The altered intracellular vesicular trafficking that results from decreased expression of core endocytosis or autophagy pathway components promotes proliferation without regulating differentiation. Autophagic flux controls the growth signaling pathway (Egfr-Ras85D-mitogen activated protein kinase), and in the mutants, activated Egfr is stabilized and recycled to the cell surface by the endosomal pathway, rather than autophagy-mediated degradation. This suggests that autophagy acts as a tumor suppressor in the early stages of CRC progression preventing the transition from early-stage cancer stem cells to later-stage cancer.

A *Drosophila* ISC model of stem cell tumors that expresses a human gain-of-function RAF allele (that hyperactivates downstream oncogenic MAPK pathway) results in the formation of large tumors.[65] This model has been used in drug screening and has identified several compounds that inhibit tumor formation including rapamycin, an mTORC1 inhibitor, that can stimulate autophagy.

While not an established disease model, the degradation of the *Drosophila* larval midgut during metamorphosis presents a novel system to understand the regulation of autophagy-dependent cell death.[26] In response to the steroid hormone, ecdysone, the expression of many *Atg* genes is upregulated in an Ecdysone receptor dependent manner and autophagy is induced at the onset of larval midgut degradation.[28,66] The degradation of the larval midgut is dependent on autophagy, as inhibition of autophagy significantly delays degradation.[28] Compared to the requirement of the autophagy machinery induced in response to starvation, only a subset of the components are required for autophagy-dependent midgut degradation.[67] This tissue then provides a unique model to understand the role of autophagy-dependent cell death.

These models will provide unique opportunities to combine the power of *Drosophila* genetics and the simplicity of the gut system to study the role of autophagy in tumor formation. Not only will this serve to further the understanding of the contribution of autophagy in genetically complex tumors but also the role of autophagy in premalignant stages and to examine early events that may contribute to tumorigenesis.

2.4 Cancer cachexia in *Drosophila*

Beyond the cooperative interactions between the tumor and the tumor microenvironment there are also complex interactions between the tumor and host at distant tissues. The ability of the tumor to promote metabolic dysfunction in distant tissues/organs can result in the dramatic atrophy of adipose tissue and skeletal muscle causing cancer cachexia. Cancer cachexia is a complex metabolic disorder characterized by alterations in insulin signaling, inflammatory and immune responses, and muscle homeostasis.[68,69] In cancer patients, cachexia can contribute to morbidity and mortality due to loss of body weight (due to loss of muscle and fat) and a decline in overall health.[69–71] Cachexia is a significant problem for cancer treatment yet the causes of cachexia and correlation between its severity and tumor burden are poorly understood with limited treatment options available.[68,70,71] A greater understanding of the molecular mechanisms driving cachexia is required to identify therapeutic interventions that can prevent/reverse the condition.

Recent studies have identified autophagy as one of the mechanisms of tissue wasting during cancer cachexia.[72–74] There is a correlation between the ability of tumors to systemically induce autophagy (dependent on IL-6 secreted by tumor cells) with the potential to promote cachexia.[74] By utilizing sera from cancer patients with weight loss and conditioned media from cancer cells that induce autophagy in cell culture, IL-6 was identified as the autophagy-inducing activity.[74] Furthermore, in a *Drosophila* Ras^{V12} *scrib* mutant cooperative tumorigenesis model, tumor growth was shown to be supported by local and systemic autophagy requiring IL-6–JAK–STAT cytokine signaling pathways.[48] In a mouse model of cachexia, BNIP3 (BCL2 and adenovirus E1B 19-kDa-interacting protein 3), a Bcl-2 family member that is a regulator of autophagy, is induced during muscle wasting.[75] During muscle atrophy in mice, *Atg* genes are upregulated as well as genes involved in the regulation of autophagy, including *Bnip3*, and *Bnip3l*.[76] The knockdown of *Bnip3* markedly reduced the induction of autophagosome formation in skeletal muscle following fasting.[76] However, it is important to note that while the induction of autophagy by BNIP3 can promote cell survival, in other context it is associated with autophagy during cell death,[77] again highlighting the context-dependence of functions of autophagy.

In *Drosophila* adults, tumors have been found to promote wasting in distant tissues, similar to a cachexia-like response.[78,79] Two independent models of cancer cachexia identified Imaginal morphogenesis protein-Late 2 (ImpL2, an insulin growth factor binding protein) secreted from tumors that mediates organ wasting. Using the Ras^{V12} $scrib^{-/-}$ tumor model, the tumor-host interaction was examined by transplantation of malignant or benign tumors into adult hosts. The malignant tumors, but not benign ones, induced tissue wasting in the ovary, fat body and muscle.[78] In an adult gut tumor model, expression of activated Yorkie (Yap1 orthologue) causes over-proliferation of intestinal cells resulting in altered metabolism and systemic wasting in organs including the ovary, fat body, and muscle.[79] The ablation of *Impl2* in theses tumors partially rescued the wasting phenotypes observed in the distant tissues, without affecting tumor growth. The partial rescue suggests that an additional factor(s) involved in tumor-host interaction may also contribute to organ wasting.

The establishment of *Drosophila* models where tumors induce cachexia-like phenotypes, have provided insights into the mechanisms whereby an insulin antagonist secreted by the tumor drives metabolic changes and wasting in distant tissues. These models now present an opportunity to address the contributions of autophagy to tumor-host interactions to study tumor-induced organ wasting.

3. Neurodegenerative diseases

Neurodegenerative diseases are distinguished by aberrant protein products that accumulate, aggregate and ultimately are thought to cause the death of specific neuronal cell types. Given the post-mitotic nature of neurons and the high metabolic demands for correct function, they are reliant on effective autophagic-lysosomal clearance of dysfunctional proteins to maintain homeostasis and ward off degenerative effects throughout their lifespan. Genetic studies in mice have confirmed an essential role for autophagy in maintaining healthy neurons since deficiency of *Atg5* in neural cells[80] or *Atg7* in Purkinje cells[81] causes progressive neurodegenerative phenotypes. Conversely, ubiquitous expression of Atg5 in mice enhances levels of autophagy and results in an increase in lifespan further supporting a role for autophagy in promoting neuronal health.[82] With age being a significant risk factor for neurodegenerative disease,[83] increasing the levels of autophagy throughout lifespan may prevent the build-up of pathogenic proteins and subsequent development of neurodegenerative diseases. However, for this to be effective a much better understanding of the regulation of autophagy, particularly during ageing, is required such that targeted approaches can be developed. *Drosophila* presents as an ideal model system for these analyses as they have been used extensively to model various human neurodegenerative diseases[84,85] and show an age-dependent decrease in the expression level of autophagy genes.

3.1 Modeling neurodegeneration in *Drosophila*

Despite the fact that many human neurodegenerative disease genes are conserved in *Drosophila*, ectopic expression of the human gene (with or without disease-causing mutations) is commonly used to study the effects of alterations to cellular protein homeostasis mechanisms. Neuronal-specific expression of human disease genes has provided important insight into disease-related mechanisms. A significant advantage of studies in this model organism is that neuronal cells are present in the context of the whole organism where they are in contact with other cells, including glial cells. *Drosophila* allows for the analysis of non-cell autonomous analyses where expression in all neurons (or particular neuronal sub-types of interest) can be analyzed for any effects on surrounding cell types. Conversely, ectopic expression in glial cells can be readily assayed for any deficits in neuronal function supporting communication between these cell types. Subset-specific expression of glial cells is also now available to further delineate the functionally diverse cell types that are involved.[86] In addition to these gain-of-function approaches, neurodegenerative phenotypes have also been observed in *Drosophila* with a deficiency of various conserved human disease genes via RNAi mediated knockdown or genetic loss-of-function approaches. In many cases there is redundancy in the human disease genes where a single gene in *Drosophila* can highlight fundamental processes relevant to disease progression. In addition to the vast array of genetic resources available to construct models of human genetic disease in *Drosophila*, their lifespan of just ~80 days (compared to ~80 years in humans) and conservation of the pathways that contribute to aging make it a powerful model for the analysis of mechanisms contributing to neurodegeneration that are relevant for human disease and essential for the design of novel therapeutic approaches.[87]

A role for autophagy in maintaining neuronal homeostasis has been further investigated in *Drosophila*, as neurodegenerative phenotypes have been associated with decreased activity of individual components of the conserved autophagy pathway. Decreased expression of *Atg5*,[88] *Atg6* or *Pi3K59F*,[89] *Atg7*,[90] *Atg17/Fip200*,[91] *Atg8a*,[92] *Atg16*,[93] *Syntaxin17*,[94] as well as *Atg5* or *Atg9*[95] all affect neuronal health and function (Table 1). Modulators of the autophagy pathway have also been identified in genetic screens for contributors to neurodegeneration, in addition to other conserved components of basic cellular functions including protein trafficking, mitochondrial function as well as lipid and iron metabolism.[96–98] Thus, studies in *Drosophila* have confirmed and extended our understanding of the contribution that

Table 1 Autophagy genes that regulate neuronal function in *Drosophila*.

Autophagy gene	**Neuronal phenotype**	**Other effects**	**References**
Atg5 (RNAi in all neuronal cells or mushroom bodies)	Decreased memory function	Non-cell autonomous effects	95
Atg5 (null mutant)	Climbing defect, increased Ref(2)P, increased TUNEL	Rescued by WT not E122D Atg5	88
Atg6 (RNAi in dopaminergic neurons)	Decreased lifespan and climbing defect	Ectopic Buffy (pro-survival Bcl 2) enhances phenotype	89
Pi3K59F/Vps34 (RNAi in dopaminergic neurons)	Decreased lifespan and climbing defect	Ectopic Buffy suppresses phenotype	89
Atg7 (null mutant)	Decreased lifespan, climbing defect and ubiquitin aggregates	Viable, developmental delay, increased stress sensitivity	90
Atg8a (hypomorphic mutants)	Decreased lifespan and ubiquitin aggregates	Atg8a expression increases lifespan	92
Atg9 (RNAi in all neuronal cells or mushroom bodies)	Decreased memory function	Non-cell autonomous effects	95
Atg16 (null mutant)	Decreased lifespan and climbing defect	Delayed ethanol-induced sedation response	93
Atg17/Fip200 (null allele)	Climbing defect, brain vacuolization and ubiquitin aggregates	No axonal defects	91
Syntaxin 17 (mutants)	Climbing defect, early lethality, increased Atg8a autophagosomes and Ref(2) P aggregates		94

individual components of the autophagy machinery make to the maintenance of neuronal homeostasis. *Drosophila* presents as an ideal model to further investigate the fundamental contribution of autophagy pathway components to neurodegenerative diseases, often termed more broadly as dementias. In particular, studies have already shown a role for autophagy in disease mechanisms for the late-onset dementias including Alzheimer's disease (AD), Parkinson's disease (PD), expanded repeat diseases (including Huntington's disease), as well as some early-onset dementias such as lysosomal diseases (Fig. 1).

3.2 Alzheimer's disease

Alzheimer's disease (AD) is the most common form of dementia and is characterized by an age-related decline in cognitive abilities. Clinical diagnosis for AD involves assessment of cognitive abilities together with determining the level of deposition of amyloid plaques and/or tangles of the Tau protein in the brain. Human genetic studies of rare inherited forms of AD have revealed mutations in the genes for amyloid precursor protein (APP), Presenilin1 (PSEN1) or PSEN2.[99] However, the majority of AD cases are sporadic where the causative agent is not known. Genome-wide association studies (GWAS) have identified mutations in candidate risk genes that could contribute to sporadic AD.[100,101] The most significant of these is *ε4* variant of *apolipoprotein E (APOE)* and others include genes encoding components of endocytosis, innate immunity as well as metabolic pathways. Although the cause of AD remains controversial, there is clear evidence for autophagic disruption affecting cellular trafficking, in particular in the endolysosomal system.[102] However, there is conflicting data as to whether autophagy is protective or pathogenic in AD.[103] Since many of the autophagy genes, as well as those identified in inherited and sporadic AD, are conserved in *Drosophila*, it provides an ideal system to determine the contribution of autophagy to AD pathology.

Models for AD in *Drosophila* include ectopic expression of amino acids 1–42 of human amyloid-β (Aβ42) where Aβ42 is the product of amyloidogenic processing of APP by β-secretase (BACE1) then the γ-secretase complex in which PSEN is the catalytic component. Ectopic expression of Aβ42 has been shown to result in a neurodegenerative phenotype associated with aging and accumulation of Aβ42 in neuronal vesicles.[83,104,105] This was accompanied by autophagic disruption, and it was proposed that autophagy played a protective role at early stages and a pathogenic role at later stages.[105] Thus, it would seem that autophagy has a clear contribution to Aβ42 neuronal toxicity and possibly may have opposing contributions at different stages of life. However, generating the constructs for ectopic Aβ42 expression requires inclusion of an exogenous secretory signal sequence and the choice of peptide sequence has been shown to affect phenotypic outcomes.[106] Since processing of APP occurs within the endolysosomal network, an alternate approach is to ectopically express both human APP and BACE1 such that the endogenous subcellular localization of fragments better reflects the endogenous situation.[107]

There is significant crosstalk between the endosomal pathway where APP processing takes place and the autophagy pathway as they both converge at the lysosome. The master regulator of lysosomal biogenesis, Transcription factor EB (TFEB) has been shown to be reduced in AD.[108] Conversely it has also been shown that increased levels of neuronal TFEB lead to a reduction in toxicity in a mouse model for AD.[109] *Mitf* is the single gene from this family of transcription factors present in *Drosophila* and it has been shown to regulate lysosomal activity thus providing an ideal system to determine its contribution to AD pathogenesis.[110]

In addition to studies of APP and its various processing products, ectopic expression of human Tau both as wild-type as well as an aggregate prone mutated form is also commonly used in *Drosophila*.[111] Ectopic expression of Tau in the eye results in phenotypes that have formed the basis for genetic modification analyses of GWAS candidate risk genes, some of which are able to modify the Tau phenotypes providing support for a role in Tau toxicity.[112–114] One of these candidate risk genes is phosphatidylinositol binding clathrin assembly protein (PICALM) which is conserved in *Drosophila* as Like-AP180 (Lap). Decreased levels of Lap in all neuronal cells has been shown to result in increased levels of Atg8a indicating elevated levels of autophagy as well as causing increased accumulation of ectopically expressed Tau.[115] Thus using *Drosophila* we can gain insights into the mechanisms that regulate autophagy and contribute to the clearance of Tau and Amyloid. This will allow for the investigation into ways in which these processes can be enhanced in an *in vivo* context to identify potentials avenues for therapeutic intervention.

3.3 Parkinson's disease

Parkinson's disease (PD) is the most common neurodegenerative movement disorder, characterized by progressive degeneration of dopaminergic neurons in the substantia nigra which results in the classical movement defects as well as other neurodegenerative symptoms. Parkinson's disease pathology is observed as intraneuronal inclusions of accumulated α-synuclein known as Lewy bodies. Most cases of PD are sporadic, however, some familial mutations have been identified providing important insight into the disease mechanism. Autosomal dominant mutations have been identified in *α-synuclein* and *Leucine-rich repeats kinase 2 (LRRK2)* in addition to autosomal recessive mutations in *Parkin, Pink1* and *DJ-1* which have all been linked to autophagy.[116]

Although α-synuclein is not conserved in *Drosophila*, ectopic expression of the human protein has been shown to cause loss of dopaminergic (DA) neurons as well as a decline in locomotor activity.[117] Ectopic expression of α-synuclein in mice leads to decreased levels of autophagy and neuronal expression in *Drosophila* was shown to enhance toxicity of mutant huntingtin protein, an indirect measure of autophagy.[118] In addition, Spermidine which has been shown to increase autophagy levels has been shown to provide protection against α-synuclein neurotoxicity in *Drosophila*.[119] Increased expression of human LAMP2 in *Drosophila* neurons also leads to elevated levels of Atg5, increased Atg8 puncta and reduced toxicity of ectopically expressed α-synuclein.[120] Thus there is evidence of a protective role for autophagy in α-synuclein induced neuronal toxicity.

LRRK2 encodes a large multi domain protein with leucine-rich repeats, a kinase domain and a GTPase domain and mutations in LRRK2 are the most common genetic determinants of PD. Ectopic expression of wild-type or the constitutively active (G2019S) form of LRRK2 have been shown to regulate autophagy via calcium dependent signaling pathway in HEK293T cells.[121] LRRK2 is conserved as a single *lrrk* gene in *Drosophila* and functional analysis of mutant alleles has revealed endolysosomal and autophagic defects.[122] In addition, ectopic expression of a mutated form of human LRRK2 (G2019S) specifically in dopaminergic neurons is a commonly used as a PD model in *Drosophila*.[123,124] These models can be used as a basis for screening PD candidate risk loci with respect to age-dependent neurodegeneration and autophagy.[125,126]

Recessive mutations in *Parkin* and *Pink* highlight a role for mitochondrial homeostasis in preventing PD. *Pink1* encodes a mitochondrial Ser/Thr kinase which is relocated to the outer mitochondrial membrane upon damage where it subsequently recruits Parkin. *Parkin* encodes a ubiquitin ligase which then is activated and results in recognition by an autophagy adaptor for delivery of the mitochondria to the lysosome for degradation in a specialized form of autophagy termed mitophagy. It has been shown that neurodegeneration observed in *pink* or *parkin* mutants can be suppressed by ectopic expression of Atg1 and this was shown to be accompanied by an increase in mitochondrial fission.[127] Thus the removal of damaged mitochondria by mitophagy has a clear contribution to the pathogenesis of PD.

3.4 Expanded repeat neurodegenerative diseases

The family of expanded repeat neurodegenerative diseases includes the polyglutamine (PolyQ) diseases where a tandem CAG repeat sequence in the coding region of the affected genes results in increases in copy number beyond a pathogenic threshold. The most common of these is Huntington's disease and the family also includes a number of Spinocerebellar Ataxia movement disorders. There are many models for toxicity resulting from the ectopic expression of polyQ tracts either alone or within disease genes in *Drosophila*.[128]

Autophagy induction by rapamycin reduces neurodegenerative phenotypes in flies expressing polyQ containing huntingtin (Htt) disease protein (Q120) thus supporting a protective role for autophagy in these diseases.[129] In addition, increased expression of Atg1 was shown to suppress the rough eye toxicity observed with Htt (Q120) also providing support for autophagy contributing to polyQ toxicity.[130] An increased level of autophagy was observed with ectopic expression of a different pathogenic polyQ containing protein, Ataxin3 (Q78) in flies and this was further enhanced with reduced levels of Atg5 thus providing genetic confirmation for the contribution of autophagy to prevention of toxicity.[131] In addition the autophagy adaptor p62 is required for reducing retinal degeneration in Ataxin 3 (Q78) expressing flies thus providing further support for autophagy playing a protective role in these disease models.[132]

However, not all expanded repeat sequences fall in the coding region of the affected genes and also not all are comprised of CAG triplet repeats thus polyQ toxicity is not the sole factor responsible for these diseases which share mutation mechanism, similar pathogenic thresholds and many of which are clinically indistinguishable.[133] Repeat containing double stranded RNA ($rCAG.rCUG_{\sim 100}$) was shown to be pathogenic when expressed during eye development and this toxicity was shown to be enhanced by decreased expression of *Atg1* or *Atg18*.[134] This suggests that it is not only polyQ containing protein products that cause toxicity and the role of autophagy is essential for a much broader range of pathogenic agents in this family of expanded repeat neurodegenerative diseases.

4. Lysosomal diseases

Lysosomal diseases are a group of debilitating inherited metabolic diseases that result from a deficit in lysosomal function that leads to aberrant storage of various cellular components and cumulative cellular dysfunction. Given the dependence of autophagy completion on lysosomal fusion for degradation of cellular cargo, autophagy has a significant role to play in these diseases. In most cases the mutated lysosomal disease genes encode specific lysosomal hydrolases that contribute to the degradation of defined molecules within the cell. Deficit of these enzymes leads to accumulation and storage of incomplete metabolic breakdown products. In mouse lysosomal disease models, the autophagic delivery of bulk cytosolic contents to the lysosome is impaired.[135] The mechanism of lysosomal dysfunction leading to impaired autophagy in these diseases presents as a suitable target for investigating therapeutic intervention. Conservation of core catabolic processes and pathways in *Drosophila* allows for a better understanding of the role for autophagy in these lysosomal diseases, some of which are responsible for childhood dementias.

The mucopolysaccharidoses (MPS) are characterized by the loss of a specific enzyme required for the stepwise degradation of various glycosaminoglycans. It is interesting that although all cells are affected in these diseases there is a subset for which the brain is the most affected and results in the development of devastating forms of childhood dementia. One of these is Sanfilippo Type A or MPS IIIA where a mutation disrupts the sulfamidase enzyme (SGSH) which is required for breakdown of heparan sulfate. In *Drosophila* reduced activity of SGSH specifically in neuronal cells results in a locomotor defect that was enhanced by decreased activity of *Atg1* or *Atg18*.[136] In addition, the cellular localization of the Ref(2)P/p62 reporter is disrupted further supporting autophagic disruption in MPS IIIA. In the case of MPS Type VII, mutation in beta-glucuronidase is responsible, and a mutant *Drosophila* model showed a neuropathological defect in addition to neuromuscular degeneration and locomotor defect.[137] These phenotypes were shown to be attenuated after treatment with resveratrol, an inducer of autophagy.

Gaucher disease results from a deficit of the glucocerebrocidase (GBA) lysosomal enzyme responsible for degradation of glucocerebrocide molecules in the cell. Knockout of the two endogenous conserved *GBA* orthologues in *Drosophila* revealed autophagic defects in neuronal tissues that are rescued by treatment with rapamycin a well-known inducer of autophagy.[138] Interestingly, GBA1 was also found to be important for the developmental autophagy-dependent cell death in *Drosophila* midgut, a function that appears to be conserved in mammals.[139,140] Thus, GBA is important in different cellular contexts for autophagy regulation associated with disease.

Mucolipidosis Type IV results from recessive mutations in the transient receptor potential calcium channel mucolipin (TRPML) that localizes to late endosomes and lysosome. Loss of function mutations in the conserved *Drosophila trpml* gene result in locomotor dysfunction, in addition to increased staining with ubiquitin aggregates and autophagy markers.[141] Further investigation established that Trpml is required for fusion between the late endosome and the lysosome and that phenotypes could be ameliorated by decreased TOR activity.[142]

Thus, there are a number of lysosomal diseases where the resultant childhood dementia has now been modeled successfully in *Drosophila* and an autophagic contribution to disease has been identified. In addition, there are similarities emerging between childhood and adult-onset dementias in terms of autophagic disruption. There is a need for a greater understanding of the mechanisms by which autophagy contributes to these diseases such that successfully targeted interventions can be developed.

Together these data describe contributions of autophagic disruption to neurodegeneration in AD, PD, expanded repeat and lysosomal diseases. There are other examples where increased expression of autophagy pathway components can reduce symptoms of neurodegeneration. TARDBP/TDP-43 (TAR DNA binding protein) has been shown to be pathogenic in amyotrophic lateral sclerosis (ALS) and frontotemporal dementia (FTD) and it has been observed that increased levels of Atg7 conferred protection in *Drosophila* disease models.[143] In addition, multiple studies have determined the effect of various autophagy-inducing compounds on several neurodegenerative disease models. However, there are multiple cellular pathways that are affected by alterations to autophagy, highlighting the advantage of *Drosophila* as a whole animal model. There are other possible contributing factors to autophagy since increased expression of Ref(2)P/p62 during midlife has been shown to increase lifespan in *Drosophila* and may also facilitate more effective removal of pathogenic proteins.[144] In addition, it has been shown in *Drosophila* that reduced autophagy induces immune activation which is a hallmark of many neurodegenerative diseases in humans and may be a significant contributing factor to neuronal decline.[145,146]

5. Infectious disease

Autophagy plays an important role in limiting the replication of a number of pathogens, including bacteria and viruses. The selective autophagic removal of bacteria and viruses, termed xenophagy, is an evolutionarily conserved immune response. The importance of xenophagy in human physiology is exemplified by inherited defects in autophagy-relevant genes that predispose to infectious disease.[2] Several bacteria and viruses have evolved strategies to avoid their xenophagic degradation by actively inhibiting the hosts autophagic responses.[147,148]

Autophagy is induced in macrophages following *Mycobacterium tuberculosis* exposure. The host resistance to *M. tuberculosis* infection requires the ubiquitin ligase Smurf1 that functions together with Parkin in selective bacterial autophagy.[149] Macrophage-specific *Atg5* conditional mouse knockouts, are defective for autophagy and have increased susceptibility to *M. tuberculosis*[150,151] and *Listeria monocytogenes* infection.[152] Furthermore, the genetic inhibition of autophagy in *Atg5* conditional knock out mice also makes them more susceptible to lethality from Sindbis virus infection.[153] Importantly, the pharmacological stimulation of autophagy has been shown to be sufficient to reduce the mortality of mice infected with the mosquito-borne diseases Chikungunya or West Nile virus.[154] Hence, there is therapeutic potential for autophagy activation to be utilized in the treatment of pathogens and antibiotic-resistant infections. While *Drosophila* lack an adaptive immune system, innate immunity mechanisms (first discovered in *Drosophila*) are conserved between flies and mammals.[155] Studies in *Drosophila* have demonstrated that autophagy is an essential host response against bacteria and viruses *in vivo*, thus providing a genetically amenable model system to study autophagy in innate immunity.[155,156]

5.1 Bacterial infection

Similar to mammals, in *Drosophila* the nuclear factor-κB (NF-κB) pathways, Toll and Imd, are essential for antibacterial and antifungal responses by regulating the production of antimicrobial peptides (AMPs).[157,158] Studies have revealed that autophagy also functions to limit bacterial infection in *Drosophila*.[159,160] In *Drosophila* hemocytes (macrophage-like cells) infection with *L. monocytogenes* induces autophagy to restrict bacterial replication in the cytoplasm. In adults challenged with *L. monocytogenes* infection the knockdown of *Atg5* in hemocytes results in increased bacterial replication and increased lethality, and the induction of autophagy by rapamycin treatment or Atg1 expression reduces the bacterial load.[160] The induction of autophagy in response to *L. monocytogenes* requires the PGN recognition protein family member PGRP-LE.[160] The *Vacuolar protein sorting 15 (Vps15)* gene (also known as *immune response deficient 1, ird1*) encodes a serine/threonine kinase that is a member of the Vps34/class III PI3-Kinase complex that functions in autophagy.[161] Mutant *ird1* animals have altered Imd and Toll pathway activity, dysregulated AMP expression and are more susceptible to bacterial infection (both *Escherichia coli* and *Micrococcus luteus*).[159] Furthermore, the ubiquitin ligase Parkin that functions in mitophagy is required to target pathogens such as *M. tuberculosis* in both *Drosophila* and humans.[162]

5.2 Viral infection

The major immune defense against viral infection in *Drosophila* is the RNA interference pathway, with additional roles for the evolutionarily conserved immune signaling pathways Toll, Imd, and JAK-STAT.[163–166] In response to some viruses, autophagy has been shown to restrict infection. The viruses are directly sensed by Toll 7 in the infected cells to induce autophagy,[167–169] dependent on PI3K and Tor signaling.[168]

There are several *Drosophila* models of viral infection including viruses that cause disease in humans. Recently a *Drosophila* model was established to study the innate immune responses to Zika virus infection.[170] In the *Drosophila* model, Zika virus infection of the brain causes Imd-dependent upregulation of *Stimulator of interferon genes, Sting*. An evolutionarily conserved function of STING (TMEM173) is the detection of pathogen cyclic dinucleotides to activate antimicrobial innate immune signaling pathways.[171,172] Mammalian studies have identified a role for STING in antibacterial autophagy.[173] In response to Zika virus infection increased Sting promotes induction of autophagy to limit the infection, as loss of autophagy increases infection and mortality, while pharmacological induction of autophagy protects from disease.[170] These models will be important to understand the role of autophagy as an evolutionarily conserved component of the antiviral and antimicrobial pathways.

6. Using *Drosophila* models for drug discovery

The modulation of autophagy in the pathological context has significant potential for therapeutic intervention, with great investment in developing both pharmacological inhibitors and inducers of autophagy.[12,148,174] However, given the context-specific functions of autophagy it will be important to determine whether it will be protective or pathogenic by promotion of cell survival or cell death. There are multiple well-defined molecular targets that are suitable to either suppress or enhance autophagy. Given the non-autophagic additional functions of many *Atg* gene products it is imperative to establish appropriate models to evaluate suitable targets for modulation.[175] It still remains to be established whether global modulation of autophagy can be used for treatment or if disease cells/tissue specific treatment is required.

Therapeutic drug discovery requires a time- and cost-effective platform to screen chemical libraries to find an affinity for a specific protein or biological activity that is characteristic of a particular disorder/disease. Importantly, even when the molecular mechanism for the disease is poorly understood *Drosophila* can be used for effective phenotypic screening. This also has the advantage over *in vitro* studies of providing a more physiologically relevant context where small molecules are screened for their effect on the phenotype in cells, tissue and whole animals.[176,177] *Drosophila* models can be used in early-stage drug discovery and for *in vivo* validation of identified compounds by assessment of efficacy and toxicity. Some of the models described above, have been successfully used in chemical screening approaches.

Chemical screens have been successfully performed using *Drosophila* models of cancer.[177,178] The pioneering study using a *Drosophila* model of multiple endocrine neoplasia type 2 (MEN2), that expresses Ret kinase, was employed to screen for kinase inhibitors, which identified AD57. Treatment with AD57 rescued oncogenic Ret-induced lethality, whereas related Ret inhibitors imparted reduced efficacy and enhanced toxicity.[179,180] Furthermore, toxicity due to the inhibition of TOR was overcome by tailoring the drug such that it no longer bound TOR. The optimized compounds improve the survival of the MEN2 model *Drosophila* adults as well as the xenograft growth of human MEN2 cells. Whole animal screening of compounds including multiple FDA-approved drugs for the ability to target human Raf oncogene expression in ISC in *Drosophila* adult that causes intestinal tumors identified several compounds that that inhibited tumor growth, including rapamycin.[65] This study demonstrated the efficacy of the drug treatment on preventing tumor growth and also the potential side effect of promoting the proliferation of wild type cells.

In addition to cancer models, several chemical screens have been performed using other *Drosophila* models of disease.[176,177] While the genetic basis contributing to several neurological diseases has been described, the identification of effective treatments has been challenging. There have been numerous screens using *Drosophila* models of neurodegenerative diseases including Fragile X syndrome, Alexander disease, Friedreich's ataxia that have aided in identification of potential therapeutics.[181–183] In addition, other *Drosophila* models of disorders used in drug screens include polycystic kidney disease,[184] seizure disorders,[185,186] and obesity.[187]

With the technological advancements in personalized medicine, the ability to generate "*Drosophila* avatars" of human diseases especially in the case of modeling patient-specific tumors, will provide the capacity to undertake whole animal drug screens and investigate novel drug combinations. This can also be combined with "lifestyle" factors, such as diet. In the search for novel therapeutic compounds for disease treatment, *Drosophila* has now emerged as a highly suitable model organism to identify such compounds as well as testing efficacy and toxicity.

7. Conclusion

With alterations in autophagic flux identified in the pathology of many human diseases, it is important to understand how this contributes to pathogenesis. This will be essential to determine where and when modulation of autophagy will have the greatest benefit. Whether induction of autophagy will be protective or contribute to pathogenesis; if systemic or local modulation is required in diseased cells/tissue; and if modulation of autophagy should be acute or prolonged. A combined understanding of all these factors will contribute to determining the therapeutic potential of modulation of autophagy.

Model organisms provide powerful tools for understanding the molecular and genetic contribution of autophagy in human disease and for drug discovery. *Drosophila* disease models will continue to provide a powerful genetic tool, with a rapid pace of discovery and affordability that will complement the advancement in personalized medicine (such as organoids and patient derived xenografts). One of the greatest strengths in using *Drosophila* models of disease is the capacity to study the complex *in vivo* biology, including the host interaction with pathogenic cells/tissues/organs that has made diseases difficult to treat. With the growing advancement in new technologies, many areas of biomedical research are likely to benefit from findings in *Drosophila* research. Furthermore, with the ability to generate patient-specific humanized *Drosophila*, this will enable the rapid screening of therapeutics *in vivo* that can then be translated to the patient.

Acknowledgments

The *Drosophila* work in DD and SK laboratory has been supported by the National Health and Medical Research Council (NHMRC) of Australia Project Grant (1124490), an Australian Research Council (ARC) Discovery Grant (DP170100623) and a NHMRC Senior Principal Research Fellowship (1103006). LO'K has been supported by the Hopwood Centre for Neurobiology, South Australian Health and Medical Research Institute (SAHMRI).

Conflict of interest

The authors declare no conflict of interest regarding the publication of this paper.

References

1. Dikic I, Elazar Z. Mechanism and medical implications of mammalian autophagy. *Nat Rev Mol Cell Biol*. 2018;19:349.
2. Levine B, Kroemer G. Biological functions of autophagy genes: a disease perspective. *Cell*. 2019;176:11–42.
3. Mizushima N. A brief history of autophagy from cell biology to physiology and disease. *Nat Cell Biol*. 2018;20:521–527.
4. Galluzzi L, Baehrecke EH, Ballabio A, et al. Molecular definitions of autophagy and related processes. *EMBO J*. 2017;36:1811–1836.
5. Kaushik S, Cuervo AM. Chaperone-mediated autophagy: a unique way to enter the lysosome world. *Trends Cell Biol*. 2012;22:407–417.
6. Liang XH, Jackson S, Seaman M, et al. Induction of autophagy and inhibition of tumorigenesis by beclin 1. *Nature*. 1999;402:672.
7. Rea SL, Majcher V, Searle MS, Layfield R. SQSTM1 mutations—bridging Paget disease of bone and ALS/FTLD. *Exp Cell Res*. 2014;325:27–37.
8. Geisler S, Holmström KM, Skujat D, et al. PINK1/Parkin-mediated mitophagy is dependent on VDAC1 and p62/SQSTM1. *Nat Cell Biol*. 2010;12:119.
9. Cadwell K, Liu JY, Brown SL, et al. A key role for autophagy and the autophagy gene Atg16l1 in mouse and human intestinal Paneth cells. *Nature*. 2008;456:259.
10. Liu Y, Shoji-Kawata S, Sumpter Jr RM, et al. Autosis is a Na+,K+-ATPase-regulated form of cell death triggered by autophagy-inducing peptides, starvation, and hypoxia-ischemia. *Proc Natl Acad Sci U S A*. 2013;110:20364–20371.
11. Xie C, Ginet V, Sun Y, et al. Neuroprotection by selective neuronal deletion of Atg7 in neonatal brain injury. *Autophagy*. 2016;12:410–423.
12. Rubinsztein DC, Codogno P, Levine B. Autophagy modulation as a potential therapeutic target for diverse diseases. *Nat Rev Drug Discov*. 2012;11:709.
13. St Johnston D. Using mutants, knockdowns, and transgenesis to investigate gene function in Drosophila. *Wiley Interdiscip Rev Dev Biol*. 2013;2:587–613.
14. Brand AH, Perrimon N. Targeted gene expression as a means of altering cell fates and generating dominant phenotypes. *Development*. 1993;118:401–415.
15. Duffy JB. GAL4 system in Drosophila: a fly geneticist's Swiss army knife. *Genesis*. 2002;34:1–15.
16. McGuire SE, Le PT, Osborn AJ, Matsumoto K, Davis RL. Spatiotemporal rescue of memory dysfunction in Drosophila. *Science*. 2003;302:1765–1768.
17. Lai SL, Lee T. Genetic mosaic with dual binary transcriptional systems in Drosophila. *Nat Neurosci*. 2006;9:703–709.

18. Potter CJ, Tasic B, Russler EV, Liang L, Luo L. The Q system: a repressible binary system for transgene expression, lineage tracing, and mosaic analysis. *Cell.* 2010;141:536–548.
19. Şentürk M, Bellen HJ. Genetic strategies to tackle neurological diseases in fruit flies. *Curr Opin Neurobiol.* 2018;50:24–32.
20. Bier E. Drosophila, the golden bug, emerges as a tool for human genetics. *Nat Rev Genet.* 2005;6:9.
21. Reiter LT, Potocki L, Chien S, Gribskov M, Bier E. A systematic analysis of human disease-associated gene sequences in Drosophila melanogaster. *Genome Res.* 2001;11: 1114–1125.
22. Bellen HJ, Wangler MF, Yamamoto S. The fruit fly at the interface of diagnosis and pathogenic mechanisms of rare and common human diseases. *Hum Mol Genet.* 2019;28:R207–R214.
23. Bangi E, Ang C, Smibert P, et al. A personalized platform identifies trametinib plus zoledronate for a patient with KRAS-mutant metastatic colorectal cancer. *Sci Adv.* 2019;5:eaav6528.
24. Mulakkal NC, Nagy P, Takats S, Tusco R, Juhasz G, Nezis IP. Autophagy in Drosophila: from historical studies to current knowledge. *Biomed Res Int.* 2014;2014:273473.
25. Nagy P, Varga Á, Kovács AL, Takáts S, Juhász G. How and why to study autophagy in Drosophila: it's more than just a garbage chute. *Methods.* 2015;75:151–161.
26. Denton D, Kumar S. Autophagy-dependent cell death. *Cell Death Differ.* 2019;26: 605–616.
27. Arakawa S, Tsujioka M, Yoshida T, et al. Role of Atg5-dependent cell death in the embryonic development of Bax/Bak double-knockout mice. *Cell Death Differ.* 2017;24:1598–1608.
28. Denton D, Shravage B, Simin R, et al. Autophagy, not apoptosis, is essential for midgut cell death in Drosophila. *Curr Biol.* 2009;19:1741–1746.
29. Nassour J, Radford R, Correia A, et al. Autophagic cell death restricts chromosomal instability during replicative crisis. *Nature.* 2019;565:659.
30. Bishop E, Bradshaw TD. Autophagy modulation: a prudent approach in cancer treatment? *Cancer Chemother Pharmacol.* 2018;82:913–922.
31. Brumby AM, Richardson HE. Using Drosophila melanogaster to map human cancer pathways. *Nat Rev Cancer.* 2005;5:626.
32. Xu T, Rubin GM. Analysis of genetic mosaics in developing and adult Drosophila tissues. *Development.* 1993;117:1223–1237.
33. Lee T, Luo L. Mosaic analysis with a repressible cell marker (MARCM) for Drosophila neural development. *Trends Neurosci.* 2001;24:251–254.
34. Richardson HE, Portela M. Modelling cooperative tumorigenesis in Drosophila. *Biomed Res Int.* 2018;2018:4258387.
35. Pylayeva-Gupta Y, Grabocka E, Bar-Sagi D. RAS oncogenes: weaving a tumorigenic web. *Nat Rev Cancer.* 2011;11:761–774.
36. Dimauro T, David G. Ras-induced senescence and its physiological relevance in cancer. *Curr Cancer Drug Targets.* 2010;10:869–876.
37. Bilder D, Li M, Perrimon N. Cooperative regulation of cell polarity and growth by Drosophila tumor suppressors. *Science.* 2000;289:113–116.
38. Brumby AM, Richardson HE. Scribble mutants cooperate with oncogenic Ras or Notch to cause neoplastic overgrowth in Drosophila. *EMBO J.* 2003;22:5769–5779.
39. Bilder D, Perrimon N. Localization of apical epithelial determinants by the basolateral PDZ protein Scribble. *Nature.* 2000;403:676–680.
40. Igaki T, Pagliarini RA, Xu T. Loss of cell polarity drives tumor growth and invasion through JNK activation in Drosophila. *Curr Biol.* 2006;16:1139–1146.

41. Pagliarini RA, Xu T. A genetic screen in Drosophila for metastatic behavior. *Science*. 2003;302:1227–1231.
42. Uhlirova M, Bohmann D. JNK- and Fos-regulated Mmp1 expression cooperates with Ras to induce invasive tumors in Drosophila. *EMBO J*. 2006;25:5294–5304.
43. Wu M, Pastor-Pareja JC, Xu T. Interaction between Ras(V12) and scribbled clones induces tumour growth and invasion. *Nature*. 2010;463:545–548.
44. Galluzzi L, Pietrocola F, Bravo-San Pedro JM, et al. Autophagy in malignant transformation and cancer progression. *EMBO J*. 2015;34:856–880.
45. White E. Deconvoluting the context-dependent role for autophagy in cancer. *Nat Rev Cancer*. 2012;12:401–410.
46. Manent J, Banerjee S, de Matos Simoes R, et al. Autophagy suppresses Ras-driven epithelial tumourigenesis by limiting the accumulation of reactive oxygen species. *Oncogene*. 2017;36:5576.
47. Chi C, Zhu H, Han M, Zhuang Y, Wu X, Xu T. Disruption of lysosome function promotes tumor growth and metastasis in Drosophila. *J Biol Chem*. 2010;285: 21817–21823.
48. Katheder NS, Khezri R, O'Farrell F, et al. Microenvironmental autophagy promotes tumour growth. *Nature*. 2017;541:417.
49. Paiardi C, Mirzoyan Z, Zola S, et al. The stearoyl-CoA desaturase-1 (Desat1) in Drosophila cooperated with Myc to induce autophagy and growth, a potential new link to tumor survival. *Genes (Basel)*. 2017;8:131.
50. Nagata R, Nakamura M, Sanaki Y, Igaki T. Cell competition is driven by autophagy. *Dev Cell*. 2019;51:99–112.
51. Warr CG, Shaw KH, Azim A, Piper MDW, Parsons LM. Using mouse and Drosophila models to investigate the mechanistic links between diet, obesity, type II diabetes, and cancer. *Int J Mol Sci*. 2018;19:4110.
52. Ayyaz A, Jasper H. Intestinal inflammation and stem cell homeostasis in aging Drosophila melanogaster. *Front Cell Infect Microbiol*. 2013;3:98.
53. Fox DT, Spradling AC. The Drosophila hindgut lacks constitutively active adult stem cells but proliferates in response to tissue damage. *Cell Stem Cell*. 2009;5:290–297.
54. Jiang H, Edgar BA. Intestinal stem cell function in Drosophila and mice. *Curr Opin Genet Dev*. 2012;22:354–360.
55. Lemaitre B, Miguel-Aliaga I. The digestive tract of Drosophila melanogaster. *Annu Rev Genet*. 2013;47:377–404.
56. Takashima S, Mkrtchyan M, Younossi-Hartenstein A, Merriam JR, Hartenstein V. The behaviour of Drosophila adult hindgut stem cells is controlled by Wnt and Hh signalling. *Nature*. 2008;454:651.
57. Wood LD, Parsons DW, Jones S, et al. The genomic landscapes of human breast and colorectal cancers. *Science*. 2007;318:1108–1113.
58. Cordero J, Vidal M, Sansom O. APC as a master regulator of intestinal homeostasis and transformation: from flies to vertebrates. *Cell Cycle*. 2009;8:2927–2932.
59. Lee W-C, Beebe K, Sudmeier L, Micchelli CA. Adenomatous polyposis coli regulates Drosophila intestinal stem cell proliferation. *Development*. 2009;136:2255–2264.
60. Martorell O, Merlos-Suarez A, Campbell K, et al. Conserved mechanisms of tumorigenesis in the Drosophila adult midgut. *PLoS One*. 2014;9 e88413.
61. Bangi E, Murgia C, Teague AG, Sansom OJ, Cagan RL. Functional exploration of colorectal cancer genomes using Drosophila. *Nat Commun*. 2016;7 13615.
62. Nagy P, Sándor GO, Juhász G. Autophagy maintains stem cells and intestinal homeostasis in Drosophila. *Sci Rep*. 2018;8:4644.
63. Zhang P, Holowatyj AN, Roy T, et al. An SH3PX1-dependent endocytosis-autophagy network restrains intestinal stem cell proliferation by counteracting EGFR-ERK signaling. *Dev Cell*. 2019;49:574–589.e575.

64. Zhang P, Holowatyj AN, Ulrich CM, Edgar BA. Tumor suppressive autophagy in intestinal stem cells controls gut homeostasis. *Autophagy*. 2019;15:1668–1670.
65. Markstein M, Dettorre S, Cho J, Neumuller RA, Craig-Muller S, Perrimon N. Systematic screen of chemotherapeutics in Drosophila stem cell tumors. *Proc Natl Acad Sci U S A*. 2014;111:4530–4535.
66. Denton D, Xu T, Dayan S, Nicolson S, Kumar S. Dpp regulates autophagy-dependent midgut removal and signals to block ecdysone production. *Cell Death Differ*. 2019;26:763–778.
67. Xu T, Nicolson S, Denton D, Kumar S. Distinct requirements of autophagy-related genes in programmed cell death. *Cell Death Differ*. 2015;22:1792–1802.
68. Petruzzelli M, Wagner EF. Mechanisms of metabolic dysfunction in cancer-associated cachexia. *Genes Dev*. 2016;30:489–501.
69. Porporato P. Understanding cachexia as a cancer metabolism syndrome. *Oncogenesis*. 2016;5:e200.
70. Argilés JM, Busquets S, Stemmler B, López-Soriano FJ. Cancer cachexia: understanding the molecular basis. *Nat Rev Cancer*. 2014;14:754.
71. Fearon K, Arends J, Baracos V. Understanding the mechanisms and treatment options in cancer cachexia. *Nat Rev Clin Oncol*. 2013;10:90.
72. Aversa Z, Pin F, Lucia S, et al. Autophagy is induced in the skeletal muscle of cachectic cancer patients. *Sci Rep*. 2016;6:30340.
73. Penna F, Costamagna D, Pin F, et al. Autophagic degradation contributes to muscle wasting in cancer cachexia. *Am J Pathol*. 2013;182:1367–1378.
74. Pettersen K, Andersen S, Degen S, et al. Cancer cachexia associates with a systemic autophagy-inducing activity mimicked by cancer cell-derived IL-6 trans-signaling. *Sci Rep*. 2017;7:2046.
75. Asp ML, Tian M, Wendel AA, Belury MA. Evidence for the contribution of insulin resistance to the development of cachexia in tumor-bearing mice. *Int J Cancer*. 2010;126:756–763.
76. Mammucari C, Milan G, Romanello V, et al. FoxO3 controls autophagy in skeletal muscle in vivo. *Cell Metab*. 2007;6:458–471.
77. Zhang J, Ney PA. Role of BNIP3 and NIX in cell death, autophagy, and mitophagy. *Cell Death Differ*. 2009;16:939–946.
78. Figueroa-Clarevega A, Bilder D. Malignant Drosophila tumors interrupt insulin signaling to induce cachexia-like wasting. *Dev Cell*. 2015;33:47–55.
79. Kwon Y, Song W, Droujinine IA, Hu Y, Asara JM, Perrimon N. Systemic organ wasting induced by localized expression of the secreted insulin/IGF antagonist ImpL2. *Dev Cell*. 2015;33:36–46.
80. Hara T, Nakamura K, Matsui M, et al. Suppression of basal autophagy in neural cells causes neurodegenerative disease in mice. *Nature*. 2006;441:885–889.
81. Komatsu M, Waguri S, Chiba T, et al. Loss of autophagy in the central nervous system causes neurodegeneration in mice. *Nature*. 2006;441:880–884.
82. Pyo JO, Yoo SM, Ahn HH, et al. Overexpression of Atg5 in mice activates autophagy and extends lifespan. *Nat Commun*. 2013;4:2300.
83. Omata Y, Lim YM, Akao Y, Tsuda L. Age-induced reduction of autophagy-related gene expression is associated with onset of Alzheimer's disease. *Am J Neurodegener Dis*. 2014;3:134–142.
84. Jaiswal M, Sandoval H, Zhang K, Bayat V, Bellen HJ. Probing mechanisms that underlie human neurodegenerative diseases in Drosophila. *Annu Rev Genet*. 2012;46: 371–396.
85. McGurk L, Berson A, Bonini NM. Drosophila as an in vivo model for human neurodegenerative disease. *Genetics*. 2015;201:377–402.

86. Kremer MC, Jung C, Batelli S, Rubin GM, Gaul U. The glia of the adult Drosophila nervous system. *Glia*. 2017;65:606–638.
87. Piper MDW, Partridge L. Drosophila as a model for ageing. *Biochim Biophys Acta Mol Basis Dis*. 2018;1864:2707–2717.
88. Kim M, Sandford E, Gatica D, et al. Mutation in ATG5 reduces autophagy and leads to ataxia with developmental delay. *Elife*. 2016;5:e12245.
89. M'Angale PG, Staveley BE. Inhibition of Atg6 and Pi3K59F autophagy genes in neurons decreases lifespan and locomotor ability in Drosophila melanogaster. *Genet Mol Res*. 2016;15:gmr15048953.
90. Juhasz G, Erdi B, Sass M, Neufeld TP. Atg7-dependent autophagy promotes neuronal health, stress tolerance, and longevity but is dispensable for metamorphosis in Drosophila. *Genes Dev*. 2007;21:3061–3066.
91. Kim M, Park HL, Park HW, et al. Drosophila Fip200 is an essential regulator of autophagy that attenuates both growth and aging. *Autophagy*. 2013;9:1201–1213.
92. Simonsen A, Cumming RC, Brech A, Isakson P, Schubert DR, Finley KD. Promoting basal levels of autophagy in the nervous system enhances longevity and oxidant resistance in adult Drosophila. *Autophagy*. 2008;4:176–184.
93. Varga K, Nagy P, Arsikin Csordas K, Kovacs AL, Hegedus K, Juhasz G. Loss of Atg16 delays the alcohol-induced sedation response via regulation of Corazonin neuropeptide production in Drosophila. *Sci Rep*. 2016;6:34641.
94. Takats S, Nagy P, Varga A, et al. Autophagosomal Syntaxin17-dependent lysosomal degradation maintains neuronal function in Drosophila. *J Cell Biol*. 2013;201:531–539.
95. Bhukel A, Beuschel CB, Maglione M, et al. Autophagy within the mushroom body protects from synapse aging in a non-cell autonomous manner. *Nat Commun*. 2019;10 1318.
96. Deal SL, Yamamoto S. Unraveling novel mechanisms of neurodegeneration through a large-scale forward genetic screen in Drosophila. *Front Genet*. 2018;9:700.
97. Haelterman NA, Jiang L, Li Y, et al. Large-scale identification of chemically induced mutations in Drosophila melanogaster. *Genome Res*. 2014;24:1707–1718.
98. Yamamoto S, Jaiswal M, Charng WL, et al. A drosophila genetic resource of mutants to study mechanisms underlying human genetic diseases. *Cell*. 2014;159:200–214.
99. Karch CM, Cruchaga C, Goate AM. Alzheimer's disease genetics: from the bench to the clinic. *Neuron*. 2014;83:11–26.
100. Gao S, Casey AE, Sargeant TJ, Makinen VP. Genetic variation within endolysosomal system is associated with late-onset Alzheimer's disease. *Brain*. 2018;141:2711–2720.
101. Lambert JC, Ibrahim-Verbaas CA, Harold D, et al. Meta-analysis of 74,046 individuals identifies 11 new susceptibility loci for Alzheimer's disease. *Nat Genet*. 2013;45:1452–1458.
102. Cataldo AM, Petanceska S, Terio NB, et al. Abeta localization in abnormal endosomes: association with earliest Abeta elevations in AD and Down syndrome. *Neurobiol Aging*. 2004;25:1263–1272.
103. Nilsson P, Saido TC. Dual roles for autophagy: degradation and secretion of Alzheimer's disease Abeta peptide. *Bioessays*. 2014;36:570–578.
104. Ling D, Magallanes M, Salvaterra PM. Accumulation of amyloid-like Abeta1-42 in AEL (autophagy-endosomal-lysosomal) vesicles: potential implications for plaque biogenesis. *ASN Neuro*. 2014;6: e00139.
105. Ling D, Salvaterra PM. A central role for autophagy in Alzheimer-type neurodegeneration. *Autophagy*. 2009;5:738–740.
106. Allan K, Perez KA, Barnham KJ, Camakaris J, Burke R. A commonly used Drosophila model of Alzheimer's disease generates an aberrant species of amyloid-beta with an additional N-terminal glutamine residue. *FEBS Lett*. 2014;588:3739–3743.

107. O'Keefe L, Denton D. Using Drosophila models of amyloid toxicity to study autophagy in the pathogenesis of Alzheimer's disease. *Biomed Res Int*. 2018;2018 5195416.
108. Wang H, Wang R, Xu S, Lakshmana MK. Transcription factor EB is selectively reduced in the nuclear fractions of Alzheimer's and amyotrophic lateral sclerosis brains. *Neurosci J*. 2016;2016:4732837.
109. Xiao Q, Yan P, Ma X, et al. Neuronal-targeted TFEB accelerates lysosomal degradation of APP, reducing Abeta generation and amyloid plaque pathogenesis. *J Neurosci*. 2015;35:12137–12151.
110. Bouche V, Espinosa AP, Leone L, Sardiello M, Ballabio A, Botas J. Drosophila Mitf regulates the V-ATPase and the lysosomal-autophagic pathway. *Autophagy*. 2016;12:484–498.
111. Wittmann CW, Wszolek MF, Shulman JM, et al. Tauopathy in Drosophila: neurodegeneration without neurofibrillary tangles. *Science*. 2001;293:711–714.
112. Chapuis J, Hansmannel F, Gistelinck M, et al. Increased expression of BIN1 mediates Alzheimer genetic risk by modulating tau pathology. *Mol Psychiatry*. 2013;18: 1225–1234.
113. Dourlen P, Fernandez-Gomez FJ, Dupont C, et al. Functional screening of Alzheimer risk loci identifies PTK2B as an in vivo modulator and early marker of Tau pathology. *Mol Psychiatry*. 2017;22:874–883.
114. Shulman JM, Imboywa S, Giagtzoglou N, et al. Functional screening in Drosophila identifies Alzheimer's disease susceptibility genes and implicates Tau-mediated mechanisms. *Hum Mol Genet*. 2014;23:870–877.
115. Moreau K, Fleming A, Imarisio S, et al. PICALM modulates autophagy activity and tau accumulation. *Nat Commun*. 2014;5:4998.
116. Hewitt VL, Whitworth AJ. Mechanisms of Parkinson's disease: lessons from Drosophila. *Curr Top Dev Biol*. 2017;121:173–200.
117. Feany MB, Bender WW. A Drosophila model of Parkinson's disease. *Nature*. 2000;404:394–398.
118. Winslow AR, Chen CW, Corrochano S, et al. Alpha-Synuclein impairs macroautophagy: implications for Parkinson's disease. *J Cell Biol*. 2010;190:1023–1037.
119. Buttner S, Broeskamp F, Sommer C, et al. Spermidine protects against alpha-synuclein neurotoxicity. *Cell Cycle*. 2014;13:3903–3908.
120. Issa AR, Sun J, Petitgas C, et al. The lysosomal membrane protein LAMP2A promotes autophagic flux and prevents SNCA-induced Parkinson disease-like symptoms in the Drosophila brain. *Autophagy*. 2018;14:1898–1910.
121. Gomez-Suaga P, Luzon-Toro B, Churamani D, et al. Leucine-rich repeat kinase 2 regulates autophagy through a calcium-dependent pathway involving NAADP. *Hum Mol Genet*. 2012;21:511–525.
122. Dodson MW, Leung LK, Lone M, Lizzio MA, Guo M. Novel ethyl methanesulfonate (EMS)-induced null alleles of the Drosophila homolog of LRRK2 reveal a crucial role in endolysosomal functions and autophagy in vivo. *Dis Model Mech*. 2014;7:1351–1363.
123. Liu Z, Wang X, Yu Y, et al. A Drosophila model for LRRK2-linked parkinsonism. *Proc Natl Acad Sci U S A*. 2008;105:2693–2698.
124. Ng CH, Mok SZ, Koh C, et al. Parkin protects against LRRK2 G2019S mutant-induced dopaminergic neurodegeneration in Drosophila. *J Neurosci*. 2009;29: 11257–11262.
125. Lavoy S, Chittoor-Vinod VG, Chow CY, Martin I. Genetic modifiers of neurodegeneration in a Drosophila model of Parkinson's disease. *Genetics*. 2018;209: 1345–1356.
126. Nagoshi E. Drosophila models of sporadic Parkinson's disease. *Int J Mol Sci*. 2018;19: E3343.

127. Ma P, Yun J, Deng H, Guo M. Atg1 mediated autophagy suppresses tissue degeneration in pink1/parkin mutants by promoting mitochondrial fission in Drosophila. *Mol Biol Cell.* 2018;29:3082–3092. mbcE18040243.
128. McLeod CJ, O'Keefe LV, Richards RI. The pathogenic agent in Drosophila models of 'polyglutamine' diseases. *Hum Mol Genet.* 2005;14:1041–1048.
129. Ravikumar B, Vacher C, Berger Z, et al. Inhibition of mTOR induces autophagy and reduces toxicity of polyglutamine expansions in fly and mouse models of Huntington disease. *Nat Genet.* 2004;36:585–595.
130. Wang T, Lao U, Edgar BA. TOR-mediated autophagy regulates cell death in Drosophila neurodegenerative disease. *J Cell Biol.* 2009;186:703–711.
131. Bilen J, Bonini NM. Genome-wide screen for modifiers of ataxin-3 neurodegeneration in Drosophila. *PLoS Genet.* 2007;3:1950–1964.
132. Saitoh Y, Fujikake N, Okamoto Y, et al. p62 plays a protective role in the autophagic degradation of polyglutamine protein oligomers in polyglutamine disease model flies. *J Biol Chem.* 2015;290:1442–1453.
133. Richards RI, Samaraweera SE, van Eyk CL, O'Keefe LV, Suter CM. RNA pathogenesis via Toll-like receptor-activated inflammation in expanded repeat neurodegenerative diseases. *Front Mol Neurosci.* 2013;6:25.
134. Samaraweera SE, O'Keefe LV, Price GR, Venter DJ, Richards RI. Distinct roles for Toll and autophagy pathways in double-stranded RNA toxicity in a Drosophila model of expanded repeat neurodegenerative diseases. *Hum Mol Genet.* 2013;22: 2811–2819.
135. Settembre C, Fraldi A, Jahreiss L, et al. A block of autophagy in lysosomal storage disorders. *Hum Mol Genet.* 2008;17:119–129.
136. Webber DL, Choo A, Hewson LJ, et al. Neuronal-specific impairment of heparan sulfate degradation in Drosophila reveals pathogenic mechanisms for Mucopolysaccharidosis type IIIA. *Exp Neurol.* 2018;303:38–47.
137. Bar S, Prasad M, Datta R. Neuromuscular degeneration and locomotor deficit in a Drosophila model of mucopolysaccharidosis VII is attenuated by treatment with resveratrol. *Dis Model Mech.* 2018;11:dmm036954.
138. Kinghorn KJ, Gronke S, Castillo-Quan JI, et al. A Drosophila model of neuronopathic Gaucher disease demonstrates lysosomal-autophagic defects and altered mTOR signalling and is functionally rescued by rapamycin. *J Neurosci.* 2016;36: 11654–11670.
139. Dasari SK, Bialik S, Levin-Zaidman S, et al. Signalome-wide RNAi screen identifies GBA1 as a positive mediator of autophagic cell death. *Cell Death Differ.* 2017;24: 1288–1302.
140. Dasari SK, Schejter E, Bialik S, et al. Death by over-eating: the Gaucher disease associated gene GBA1, identified in a screen for mediators of autophagic cell death, is necessary for developmental cell death in Drosophila midgut. *Cell Cycle.* 2017;16:2003–2010.
141. Venkatachalam K, Long AA, Elsaesser R, Nikolaeva D, Broadie K, Montell C. Motor deficit in a Drosophila model of mucolipidosis type IV due to defective clearance of apoptotic cells. *Cell.* 2008;135:838–851.
142. Wong CO, Li R, Montell C, Venkatachalam K. Drosophila TRPML is required for TORC1 activation. *Curr Biol.* 2012;22:1616–1621.
143. Donde A, Sun M, Jeong YH, et al. Upregulation of ATG7 attenuates motor neuron dysfunction associated with depletion of TARDBP/TDP-43. *Autophagy.* 2019;1–11. https://doi.org/10.1080/15548627.2019.1635379.
144. Aparicio R, Rana A, Walker DW. Upregulation of the autophagy adaptor p62/ SQSTM1 prolongs health and lifespan in middle-aged Drosophila. *Cell Rep.* 2019;28:1029–1040.e1025.

145. Richards RI, Robertson SA, O'Keefe LV, et al. The enemy within: innate surveillance-mediated cell death, the common mechanism of neurodegenerative disease. *Front Neurosci*. 2016;10:193.
146. Shukla AK, Spurrier J, Kuzina I, Giniger E. Hyperactive innate immunity causes degeneration of dopamine neurons upon altering activity of Cdk5. *Cell Rep*. 2019;26:131–144.e134.
147. Dong X, Levine B. Autophagy and viruses: adversaries or allies? *J Innate Immun*. 2013;5:480–493.
148. Galluzzi L, Bravo-San Pedro JM, Levine B, Green DR, Kroemer G. Pharmacological modulation of autophagy: therapeutic potential and persisting obstacles. *Nat Rev Drug Discov*. 2017;16:487–511.
149. Franco LH, Nair VR, Scharn CR, et al. The ubiquitin ligase Smurf1 functions in selective autophagy of Mycobacterium tuberculosis and anti-tuberculous host defense. *Cell Host Microbe*. 2017;21:59–72.
150. Castillo EF, Dekonenko A, Arko-Mensah J, et al. Autophagy protects against active tuberculosis by suppressing bacterial burden and inflammation. *Proc Natl Acad Sci U S A*. 2012;109:E3168–E3176.
151. Watson RO, Manzanillo PS, Cox JS. Extracellular M. tuberculosis DNA targets bacteria for autophagy by activating the host DNA-sensing pathway. *Cell*. 2012;150:803–815.
152. Zhao Z, Fux B, Goodwin M, et al. Autophagosome-independent essential function for the autophagy protein Atg5 in cellular immunity to intracellular pathogens. *Cell Host Microbe*. 2008;4:458–469.
153. Orvedahl A, MacPherson S, Sumpter Jr R, Talloczy Z, Zou Z, Levine B. Autophagy protects against Sindbis virus infection of the central nervous system. *Cell Host Microbe*. 2010;7:115–127.
154. Shoji-Kawata S, Sumpter R, Leveno M, et al. Identification of a candidate therapeutic autophagy-inducing peptide. *Nature*. 2013;494:201–206.
155. Buchon N, Silverman N, Cherry S. Immunity in Drosophila melanogaster—from microbial recognition to whole-organism physiology. *Nat Rev Immunol*. 2014;14:796–810.
156. Bier E, Guichard A. Deconstructing host-pathogen interactions in Drosophila. *Dis Model Mech*. 2012;5:48–61.
157. Ferrandon D, Imler JL, Hetru C, Hoffmann JA. The Drosophila systemic immune response: sensing and signalling during bacterial and fungal infections. *Nat Rev Immunol*. 2007;7:862–874.
158. Lemaitre B, Nicolas E, Michaut L, Reichhart JM, Hoffmann JA. The dorsoventral regulatory gene cassette spatzle/Toll/cactus controls the potent antifungal response in Drosophila adults. *Cell*. 1996;86:973–983.
159. Wu J, Randle KE, Wu LP. ird1 is a Vps15 homologue important for antibacterial immune responses in Drosophila. *Cell Microbiol*. 2007;9:1073–1085.
160. Yano T, Mita S, Ohmori H, et al. Autophagic control of listeria through intracellular innate immune recognition in drosophila. *Nat Immunol*. 2008;9:908–916.
161. Anding AL, Baehrecke EH. Vps15 is required for stress induced and developmentally triggered autophagy and salivary gland protein secretion in Drosophila. *Cell Death Differ*. 2015;22:457–464.
162. Manzanillo PS, Ayres JS, Watson RO, et al. The ubiquitin ligase parkin mediates resistance to intracellular pathogens. *Nature*. 2013;501:512–516.
163. Costa A, Jan E, Sarnow P, Schneider D. The Imd pathway is involved in antiviral immune responses in Drosophila. *PLoS One*. 2009;4: e7436.
164. Dostert C, Jouanguy E, Irving P, et al. The Jak-STAT signaling pathway is required but not sufficient for the antiviral response of drosophila. *Nat Immunol*. 2005;6:946–953.
165. Merkling SH, van Rij RP. Beyond RNAi: antiviral defense strategies in Drosophila and mosquito. *J Insect Physiol*. 2013;59:159–170.

166. Zambon RA, Nandakumar M, Vakharia VN, Wu LP. The toll pathway is important for an antiviral response in Drosophila. *Proc Natl Acad Sci U S A*. 2005;102:7257–7262.
167. Moy RH, Gold B, Molleston JM, et al. Antiviral autophagy restricts Rift Valley fever virus infection and is conserved from flies to mammals. *Immunity*. 2014;40:51–65.
168. Nakamoto M, Moy RH, Xu J, et al. Virus recognition by Toll-7 activates antiviral autophagy in Drosophila. *Immunity*. 2012;36:658–667.
169. Shelly S, Lukinova N, Bambina S, Berman A, Cherry S. Autophagy is an essential component of Drosophila immunity against vesicular stomatitis virus. *Immunity*. 2009;30: 588–598.
170. Liu Y, Gordesky-Gold B, Leney-Greene M, Weinbren NL, Tudor M, Cherry S. Inflammation-induced, STING-dependent autophagy restricts Zika virus infection in the Drosophila brain. *Cell Host Microbe*. 2018;24:57–68.e53.
171. Margolis SR, Wilson SC, Vance RE. Evolutionary origins of cGAS-STING signaling. *Trends Immunol*. 2017;38:733–743.
172. Martin M, Hiroyasu A, Guzman RM, Roberts SA, Goodman AG. Analysis of Drosophila STING reveals an evolutionarily conserved antimicrobial function. *Cell Rep*. 2018;23:3537–3550.e3536.
173. Moretti J, Roy S, Bozec D, et al. STING senses microbial viability to orchestrate stress-mediated autophagy of the endoplasmic reticulum. *Cell*. 2017;171:809–823.e813.
174. Rubinsztein DC, Bento CF, Deretic V. Therapeutic targeting of autophagy in neurodegenerative and infectious diseases. *J Exp Med*. 2015;212:979–990.
175. Galluzzi L, Green DR. Autophagy-independent functions of the autophagy machinery. *Cell*. 2019;177:1682–1699.
176. Strange K. Drug discovery in fish, flies, and worms. *ILAR J*. 2016;57:133–143.
177. Su TT. Drug screening in Drosophila; why, when, and when not? *Wiley Interdiscip Rev Dev Biol*. 2019;8 e346.
178. Yadav AK, Srikrishna S, Gupta SC. Cancer drug development using drosophila as an in vivo tool: from bedside to bench and back. *Trends Pharmacol Sci*. 2016;37:789–806.
179. Dar AC, Das TK, Shokat KM, Cagan RL. Chemical genetic discovery of targets and anti-targets for cancer polypharmacology. *Nature*. 2012;486:80.
180. Vidal M, Wells S, Ryan A, Cagan R. ZD6474 suppresses oncogenic RET isoforms in a Drosophila model for type 2 multiple endocrine neoplasia syndromes and papillary thyroid carcinoma. *Cancer Res*. 2005;65:3538–3541.
181. Chang S, Bray SM, Li Z, et al. Identification of small molecules rescuing fragile X syndrome phenotypes in Drosophila. *Nat Chem Biol*. 2008;4:256.
182. Seguin A, Monnier V, Palandri A, et al. A yeast/Drosophila screen to identify new compounds overcoming frataxin deficiency. *Oxid Med Cell Longev*. 2015;2015: 1–10. 565140.
183. Wang L, Hagemann TL, Messing A, Feany MB. An in vivo pharmacological screen identifies cholinergic signaling as a therapeutic target in glial-based nervous system disease. *J Neurosci*. 2016;36:1445–1455.
184. Hofherr A, Wagner CJ, Watnick T, Köttgen M. Targeted rescue of a polycystic kidney disease mutation by lysosomal inhibition. *Kidney Int*. 2016;89:949–955.
185. Stilwell GE, Saraswati S, Littleton JT, Chouinard SW. Development of a Drosophila seizure model for in vivo high-throughput drug screening. *Eur J Neurosci*. 2006;24:2211–2222.
186. Streit AK, Fan YN, Masullo L, Baines RA. Calcium imaging of neuronal activity in Drosophila can identify anticonvulsive compounds. *PLoS One*. 2016;11: e0148461.
187. Gasque G, Conway S, Huang J, Rao Y, Vosshall LB. Small molecule drug screening in Drosophila identifies the 5HT2A receptor as a feeding modulation target. *Sci Rep*. 2013;3 srep02120.

Index

Note: Page numbers followed by "*f*" indicate figures and "*t*" indicate tables.

D

L

M

R

S

T

FSC
www.fsc.org
MIX
Papier | Fördert gute Waldnutzung
FSC® C083411

Kolibri 360 GmbH
Ferdinand-Jühlke-Straße 7
99095 Erfurt, Deutschland
produktsicherheit@kolibri360.de